MEMORIAL VOLUME FOR

TAI TSUN WU

A Pioneer of Mathematical and Theoretical Physics

MEMORIAL VOLUME FOR
TAI TSUN WU

A Pioneer of Mathematical and Theoretical Physics

Editors

Sau Lan Wu University of Wisconsin-Madison, USA
Hung Cheng Massachusetts Institute of Technology, USA
Wasikul Islam University of Wisconsin-Madison, USA
Barry McCoy Stony Brook University, USA
John Myers Harvard School of Engineering and Applied Sciences, USA
Per Osland University of Bergen, Norway
Shaojun Sun University of Wisconsin-Madison, USA

World Scientific

NEW JERSEY • LONDON • SINGAPORE • BEIJING • SHANGHAI • HONG KONG • TAIPEI • CHENNAI • TOKYO

Published by

World Scientific Publishing Co. Pte. Ltd.
5 Toh Tuck Link, Singapore 596224
USA office: 27 Warren Street, Suite 401-402, Hackensack, NJ 07601
UK office: 57 Shelton Street, Covent Garden, London WC2H 9HE

Library of Congress Control Number: 2025945515

British Library Cataloguing-in-Publication Data
A catalogue record for this book is available from the British Library.

MEMORIAL VOLUME FOR TAI TSUN WU
A Pioneer of Mathematical and Theoretical Physics

ISBN 978-981-98-2190-7 (hardcover)
ISBN 978-981-98-2191-4 (ebook for institutions)
ISBN 978-981-98-2192-1 (ebook for individuals)

For any available supplementary material, please visit
https://www.worldscientific.com/worldscibooks/10.1142/14546#t=suppl

Desk Editor: Carmen Teo Bin Jie

Typeset by Stallion Press
Email: enquiries@stallionpress.com

Contents

Chapter 1: A Letter to Tai Tsun Wu from His Wife Sau Lan Wu, Tai Tsun Wu's Life Story and His Career Advancement 1

(1) A Letter to Tai Tsun Wu from his wife Sau Lan Wu 1

(2) Reflections on the Childhood and Formative Years of Tai Tsun Wu 3

(3) Celebrating the Legacy of Tai Tsun Wu, a Brilliant Theoretical Physicist 7

Chapter 2: Selected Photos of Tai Tsun Wu in His Happy Days 17

Chapter 3: Remembrances from Tai Tsun Wu's Family, Friends, and Collaborators 41

(1) Remembrances from Family Members 41

(2) Remembrances from Friends and Collaborators 46

Chapter 4: Celebration of Tai Tsun Wu's 60th birthday at CERN in 1994 — A Fun Dialog and Quiz edited by Conrad Newton and Per Osland 67

Chapter 5: Tai Tsun Wu's CV and Published Obituary, and Harvard Faculty of Arts and Sciences Memorial Minutes 75

(1) Curriculum Vitae of Tai Tsun Wu 75

(2) Obituary published in Physics Today 79

(3) Obituary published in Harvard Physics department website 83

(4) Abridged Memorial Minute presented at the meeting of the Faculty of Arts and Sciences, Harvard University, October 7, 2025 — Hung Cheng, Sheldon Glashow, John Hutchinson, and Arthur Jaffe 84

(5) Full Version of Memorial Minute presented at the meeting of the Faculty of Arts and Sciences, Harvard University, October 7, 2025 — Hung Cheng, Sheldon Glashow, John Hutchinson, and Arthur Jaffe 86

Chapter 6: Electromagnetic Theory (66 papers)........ 91

Chapter 7: Yang–Mills Theory & Monopoles (23 papers) 123

Chapter 8: Mathematical & Theoretical Physics in High Energy Physics (131 papers) 133

Chapter 9: Quantum Field Theory and Statistical Physics (40 papers) 187

Chapter 10: Particle Physics Phenomenology and Higgs Physics (132 papers) 209

Chapter 11: Concept of the Basic Standard Model and a Relation between the Three Gauge Coupling Constants (5 papers) 267

Chapter 12: Books (6 books) 273

Chapter 13: Selected Publications of Tai Tsun Wu (11 papers) 281

Chapter 1
A Letter to Tai Tsun Wu from His Wife Sau Lan Wu, Tai Tsun Wu's Life Story and His Career Advancement

(1) A Letter to Tai Tsun Wu from his wife Sau Lan Wu

To my dearest husband Tai Tsun Wu
From your wife Sau Lan Wu

TT, where are you? Can you hear me calling you? I have your ashes in an urn in your bedroom of our apartment. I say good morning to you every morning, and I wish you good night in the evening. I talk to you every day, sharing with you the stories of my daily life — what I have done, where I have been, and what has been in my thoughts.

I am trying to gain strength so that when we meet in heaven, I am strong enough to take care of you. I go to the Stanford Memorial Church every week to talk to you, praying to God for both of us. I remember the song "I walk with God". When I walk alone, I walk with you and God.

Dear TT, this is your wife writing to you. TT, you are the smartest person I know. You told me that when you were a kid, you went to a bookstore, read the books, then you could memorize the contents of the books without needing to buy them.

You used to tell me when you were doing calculations, you usually had 2000 mathematical terms in your head. You were a genius and a great scientist and had a great mind. You did not care about joining the bandwagon nor getting instant fame. Your final publication of your proposal of the Basic Standard Model, which you were most proud of, was published a year before your death at the age of 89. Between the publication of this paper (November 2023) and the time soon after your death (July 19, 2024), this publication has been downloaded 12,921 times.

You were extremely supportive of my career. You told me that if we were going to have children, you would take care of the children so I could pursue my research. With great regret, I did not take your offer, otherwise, we would have the brightest children with us.

We spent many wonderful years together at CERN, Geneva, Switzerland, inspiring and motivating each other in our work. Every Sunday, we had a wonderful walk together to the church Paroisse de Saint-Julien in Meyrin, where I prayed for both of us. In 2019, due to the Covid pandemic, we moved to Cambridge, MA, where you continued teaching at Harvard. Again, every day, we held our hands and enjoyed tremendously walking up the hill to the Harvard Observatory of Garden street.

After your retirement, we moved to the Vi community at Palo Alto, California on April 12, 2022 next to Stanford. For three months, we had memorable times walking every day around the perimeter of Vi. I still remember each tree that you admired. I call our walking path the 'TT memorial path'. Our daily walk stopped in July 2022 when you broke your hips and since then I took care of you every day at home.

On July 10, 2024, I took you to Stanford Hospital for tests but you never came back home. From July 10 on, I stayed with you day and night by your side in the hospital. Up to 2 days before you passed away on July 19, we both were confident that we would go home together soon.

On July 17, things had a difficult turn and you were partially unconscious with me holding your hand and whispering to you. I watched your last breath and you passed away at 10:48 pm on July 19 peacefully without realizing you were leaving me behind.

TT, I love you every minute of every day. My tears have dried up, but my heart still cries out for you. TT, where are you? Do you hear me calling you? I am forever with you. When I walk alone, I walk with you and God.

(2) Reflections on the Childhood and Formative Years of Tai Tsun Wu

Hwa Tsang, Tai Tsun Wu's cousin

Family Background and Early Years

Tai Tsun Wu (吴大峻) was born on Dec 1, 1933 in Shanghai, China. His father, King Ching Wu (吳錦慶), earned a Bachelor's degree in Civil Engineering from one of the colleges of the University of London, England, and later obtained a Master's degree in Civil Engineering from Cornell University, Ithaca, New York. His mother, Wei Van Tsang (張緯文), received both her Bachelor's and Master's degrees in Physics from Cornell University, where the two met.

Born prematurely, Tai Tsun weighed just over three pounds at birth. Thanks to loving care, he grew into a healthy, happy child. His parents brought him a wonderful younger brother, Tai Te Wu (吴大德).

Childhood in Shanghai

Tai Tsun, a precocious and lively child, was homeschooled until the fall of 1942, when he entered the first year of junior high at the Second Affiliated Middle School of Datong University (上海大同大学附中二院, equivalent to 7th grade, 初中一年级).

Throughout his childhood, Tai Tsun's closest companions were his brother Tai Te and his maternal cousins Tung Tsang (張通, born 1932) and Hwa Tsang (張華, born 1937). Together, they played Mahjong (麻将), Gomoku (五子棋), Chinese checkers (跳棋), Chinese chess (象棋), Go (围棋), and various card games like Go Fish, Hearts, and Bridge. The older boys (Tung, Tai Tsun, and Tai Te) often ganged up on young Hwa during games, but she never minded — she was simply delighted to be included. An avid ping pong player, Tai Tsun also developed an early love for reading, immersing himself in Chinese classical novels such as *Romance of the Three Kingdoms* (三国演义), *Journey to the West* (西游记), *Water Margin* (水浒传), and *Legend of the Swordsmen of Shu Mountain* (蜀山剑侠传), as well as other martial arts tales (武侠小说).

Formative Years in Rigorous Mathematics Training

The mathematics curriculum at Datong University (大同大学) and its affiliated middle school (附中) was notably rigorous and advanced for its time. I vividly remember one particular challenge presented by a teacher fondly nicknamed "Old Master Mei" (梅老爷) during the second-year Geometry class in the fall semester of my third year of junior high (初三). On the first day of that term, he posed a problem that became a local legend among the students:

In triangle ABC, the lengths of the angle bisectors of angles B and C are equal. Prove that angle B = angle C. Provide a direct geometric proof using supplemental lines, not an indirect argument or algebraic calculation.

This problem was tough but highly admired by all students, and it sparked an intellectual excitement in us that lasted well beyond that semester. Over 60 years later, I still remember how difficult it was to solve, and out of curiosity, I reached out to Tai Tsun, my long-time friend, to revisit the solution. He had the patience to reflect deeply on this puzzle, and gradually, after all these years, we pieced together the intricate proof once more.

Departure from Shanghai for Hong Kong

On 27th April 1949, as the People's Liberation Army prepared to take Shanghai after a key victory, the Wu family and cousin Tung Tsang (張通) boarded a boat for Hong Kong, then a British Crown Colony. On this journey, they possibly shared a boat with Du Yuesheng (杜月笙), a prominent Shanghai figure. Other members of the Tsang family had left for Hong Kong earlier in 1949. Tai Tsun was almost finishing his first year in College at Shanghai Datong University studying electrical engineering. We all felt sadness at leaving Datong University and its affiliated middle school, where we had received excellent science and math education and formed close friendships.

After the family relocated to Hong Kong, they continued their close-knit traditions. Outings, walks, movie trips, and even collecting tram tickets (电车票) were among their favorite activities. The Wu family lived at No. 404 King's Road (英皇道404号) in the North Point area (北角). And the Tsang family including Hwa's grandparents, my second uncle's family, another uncle's child, together with my grandfather's two nursing aides, and other maids & cook, lived a block away at No. 324 King's Road (英皇道324号). I would visit the Wu household two or three times a week.

Mr. Wu taught Civil Engineering at the University of Hong Kong, and Mrs. Wu taught Physics at St. Paul's Coeducational College (圣保罗男女校), later moving on to the University of Hong Kong.

The Path from Hong Kong to America

At that time, Hong Kong had only one university, and competition for admission was fierce. Despite scoring perfectly on the entrance exam, Tai Tsun was denied admission to the University of Hong Kong because he was neither a Hong Kong native nor a graduate of an approved Hong Kong high school. While we were disheartened, Tai Tsun remained undeterred. Throughout his life, Tai Tsun nurtured an insatiable passion for learning. With a meticulous mind and a thoughtful approach, he absorbed new concepts effortlessly. Though he never enrolled at the University of Hong Kong, he independently mastered a wide range of subjects — and relished the freedom to ponder their deeper connections. In this self-directed journey, he truly thrived.

Meanwhile, his uncle, Ngeu Faung Tsang (張藕舫), became a visiting assistant professor at the University of Minnesota in December 1949 and helped Tai Tsun secure admission to the university's Electrical Engineering program. Tai Tsun's family was highly supportive of his academic journey and in this admission process. Tai Tsun's cousin, Tung Tsang (張通), having earned a bachelor's degree in Chemical Engineering from Datong University, also began graduate studies at Minnesota in Fall 1949. A world of new opportunities in the United States was calling Tai Tsun to pursue his dreams and embark on a remarkable new journey.

A New Life in the United States of America

On November 4, 1950, Tai Tsun arrived in San Francisco and visited Professor Hu Dunfu (胡敦復), a close family friend and mentor who was also the founder and president of Datong University. After a brief stay in Seattle, Tai Tsun proceeded to the University of Minnesota for his undergraduate studies. Though the three cousins lived at different addresses and arrived at different times, they were thrilled to be in the same city, sharing the joy of being together in a new country. Tai Tsun lived a frugal life in America, often subsisting on rice and canned vegetables, particularly sweet peas, which he grew fond of them for life. Despite the financial hardships, his perseverance led to academic success. Eventually, my father sponsored Mr. and Mrs. Wu's immigration to the United States, where they taught at the University of Arkansas before retiring in Oklahoma City.

As an undergraduate at the University of Minnesota from January 1951 to June 1953, Tai Tsun excelled in all his classes, earning a

Bachelor's degree in electrical engineering. On March 23, 1953, he took The William Lowell Putnam Mathematical Competition, the preeminent mathematics competition for undergraduate college students in the United States and Canada. That year, he achieved one of the top five scores, becoming a Putnam Fellow and earning a fellowship to study at Harvard University. He earned his Ph.D. at Harvard in 1956 and remained there as a Junior Fellow, later holding various distinguished faculty positions until his retirement in 2021 — a remarkable academic career that spanned nearly 70 years.

Tai Tsun passed away in 2024. The world lost a brilliant and dedicated scientist. I lost a dear cousin and friend. I miss his kindness, wisdom, energy, and humor. I am grateful for his life and treasure his memory.

(3) Celebrating the Legacy of Tai Tsun Wu, a Brilliant Theoretical Physicist

Wasikul Islam & Sau Lan Wu

Tai Tsun Wu has been a distinguished figure in theoretical physics and a longtime professor at Harvard University. Wu's journey from an exceptional young scholar to a globally respected physicist was marked by brilliance, dedication, and groundbreaking contributions. Wu's contributions in physics spanned a remarkable range of fields, from antenna theory and statistical mechanics to quantum field theory and high-energy particle physics. His work advanced fundamental science and bridged the gap between physics and mathematics, leaving a lasting legacy in both disciplines.

Early Life and Early Career Days

Born on December 1, 1933, in Shanghai, China, Tai Tsun Wu's intellectual journey began in a world of upheaval and transformation. He was home-schooled by his mother, who held a Master's degree in Physics from Cornell University — until he attended high school in

Shanghai. He moved to the United States in November 1950, where he pursued electrical engineering at the University of Minnesota. Wu's exceptional talent in mathematics became evident early on; in 1953, he was named a Putnam Fellow, one of the most prestigious honors in the mathematical community. His academic trajectory took a decisive turn when he joined Harvard University as a graduate student, earning his Ph.D. in applied physics in 1956 at the age of 22, under the guidance of Professor Ronold King. Wu's doctoral thesis on high-frequency scattering and antenna theory established him as a prodigious talent in electromagnetic wave theory. His early work on antenna theory laid the groundwork for his later explorations into more abstract and profound areas of physics. After graduation, he was named a Junior Fellow in 1956 in the Harvard Society of Fellows, a position that allowed him the freedom to pursue research anywhere. He went to the Institute for Advanced Study in Princeton, where he was inspired by and collaborated with Professor C.N. Yang and Professor T.D. Lee (Nobel Laureates in 1957). Wu's ability to seamlessly transition between applied and theoretical physics was a hallmark of his career, reflecting his deep understanding of both the practical and conceptual dimensions of science. Wu joined the Harvard faculty in 1959, beginning a career that would span over six decades. Later, in 1966, he was appointed the Gordon McKay Professor of Applied Physics. In 1994, he was offered a joint appointment at the Harvard Physics Department to become a Professor of Physics.

Early Contributions to Statistical Mechanics

In the 1960s, Wu turned his attention to statistical mechanics, collaborating with Barry McCoy and Craig Tracy on the two-dimensional Ising model. Their work led to the exact computation of spin correlation functions, revealing how rotational invariance emerges near critical points. This discovery, which involved Painlevé transcendental functions, not only solved a long-standing problem in statistical mechanics but also opened new avenues in the study of integrable systems.

Breakthrough in High-Energy Scattering: Predicting Rising Cross-Sections

One of Wu's most remarkable achievements came through his collaboration with Hung Cheng at MIT, where they explored the high-energy behavior of particle scattering. By summing large classes of Feynman diagrams, they predicted a rise in proton-proton total cross-sections — a bold proposition that challenged prevailing expectations at the time, which assumed the cross-sections remained flat. Their theoretical prediction was experimentally confirmed in the 1970s by the ISR experiment at CERN, marking a milestone in particle physics. This work offered foundational insights into diffraction scattering and multiple hadron production, contributing significantly to our understanding of high-energy collisions and laying the groundwork for some of the future developments in quantum chromodynamics.

The Wu–Yang Monopole and the Wu–Yang Dictionary

One of Wu's most significant contributions emerged from his collaboration with C.N. Yang (Nobel Laureate) in the mid-1970s, where they revolutionized the mathematical treatment of magnetic monopoles. Drawing on differential geometry and fiber bundle theory, Wu and Yang, in a paper titled "*Dirac Monopole Without Strings: Monopole Harmonics*", introduced "monopole harmonics," a refined generalization of spherical harmonics that replaced the earlier concept of singular Dirac strings. This innovative framework described particle wave functions in a field of magnetic monopole using smooth functions within appropriate coordinate patches, connected across patches by gauge transformations. Their work not only provided a more rigorous foundation for understanding monopoles but also offered powerful insights into topological gauge field configurations in higher dimensions. Wu's work with C.N. Yang, titled "*Concept of Nonintegrable Phase Factors and Global Formulation of Gauge Fields*", remains one of his most famous works so far, as it has been cited more than 1,184 times to date. The so-called "Wu–Yang dictionary" became a crucial work in theoretical physics, linking concepts in gauge theory and differential geometry, inspiring decades of interdisciplinary research.

Anharmonic Oscillators and Perturbation Theory

Wu's work with Carl M. Bender on the anharmonic oscillator is another highlight of his career. Their 1969 paper, "*Anharmonic Oscillator*", explored the oscillator's unusual behavior, revealing its divergent perturbation series and complex analytic structure, offering insights relevant to λϕ4 field theory. In 1973, a follow-up work, "*Anharmonic Oscillator. 2: A Study of Perturbation Theory in Large Order*", extended the previous work while exploring the behavior of quantum systems beyond the harmonic approximation. Using the Wentzel–Kramers–Brillouin (WKB) analysis and difference-equation methods, they successfully determined the asymptotic behavior of the Rayleigh–Schrödinger coefficients for both Wick-ordered and non-Wick-ordered oscillators. Their theoretical predictions aligned remarkably well with numerical results from computer studies extending to perturbation theory's 150th order. This breakthrough provided a deeper understanding of perturbation theory's divergence behavior and established a critical framework for analyzing large-order effects in quantum systems.

Bremsstrahlung Processes in Gauge Theories

Wu's contributions to quantum electrodynamics (QED) and gauge theories are exemplified by his influential work on bremsstrahlung processes, conducted in collaboration with Frits A. Berends, R. Gastmans, and others. Their landmark papers — "*Single Bremsstrahlung Processes in Gauge Theories*" (1981) and "*Multiple Bremsstrahlung in Gauge Theories at High Energies*" (1981) — significantly advanced the understanding of radiation processes in high-energy particle collisions.

In the first paper, Wu and his collaborators demonstrated that single hard bremsstrahlung cross-sections exhibit remarkable simplicity in the ultrarelativistic limit. They showed that these cross-sections can be elegantly factorized into two components: one associated with the lowest-order elastic process and another linked to the infrared factor describing soft quantum emission. This clear and systematic formulation offered practical insights into the behavior of gauge theories in high-energy regimes.

The second paper extended this work to multiple bremsstrahlung processes, providing a comprehensive formalism for QED at high energies. By introducing helicity states for both fermions and gauge particles — a powerful technique that simplifies calculations — Wu and his colleagues achieved automatic cancellations between complex Feynman diagrams. This approach not only streamlined theoretical predictions but also improved the precision of experimental analyses at particle accelerators like CERN.

Cited 548 and 453 times, respectively, these papers became foundational references in the field, shaping both theoretical advancements and experimental methodologies in high-energy physics.

CP Violation and High-Energy Scattering

Wu's early collaboration with C. N. Yang on CP violation in the decay of neutral kaons ("*Phenomenological Analysis of Violation of CP Invariance in Decay of K0 and anti-K0*", 1964) was another significant contribution to particle physics. This influential paper played a pivotal role in advancing the understanding of CP symmetry violations in the Standard Model of Particle Physics.

In another notable work, Wu and Yang explored the behavior of particles in high-energy scattering processes in their 1965 paper, "*Some Speculations Concerning High-Energy Large Momentum Transfer Processes*". In this study, they proposed that the rapid decrease in differential cross-sections at large scattering angles with increasing energy could arise from a mechanism independent of the method of excitation. This novel hypothesis suggested that certain fundamental properties of particle interactions might transcend the specifics of individual processes, pointing to deeper underlying principles in quantum field theory. By highlighting potential universal behaviors in large-momentum transfer processes, their work laid the groundwork for future studies in perturbative QCD and the scaling behavior observed in deep inelastic scattering.

Tai Tsun Wu's Contributions to Higgs Physics

Tai Tsun Wu has made significant contributions to Higgs physics through theoretical advancements and early mass predictions for the Higgs boson and the top quark. His research spans various topics within quantum field theory, electroweak theory, and collider physics. Below are some key highlights of his impactful work:

a. Relativistic Decay Calculations

In his 1988 paper "*Relativistic Theory of the Decay of Y Into Higgs + Photon*", Wu, alongside collaborators, explored the decay of upsilon (Y) into a Higgs boson and a photon. By employing the Bethe–Salpeter equation to calculate vertex functions, Wu accounted for relativistic effects and found that the decay rate was significantly smaller than predicted by non-relativistic calculations. This contribution enhanced the understanding of Higgs production in specific quarkonium systems.

b. Electroweak Theory and Mass Predictions

Wu's series of papers in the early 1990s made pivotal contributions to electroweak theory. In "*Parameters in the Electroweak Theory: Masses of the Top Quark and the Higgs Boson*", he derived mass relations for the Higgs boson and top quark. From these relations, he predicted the Higgs boson mass as approximately 190 GeV and the top quark mass as 120 GeV via the absence of quadratic divergences in the Standard Model.

c. Two-Higgs Doublet Model

Wu's 1994 paper, "*Mass Relations in the Two Higgs Doublet Model*", extended his theoretical framework to a scenario involving two Higgs doublets. By setting quadratic divergences to zero, he derived four key mass relations crucial for understanding mass hierarchies and the behavior of scalar particles in extended Higgs sectors.

d. Higgs Production and Detection

Wu also contributed significantly to predicting the feasibility of Higgs detection at colliders. In "*Can the Higgs boson be detected at the Fermilab Tevatron collider?*", he analyzed production cross-sections and demonstrated that upgraded luminosity conditions could enable Higgs detection even at lower-energy colliders like the Tevatron.

In "*Production of Isolated Higgs Particle at the Large Hadron Collider*", Wu applied relativistic quantum gauge field theory to predict Higgs particle kinematics at the LHC. His innovative concept of treating the LHC as a "Pomeron Collider" provided fresh insights into hadronic interactions at high energies.

e. Higgs Decay Mechanisms

Wu's 2011 papers, "*Higgs Decay into Two Photons, Revisited and Higgs Decay $H \to \gamma\gamma$*" through a W Loop, addressed discrepancies in the theoretical calculations of the Higgs $\to \gamma\gamma$ decay process. By avoiding dimensional regularization, Wu presented a more straightforward calculation that aligned with the decoupling theorem. This result challenged the earlier conventional approach and offered a preferred theoretical framework for the decay mechanism.

Tai Tsun Wu's Final Paper: "Concept of the Basic Standard Model and a Relation Between the Three Gauge Coupling Constants"

Tai Tsun Wu's final paper at the age of 90, "*Concept of the Basic Standard Model and a Relation Between the Three Gauge Coupling Constants*" (*Int. J. Mod. Phys. A* **38**, 2350128), published in November 2023, proposes a significant revision to the Standard Model of particle physics. Motivated by the desire to simplify the model by reducing the number of parameters, Wu introduced a new theoretical framework, which he called the Basic Standard Model. This model streamlines the Higgs mechanism by limiting its coupling to only the four heaviest elementary particles — the top quark, the Higgs boson, the Z boson, and the W boson. These particles, each with masses around 100 GeV/c^2, directly couple to the Higgs field, while the masses of the remaining 11 massive particles are generated through radiative corrections.

By reducing the number of interaction terms in the Lagrangian, Wu's new model offers an elegant yet theoretically sound approach to mass generation. A key feature of the Basic Standard Model is the derivation of a novel relation between the three gauge coupling constants of the Standard Model's symmetry group U(1)×SU(2)×SU(3). This prediction, made entirely on theoretical grounds and without experimental input, turned out to be in agreement with observed experimental results, providing support for the model's validity.

In addition to this novel prediction, Wu's work aimed to address the puzzling mass of the bottom quark, which differs significantly from the masses of the W, Z, Higgs, and top quark. Wu was particularly intrigued by this discrepancy and hoped that his model would offer insights into the mechanism behind the bottom quark's mass. Published in November 2023, Wu's paper attracted significant attention, with over 12,921 downloads by the time of his passing in July 2024. Wu took immense pride in this work, viewing it as a crucial step toward a deeper understanding of mass generation in particle physics. Wu expressed hope that future theorists would build on his Basic Standard Model to explore unresolved questions and continue his legacy in the field of theoretical physics.

A Life of Collaboration and Mentorship

Wu's scientific achievements were matched by his dedication to mentorship and collaboration. He worked extensively with institutions such as CERN, DESY, the Institute for Advanced Study at Princeton, Rockefeller University, and Utrecht University, fostering a global network of scientific inquiry. His collaborations were not limited to his peers; he also nurtured the careers of countless students and junior researchers, many of whom went on to make significant contributions to physics.

Honors and Legacy

Throughout his distinguished career, Tai Tsun Wu received numerous accolades that underscored his significant contributions to physics. Notable honors included the Dannie Heineman Prize for Mathematical Physics (1999), the Alexander von Humboldt U.S. Senior Scientist Award (1985–86), and the IEEE EMC Society Transactions Prize Paper Award (1983). He was a fellow of the American Academy of Arts and Sciences (1977–2024) and a member of Academia Sinica in Taiwan, alongside memberships in prestigious organizations such as Eta Kappa Nu, Tau Beta Pi, and Sigma Xi. Wu's extensive body of work comprised over 400 research papers and six influential books, including *The Scattering and Diffraction of Waves* (1959), *The Two-Dimensional Ising Model* (1973), and *Expanding Protons: Scattering at High Energies* (1987), leaving a lasting impact on theoretical physics.

The Tale of a Very Personal Equation

Tai Tsun Wu's most impactful equations may not have always been written on a chalkboard — one of his most remarkable "equations" unfolded unexpectedly in 1962 when he met Sau Lan Yu, a summer student at Brookhaven National Laboratory from Vassar College, while he was a visiting physicist from Harvard University. One day, while looking for physicist Jack Steinberger (Nobel Laureate 1988), Sau Lan crossed paths with Tai Tsun Wu, Steinberger's office mate, who kindly informed her that Steinberger was back. What seemed routine turned out to be fate. Later, as a Harvard graduate student, Sau Lan often studied at the library, where a Harvard librarian decided to play a matchmaker and brought the two together. The couple married in June 1967 at Harvard Memorial Church, where Nobel Laureate C.N. Yang gracefully escorted the bride, Sau Lan, down the aisle, forging a deep personal and professional bond that spanned the next six decades. With each other's unwavering love and support, Sau Lan Wu carved out a successful career in experimental physics, while Tai Tsun Wu distinguished himself as a respected theoretical physicist, proving that some of life's most meaningful equations are written with love.

References

1. Tai Tsun Wu. *Physics Today*, August 2024. https://pubs.aip.org/physicstoday/online/43750/Tai-Tsun-Wu.
2. Tai Tsun Wu (1933–2024). *Harvard University, Department of Physics*, August 2024. https://www.physics.harvard.edu/news/tai-tsun-wu-1933-2024.
3. Tai Tsun Wu. *Wikipedia: The Free Encyclopedia*, July 23, 2024. https://en.wikipedia.org/wiki/Tai_Tsun_Wu.
4. Wu–Yang Dictionary. *Wikipedia: The Free Encyclopedia*, June 2024. https://en.wikipedia.org/wiki/Wu%E2%80%93Yang_dictionary.
5. Wu–Yang Monopole. *Wikipedia: The Free Encyclopedia*, May 2024. https://en.wikipedia.org/wiki/Wu%E2%80%93Yang_monopole.

Chapter 2
Selected Photos of Tai Tsun Wu in His Happy Days

Note: The years for some of the photos are approximate.

Photo (1): Tai Tsun Wu, the teenager.

Photo (2): Happy Tai Tsun Wu when he submitted his Ph.D. thesis to Harvard University; in 1956.

Photo (3): Tai Tsun Wu received his Ph.D. degree from Harvard University at the age of 22; in 1956.

Photo (4): Tai Tsun Wu and Sau Lan Wu met at the Harvard University campus; in 1965.

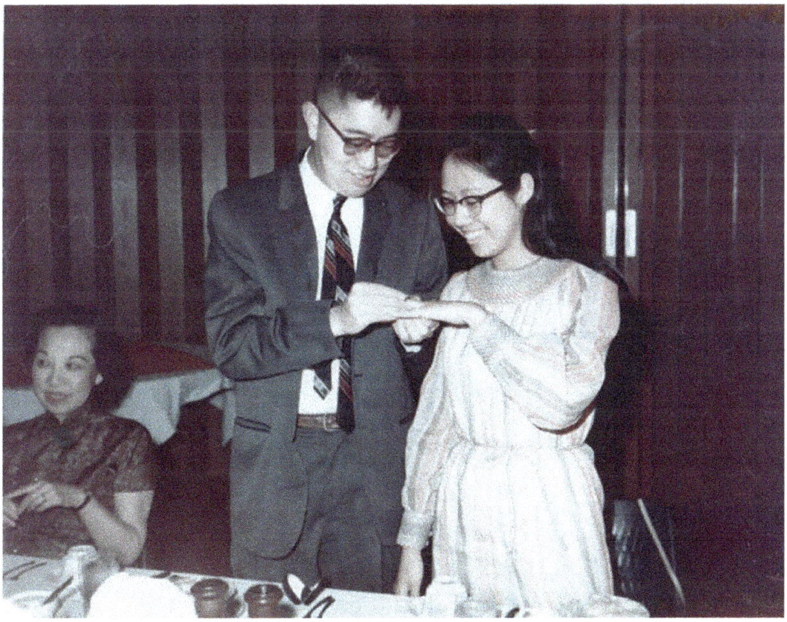

Photo (5): The engagement of Tai Tsun Wu and Sau Lan Wu in 1965.

Photo (6): Tai Tsun Wu with his mother Wei Van Tsang and father King Ching Wu; in 1965.

Photo (7): Tai Tsun Wu and his lifelong collaborator Hung Cheng; in 1965.

Photo (8): Tai Tsun Wu and his collaborator Barry McCoy by the Charles River, Boston, MA; in 1966.

Photo (9): C. N. Yang (Nobel Laureate 1957), presenting as the father of the bride walking down the aisle at the wedding of Tai Tsun Wu and Sau Lan Wu on June 18, 1967 at Harvard Memorial Church, Cambridge, MA.

Photo (10): The wedding of Tai Tsun Wu and Sau Lan Wu on June 18, 1967 at Harvard Memorial Church, Cambridge, MA.

Photo (11): Tai Tsun Wu's parents at the wedding on June 18, 1967 — from left to right: C. N. Yang, Tai Tsun Wu, Sau Lan Wu, Tai Tsun Wu's mother Wei Van Tsang, and Tai Tsun Wu's father King Ching Wu.

Photo (12): Tai Tsun Wu and Sau Lan Wu with their wedding cake on June 18, 1967.

Photo (13): Happy Married Couple on June 18, 1967.

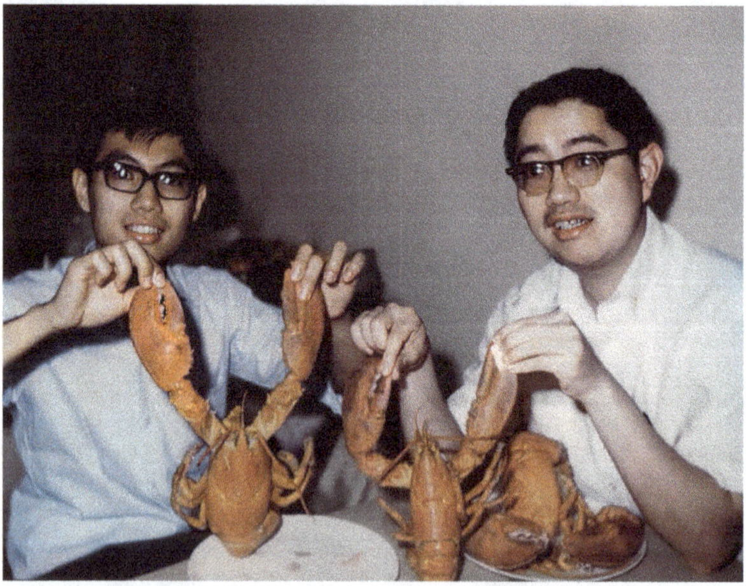

Photo (14): Tai Tsun Wu and his brother-in-law Ming Lun Yu in Tai Tsun Wu's apartment, Cambridge, MA, enjoying Maine lobsters; in 1973.

Photo (15): A photo from the cover of DESY Journal of October 1983. Tai Tsun Wu and his collaborator Harry Lehmann (Universität Hamburg, Germany) were discussing a formula of their research work. Tai Tsun Wu was pointing to the formula he wrote on the blackboard at the DESY auditorium; in 1983.

Photo (16): Vacation in Hawaii, USA — from left to right: Tai Tsun Wu, Sau Lan Wu, and Tai Tsun's mother-in-law Ying Lai; in 1985.

Photo (17): Tai Tsun Wu and Sau Lan Wu looked serious; in 1986.

Photo (18): A day in Zermatt, Switzerland; in 1987.

Photo (19): Tai Tsun Wu sailing in Geneva Lake with Sau Lan Wu's colleague Haimo Zobernig (University of Wisconsin) and Tai Tsun's mother-in-law Ying Lai; in 1990.

Photo (20): Tai Tsun Wu and Sau Lan Wu went to Chamonix, France. They took the cable car to Plan de l'Aiguille and went out for a walk. Tai Tsun twisted his ankle and could not walk back to the cable car. They called emergency services for help. A helicopter came to take Tai Tsun to the hospital in Geneva. As shown in the photo, a Fracture Walker Boat was mounted to stabilize Tai Tsun's foot, ankle and calf (red in the picture), before he was lifted into the helicopter. A beautiful glacier was visible in the background; in 1990.

Photo (21): At the celebration at CERN on the 60th birthday of Tai Tsun Wu in 1994 — Cecilia Jarlskog (Lund University, Sweden) and Tai Tsun Wu; in 1994.

Photo (22): At the celebration at CERN on the 60th birthday of Tai Tsun Wu with his colleagues — from left to right: Sau Lan Wu, C. N. Yang, Cecilia Jarlskog (Lund University, Sweden), and Tai Tsun Wu; in 1994.

Photo (23): At the celebration at CERN on the 60th birthday of Tai Tsun Wu with his colleagues — from left to right: Tai Tsun Wu, Jack Steinberger (Nobel Laureate 1988), and C. N. Yang (Nobel Laureate 1957); in 1994.

Photo (24): At the celebration at CERN on the 60th birthday of Tai Tsun Wu with his collaborators — from left to right: Raymond Gastmans (Katholieke Universiteit Leuven, Belgium), Harry Lehmann (Universität Hamburg, Germany), Tai Tsun Wu, and Per Osland (University of Bergen, Norway); in 1994.

Photo (25): At the celebration at CERN on the 60th birthday of Tai Tsun Wu with his collaborators — from left to right: Per Osland (University of Bergen, Norway), Tai Tsun Wu, Sau Lan Wu, and Cecilia Jarlskog (Lund University, Sweden); in 1994.

Photo (26): At the celebration at CERN on the 60th birthday of Tai Tsun Wu with his collaborators — from left to right: Per Osland (University of Bergen, Norway), Tai Tsun Wu, Cecilia Jarlskog (Lund University, Sweden), Sau Lan Wu, Alexander Grossmann (Centre de Physique Théorique II, CNRS, Marseille, France), Maurice Jacob (CERN), and Raymond Gastmans (Katholieke Universiteit Leuven, Belgium); in 1994.

Photo (27): Tai Tsun Wu at the retirement party of his thesis advisor and collaborator Ronold W. P. King (Harvard University); in 2005.

Photo (28): What are you trying to solve, Raymond? Three physicists, one blackboard, and a universe full of unsolved questions. Tai Tsun Wu, Per Osland (University of Bergen, Norway), and Raymond Gastmans (Katholieke Universiteit Leuven, Belgium) diving deep into the discussions, tackling some puzzling problems at Tai Tsun's Theory office at CERN; in 2012.

Photo (29): From left to right: Tai Tsun Wu, François Englert (Université Libre de Bruxelles, Belgium), and Sau Lan Wu at the 2013 Physics Nobel Prize celebration at CERN for Higgs and Englert; in 2013.

Photo (30): Tai Tsun Wu and his collaborator Raymond Gastmans (Katholieke Universiteit Leuven, Belgium) at CERN auditorium. Tai Tsun Wu used the blackboard to work out their research formulas; in 2014.

Photo (31): Tai Tsun Wu and Sau Lan Wu at the ATLAS detector in the underground tunnel of LHC; in summer of 2014.

Photo (32): Tai Tsun Wu and Sau Lan Wu in Parc de Budé, Geneva, Switzerland; in 2016.

Photo (33): Tai Tsun Wu and Sau Lan Wu in Balexert Shopping Center, Geneva, Switzerland; in Christmas 2019.

Photo (34): Tai Tsun Wu with his happy family members — from left to right: Tai Tsun Wu, Sau Lan Wu, niece Elaine Yu, mother-in-law Ying Lai, sister-in-law Lynne Yu, and niece Joyce Yu; in 1982.

Photo (35): Tai Tsun Wu with his happy family members — from left to right: in the front row: niece Joyce Yu, mother-in-law Ying Lai, and sister-in-law Lynne Yu, in the back rows: niece Elaine Yu, Tai Tsun Wu and brother-in-law Ming Lun Yu; in 1994.

Photo (36): Family excursion — from left to right: niece Elaine Yu, Tai Tsun Wu, mother-in-law Ying Lai, brother-in-law Ming Lun Yu, sister-in-law Lynne Yu, niece Joyce Yu, and Sau Lan Wu; in 1999.

Photo (37): Tai Tsun Wu with his happy family members — from left to right: mother-in-law Ying Lai, brother-in-law Ming Lun Yu, Tai Tsun Wu and Sau Lan Wu; in 2004.

Photo (38): On the wedding day of Joyce Yu — from left to right: brother-in-law Ming Lun Yu, sister-in-law Lynne Yu, niece Joyce Yu, Sau Lan Wu and Tai Tsun Wu; in 2016.

Sau Lan Wu would like to thank Evelyn Mietzner for working with her in producing the wonderful set of photos.

Chapter 3
Remembrances from Tai Tsun Wu's Family, Friends, and Collaborators

(1) Remembrances from Family Members

(a) From Ming Lun Yu and Lynne Yu (Brother-in-law and Sister-in-law)

It is well-known that Professor Tai Tsun ("TT") Wu was an intelligent and serious physicist. He made invaluable contributions to the understanding of many areas of physics including statistical physics and elementary particle physics. On the other hand, few people had seen his "softer" side where he enjoyed Mother Nature and had fun.

I am Ming Lun Yu, his brother-in-law. My wife Lynne Alison Yu and I had a lot of fun with the beloved "TT". In the earlier days when I was a graduate student, "TT" and Sau Lan came up with the idea of driving from Los Angeles all the way to northern California to see the natural wonders and visit some universities along the way. The over 1000-mile round trip took us through the beautiful coastal Highway 1, saw the Redwood National Park and the unique Lassen Volcanic National Park, and visited UC Berkeley and UC Santa Barbara along the way. The success of the trip precipitated future family trips to Disneyland and Disneyworld, Hawaii islands, and Western Europe when "TT" and Sau Lan were doing research at CERN. But probably very few of you would have imagined that "TT" took a rafting trip along the Colorado River all by himself in the 1980s.

Many of your probably noticed that "TT" and Sau Lan never had a TV in their home and never went to movie theaters. But when "TT" was visiting us, we succeeded to "corrupting" him to go to see movies and watch TV. Our two daughters had a lot of good time with "Uncle TT". They will describe that in their separate writeup.

We treasure the memories of a fun loving "TT".

"If TT took your side of the debate, you would definitely win" — Lynne Yu

(b) From Joyce Yu
(Niece, daughter of Ming Lun Yu and Lynne Yu)

We called him "Uncle TT."

My memories of being his niece throughout childhood, my teenage years, and into adulthood are ones I'll always cherish. As kids, my sister and I would get so excited whenever Uncle TT came to visit. He loved his nieces, and it showed. His kindness felt easy and effortless — even a bit youthful himself.

As the years passed, I'd catch him on speakerphone and call out, "Hi TT!" and he'd shout back with joy, "HI JOYCE!!!"

We always kept things light and simple. He was generous with his warmth, and especially his smiles.

I will always remember him fondly and feel grateful to have had an uncle who was a good man. We all miss him very much.

(c) From Elaine Yu
(Niece, daughter of Ming Lun Yu and Lynne Yu)

As my uncle's youngest niece, TT always struck me as a unique character and was consistently kind to me. He of course was known as a titan in the world of theoretical physics, but I got to see him as our sweet uncle.

When my sister and I were children during our hyper-energetic shenanigans age, he was patient and willing to play with us. I recall my sister and I playing a game of "hostage" where we tied our uncle to a chair in our room with jump rope, turned out the lights, closed the door and ran away giggling. TT happily played along with no complaints; my parents found him tied up in the dark later.

TT also never talked down to me as a child and always talked up to me as an adult. Occasionally he prodded my mind with physics questions he was working on, and would ask me things like "Elaine, how to calculate the shadow of nuclei?" with complete seriousness, wondering if somehow my young mind might come up with something he hadn't thought about. Naturally I had no idea, but he forced me to think about it anyway.

He bought me his favorite books like Sir Arthur Conan Doyle's Sherlock Holmes series when I was teenager as they were his favorite books.

TT was also a math and physics genius with a montropic mind. He was the only person that I knew that could do complex math in his head, constantly working out calculation problems, *anytime, anywhere*.

It was also never clear from his facial expressions that he was deeply embroiled in a math or physics problem. I remember approaching him when he was resting on the couch, where he appeared half asleep, and I informed him "TT, there's lunch," and he would reply "Wait, I'm working on a problem." Generally, if he ever spent a long time in a bathroom, I assumed he was finding a quiet place to hide to think about a calculation.

I distinctly remember arriving at the Fremont BART train station parking lot, walking towards the station, and a family member asking TT if he could remember our parking stall number so we could pay inside. He replied, "Don't interrupt me, don't introduce a new number, I've been working on this calculation in my head."

I remember TT telling me he once considered focusing on a career in Chemistry when he was in school, but he was too clumsy to pursue such work.

Despite achieving financial success at Harvard, TT also had simple tastes and practically no need for material things, a quality I greatly admire. He was one of those people that seemed completely nourished by ideas and intellectual conversation. He also loved to debate (mostly about physics) and often liked to play devil's advocate for the sake of challenging your assumptions.

He also liked to wear the same thing everyday, which according to my memory did not change throughout the decades of knowing him. He wore a light colored camel trench coat, baby blue knitted sweater Sau Lan's mother, white shirt and slacks. If the sweater had holes, my grandmother would patch it up. The only time he would buy new clothes is if his clothes were truly disintegrating.

As a child I remember visiting his home in Boston when he was working at Harvard and only seeing peanut butter in his refrigerator. He loved peanut butter. I once asked him how he got so tall when he was growing up in Shanghai and he said he ate many eggs everyday.

Even during his last years during his battle with cancer, he was still working on physics problems. His one other pleasure according to Sau Lan was reading kung fu novels in Chinese that caused him to stay up late at night.

Perhaps the only thing he loved more than physics was Sau Lan. He was always brimming with pride to be her husband and supportive of her career.

TT was generous and kind, inquisitive and it was very clear he loved our family. He is sorely missed.

(d) Patrick Tai 戴振樑 (family friend)

I met T.T., strangely enough, when we both visited our dentist in Cambridge, MA. Coincidentally, we had the same dentist, and we both were there for our dental appointment!

We laughed at the coincidence. In the pleasant mood, T.T. humorously pointed to a parked car and said: "This is a Ford. … It's strange that they could not decide whether it is a 'F' or a 'D'!"

We all thought it was a very funny joke. I liked T.T. right away.

Afterwards, I met T.T. several times, sometimes with you. Maybe we tried out your skill in steaming Chinese dumplings using steamers. Sometimes we did that in T.T.'s Faculty Residence in Cambridge. We enjoyed ourselves.

I am sure it could be written better, but the addition note that I might add to the Memorial of T.T. may go like this:

I met Professor T.T. Wu outside our (same) dentist's office. He was leaving and was in a good mood. He pointed to a car parked there and commented: "It is a Ford. They could not decide whether it is 'F' or 'D'!"

Leaving a dental operation in a good mood was strange enough; but making a joke labeled him as a real humorous person. I liked him immediately!

(2) Remembrances from Friends and Collaborators

(a) Carl M. Bender
(Washington University in St. Louis)

It was almost midnight and I was in my pajamas when there was a knock on the door. It was Tai and he was wearing a broad smile. "Let's go to my office. I have an idea and we have eight hours to work." There are not many thesis advisors who are truly enthusiastic about the thesis research of their graduate students. Tai was rather unusual, to say the least.

While I was Tai's graduate student, Tai almost never taught me facts. I already knew the equations and formulas that we were studying. But Tai was a teacher in the profoundest sense of the word. He demonstrated how to think. He showed me how to make friends and shake hands with a problem, how to disconnect from a nonproductive approach, how to think in novel ways, how to make an indirect attack on a difficult problem that doesn't yield to a conventional analysis.

When I made advances of my own, Tai's praise was warm, effusive, and sincere, and his influence has lasted for my entire life. When faced with a difficult analysis, I still begin with the question, "What would Tai try in this situation, how would he approach this problem; what would he try first, and what would he do if this failed." It is rare indeed for a graduate student to have such a mentor.

All together, Tai and I co-authored six influential papers, two while I was Tai's graduate student. We examined perturbation theory in great depth. We explained why perturbation series diverge (Bender–Wu singularities) and we created a huge new area of theoretical research activity (large-order behavior of perturbation theory). Over the years Tai transformed from my enthusiastic mentor to my warm and caring colleague. I am extraordinarily lucky to have known and worked with him.

(b) Hong-Mo Chan
(UK Research and Innovation: Science and Technology Facilities Council (STFC))

I first met TT when I went to the Princeton Institute from Hong Kong. We were both young, of nearly the same age. I was a fresh graduate from Britain, but obliged by a grant to teach back in Hong Kong for a couple of years, but TT had already gained then a name for himself. My first memory of him is a call past midnight (1 am to be exact) asking me to join him in a nearby café for ice cream. This was a unwelcome request for me, for being a light sleeper all my life, I have always kept a rather strict schedule. But I appreciated the friendly intention and went along. And we have been friends ever since, though meeting only infrequently when our trajectories crossed.

Though friends, we have never worked together, hence the absence of any links to common work, as suggested by the editors. We differed too widely in interest and taste. And also, knowing him to be far superior in both knowledge and competence, I feared being reduced to just a follower if I worked with him.

Even so, TT has now left me an example to follow. The news of his decease, much belated, came to me only recently from Sau Lan, because I myself am in a care home detached from usual communications. It was a surprise, but surprisingly not of great sadness, for both of us are now of such an age that the end, we know, can come any day. However, we are searchers and it is to us both a vocation and a duty to seek the truth, although the truth, as TT said the last time we met in Singapore, may be revealed only after we are gone. You TT have bravely remained in your post seeking the truth until the very end. I applaud your dedication, and shall try to do the same.

(c) Ken Chen (Sandia National Laboratories, Albuquerque, New Mexico, USA)

I first met Professor Wu when he visited the Air Force Research Laboratory at Kirtland Air Force Base in Albuquerque, New Mexico in the early 1980's. He stayed with my family and we developed a lasting friendship for many decades. Initially, he helped review generation and coupling effects of Nuclear Electromagnetic Pulses, but these discussions led to great progress in our understanding of other problems, notably electromagnetic coupling to long lines, grazing incident on power lines, and buried long-line coupling.

Tai was a special influence in my life, especially how he picked problems to solve, much like da Vinci selecting the Mona Lisa as a subject:

"It was in the nineteen sixties when I had a chance encounter with Robert Woodward, a professor of chemistry at Harvard specializing in organic chemistry. I had known him slightly before, and this was the only significant conversation that had occurred between us. But this conversation had a major effect on my career. This is undoubtedly the major advantage of being at a first-rate university — the chance encounter with first-rate minds. Here is a very rough reconstruction of that conversation.

I opened this conversation by asking him the following question: 'You have a reputation of always knowing which problem is the most important one at any moment, and then working on that problem. How do you figure out which problem is the most important?' "

RW: 'Nobody can know which problem (in organic chemistry) is the most important, and I certainly do not.'

Me: 'In this case, how do you choose which problem to work on?'

RW: 'This is a better-defined question: within reason, I always choose to work on the most difficult problem.'

He then thought for a moment, and then added: 'After some time, the less difficult problems are all solved, but I am still working on my most difficult problem. It was then that the most difficult problem that I am still working on becomes the most important problem to solve.'

Even with such a great career he longed to answer great questions after retirement from Harvard. Our conversations on Yang–Mills non-Abelian gauge theory captured this most poignantly:

"I was still a student in the mid-fifties and thus missed these important developments at that time. And you were still a teenager at that time. If there is a major development coming in the near future as I am expecting, I would not like to miss it again."

(d) Ekaterina Christova
(Institute for Nuclear Research and Nuclear Energy in the Bulgarian Academy of Science, Sofia, Bulgaria)

In Memoriam: T.T. Wu (1933–2024)

With sadness and profound respect, we remember Dr. Tai Tsun Wu for his pioneering research in particle physics, and particularly for his key role in the development of weak interactions and our understanding of the Standard Model (SM) of particle physics.

One of his most notable and lasting contributions came in his study of the Higgs boson and in understanding its decay into two photons (2γ) — a key signature in the discovery of the Higgs boson at the Large Hadron Collider in 2012.

In 2014 T.T. Wu discussed this decay with Dr. Ivan Todorov (IT), the head of the theory division in our Institute. As the photon is a massless vector field, it does not interact directly with the Higgs boson but via W-loops. There are many calculations of the amplitude of this decay, but with controversial results, both in unitary gauge, that involves only physical fields, and in the commonly used R\xi-gauge, when one deals with divergent integrals and some kind of regularization is always applied (usually DimReg. is used). Thus, the problem seems to originate from ambiguities of the chosen regularization. To avoid this, T.T. Wu and IT suggested to apply the dispersion method, its advantage being that it deals only with finite quantities and avoids the ambiguities with regularization.

In the dispersion method one starts with the absorptive (imaginary), absolutely convergent part of the amplitude, the full amplitude is obtained by the dispersion integral over the absorptive amplitude. But again we face uncertainties — though convergent, the dispersion integral determines the amplitude up to an additive constant and a boundary condition is needed to fix it. As I already had some experience with Cutkosky rule for obtaining the absorptive amplitude, IT suggested that we do this calculation together. In the SM there is no coupling of Higgs boson to photons, and it seems natural to apply the unsubtracted dispersion integral, thus we obtain [1] an earlier result of T.T. Wu and collaborators. However, using the same dispersion integral but with subtraction, the subtraction constant determined by the Goldstone equivalence theorem that determines the Lagrangian in the limit of zero mass of W^\pm, the result [2] coincides with calculations in R\xi-gauge with DimReg. These results are published in 2 papers, both initiated by the discussions of T.T. Wu with IT.

This is just one example of Dr. Wu's spirit and influence on research in Physics and I'm happy to have been involved in it.

References

[1] E. Christova and I. Todorov, *Bulg. J. Phys.* **42**, 296 (2015), arXiv: 1410.7061.

[2] I. Boradjiev, E. Christova and H. Eberl, *Phys. Rev. D* **97**, 073008 (2018), arXiv: 1711.07298.

(e) Remembrances of Tai Tsun Wu — Robert Coquereaux (Physicien théoricien au Centre de Physique Théorique (CPT), Marseille, France)

I met Tai for the first time in Marseille, probably in the spring of 1979. He was visiting his friend and colleague Alex Grossman who was working at the Centre de Physique Théorique, in Luminy. At that time I was a young theoretical physicist who had just completed his PhD, also at CPT, and I had been lucky enough to get a research position at CNRS, in France, just after my thesis. We had a few discussions, on physics of course, and I remember that Tai kindly asked me to explain to him what I had been working on. It was a short visit, and Tai, who was a known figure in the world of particle physics, was very busy. However, a short time after Tai's visit, Alex disembarked into my office and said: "Robert, I just had a chat with Tai. He invites you for one year at Harvard". I have to confess that I was a little dumbfounded because I had not officially applied for anything but I was of course delighted at the idea of spending a full year in Cambridge. I could manage to obtain a one year leave of absence from CNRS, so that I could accept Tai's kind offer, and on the first days of 1980, together with my wife and two young children, we landed in Boston.

This was a wonderful year. With the help of Tai and Margaret, I readily got used to this New World and to this prestigious institution. Several times a week, after his class, Tai was looking for me, in an office located in Jefferson, or I was joining him at Gordon McKay, to have lunch together in Harvard Square, Then Tai would have tea, or rather a cup of hot water, an "infinitely weak tea", as he used to call it, and I had a coffee. Sometimes we played Go, at lunchtime, but he always won! For some reason, we did not publish any joint paper, but we had countless conversations, on all kinds of topics, mostly particle physics, but also statistical mechanics or mathematical physics. We were already involved in distinct projects, and Tai had been very clear from the start: I should follow my own path, even if it was not going in the same direction as his. I nevertheless suspect that, without openly criticizing, he was disappointed from the fact that I had a tendency to be too formal, preferring the beauty of mathematical constructions to its actual physical applications. On the contrary, Tai was never losing sight of physics, even if he was using rather abstract methods from algebra or analysis. At that time, Tai used to wake up very early, and he started his days by checking (correcting!) several pages of the Bateman Manuscript Project — a five-volume compilation of the mathematical theory of special functions — to warm up his brain cells! He was then ready for a new day of research and teaching at the university. Tai used to come every day to the institute, but sometimes his wife, Sau Lan, who was already a famous experimental physicist and who was not often in Cambridge, as she was travelling around the world, was going through Boston; on such occasions Tai was telling me: "Robert, tomorrow I shall not come to the university: my wife is visiting!" Thanks to Tai, this year 1980 was, for me, truly exceptional. Independently of our interactions, I could meet several other grand figures of science who were luminaries at Harvard University at that time. I could talk to them, and benefit from their advice and teachings. I shall never forget.

One page is of course too short for collecting more memories but I can mention that I spent another year with Tai in Harvard. This was in 1987. Again I was there with my wife and two kids. Again this was a wonderful year, and I once more had the opportunity to witness the eclecticism of Tai's mind, easily jumping from quantum field theory to mechanics, from particle physics to the Ising model, while writing research articles on nucleon polarisation or on antenna properties. …Tai was at the same time giving lectures in Harvard on the propagation of sonic or electromagnetic waves in the oceans, taking into account pressure, temperature gradients, salinity, or even waves or swells. Although it was totally outside of my research interests, I decided to follow his class, and this was an unexpected and nice experience. Later we mostly met at CERN, where Tai had an office and spent several months each year, I too would occasionally visit this institute. Each time we had nice conversations, for instance on the LHC, and more than once Tai tried to convince me of going back to the phenomenology of particle physics (he failed). I have fond memories of Tai Tsun Wu, especially those connected with my first visit to the States. He gave me the opportunity of sharing the activities of Harvard University scientists, of discovering new landscapes, and of meeting several bright minds. I am deeply grateful to Tai for all of this. It has been a nice period in my life, probably also because I was younger then!

(f) George Fikioris
(School of Electrical and Computer Engineering, National Technical University of Athens, Greece)

Tai was my Ph.D. advisor at Harvard. Here, I will describe the story behind [1], which I co-wrote with Tai after I received my Ph.D. in 1993. The paper [1] pertains to unphysical numerical solutions of a very popular integral equation in the theory of linear antennas.

One day in 1994, I visited Harvard to spend the day with Tai. That morning, I was aware of the existence of unphysical solutions, but I could not adequately explain, predict, or avoid them. So soon after arriving, I asked Tai a vague question about the solutions. Although he acknowledged my question and told me it was interesting, he did not immediately answer.

We spent the entire day talking about unrelated problems, had lunch in Harvard Square, and returned to the Gordon McKay Lab late in the afternoon. When our other discussions winded down, I said, "What about the question I asked you this morning?" He immediately replied "I'll tell you how to answer that question," and started writing on the whiteboard. I listened carefully, taking notes.

A few hours later, I had a more-or-less complete answer to my morning question and the foundations for [1] had been laid. Tai's key idea was to apply a numerical method to the linear antenna of infinite length, for which the "numerical solution" could be expressed analytically. The understanding thus obtained easily carried over to the realistic (finite-length) antenna, for which no such analytical study was possible. But what Tai wrote on the whiteboard was much more than an original and foundational idea; his writings included nontrivial, illuminating calculations that he carried out effortlessly, as if he was copying from notes.

There were no such notes. As far as I know, Tai became aware of the problem when I asked him my question that morning. We had spent the entire day together — talking about unrelated topics — so I am sure he did not write anything down. It seems that he solved the problem and performed the necessary calculations subconsciously.

Tai was the best advisor I could have, and a very dear friend who guided me in many aspects of my life. My father John (who passed away in 2012) felt the same way as Tai was his advisor, too. It is my privilege to be the second generation Wu/King student.

Reference

[1] G. Fikioris and T. T. Wu, "On the application of numerical methods to Hallen's equation," *IEEE Trans. Antennas Propagat.*, vol. 49, no. 3, pp. 383–392, March 2001.

(g) Raymond Gastmans (Institute for Theoretical Physics, University of Leuven, Belgium)

As an outcome of the 1978 LEP Summer Study in Les Houches, France, TTWu and I worked out the high energy limit of the cross section formula for the process $e^{+}+e^{-} \rightarrow e^{+}+e^{-}+\gamma$. Its knowledge was essential for an accurate determination of the luminosity of the electron-positron collider LEP at CERN. After a lot of calculations, the formula was found to be sur-surprisingly simple, which made us believe there had to be a much better way to derive the result. Together with the colleagues Patrick De Causmaecker and Walter Troost from the University of Leuven, we developed such a procedure, which is presently known as the helicity method. Until then, cross sections were obtained by summing all Feynman diagrams and squaring the sum. In the helicity method, the cross section is the sum of the squares of individual helicity amplitudes, which are easy to derive. Because of its simplicity, the helicity method is now widely used to study high energy collisions, even if they are described by extremely large numbers of Feynman diagrams.

I learned from Tai that (i) it pays to question 'generally accepted ideas', and (ii) it pays to work hard. I am very grateful to him for those lessons.

(h) Vassilios Houdzoumis
(School of Electrical and Computer Engineering, National and Technical University of Athens, Athens, Greece)

"In addition to his exceptional scientific stature, I will remember T.T. Wu for his fatherly advice.

In correspondence with me, his style had always been witty and unconventional; thus delightful."

(i) Choo Hiap
(National University of Singapore, Singapore)

"I first met Tai Tsun in person at the China conference in Jan 1980 (1980 年广州粒子物理理论讨论会) although I learned much of his famous work with Prof. C.N. Yang much earlier, the Wu–Yang ansatz 1967, the global formulation of Wu–Yang monopole in 1975. My first impression then, he smiled heartily, childlike, he was warm and sincere.

I really got to know Tai Tsun much better in 2006 when he was invited as the LKY visitor under the Lee Kuan Yew Distinguished Visitors Programme, visiting Singapore universities, 28 August–7 September 2006; during that week we met daily. He was very kind and generous to me, introduced and explained several frontier topics to work on. In discussing physics with him, he would first ask what I would say, and then he would explain to me the deeper way of analysing the problem.

I also remember he explained to me about an antenna design by him for the aircraft. He also visited some of the local bookshops and bought several Chinese classics and thence I realized he also enjoyed reading Chinese literature and philosophy.

Subsequently we met whenever he visited Singapore, e.g. in 2007, Conference in Honour of C.N. Yang's 85th Birthday, Singapore, 31 Oct–03 Nov 2007.

We maintain email correspondences until Feb 2024. I missed Tai Tsun dearly.

(j) Manja Klemencic and Khaled El Rouayheb (Harvard University, USA)

Remembering Tai Tsun Wu, a neighbor and a friend.

In the early 2010s, we would walk through Robinson Street, past Tai's house, every day on the way to and back from Montessori school in Cambridge. It was on one of these walks that we first met Tai. We struck up friendly, neighborly conversation. Soon, our children would recognize Tai from afar and greet him. We remember him in his elegant beige coat, freshly pressed shirts, and thick glasses. Tai loved talking to the children. With our son, he discussed chess. Tai said that he played only once with his supervisor, and that he lost in four moves with the Scholar's Mate. When our daughter got new glasses for being shortsighted, he showed her his own glasses and reassured her that he had them all his grown life and still sees well.

A few years later, we turned to Tai and his neighbor Nur Yalman when deciding whether to purchase a house on sale at Robinson Street. Tai and Nur's houses were of the same Bauhaus-style, cute and functional with large windows, that Harvard University built on Robinson Street for its faculty in the late 40s and early 50s. They both encouraged us to purchase the house for which we are still grateful. After we purchased the house and moved in, our friendship with Tai strengthened. For his 80th birthday, his colleagues invited him to a Chinatown restaurant in Boston and got him a huge cake. He brought the cake to us since soon after his birthday on 1st December he was travelling back to Geneva. He joked that his house is very comfortable without him since he always left heating rather high to make sure that pipes didn't freeze over the winter.

It was few years later that we finally met Tai's wife Sau Lan. Tai spoke to us about Sau Lan often and lovingly. His three months of separation from her every academic year were evidently difficult for him. He was so happy when Sau Lan joined him in Cambridge. When she came, she helped renovate the house and make it more suitable for full-year residence. Meeting Sau Lan, I understood why Tai admired and loved her so much. She is a remarkable woman, an accomplished academic, with a strong practical sense and a very strong bond to her husband. Sau Lan and Tai engaged our children to run small errands for them: shovel snow, pick up the newspapers when away, prepare the bins for weekly trash pickup. These were their first ever jobs, and their first monthly earnings, which they still remember fondly, ten years later.

Tai and Sau Lan returned permanently to Cambridge from Geneva with 6 large suitcases in March of 2020 at the height of COVID-19. These were challenging times. The Cambridge house was a good place to isolate as it allowed them to walk around and up the stairs. We were sharing news and tips on managing the hard times. With Sau Lan, and Tai next to her, we talked every day. Our own parents are abroad — Denmark, Slovenia, and Lebanon. Tai and Sau Lan were, in a sense, the only family we had in Cambridge. We helped each other as much as we could. Since Tai and Sau Lan kept in isolation, I offered to help them pack up Tai's office at Harvard when he decided to retire. I arranged for movers and organized his books, papers, and even two dear Harvard office chairs to be shipped to California. It was emotional to say goodbye when they decided to move to California. But the move was obviously the right decision for them since they were moving to a beautiful, spacious new apartment in superb senior living complex in Palo Alto. We took care of their house in Cambridge, but it didn't make sense for them to keep the house given that they were not planning to return. I spent more than six months with Sau Lan and Tai organizing their belongings in Cambridge house and preparing them for shipment to California. It was hard work and emotionally draining for them to decide on items that have been in the house over 40 years since Tai purchased it from Harvard University. With Tai, we went through every book in his study and all the bookshelves of notes and materials. It was very touching seeing Tai decide, item for item, which of his belongings he wanted to be sent over. Sau Lan and I talked regularly doing the process, sometimes more than once a day, though during our conversations she would often turn to Tai to ask whether he wanted something to be sent or not. I particularly remember a beautiful walnut table from Denmark which Tai worked on in his study and which he was very happy to receive. I also made sure to send a framed article in New York Times which mentioned Tai and his work. The shipment was sent in May 2023, just before we left for Europe for our sabbatical leave. Sadly, we never saw Tai in person after they left Cambridge for California.

We saw him, on FaceTime, for the last time a few hours before he died in a Stanford hospital. Sau Lan called us to allow us to say goodbye, for which we are for ever grateful. We cherish our memories of Tai the way we knew him during our Cambridge years. We can still see him vividly, in our mind's eye, as he was when we first met him on our walks to our children's school: dressed elegantly, walking in a careful, stately manner, an elderly gentleman who spoke kindly to our children, mentioned his wife Sau Lan with adoration, and whose astounding intelligence and life wisdom was apparent after just a few minutes of conversation.

Dr Manja Klemencic and Professor Khaled El Rouayheb with Ajda and Rami.

(k) James C. Liao
(President of ACADEMIA SINICA, Taiwan, R. O. C.)

中央研究院
ACADEMIA SINICA
128 Academia Road, Section 2
Taipei 115, Taiwan, R.O.C.

廖俊智 院長
James C. Liao, Ph.D.
President
Tel:+886-2-2789-9400
Fax:+886-2-2785-3852
http://www.sinica.edu.tw
E-mail:liaoj@gate.sinica.edu.tw

February 11, 2025

To the Family of Dr. Tai Tsun Wu:

On behalf of my colleagues at Academia Sinica, I wish to express our profound sorrow upon belatedly learning of the passing of Dr. Tai Tsun Wu, one of our most renowned Academicians and a well-loved colleague and friend. We extend our heartfelt condolences to your family for your loss and are deeply saddened by this news.

Dr. Wu dedicated his life's work to physics. A profound scholar, he made significant contributions to various areas, including antenna theory, statistical physics, and perturbative quantum field theory. His groundbreaking research and innovative theories garnered him widespread recognition in the academic community. In his later years, his passion for academia remained undiminished. He devoted his time to research and to mentoring the next generation of scholars.

His unstinting efforts were truly exemplary. During his distinguished career, Dr. Wu was elected Member of the American Academy of Arts and Sciences. He also received numerous honors, including the Dannie Heineman Prize for Mathematical Physics of the American

Institute of Physics and the American Physical Society and the Humboldt Foundation Prize, garnering well-deserved acclaim.

Please know that our thoughts are with you. The memory of Dr. Wu's exceptional character will endure at Academia Sinica, and we will draw strength from his many achievements. May he rest in peace and may his contributions forever receive the honor they deserve.

Sincerely,

James C. Liao, PhD
President

(l) Dionisios Margetis
(University of Maryland, College Park, MD, USA)

Tai Tsun Wu was my doctoral advisor. I was his last Ph.D. student at Harvard. He was my mentor and friend. Our close interaction started during my first week as a graduate student, and continued until a couple of months before Tai's passing. Tai taught me to tackle problems by wrestling with them fiercely, as if my life depended on this; and to keep believing in myself even when I fail.

There are many memorable vignettes of my experience in working with Tai. I will write about two of these here.

When I first arrived at Harvard, I was eager to impress my advisor with my knowledge of applied mathematics. In our first meeting, Tai kindly asked me to go to dinner with him at a Chinese restaurant on Beacon Street. His next question was whether I knew how to use chopsticks. I hesitated, and almost whispered something like "Not so well". Tai, in a half-serious manner, said "You will never graduate if you don't learn how". During dinner, Tai asked me to explain to him what I had learned about the Wiener-Hopf method from coursework or on my own. I made the attempt to summarize the main points from Noble's book. "I don't understand", Tai rebuked immediately. After a few more attempts of mine, and Tai's sharp criticisms of them, he finally remarked "You seem to know a lot. From now on you need to understand."

Three years later, I was working on a problem related to superdirective current distributions in electromagnetism, as a part of my thesis. Roughly speaking, highly oscillatory currents can generate arbitrarily narrow radiation patterns. Seeds of this idea can be traced back to Oseen's old hypothesis for the "Einstein needle radiation". For a circular geometry in 2D, the formulation led to an infinite series involving Bessel functions for the current. I tried to evaluate this series asymptotically for a large radius. This looked like a horribly tricky task to me. I kept trying for days. One day, I mentioned to Tai my slow progress. "You are too impatient", Tai responded and asked me to remind him of the specifics. He suggested that we go to lunch together. During lunch, we discussed numerous topics unrelated to the subject of my research, e.g., my progress on dating girls and the political situation in Greece. When we returned to the Gordon McKay Lab, Tai took a nap in a big chair. After about half an hour or a little longer, he stood up smiling. "I can tell you how to simplify the series", he said. Tai started to write on a whiteboard with great detail. It was obvious to me that he had not kept or written any notes on this previously. I concluded that he had thought about the solution of the problem during his short napping. His argument involved a delicate and extremely technical decomposition of each term of the series, with a subsequent conversion of the series to integrals. Tai unveiled the decomposition in an intuitive fashion, like the artist who paints a beautiful picture. Later on, I verified that the approximation derived this way was in excellent agreement with numerics.

Dionisios Margetis,
Professor of Mathematics,
University of Maryland, College Park

(m) John Myers
(Harvard University, Cambridge, MA, USA)

Tai was so very alive. Over a long afternoon, we might sit silent for half an hour, waiting for an idea to come — a trick he taught me — then have a loud argument, and then buy and devour a half gallon of his favorite Mississippi Mud ice cream. I first encountered Tai in 1960 when he was teaching a special topics course, nominally in electromagnetism. What I best remember of that class had more to do with quantum theory: Tai spent the class hour at the blackboard, on which we could see him working on a problem that he currently cared about, namely the convergence or lack of it of the Born approximation.

Two years later, Tai played a crucial part in my life. He and I had done our graduate work under Prof. R. W. P. King, a world expert on antennas. When I turned in a draft of my thesis to Prof. King, he gave it to Tai to evaluate. Without Tai's approval, that would have been the end of that thesis.

Years after graduation, I came back to Harvard to collaborate with Tai in many ventures in electromagnetism and a few in particle physics. Unlike electromagnetism, which is essentially defined by the equations of Maxwell established in the 19th century, particle physics is open and, in the word borrowed by Tai from C. N. Yang, "muddy." Tai and Hung Cheng overturned the habit of relying on the first few orders of perturbations theory to sum selected Feynman diagrams to all orders, arriving at their famous prediction that the cross section of proton-proton scattering would continue to increase with energy, without limit [H. Cheng and T. T. Wu, *Phys. Rev. Lett.* **24** (1970) 1456]. Ever increasing energies of collider experiments confirmed this.

In 2011, Tai, with Raymond Gastmans and Sau Lan Wu, again broke a habit of theoretical particle physics, the habit of ignoring the strict mathematics that distinguishes integral of a limit from the limit of an integral. They showed how the failure to make that distinction leads to wrong answer in the decay of a Higgs particle to two photons. As if that wasn't enough, more drama came in 2023 with the publication that drew on both the need to sum selected Feynman diagrams to all orders and the need to distinguish a limit of an integral from the integral of a limit, culminating in Tai's and Sau Lan's "Basic Standard Model" that reduces the number of free parameters from 25 to 6 [T. T. Wu and Sau Lan Wu, *International Journal of Modern Physics A*, Vol. 38, No. 28 (2023) 2350128].

For over sixty years I have looked to Tai as a friend, as a mentor, and as a model of courage in science.

(n) Per Osland
(University of Bergen, Norway)

I owe my career as a physicist to Tai. In the 1980s the openings at Universities in Norway were very scarce. After some years as a postdoc at NORDITA and at Harvard (with Roy Glauber) I came to DESY in 1982. Tai was on sabbatical there, since Sau Lan was at that time working in TASSO. We started working together on magnetic monopoles, solving the Dirac equation for Kazama–Yang bound states. This work relied on a lot of properties of various special functions described in the famous Bateman project books. Tai was a great expert on this. At the end of my two years at DESY there was still no opening in Norway, but Tai generously offered me a postdoc position at Harvard. At Harvard we continued our study of monopoles, and also wrote a couple of papers on ladder diagrams in phi-cubed theory. I stayed at Harvard until offered a position in Bergen in 1987.

Our collaboration continued after I moved to Bergen. Sau Lan was by then in ALEPH, and preparing for the LHC. When not teaching at Harvard, Tai was spending most of his time at CERN. So our collaboration continued, by Tai coming up from CERN, or I visiting him there. Our topics included a study of the non-linear

Boltzmann equation, divergences in the electroweak theory, Higgs phenomenology and analytic studies of neutrino propagation and mixing. He taught me a lot of very useful mathematics.

Tai was very generous, in identifying co-authors, giving credit, and in insisting on paying restaurant bills. I have many fond memories of working sessions at large blackboards (but first they had to be carefully washed), or him seated in a good chair after a good meal. For really deep thinking, he would then sometimes take off his socks.

(o) Jean-Marc Richard

(Institut de Physique des 2 Infinis de Lyon, Université Claude Bernard (Lyon 1), France)

This note contains a brief recollection of my interaction with Tai T. Wu.

Starting from the early 90s, I had the chance to meet Tai regularly at CERN and benefit from long discussions with him on many physics issues, often finishing in a French restaurant across the border, where we talked on a wide variety of topics, including French politics that Tai followed rather closely.

Tai had a deep expertise in many fields, but during most of the recent years, he concentrated his efforts mainly on the dynamics associated with the Higgs boson. André Martin and I nevertheless convinced him to investigate together some problems of mathematical physics into which he plunged with relish and energy. The first one was the domain of stability of three charges in the space of their mass ratio [1–3]. The second one, triggered by Jean-Louis Basdevant, was about the comparison of N-body and (N − 1)-body energies, a problem encountered in several domains such as nuclear physics, quark models and stability of matter [4, 5]. The third one, on which Nick Khuri collaborated, concerned the low-energy behavior of scattering amplitude in 2-dimensional quantum mechanics [6]. In each case, Tai astonished us by his enthusiasm and rigor, and he was really the driving force of the team.

References

[1] A. Martin, J.-M. Richard, and T. T. Wu. "Stability of three-unit-charge systems". In: *Phys. Rev. A* **46**.7 (Oct. 1992), pp. 3697–3703. Doi: 10.1103/PhysRevA.46.3697.

[2] A. Martin, J.-M. Richard, and T. T. Wu. "Stability of systems of three arbitrary charges: General properties". In: *Phys. Rev. A* **52**.4 (Oct. 1995), pp. 2557–2563. Doi: 10.1103/PhysRevA.52.2557.

[3] A. Krikeb, A. Martin, J.-M. Richard, and T.-T. Wu. "Stability Domain of Systems of Three Arbitrary Charges". In: *Few-Body Systems* **29**.4 (Dec. 2000), pp. 237–257. Doi: 10.1007/s006010070003.

[4] J.-L. Basdevant, A. Martin, J.-M. Richard, and T. T. Wu. "Optimized lower bounds in the three-body problem". In: *Nuclear Physics B* **393**.1 (Mar. 1993), pp. 111–125. Doi: 10.1016/0550-3213(93)90240-P.

[5] A. Benslama, A. Metatla, A. Bachkhaznadji, S. R. Zouzou, A. Krikeb, J.-L. Basdevant, J.-M. Richard, and T. T. Wu. "Optimized Lower Bound for Four-Body Hamiltonians". In: *Few-Body Systems* **24**.1 (May 1998), pp. 39–54. Doi: 10.1007/s006010050075.

[6] N. N. Khuri, A. Martin, J.-M. Richard, and T. T. Wu. "Low-energy potential scattering in two and three dimensions". In: *Journal of Mathematical Physics* **50**.7 (July 2009), pp. 072105–072105. Doi: 10.1063/1.316780to3. arXiv: 0812.4054 [math-ph].

Chapter 4
Celebration of Tai Tsun Wu's 60th birthday at CERN in 1994 — A Fun Dialog and Quiz edited by Conrad Newton and Per Osland

**** WHU'S Wu? A Roast in Honor of TT Wu's 60th Birthday ****

General Characteristics

Tai has been described by a famous physicist as which of the following?
A> The most persevering physicist
B) The best preserved physicist
C) The most perverse physicist

Note: I think answer (c) is almost as good as (a). The idea for this question came about because both Sau Lan and Raymond misread the title of the colloquium given by CN Yang, entitled "TT Wu, the most focused and persevering physicist I know". Sau Lan read it as (b), and Raymond read it as (c). In justice to them, it should be said that "persevering" was misspelled.

How does Tai signal his presence?
A) By loud coughing and blowing his nose
B) By whistling "somewhere over the rainbow" while he works
C> By singing Simon and Garfunkel's "Cecilia, you're breaking my heart"

Note: Clearly, the correct answer is (a), but Cecilia Jarlskog was sitting next to Tai at dinner, and so he had no choice but to answer with (c). Actually, coughing is not an appropriate description — “snorting” would be better — but we did not want to offend too many people’s sensibilities.

Food

When Tai goes to a restaurant for dinner, he considers which of the following issues:

A) The quality of the linens and the silverware
B) The charm of the hostess
C> The number of calories per dollar

Note: Tai is very fond of the waitress at the China Inn, but clearly (c) is the correct answer. Answer (a) is a joke.

When Tai goes for lunch at the CERN cafeteria, he prefers to eat:

A) Beef Kidney
B) Frogs’ Legs with ketchup, when available
C> Nut cake with red jelly

Note: Tai has gout, and therefore he cannot eat internal organs, especially kidney and liver. Last year, I was responsible for an attack of gout, because I did not realize that the French word “rognons” is a food word for “kidney”. Tai spent the next two weeks walking around CERN in bathroom slippers, because his big toe was too swollen for him to wear shoes.
Answer (b) is a joke which should be appreciated especially by the Europeans. At the dinner, Tai was presented with item (c) as a gift.

Entertainment

Tai’s favorite reading is:

A) Agatha Christie novels
B> The Bateman manuscript
C) The Herald Tribune

Note: All of the above were good answers. At any given moment, Tai could probably answer (a), (b), or (c). However, in the long term, (b) is the correct answer.

Tai prefers which of the following movies?
A> The Wizard of Oz
B) Terminator II: Judgement Day
C) Home Alone

Note: Answer (b) is a joke. Answer (c) is less good than answer (a).

Tai's favorite hobby is:
A) Collecting old cars (one at a time)
B) Gardening
C> Electromagnetism

Note: Answer (a) refers to Tai's habit of buying old cars. Answer (b) is clearly a joke — just look at his lawn. Joachim Bartels once asked Tai how he found time to pursue questions in electromagnetism, in addition to his research in high energy physics. Tai's response: "Some people have gardening. I have electromagnetism."

Tai prefers which of the following water sports:
A) Whitewater rafting down the Grand Canyon
B> Whale watching in Boston Harbor
C) Soaking in the hot tub

Note: Someone told us that Tai once went whitewater rafting with Barry McCoy. Tai said it was not true that he went through the Grand Canyon, but perhaps he was elsewhere on the Colorado river. To his close friends, it was obvious that the answer should be (c), but Tai was apparently too embarrassed to answer honestly. He was at first reluctant to answer, and said that it depended on the time of year, but we said he was only allowed to give one answer.

Professional Life

What are the most important tools for a long calculation?

A) Several dozen lead pencils
B> Several gallons of Breyer's vanilla ice cream
C) Several boxes of Kleenex

Note: Tai does not use pencils, only pens, so (a) is incorrect. (c) is not incorrect, but on balance (b) is more important.

When Tai needs to concentrate on a problem, he

A) Swings his pen back and forth between two fingers
B) Picks his nose
C> Paces up and down

Note: Truly, any of the above answers was correct. Sau Lan, who formulated the question and the answers, thought that the answer should be (a). Maurice Jacob replied without hesitation that the answer was (b). Answer (c) was the safest answer to give in mixed company.

When Tai is stuck on a calculation, he

A) Takes a nap
B) Takes a hot bath
C> Takes off his shoes and socks, and inspects the soles for embedded grains of sand

Note: Again, all of the above answers look good. I suppose that Sau Lan thought that answers (a) and (b) were too obvious. She says she forced Tai to reply with (c).

Where did Tai first meet Frank Yang?

A) As a young boy in Shanghai
B) In Einstein's office at the Institute for Advanced Studies
C> Within sight of the women's dormitory at Harvard

Note: All of these sound plausible. Both Wu and Yang are from Shanghai originally, and they spent time together at the Institute for Advanced Studies. Nevertheless, (c) is not only funny, but correct. Yang's brother was a friend of Tai's at Harvard when both were students, and Tai met Yang when Yang was visiting his brother. I'm not certain how close to Radcliffe they were at the time. Incidentally, Yang was named Frank after Benjamin Franklin, presumably after his scientific aptitude became apparent.

Tai is most allergic to which of the following:

A> Chalk Dust
B) Dimensional Regularization
C) Supersymmetric Partners

Note: This was definitely a question to ruffle some feathers among the guests, since supersymmetry has the status of a religion among the local population, and Tai is unquestionably a heretic. Dimensional regularization is also very popular among most high energy physicists, and Tai does not like it. To avoid making more enemies than necessary, Tai replied with (a), the correct answer from a purely medical point of view. Maurice Jacob, when he heard Tai's answer, said "You are too kind." John Ellis, one of the cult leaders of supersymmetry here at CERN, asked me about the questions afterwards, and claimed to remember that. Tai had replied with (b). I was a little nervous when he asked who had prepared the questions, since I wrote this question.

Domestic Life

Where did Tai first meet Sau Lan?

A> In Jack Steinberger's office at Brookhaven
B) Square dancing in Memorial Hall at Harvard
C) At the US consulate in Hong Kong

Note: We assumed that people would have trouble with this question, but I seem to remember that, most groups got it right. (b) sounds like the right answer for a normal couple, but Tai and Sau Lan are not a normal couple. Professor Yang gave the answer away in his talk when he said that his wife had given up teaching Tai how to dance. Parenthetically, the Yangs also gave up on teaching him how to swim.

Why did Tai lose his hair soon after he married?

A) Because Sau Lan needed it for her head

B) "A busy brain gathers no hair"

C> He only needed it as mating plumage

Note: Sau Lan has very long hair, so (a) made a pretty good joke. Answer (b) is a variation of a Dutch proverb about busy roads gathering no grass. Clearly none of the answers were serious, and so Tai chose the funniest one.

Why did Sau Lan agree to marry Tai?

A) Because he was so cute with his full head of hair

B> Because he said that she did not have to cook

C) Because he promised to do the income taxes for both her and her mother

Note: Answer (a) is supposed to be a joke — Sau Lan made up this question. Although it is true that Tai does the taxes, answer (b) is best. Sau Lan said that answer (b) was in fact part of their agreement. Since Tai's mother cannot cook, he does not seem to mind. To be fair, it should be pointed out that Sau Lan eats nothing but water cress, and Tai is a bottomless pit. Fortunately, Sau Lan's mother is a FANTASTIC cook, and she spends a lot of time visiting at CERN.

Why doesn't Tai do any cooking?

A> Because the eggs always fly out of the pan

B) Because Sau Lan needs the space in the kitchen for the Xerox machine

C) Because that doesn't leave enough time for eating and napping

Note: All of the answers are pretty good. Answer (a) is based on one of Tai's stories, according to which he was boiling eggs, left the kitchen, and returned to find two blackened eggs. The strange part is that one of the eggs ended up on the floor. Tai was rather proud of this story, and not at all embarrassed by it. I was surprised to discover that Sau Lan does have a photocopier at home (a big one!), but it is not in the kitchen.

The reason Tai has a dirty raincoat:
A) His wife refuses to wash it
B) He refuses to go out in the rain with it
C> Washing destroys the raincoat's impermeability to water

Note: Sau Lan suggested this question, and the answers. Apparently her mother always asks why Tai's raincoat is dirty, but Tai adamantly refuses to let them wash it. This question tied in nicely with Professor Yang's talk, in which he compared Tai to the TV detective Columbo, who always wears a dirty raincoat, and who always gets his man, despite his apparently clumsy manner by being relentlessly focused and persevering.

OTHER

Tai' s greatest calculation is:
A) His calculation of his mother-in-law's 1993 income taxes
B) The solution of the Chocolate Icing Model
C> The discovery of the rising cross section as a function of total-calorie intake

Note: Answers (b) and (c) are somewhat humorous descriptions of real-life problems Tai has solved. Professor Yang in his talk had made it clear that answer (c) was correct, and Tai seemed to agree. Yang described the discovery as an intuitive leap based on the knowledge gained from long calculations, calculations other people could not have even dreamed of doing. "No one believed them!" he said, referring to Cheng and Wu. He confessed that even he, who was their friend, was skeptical of their claims, all of which were later verified by experiment. Professor Yang went on to say that their paper ought to be included in the 90th anniversary edition of the Physical Review, as one of the ninety great papers published in that journal.

When Tai's briefcase was stolen from the Geneva airport and later recovered, which items were missing?

A) His tax documents, and Sau Lan's, and his mother-in-law's
B) Three hundred Swiss Francs, and a pocket calculator
C> An eight hundred page top- and Higgs-mass calculation

Note: The astonishing part about this question is that all of the items mentioned above were in the briefcase (except that only his own tax documents were there), and when it was recovered, only (c) was missing. Fortunately, Tai had copies of most of the notes in other places.

The official answers are marked by a > sign. The correct answers may in some cases differ from the official answers.

Chapter 5
Tai Tsun Wu's CV and Published Obituary, and Harvard Faculty of Arts and Sciences Memorial Minutes

(1) Curriculum Vitae of Tai Tsun Wu

Date of Birth: December 1, 1933
Place of Birth: Shanghai, China
Citizenship: U.S.A.

EDUCATION:

B.S. University of Minnesota Electrical Engineering1953
S.M. Harvard University Applied Physics1954
Ph.D. Harvard University Applied Physics1956

HONORS AND FELLOWSHIPS:

Putnam Scholar, 1953
Alfred P. Sloan Foundation Fellow, 1960–1964
NSF Senior Post-Doctoral Fellow, 1966–1967
Guggenheim Fellow, 1970–1971
Fellow of the American Academy of Arts and Sciences, 1977–2024
IEEE EMC Society Transactions Prize Paper Award, 1983
Alexander von Humboldt U.S. Senior Scientist Award, 1985–1986
Dannie Heineman Prize for Mathematical Physics, American Institute of Physics, 1999 Membership in Eta Kappa Nu, Tau Beta Pi, and Sigma Xi

EMPLOYMENT:

Junior Fellow, Society of Fellows, Harvard University, 1956–1959
Assistant Professor of Applied Physics, Harvard University, 1959–1963
Associate Professor of Applied Physics, Harvard University, 1963–1966
Gordon McKay Professor of Applied Physics, Harvard University, 1966–2021
Professor of Physics (jointly), Harvard University, 1994–2021
Honorary Fellow, University of Wisconsin–Madison, 2021–2024

OTHER EXPERIENCES:

Member, Institute for Advanced Study, Princeton, NJ, 1958–1959, 1960–1961, 1962–1963
Visiting Professor, Rockefeller University, 1966–1967
Scientific Associate, DESY, Hamburg, Germany, 1970–1971, 1982–1983
Kramers Professor, Rijksuniversiteit, Utrecht, The Netherlands, 1977–1978
Scientific Associate, CERN, Geneva, Switzerland, 1977–1978, 1986–present

PUBLICATIONS:

6 books, numerous articles in books, and over 400 original papers in refereed journals (see attached selection)

SELECTED PUBLICATIONS OF TAI TSUN WU

Books

Scattering and Diffraction of Waves (with R. W. P. King), Harvard University Press, 1959; translated into Russian, 1962.

The Two-Dimensional Ising Model (with B. M. McCoy), Harvard University Press, 1973; Dover Edition, 2015.

Antennas in Matter: Fundamentals, Theory, and Applications (with R. W. P. King, G. S. Smith, and M. Owens), M.I.T. Press, 1981.

Expanding Protons: Scattering at High Energies (with H. Cheng), M.I.T. Press, 1987.

The Ubiquitous Photon: Helicity Method for QED and QCD (with R. Gastmans), Oxford University Press, 1990.

Lateral Electromagnetic Waves: Theory and Applications to Communications, Geophysical Exploration, and Remote Sensing (with R. W. P. King and M. Owens), Springer-Verlag, 1992.

All these books are based largely on the original research of the authors.

Selected Papers

1) "High-Frequency Scattering", *Phys. Rev.* **104**, 1201 (1956).

2) "High-Energy Potential Scattering", *Phys. Rev.* **108**, 466 (1957).

3) "Ground State of a Bose System of Hard Spheres", *Phys. Rev.* **115**, 1390 (1959).

4) "Some Nonequilibrium Properties of a Bose System of Hard Spheres at Extremely Low Temperatures", *Journal of Mathematical Physics* **2**, 105 (1961).

5) "Domains of Definition for Feynman Integrals over Real Feynman Parameters", *Phys. Rev.* **123**, 678 (1961).

6) "Phenomenological Analysis of Violation of CP Invariance in Decay of K0 and ¯K0", *Phys. Rev. Lett.* **13**, 380 (1964) [with C. N. Yang].

7) "Limit of Cross Sections at Infinite Energy", *Phys. Rev. Lett.* **24**, 1456 (1970) [with Hung Cheng].

8) "Impact picture of p p, anti-p p, pi+- p, and K+- p elastic scattering from 20-GeV to 5000-GeV", *Physics Letters B* **44**, 97 (1973) [with Hung Cheng and J.K. Walker].

9) "Concept of nonintegrable phase factors and global formulation of gauge fields", *Phys. Rev. D* **12**, 3845 (1975) [with C. N. Yang].

10) "Interaction of electrons, magnetic monopoles, and photons (I)", *Scintia Sinica* **21**, 317 (1978) [with T. S. Tu and C. N. Yang].

11) "Impact-picture expectations for very high-energy elastic pp and pp scattering", *Nuclear Physics B* **247**, 15 (1984) [with C. Bourrely and J. Soffer].

12) "Beamstrahlung in high energy electron-positron linear colliders with non-uniform bunches", *Physics Letters* **221B**, 203 (1989) [with M. Jacob].

13) "Quantum mechanics, quantum computing, and quantum cryptography," in A. Bokulich and G. Jaeger, *Philosophy of Quantum Information and Entanglement*, Cambridge University Press, Cambridge UK (2010) p. 247.

14) "Higgs decay $H \to \gamma\gamma$: New theoretical results and possible experimental implications", *International Journal of Modern Physics A* **30**, 1550200 (2015) [with R. Gastmans and Sau Lan Wu].

15) "Failure of the Feynman R_1 gauge for the standard model: An cxplicit example", *International Journal of Modern Physics A* Vol. 31, No. 04n05, 1650028 (2016) [with Sau Lan Wu].

16) "Comparing the R_\xi gauge and the unitary gauge for the standard model: An example", *Nuclear Physics B* **914**, 421 (2017) [with Sau Lan Wu].

17) "On the Yang–Mills non-Abelian Gauge Theory", A Festschrift in Honor of the C. N. Yang Centenary, pp. 485–500 (2022) [with Sau Lan Wu].

18) "Concept of the basic standard model and a relation between the three gauge coupling constants", *International Journal of Modern Physics A*, Vol. 38, No. 28, 2350128 (2023) [with Sau Lan Wu].

(2) Obituary published in Physics Today

Tai Tsun Wu

21 November 2024

(1 December 1933–19 July 2024)
A master of mathematical physics, Wu made groundbreaking contributions to antenna theory, statistical physics, and perturbative quantum field theory.

Written by:

Arthur M. Jaffe (Landon T. Clay Professor of Mathematics and Theoretical Science, Harvard University), and
Michael E. Peskin (Professor of Particle Physics and Astrophysics, SLAC, Stanford University)

DOI: https://doi.org/10.1063/pt.vach.hclr

Credit: Photo courtesy of Sau Lan Wu.

Tai Tsun Wu, a distinguished theoretical physicist and professor of physics and applied physics at Harvard University, passed away on 19 July 2024 in Palo Alto, California. He was 90 years old. Wu was known for his groundbreaking research in antenna theory, statistical physics, and perturbative quantum field theory.

Born on 1 December 1933, in Shanghai, China, Tai Tsun Wu came to the US in November 1950 and studied electrical engineering at the University of Minnesota. In 1953, he was a winner of the Putnam mathematical competition. Moving to Harvard, he earned a doctorate in applied physics in 1956 at the age of 22, working with Ronold King. Wu's graduate work established him as an extraordinary master of antenna theory for electromagnetic waves.

Wu joined the Harvard faculty in 1959. In 1966, he became the Gordon McKay Professor of Applied Physics; in 1994, he was also appointed a professor of physics.

Beginning in the 1960s, Wu carried out a series of tour-de-force calculations on the 2D Ising model in papers with Barry McCoy and Craig Tracy. Among other results, they computed the spin correlation function for spins at arbitrary orientation relative to the crystal axes, demonstrating explicitly how the rotationally invariant scaling theory sets in as one approaches the critical point. They discovered that the scaling functions are given in terms of Painlevé transcendental functions, opening a new line of mathematical research on integrable systems.

In a long collaboration with Hung Cheng at MIT, Wu investigated the high-energy behavior of elementary particle scattering cross sections by explicit summation of large classes of Feynman diagrams in scalar field theory and quantum electrodynamics. The results predicted the rise of the proton–proton total cross section observed in the early 1970s at the CERN Intersecting Storage Rings and were instrumental in interpreting the phenomena of diffraction scattering and multiple hadron production observed at CERN, including at the LHC.

In a series of papers with C. N. Yang in 1975–76, Wu reformulated the mathematical treatment of magnetic monopoles. Using ideas from differential geometry and the theory of fiber bundles, Wu and Yang represented the wavefunctions of particles moving in the field of a magnetic monopole in terms of "monopole harmonics," generalizing the usual spherical harmonics. These monopole harmonics are smooth in their appropriate coordinate patches and are linked across patches by gauge transformations. This formulation replaced the singular "Dirac strings" previously used to describe monopoles. These harmonics and their generalizations have become an essential tool for the study of monopoles and topological gauge field configurations in higher dimensions. Mathematicians have formulated these ideas as the "Wu–Yang dictionary" linking concepts in physics and differential geometry.

In his final years, Wu contributed to the theory of the Higgs boson. He pursued the idea that the standard model implies additional relations between the masses of its heaviest particles (W, Z, H, and t) that are not obvious from the Lagrangian and that give insight into the Higgs phenomenon. His last published work at the age of 89, "Concept of the basic standard model and a relation between the three gauge coupling constants," was published in 2023.

In recognition of his contributions, Wu received a number of honors and prizes, including the Dannie Heineman Prize for Mathematical Physics of the American Institute of Physics and the American Physical Society (1999) and the Alexander von Humboldt Prize (1985–86). He was a fellow of the American Academy of Arts and Sciences and a member of the Academia Sinica in Taiwan. Wu wrote more than 400 original papers and 6 books. Noteworthy among the books are *The Scattering and Diffraction of Waves* (1959), with R. W. P. King; *The Two-Dimensional Ising Model* (1973), with B. McCoy; and *Expanding Protons: Scattering at High Energies* (1987), with H. Cheng.

Wu was not only a brilliant scientist but also a dedicated mentor and an inspiration to young scientists. He spent significant time in collaborations at CERN, DESY, Utrecht University, the University of Bergen, the Institute for Advanced Study in Princeton, and Rockefeller University.

Wu is survived by his wife Sau Lan Wu, with whom he shared a life of intellectual and personal partnership. Tai met Sau Lan when she was a graduate student in the Harvard physics department. Sau Lan went on to become a prominent figure in experimental particle physics. She held the position of Enrico Fermi Professor at the University of Wisconsin and played a central role in the discoveries of the J/ψ particle, the gluon, and the Higgs boson. The couple was known for their close collaboration, which brought Tai Tsun Wu's theoretical insight to these experimental advances.

Tai Tsun Wu's passing is a profound loss to the scientific community and all who knew him. We hope that his legacy will continue to inspire future generations of physicists. May his memory be a source of inspiration and his contributions a guiding light for many years to come.

(3) Obituary published in Harvard Physics department website

Authored by Prof. Arthur Jaffe (Landon T. Clay Professor of Mathematics and Theoretical Science, Harvard University)

Dear Colleagues,

I learned today from Sau Lan Wu that Tai Tsun Wu passed away in Palo Alto, California, at 10:48 pm, on July 19, 2024. Tai was a formidable faculty member of both the Harvard physics department and SEAS. He was known for his groundbreaking research in both perturbative quantum field theory and in statistical physics. Tai's doctoral thesis, written with Professor Ronald King at Harvard when he was only 22 years old established him as an extraordinary master of antenna theory for electromagnetic waves.

During his subsequent productive career, Tai pursued a long collaboration with Hung Cheng at MIT. They studied the high-energy behavior of Feynman diagrams in quantum field theory, reflecting properties of renormalization theory. Tai's work on statistical mechanics models with Barry McCoy, Craig Tracy and others led to many insights, including the closed-form solution for correlation functions of the scaling limit of the Ising model in two dimensions, ostensibly an exact quantum field theory. Tai had many graduate students and collaborators. He was a regular visitor in the CERN theory group, recently working on theoretical aspects of the Higgs particle discovery. Tai Wu received the Dannie Heineman Prize for Mathematical Physics, the Humboldt Foundation Prize, and he was a member of the Academia Sinica.

Tai married Sau Lan Yu, a graduate student in the Harvard physics department and later Enrico Fermi Professor at the University of Wisconsin. Sau Lan played a central role in the discovery of the J/psi particle and In many experiments at CERN. The Wu family spent much time both in Cambridge and in Geneva. In 2023 they sold their Cambridge house and moved to Palo Alto.

Arthur Jaffe
July 23, 2024

(4) Abridged Memorial Minute presented at the meeting of the Faculty of Arts and Sciences, Harvard University, October 7, 2025 — Hung Cheng, Sheldon Glashow, John Hutchinson, and Arthur Jaffe

Tai Tsun Wu was a formidable member of both SEAS and Physics. His ground-breaking graduate research on antenna theory under Ronold W. P. King established him as a leading expert in that important field. Tai's subsequent contributions changed the course of both particle and statistical physics. His collaboration with Hung Cheng predicted rising cross-sections in hadron scattering. His work with Barry McCoy, Craig Tracy, and others led to a closed-form solution for the scaling limit of the Ising model. With C. N. Yang, he reformulated the theory of monopoles.

Tai authored over 400 publications, including six books, yet Tai shunned the limelight and the pursuit of recognition. Nevertheless, he received the Heineman Prize, the von Humboldt Prize, was elected to the American Academy of Arts and Sciences, and to the Academia Sinica.

Tai was born on December 1, 1933, in Shanghai. As a University of Minnesota undergraduate, he won the William Lowell Putnam national mathematics competition in 1953. This offers a Harvard scholarship and Tai became the first Putnam Fellow from Minnesota. He was elected a Junior Fellow at 22, became a faculty member at 25, and remained until becoming *emeritus* in 2021.

Tai was both personable and generous. He entertained many friends; one guest recalls the unusual experience of a feast including chicken feet. Another colleague arrived at Harvard without a car, and Tai insisted the newcomer drive his Dodge Dart until he returned from sabbatical.

Tai met the physics student Sau Lan Yu, and they married on June 18, 1967, in Memorial Church. Sau Lan became distinguished, discovering the J/ψ particle with Ting at Brookhaven National Laboratory, as well as leading experiments at CERN. The Wus spent much time in Cambridge and Europe, until moving to California in 2022, where Tai died July 19, 2024, in Stanford University hospital.

Respectfully submitted,

Hung Cheng (MIT)
Sheldon Glashow
John Hutchinson
Arthur Jaffe, Chair

Reference: *Hung Cheng, Sheldon Glashow, John Hutchinson, and Arthur Jaffe, "Tai Tsun Wu," abridged Memorial Minute presented at the meeting of the Faculty of Arts and Sciences, Harvard University, October 7, 2025.*

(5) Full Version of Memorial Minute presented at the meeting of the Faculty of Arts and Sciences, Harvard University, October 7, 2025 — Hung Cheng, Sheldon Glashow, John Hutchinson, and Arthur Jaffe

At a meeting of the FACULTY OF ARTS AND SCIENCES on October 7, 2025, the following tribute to the life and service of the late Tai Tsun Wu was spread upon the permanent records of the Faculty.

Tai Tsun Wu
Born: December 1, 1933
Died: July 19, 2024

Professor Tai Tsun Wu was a formidable member of both the School of Engineering and Applied Sciences (SEAS) and the Department of Physics at Harvard. At the age of 22, his ground-breaking research on antenna theory under the direction of Professor Ronold W. P. King established him as one of the leading experts in this important field. The remarkable breadth of Tai's research interests over the course of his career was underpinned by his exceptional mathematical abilities. Although he shifted the main thrust of this research to fundamental problems in physics, he continued for years to be active in solving basic electricity and magnetism problems that arise in antenna theory.

During Tai's subsequent productive career, he pursued a long collaboration with Hung Cheng, a professor at the Massachusetts Institute of Technology (MIT). Their extraordinary study of the high-energy behavior in quantum field theory illuminated properties of renormalization theory and resulted in the prediction of rising total cross-section of hadron scattering. Tai's work on statistical mechanics models with Barry McCoy, Craig Tracy, and others led to different insights, including finding a closed-form solution for correlation functions of the scaling limit of the Ising model, ostensibly an exact quantum field theory. In 1975 Tai collaborated with C. N. Yang to reformulate the theory of monopoles, leading to what is now referred to as the "Wu-Yang dictionary." Professor Wu advised many graduate students and had numerous collaborators, including John Myers.

Tai's research led to his recognition in many ways. Among other achievements, he received the Dannie Heineman Prize for Mathematical Physics and the Alexander von Humboldt Foundation Prize. He was elected to the American Academy of Arts and Sciences and to the Academia Sinica. Tai taught courses in applied mathematics and in physics in the Department of Physics and in SEAS. He was known to students and colleagues as an accessible expert on mathematical methods. Tai shunned the limelight and the pursuit of recognition; rather, he constantly focused on his research. For this reason, despite the fact that he authored over 400 publications, including six books, his work is not as widely known as it should be.

Tai Tsun Wu was born on December 1, 1933, in Shanghai, China. He came to the United States to study as an undergraduate at the University of Minnesota, where, in 1953, he won the William Lowell Putnam Mathematical Competition. This national competition for undergraduates has a Harvard connection: it was established by Elizabeth Lowell Putnam in honor of her husband and it offers a Harvard graduate school scholarship to one of the top winners each year. Tai applied to Harvard as the first Putnam Fellow from the University of Minnesota. A story that is still told is that members of the Department of Mathematics had naturally assumed that young Tai would be joining them, but he had applied to study applied physics. His doctoral thesis led to his election to the Society of Fellows at the age of 22 and to his appointment in the Harvard Faculty of Arts and Sciences at age 25, where he remained until 2021, when he became an *emeritus* professor.

Tai was fond of several local Chinese restaurants, where many friends were his guests for lunch or dinner. One of these guests recalled being feasted by Tai with a dinner that included chicken feet, an unusual experience. Tai expressed his generosity in many other ways. One new faculty member arrived at Harvard without a car just as Tai was about to leave for a sabbatical. Tai insisted that his young colleague drive his Dodge Dart until he returned.

While at Harvard, Tai became acquainted with Sau Lan Yu, a graduate student in experimental physics. They married on June 18, 1967, in the Harvard Memorial Church. Sau Lan went on to become distinguished for her role in the discovery of the J/ψ particle with Samuel Ting at Brookhaven National Laboratory, as well as for leading many experiments with the European Organization for Nuclear Research (CERN). The Wu family spent much time both in Cambridge and Europe. In 2022 they sold their Cambridge house and moved to Palo Alto, California, where Tai died in the Stanford University hospital on July 19, 2024.

Respectfully submitted,

Hung Cheng (MIT)
Sheldon Glashow
John Hutchinson
Arthur Jaffe, Chair

Portions of this Minute were previously published: Arthur Jaffe, "Tai Tsun Wu (1933–2024)," Department of Physics's website, July 23, 2024, https://www.physics.harvard.edu/news/tai-tsun-wu-1933-2024 [accessed August 11, 2025].

Reference: *This Minute was previously published by the Office of the Secretary of the Faculty of Arts and Sciences of Harvard University and The Harvard Gazette: Hung Cheng, Sheldon Glashow, John Hutchinson, and Arthur Jaffe, "Tai Tsun Wu," website of the Office of the Secretary of the Faculty of Arts and Sciences, Harvard*

University, October 8, 2025, https://officeofthesecretary.fas.harvard.edu/memorial-minutes; Hung Cheng, Sheldon Glashow, John Hutchinson, and Arthur Jaffe, "Tai Tsun Wu," The Harvard Gazette, October 8, 2025, https://officeofthesecretary.fas.harvard.edu/sites/g/files/omnuum2291/files/2025-10/Wu%20Memorial%20Minute_2025_10_07.pdf.

This memorial minute is also published in The Harvard Gazette with the title "Tai Tsun Wu, 90. Memorial Minute — Faculty of Arts and Sciences", on October 9, 2025 https://news.harvard.edu/gazette/story/2025/10/tai-tsun-wu-90/ (accessed on October 11, 2025).

The Harvard Gazette

CAMPUS & COMMUNITY

Tai Tsun Wu, 90

Memorial Minute – Faculty of Arts and Sciences

October 9, 2025 • 4 min read

Harvard University
Photo by Dylan Goodman

Chapter 6
Electromagnetic Theory (66 papers)

Note: For Chapters 4–9, citation counts up to February 25, 2025, are listed for publications with over 50 citations.

Unlike the equations of classical mechanics that had to be reorganized to conform first to relativity theory and then to the quantum revolution, the equations introduced by Maxwell that founded electromagnetic theory are basic to modern physics. In addition to his particle physics and statistical mechanics, for over more than fifty years Tai Tsun Wu applied his mathematical virtuosity and conceptual elegance to advances in electromagnetic theory. An example of this elegance is his 1961 "*Transient Response of a Dipole Antenna*," in which he provides an exact solution, obtained by identifying the transient for a finite dipole antenna with that of the exactly solvable infinite antenna. Many of his papers solved problems in antenna theory presented by Prof. R.W.P. King, and many others were with the students that he helped to educate. A selection from his extensive influential writings on electromagnetic theory is the following:

1) High-Frequency scattering, *Phys. Rev.* **104**, 1201 (1956), Tai Tsun Wu (Harvard U.), DOI: https://doi.org/10.1103/Physrev.104.1201; Citations: 59.

TAI TSUN WU†
Gordon McKay Laboratory of Applied Science, Harvard University, Cambridge, Massachusetts
(Received June 7, 1956)

In recent years, there has been a revival of interest in the phenomena of scattering at high frequencies. The simplest problems of this sort, in which the obstacle is either a circular cylinder or a sphere, are treated here. The treatment is limited to the total scattering cross sections, even though the method is by no means restricted in this way. The specific problems treated here include the scattering of a plane electromagnetic wave by a perfectly conducting circular cylinder (two possible polarizations) or a perfectly conducting sphere, the acoustic scattering by a rigid sphere, and the quantum-mechanical scattering by an impenetrable sphere.

By considering the creeping waves as defined on the universal covering space, the scattering cross sections of a circular cylinder may be expressed by asymptotic expansions for vanishingly small wavelengths. Analogous calculations yield the corresponding results for a sphere. It turns out, as expected, that the resulting expressions are accurate even for fairly large wavelengths. The first six terms of the asymptotic series are explicitly found in each case.

In conclusion, the application of the method to the determination of the approximate current distribution on the obstacle is considered briefly; also some generalizations about the scattering cross section are conjectured.

Commentary by Dr. Dionisios Margetis

This paper addresses the following intriguing and highly nontrivial question in classical electromagnetism: What is the asymptotic series for the scattering cross sections of electrically large obstacles beyond the regime of geometric optics? To address this, T. T. Wu devised a novel technique based on the Poisson summation formula. He systematically provided asymptotic expansions for the cross sections of plane-wave scattering from perfectly conducting infinite circular cylinders and spheres. His method can yield any number of terms of the series, and has given insights into the electromagnetic scattering by more complicated three-dimensional obstacles.

2) High-Energy potential scattering, *Phys. Rev.* **108**, 466(1957), Tai Tsun Wu (Harvard U.), DOI: https://doi.org/10.1103/PhysRev.108.466.

TAI TSUN WU*
Division of Engineering and Applied Physics, Harvard University, Cambridge, Massachusetts
(Received May 17, 1956; revised manuscript received June 7, 1957)

Asymptotically in the limit of high energies, successive approximations are obtained for the problem of potential scattering where the total phase shift through the potential is not small. Only the Schrödinger wave equation is treated. The method consists of first applying the stationary-phase approximation to the integral equation and then solving the resulting equation by iteration to derive the asymptotic behavior of the field. In particular, some information about the radiation field is obtained.

3) Theory of the Microstrip, *J. Appl. Phys.* **28**, 299–302 (1957), Tai Tsun Wu (Harvard U.), DOI: https://doi.org/10.1063/1.1722734.

TAI TSUN WU*
Gordon McKay Laboratory of Applied Science, Harvard University, Cambridge 38, Massachusetts
(Received October 26, 1956)

A procedure is presented for finding by successive approximation the current distribution, and thus the electromagnetic field, of the lowest mode of the microstrip transmission line. The approximation is valid provided that the strip is not too wide and the dielectric sheet not too thick. Each iteration consists of essentially solving two electrostatic problems.

As commented on by Prof. George Fikioris, Prof. Wu's pioneering work on the microstrip transmission line showed that the fields away from the line behave in a highly unusual manner. As a result, much care is required when one applies Fourier-transform methods (now often referred to as spectral-domain methods).

4) Driving Point and Input Admittance of Linear Antennas, *J. Appl. Phys.* **30**, 74–76 (1959), Tai Tsun Wu (Harvard U.), Ronold W.P. King (Harvard U.), https://doi.org/10.1063/1.1734978.

TAI TSUN WU† AND RONOLD W. P. KING
Cruft Laboratory, Harvard University, Cambridge, Massachusetts
(Received March 31, 1958)

An infinity in the input admittance of linear antennas owing to the use of an idealized delta-function generator is investigated. It is shown that the infinity may be interpreted in terms of an infinite capacitance between the two halves of the antenna. The conclusion is reached that conventionally used iterative procedures are not invalidated by difficulties with respect to the driving point.

5) High-frequency diffraction of plane waves by an infinite slit for grazing incidence, in *IRE Transactions on Antennas and Propagation*, vol. 8, no. 1, pp. 37–42, January 1960, S. R. Seshadri (Harvard U.) and T. T. Wu (Harvard U.), DOI: https://doi.org/10.1109/TAP.1960.1144804.

S. R. Seshadri and T. T. Wu
Gordon McKay Labs, Harvard University, Cambridge, MA, USA

Abstract: The scattering of plane electromagnetic waves of wave number *k* by an infinite slit of width *2a* formed by two perfectly conducting coplanar screens of zero thickness is considered. In the limit of large *ka* and at grazing incidence, the asymptotic series for the transmission cross section per unit length of the slit is evaluated up to the order(*ka*)^{−11 i 2}.

6) High-frequency diffraction of electromagnetic waves by a circular aperture in an infinite plane conducting screen, in *IRE Transactions on Antennas and Propagation*, vol. 8, no. 1, pp. 27–36, January 1960, S. R. Seshadri (Harvard U.) and T. T. Wu (Harvard U.), DOI: https://doi.org/10.1109/TAP.1960.1144810.

S. R. Seshadri and T. T. Wu
Gordon McKay Labs, Harvard University, Cambridge, MA, USA

Abstract: The scattering of plane electromagnetic wave of wave number *k* by a circular aperture of radius *a* in an infinitely conducting plane screen of zero thickness and infinite extent is considered. In the limit of large *ka* and at normal incidence, the ratio of the transmission cross section to the geometrical optical value \pi *a*^{2}, is found up to the order(*ka*)^{−5^{j}2}.

7) Electromagnetic Reflection from Sound Waves. *J. Acoust. Soc. Am.* 1 December 1960; **32** (12): 1660–1667, Hans J. Schmitt (Harvard U.) and Tai Tsun Wu (Harvard U.), DOI: https://doi.org/10.1121/1.1907984.

HANS J. SCHMITT AND TAI TSUN WU
Gordon McKay Laboratory, Harvard University, Cambridge 38, Massachusetts
(Received May 24, 1960)

The reflection of electromagnetic waves normally incident on the wavefronts of a semi-infinite standing sound wave in a liquid is discussed theoretically. Since the reflecting planes of maximum sound pressure disappear periodically, the reflected electromagnetic signal is modulated with the acoustic frequency. The modulated reflected electromagnetic wave arises (1) from the microscopic movement of the liquid surface, and (2) from the direct analog of Bragg reflection in optics. The second effect is most pronounced when the wavelength of the electromagnetic radiation in the liquid is twice the acoustic wavelength.

8) Theory of the Dipole Antenna and the Two-Wire Transmission Line, *J. Math. Phys.* **2**, 550–574 (1961), Tai Tsun Wu (Harvard U.), https://doi.org/10.1063/1.1703740; Citations: 86.

TAI TSUN WU
Gordon McKay Laboratory, Harvard University, Cambridge, Massachusetts
(Received September 19, 1960)

The properties of the dipole antenna are studied by an approximate procedure that makes use of the Wiener-Hopf integral equation. In particular, the input admittance and the radiation pattern are found. The present results thus supplement the existing theories, which are concerned mostly with short dipoles. The same procedure is then applied to several related problems. First, the back-scattering cross section of a dipole antenna is found approximately for normal incidence. Secondly, the two-wire transmission line is studied in detail by considering it to be two coupled dipole antennas. The capacitive end correction for an open end is evaluated, and the radiated power and the radiation resistance are found for a resonant section of transmission line with both ends open. Finally, the dielectric-coated antenna is considered briefly.

Commentary by Dr. Dionisios Margetis

This work introduces a (by now classic) systematic approach by the Wiener–Hopf method to the inherently difficult computation of the current distribution and radiative properties of the dipole antenna and two-wire transmission line. By first treating the problem of the semi-infinite dipole antenna exactly via a Riemann–Hilbert problem in the Fourier space, T. T. Wu proceeds to describe suitable approximations for the long thin dipole antenna. The methodology is then extended to the more demanding case with two coupled dipole antennas forming a two-wire transmission line.

9) Transient Response of a Dipole Antenna, *J. Math. Phys.* **2**, 892–894 (1961), Tai Tsun Wu (Harvard U.), DOI: https://doi.org/10.1063/1.1724237.

TAI TSUN WU
Gordon McKay Laboratory, Harvard University, Cambridge, Massachusetts
(Received May 11, 1961)

The current distributed for a dipole antenna driven by a step-function voltage is found shortly after the switch-on of the voltage.

Commentary by Dr. John Myers

In this paper he provides an exact solution for the transient behavior of a finite cyclindrical dipole antenna by using the speed of light as a limit on causal propagation to invoke the short-time behavior of the infinite antenna (John Myers).

10) Dimers on Rectangular Lattices, *J. Math. Phys.* **3**, 1265–1266 (1962), Tai Tsun Wu (Brookhaven National Laboratory), DOI: https://doi.org/10.1063/1.1703868.

TAI TSUN WU†
Brookhaven National Laboratory, Upton, New York
(Received August 17, 1962)

A simplified calculation of a known result of the dimer problem on a rectangular lattice is presented.

11) Theory of the Thin Circular Loop Antenna, *J. Math. Phys.* **3**, 1301–1304 (1962), Tai Tsun Wu (Harvard U.), DOI: https://doi.org/10.1063/1.1703875; Citations: 156.

TAI TSUN WU*
Gordon McKay Laboratory, Harvard University, Cambridge, Massachusetts
(Received April 30, 1962)

The current distribution on a thin circular loop transmitting antenna driven by a delta-function generator is determined approximately by Fourier series expansion. A difficulty encountered in previous analysis is shown to be due to an inadequate approximation.

12) Some Properties of Impedance as a Causal Operator, *J. Math. Phys.* **3**, 262–271 (1962), Tai Tsun Wu (Harvard U.), https://doi.org/10.1063/1.1703799.

TAI TSUN WU†
Gordon McKay Laboratory of Applied Science, Harvard University, Cambridge, Massachusetts
(Received September 25, 1961)

As a model for the various problems connected with domains of analyticity and dispersion relations, the concept of impedance (self-impedance) has been studied in general under the physical requirements of linearity, passivity, reproducibility, and causality. After identifying impedance functions with suitably defined positive functions, any impedance function may be decomposed to the sum of a minimum reactive part and two reactive parts, one of which is of unfamiliar nature. In the time domain, the convolution representation of the current in terms of the voltage and vice versa are obtained rigorously. As an example, a reactance obtained from the Cantor function is studied in detail. Its peculiar delta-function response is computed explicitly, and it is also used to construct a counter example to a representation by van Kampen of the S matrix of a Maxwellian field. The general results are also applied to study the Kronig-Kramers dispersion relations for dielectric constants. It is proved that one of the two relations is true under very general conditions, but the other is false in general.

13) Input Admittance of Infinitely Long Dipole Antennas Driven from Coaxial Lines, *J. Math. Phys.* **3**, 1298–1301 (1962), Tai Tsun Wu (Harvard U.), https://doi.org/10.1063/1.1703874.

TAI TSUN WU*
Gordon McKay Laboratory, Harvard University, Cambridge, Massachusetts
(Received April 30, 1962)

For an infinitely long dipole antenna driven from a coaxial line, the reflection coefficient and hence the apparent terminal admittance are determined approximately when the radii of the coaxial line are small compared with the wavelength. This result is useful because the differences of admittances for antennas with identical geometries near the driving point are less sensitive to the driving condition and hence can be found approximately by many existing theories.

14) Diffraction by a circular aperture in a unidirectionally conducting screen, in *IEEE Transactions on Antennas and Propagation*, vol. 11, no. 1, pp. 56–67, January 1963, S. R. Seshadri (Harvard U.) and T. T. Wu (Harvard U.), DOI: https://doi.org/10.1109/TAP.1963.1137969.

S. R. Seshadri and T. T. Wu
Gordon McKay Labs, Harvard University, Cambridge, MA, USA

Abstract: The diffraction of a normally incident plane electromagnetic wave with wave number k by a circular aperture of radius a in a unidirectionally conducting plane screen of zero thickness and infinite extent is considered. In the limit of large ka, the ratio of the transmission cross section to the geometrical optics value πa^{2}, is found up to the order$(ka)^{-3/2}$.

15) Transient response of linear antennas driven from a coaxial line, in *IEEE Transactions on Antennas and Propagation*, vol. 11, no. 1, pp. 17–23, January 1963, T. Wu and R. King, DOI: https://doi.org/10.1109/TAP.1963.1137980.

T. Wu and R. King
Gordon McKay Laboratory, Hamdard University, Cambridge, MA, USA

Abstract: The initial transient response of straight wires connected to coaxial lines is studied theoretically for the case where a pulse is applied to the coaxial line. The wave form of the return pulse is first found approximately for the case of a pulse of zero rise time. Since this does not correspond to any feasible experimental situation, the effect of a finite rise time is considered in detail. Numerical results are obtained for several special cases.

16) Quantization of the Electromagnetic Field, *Phys. Rev.* **129**, 1420 (1963), Tai Tsun Wu (BNL), DOI: DOI: https://doi.org/10.1103/PhysRev.129.1420.

Tai Tsun Wu†
Brookhaven National Laboratory, Upton, New York
(Received 13 September 1962)

A method of quantizing the electromagnetic field is proposed, where the underlying Hilbert space has a definite metric and the verifications of Lorentz covariance can be easily carried out with the help of Ward's identity. The usual Feynman-Dyson rules are obtained for the electron-photon system, and the longitudinal and scalar photons are found to have separately no interaction with the electron field.

17) The optical model of scattering by a composite cylinder, in *IEEE Transactions on Antennas and Propagation*, vol. 11, no. 6, pp. 703–705, November 1963, R. D. Kodis (Oxford U.) and T. T. Wu (Harvard U.), DOI: https://doi.org/10.1109/TAP.1963.1138122.

R. D. Kodis
Engineering Laboratory, Oxford University, Oxford, UK
T. T. Wu
Gordon McKay Laboratory, Harvard University, Cambridge, MA, USA

Abstract: The optical approximation to scattering by a composite cylinder includes two classes of rays: those which are reflected by the core and those which are not. It is shown here that members of the latter set may arise whenever a certain integral has a stationary phase point in a restricted parameter regime. Consideration of the restrictions in the case of forward scattering shows that relatively few core-independent rays of low order will satisfy the geometrical requirements even when a stationary phase point exists.

18) Electromagnetic Reflection from Sound Waves. II. *J. Acoust. Soc. Am.* 1 March 1964; **36**(3): 437–445, Hans J. Schmitt (Gordon

McKay Laboratory, Harvard U.) and Tai Tsun Wu (Gordon McKay Laboratory, Harvard U.), DOI: https://doi.org/10.1121/1.1918975.

HANS J. SCHMITT* AND TAI TSUN WU†

Gordon McKay Laboratory, Harvard University, Cambridge, Massachusetts
(Received 4 December 1962)

The reflection of electromagnetic waves normally incident on the surface of a liquid supporting a spherical sound wave is discussed theoretically. The effect of Bragg reflection is neglected, and only the microscopic movement of the liquid surface is taken into account. Because of the periodic nature of the sound wave, the reflected electromagnetic signal contains frequency components that correspond to both the sum and the difference of the electromagnetic and acoustic frequencies. These two frequency components have quite different properties. The problem is treated by perturbation theory, and two special cases are examined in detail.

19) The propagation of transient electromagnetic fields into a cavity formed by two imperfectly conducting sheets, in *IEEE Transactions on Antennas and Propagation*, vol. 13, no. 1, pp. 149–158, January 1965, C. Harrison (Sandia Corporation), M. Houston (Sandia Corporation), R. King (Harvard U.) and T. Wu (Harvard U.), DOI: https://doi.org/10.1109/TAP.1965.1138390.

C. Harrison and M. Houston
Sandia Corporation, NM, USA
R. King and T. Wu
Harvard University, Cambridge, MA, USA

Abstract: In this paper steady-state formulas are developed for the ratio of the resultant electromagnetic field in the screened space formed by dual, infinite, homogeneous metal plates to the incident field, and from these transfer functions the time history of the field in the cavity is computed for Gaussian-shaped input fields.

20) Electrodynamics of Intermediate Bosons and CP Noninvariance, *Phys. Rev. Lett.* **15**, 811 (1965), Tai Tsun Wu (Harvard U.), DOI: https://doi.org/10.1103/PhysRevLett.15.811.3.

21) The Cylindrical Antenna with Nonreflecting Resistive Loading, *IEEE Trans. Antennas and Propagation*, Vol. 13, 369–373 (1965), T. T. Wu (Harvard U.) and R. W. P. King (Harvard U.), DOI: https://doi.org/10.1109/TAP.1965.1138429; Citations: 373.

Tai Tsun Wu and R. W. P. King
Gordon McKay Laboratory, Harvard University,
Cambridge, Massachusetts
(Received August 10, 1964)

The distribution of current along a center-driven cylindrical antenna is obtained when the material forming the antenna is resistive. The particular case is considered when the impedance per unit length of the antenna is a function of the distance from the end. A solution is obtained specifically when the current is represented by an outward traveling wave with no reflected wave. The admittance of the antenna and the far-field pattern is determined. Field patterns are evaluated for a wide range of lengths. These are characterized by a single major lobe with a very small minor lobe structure.

Commentary by Dr. George Fikioris

T. T. Wu and the late Professor R. W. P. King used integral equations to show that a short linear antenna made of a properly designed resistive material can serve as a traveling-wave antenna; the concept — often termed "Wu–King resistivity loading" — later formed the basis of several research projects (e.g. by MITRE in the nineties) and resulted in broadband antennas for a number of Department of Defense applications.

22) Regge Poles in Quantum Electrodynamics with Massive Photons, *Phys. Rev.* **140**, B465 (1965), Hung Cheng (Harvard U.) and Tai Tsun Wu (Harvard U.), DOI: https://doi.org/10.1103/PhysRev.140.B465.

Hung Cheng*
Department of Physics, Harvard University, Cambridge, Massachusetts

AND

Tai Tsun Wu†
Gordon McKay Laboratory, Harvard University, Cambridge, Massachusetts
(Received 14 June 1965)

Starting from the Lagrangian theory of quantum electrodynamics with massive photons, we find that the lowest order radiative corrections to the fourth order "box" diagram of the Compton scattering amplitude contribute terms proportional to $(\ln t)^2$ for large t, violating the Regge behavior if uncanceled. The sixth-order ladder diagram and two others obtained from it by interchanging one of the external photon vertices with a virtual photon vertex are re-evaluated. It is found that the unwanted terms contributed by all these diagrams exactly cancel, proving that in this theory, the fermion lies on a Regge trajectory, up to the sixth-order perturbation.

23) Electrodynamics of Intermediate Bosons and CP noninvariance, *Phys. Rev. Lett.* **15**, 718 (1965), Tai Tsun Wu (Harvard U.), DOI: https://doi.org/10.1103/PhysRevLett.15.718.

Tai Tsun Wu*

Harvard University, Cambridge, Massachusetts
(Received 22 September 1965)

24) Currents, charges, and near fields of cylindrical receiving and scattering antennas, in *IEEE Transactions on Antennas and Propagation*, vol. 13, no. 6, pp. 978–979, R. King (Harvard U.) and T. T. Wu (Harvard U.), November 1965, DOI: https://doi.org/10.1109/TAP.1965.1138563.

25) The imperfectly conducting cylindrical transmitting antenna, *IEEE Transactions Antennas and Propagation* **14**, 524–534 (1966), R. W. P. King (Harvard U.) and T. T. Wu (Harvard U.), DOI: https://doi.org/10.1109/TAP.1966.1138733;Citations: 69.

R. W. P. King and T. T. Wu
Gordon McKay Laboratory, Harvard University, Cambridge, Massachusetts
(Received June 30, 1965; revised May 6, 1966)

The properties of a cylindrical antenna with a continuous ohmic resistance along its length are of interest in the design of certain types of directive broadband antennas and in the determination of the efficiency of dipole antennas. Conventionally, the contribution by ohmic resistance to the distribution of current and the impedance is contained in a particular integral that is either ignored or treated as a higher-order correction to formulas derived for perfectly conducting antennas. An alternative and more useful form has been developed in which the integral equation for the current is rearranged to permit the introduction of a complex wave number *k*. An approximate solution of this equation is then obtained in terms of the three trigonometric functions, *sin k*(*h*- |*z*|), *cos kz* - *cos kh*, and *cos 1/2k_0z* - *cos 1/2k_0h*, where *k_0* is the free-space wave number. Expressions are derived for the coefficients of these functions and for *k*. Explicit formulas are given for the distribution of current and the admittance.

Commentary by Dr. Dionisios Margetis

By the mid-60s, antenna theory was considered as an important and mature area of research within electromagnetics. Yet, some fundamental questions in this context had remained largely unanswered. This paper addresses one of the most challenging related questions known at that time, namely, the form of the current distribution of a cylindrical transmitting antenna with a continuous ohmic resistance due to dissipation along its length. The approach is nonperturbative in this work, and yields an explicit approximate solution of the underlying integral equation via the introduction of a complex wave number for the current.

26) The Electric Field Very Near a Driven Cylindrical Antenna, *Radio Science*. **1**: 353–360 (1966), R. W. P. King (Harvard U.) and T. T. Wu (Harvard U.). DOI: https://doi.org/10.1002/Rds196613353.

Ronold W. P. King
Division of Engineering and Applied Physics,
Harvard University, Cambridge, Mass. 02138, U.S.A.
Tai Tsun Wu
Division of Engineering and Applied Physics,
Harvard University, Cambridge, Mass. 02138, U.S.A.

A first-order formula is derived for the component $Ez(\rho, z)$ of the electric field very near to the surface and parallel to the axis of a perfectly conducting center-driven cylindrical antenna with radius a and length $2h$. It has the form $Ez(\rho, z) = Ez(z) \ln(\rho/a)$ where p is the radial distance from the axis of the antenna and $Ez(z)$ is a complicated expression that involves generalized sine and cosine integral functions. Zero-order results are compared with first-order values for half-and full-wave dipoles with $a/\lambda = 7.022 \times 10^{\Sigma 3}$. It is shown that zero-order values do not include large values of $Ez(\rho, z)$ near the ends and near the driving point, so that a quantitative estimate of the field near a center-driven cylindrical antenna requires a first-order evaluation.

27) An integral equation for the current in an asymmetrically driven cylindrical antenna, in *Proceedings of the IEEE*, vol. 55, no. 6, pp. 1097–1098, June 1967, S. R. Seshadri (Sylvania Electric Products) and T. T. Wu (Harvard U.), DOI: https://doi.org/10.1109/PROC.1967.5743.

S.R. Seshadri
Sylvania Electric Products, Inc., Waltham, Mass.
Tai Tsun Wu
Gordon McKay Laboratory, Harvard University, Cambridge, MA, USA

Abstract: The coupled integral equations for the transverse and the axial currents in a tubular antenna are derived for the case in which it is driven along an infinitesimally narrow circumferential gap by an EMF which is allowed to vary along the circumference. The cases where the EMF is assumed to be maintained either on the outside surface only, or on the inside surface only, or on both surfaces of the tube as in the conventional treatments of antenna theory are considered.

28) The Thick Tubular Transmitting Antenna, *Radio Science* **2**: 1061–1065 (1967), R. W. P. King (Harvard U.) and T. T. Wu (Harvard U.). DOI: https://doi.org/10.1002/Rds1967291061.

Ronold W. P. King
Gordon McKay Laboratory, Harvard University, Cambridge, Mass. 02138, U.S.A.
Tai Tsun Wu
Gordon McKay Laboratory, Harvard University, Cambridge, Mass. 02138, U.S.A.

The growing importance of electrically thick monopole antennas has suggested a re-examination of the model used in the derivation of the integral equation for the current in tubular antennas. It is shown that when such an antenna is driven from a coaxial line in the usual manner, which maintains a field along a narrow circumferential region on the outside surface of the tube (instead of along both the inner and outer surfaces), an additional term is required in the integral equation. This is negligible when the antenna is electrically thin, but of major importance when it is not. A quantitative comparison of conductances is provided over a wide range of thickness.

29) A simple formula of current in dipole antennas, in *IEEE Transactions on Antennas and Propagation*, vol. 16, no. 5, pp. 542–547, September 1968, L. Shen (University of Houston), T. T. Wu (Harvard U.) and R. King (Harvard U.), DOI: https://doi.org/10.1109/TAP.1968.1139241; Citations: 62.

L. Shen
Department of Electrical Engineering, University of Houston, Houston, TX, USA
Division of Engineering and Applied Physics, Harvard University, Cambridge, MA, USA
T. T. Wu, R. King
Division of Engineering and Applied Physics, Harvard University, Cambridge, MA, USA

Abstract: A simple and quantitatively accurate representation of the current distribution in a dipole antenna is derived. Numerical data are given and are found to be in good agreement with the experiment when h \geq 0.15\lambda.

30) Cylindrical antennas as probes for determining the properties of media, 1968 Antennas and Propagation Society International Symposium, Boston, MA, USA, 1968, pp. 61–68, R. King, S. Sandler and T. T. Wu, DOI: https://doi.org/10.1109/APS.1968.1150632.

R. W. P. King, S. S. Sandler, and T. T. Wu
Gordon McKay Laboratory, Harvard University, Cambridge, MA, USA

31) On the admittance of a monopole driven from a coaxial line, in *IEEE Transactions on Antennas and Propagation*, vol. 17, no. 6, pp. 814–817, November 1969, R. W. P. King (Harvard U.) and T. T. Wu (Harvard U.), DOI: https://doi.org/10.1109/TAP.1969.1139539.

R. W. P. King and T. T. Wu
Harvard Univ., Cambridge, MA, USA

***Abstract*—A simple formula is described for the apparent admittance $Y(ka,b/a)$ of a monopole of fixed length h and radius a when driven from a coaxial line (conductor radii b and a). If $Y(ka,b/a)$ is known at n values of ka with b/a fixed at $(b/a)_2$ and at m values of b/a with ka fixed at $(ka)_1$, it can be evaluated from the given formula for all combinations of ka and b/a in the two ranges of n and m. It is assumed that $kb \ll 1$ and $h \gg (b-a)$.**

32) The horizontal wire antenna over a conducting or dielectric half space: Current and admittance, *Radio Science* **9**: 701–709 (1974), R. W. P. King (Harvard U.), T. T. Wu (Harvard U.), and L.-C. Shen (University of Houston). DOI: https://doi.org/10.1029/Rs009I007P00701.

R. W. P. King
Gordon McKay Laboratory, Harvard University,
Cambridge, Massachusetts 02138
T. T. Wu
Gordon McKay Laboratory, Harvard University,
Cambridge, Massachusetts 02138
L.-C. Shen
Department of Electrical Engineering, University of Houston,
Houston, Texas 77004

The wave number, the distribution of current, and the admittance of a horizontal wire antenna over a homogeneous isotropic half space are determined when the complex wave number characteristic of the half space is large in magnitude compared with the real wave number of the air. The axial electric field at the surface of the half space is described. The properties of the antenna are obtained from those of the eccentrically insulated dipole in a general medium in the limit as the radius of the insulation is increased without limit while the distance of the dipole from the boundary is kept constant. Application of the new theory to the Beverage or wave antenna is discussed.

33) The dipole antenna with eccentric coating in a relatively dense medium, in *IEEE Transactions on Antennas and Propagation*, vol. 23, no. 1, pp. 57–62, January 1975, T. T. Wu (Harvard U.), L. Shen (University of Houston) and R. W. King (Harvard U.), DOI: https://doi.org/10.1109/TAP.1975.1141003.

T. T. Wu
Division of Engineering and Applied Physics, Harvard University,
Cambridge, MA, USA
L. Shen
Department of Electrical Engineering, University of Houston, Houston, TX, USA
R.W. King
Division of Engineering and Applied Physics, Harvard University,
Cambridge, MA, USA

Abstract: In a medium of high permittivity or conductivity, such as sea water, lake water, or wet earth, a dielectric coated antenna can produce a directional radiation pattern when the coating is eccentric. The current distribution on a finite or infinite dipole with such a coating is obtained. Formulas for the directivity and the radiation pattern of an infinitely long dipole with eccentric coating are also derived.

34) Analysis of Crossed Wires in a Plane-Wave Field, in *IEEE Transactions on Electromagnetic Compatibility*, vol. EMC-17, no. 4, pp. 255–265, Nov. 1975, R. W. P. King and T. T. Wu, DOI: https://doi.org/10.1109/TEMC.1975.303432.

R. W. P. King and T. T. Wu
Division of Engineering and Applied Physics, Harvard University, Cambridge, MA, USA

Abstract: The currents and charges induced in a pair of electrically thin crossed wires by a normally incident plane electromagnetic wave are derived by analytical methods. The boundary conditions at the ends and at the junction are explained. The solution of a new integro-differential equation for the currents is obtained in terms of trigonometric and integral-trigonometric functions. Depending on the electrical lengths of the crossed elements and location of their junction a variety of quite different distributions of current and charge obtain. These determine the scattered near and far fields. Graphs of computed currents and charges per unit length on the four arms of several important cases are displayed. The accurate determination of the induced currents and charges on a mathematically tractable structure-the thin-wire cross-is an early step in a study that will proceed to electrically thick cylinders, wide strips, and their junctions in crossed configurations in an effort to gain a meaningful approximate understanding of the currents and charges induced on an aircraft by an electromagnetic pulse.

35) Coupled insulated antennas in a relatively dense medium: currents and admittances, *Radio Science* **11**: 661–669 (1976), R. W. P. King (Harvard U.), T. T. Wu (Harvard U.), and L.-C. Shen (University of Houston). DOI: https://doi.org/10.1029/Rs011I008P00661.

R. W. P. King
Gordon McKay Laboratory, Harvard University, Cambridge, Massachusetts 02138
T. T. Wu
Gordon McKay Laboratory, Harvard University, Cambridge, Massachusetts 02138
L. C. Shen
Department of Electrical Engineering, University of Houston, Houston, Texas 77004

The analysis of the insulated dipole in a relatively dense medium is extended to coupled, individually insulated elements arranged in a circular array. It is shown that the transmission-line-like distributions of current are preserved for each phase sequence but with different complex wave numbers that depend not only on the electrical properties of the insulating and ambient media but also on the coupling among the elements. Special attention is given the two-element and three-element arrays.

36) A theoretical and experimental study of the ferrite rod antenna, 1977 Antennas and Propagation Society International Symposium, Stanford, CA, USA, 1977, pp. 242–245, D. V. Giri (Harvard U.), R. W. P. King (Harvard U.) and T. T. Wu (Harvard U.), DOI: https://doi.org/10.1109/APS.1977.1147716.

37) Measurement of the circuit properties of insulated linear antennas in a dissipative medium, in *IEEE Transactions on Antennas and Propagation*, vol. 25, no. 6, pp. 836–840, November 1977, K. Lee (Harvard U.), R. W. P. King (Harvard U.) and T. T. Wu (Harvard U.), DOI: https://doi.org/10.1109/TAP.1977.1141680.

Kuan Min Lee, Ronold W. P. King, Life Fellow, *IEEE* and T. T. WU

Abstract: The experimental system for measuring the electrical properties of antennas embedded in a dissipative medium is discussed. Measured input admittances of linear antennas with insulations of different sizes and dielectric constants are compared with the theoretical results. Measured current distributions of insulated antennas with the dielectric constant of the insulation comparable to that of the surrounding medium are compared with theory and are found to be in good agreement. The uncertainties in the measurement are discussed.

38) The horizontal-wire antenna over a dissipative half-space: Generalized formula and measurements, in *IEEE Transactions on Antennas and Propagation*, vol. 25, no. 6, pp. 850–854, November 1977, R. M. Sorbello, R. W. P. King, K. M. Lee, L. C. Shen and T. T. Wu, DOI: https://doi.org/10.1109/TAP.1977.1141679.

R. M. Sorbello, Member, *IEEE*, Ronold W. P. King, Life Memeber, *IEEE*,
K. M. Lee, Member, *IEEE*, L. C. Shen, Member, *IEEE*, and
T. T. Wu, Member, *IEEE*

Abstract: It is shown and verified by experiment that a horizontal resonant or traveling wave antenna placed in air close to a dense half space with the properties of lake or sea water or earth behaves like a terminated lossy transmission line. The terminal impedance is related to the radiation of the antenna into the air and the complex wavenumber and characteristic impedance are those of an infinitely long line. The complex wavenumber takes account of both dissipation in and radiation into the dense half-space.

39) Standing Waves and Notches in an EMP Simulator and Their Reduction, *IEEE Transactions On Electromagnetic Compatibility* **23**: 80–87 (1981), R. W. P. King, D. J. Blejer, and T. T. Wu. DOI: https://doi.org/10.1109/Temc.1981.303898.

RONOLD W. P. KING, FELLOW, IEEE, DENNIS J. BLEJER, MEMBER, IEEE, AND TAI TSUN WU

***Abstract*—The high standing-wave ratio in a model simulator in the midfrequency range defined by $kh = 2\pi h/\lambda \sim 2\pi$ (where h is the height of the parallel-plate region) is investigated experimentally. It is shown that the isolated deep minimum or "notch" is due to the mutual cancellation of the imaginary parts of the TM_{01} and TEM modes. The effects on the SWR and the deep minimum of changes in the magnitude and location of the terminating resistance are investigated. Also studied are the introduction of a bifurcating plate, resistive modal filters, and series aprons. It is shown that, with the proper adjustment of the last named, the "notch" can be eliminated and the SWR reduced to near two at discrete frequencies that span the entire midfrequency range. Since the low-frequency and high-frequency ranges already have low SWR's, the simulator has been made effective over the entire frequency band, in the sense that the SWR is low.**

***Key Words*—EMP simulator, standing waves, notches, SWR reduction, bifurcating plate, resistive modal filters, series aprons.**

40) The Measurement of the Electric Field Inside a Finite Dielectric Cylinder Illuminated by a Plane Wave, in *IEEE Transactions on*

Microwave Theory and Techniques, vol. 30, no. 8, pp. 1282–1286, Aug. 1982, R. Bansal, R. W. P. King and T. T. Wu, DOI: https://doi.org/10.1109/TMTT.1982.1131241.

R. Bansal, R, W. P. King, and T. T, Wu

Abstract: An experimental study of the distribution of the electric field induced inside a finite circular cylinder of water illuminated by an approximately plane electromagnetic wave is presented. The incident field was generated by using a monopole above the ground plane with a 90° corner reflector. The cylinder of water included a thin conducting tube at its center to shield the transmission lines leading to the probes. The graphs of selected measured distributions are displayed and interpreted. The measurements were earned out at 100, 300, and 600 MHz. The conductivity of the 50-cm long column of water was varied from approximately zero to 3.5 S/m. Both the amplitude and the phase of the induced electric field were measured in the experiment. Comparisons with a new theoretical solution developed by the authors are also included.

41) The propagation and reflection of lateral waves, 1983 Antennas and Propagation Society International Symposium, Houston, TX, USA, 1983, pp. 461–464, M. F. Brown (Harvard U.), R. W. P. King (Harvard U.) and T. T. Wu (Harvard U.), DOI: https://doi.org/10.1109/APS.1983.1148949.

42) The loop antenna with a cylindrical core: Theory and experiment, in *IEEE Transactions on Antennas and Propagation*, vol. 31, no. 2, pp. 225–231, March 1983, R. W. Burton (Harvard U.), R. W. P. King (Harvard U.) and T. T. Wu (Harvard U.), DOI: https://doi.org/10.1109/TAP.1983.1143052.

Robert W. Burton, Ronold W. P. King, Life Fellow, *IEEE*, and T. T. Wu

Abstract: The electromagnetic field of an electrically small loop antenna with a cylindrical core is investigated. The infinitely long cylinder may be conducting or insulating. With the help of the principle of similitude, measurements made with a small loop at a high frequency with cylinders of water with widely different conductivities are applied to a very large loop around a mountain at a very low frequency. An approximate equivalent circuit for the loop with a core is described and tested.

43) An Experimental Investigation of the Parallel-Plate EMP Simulator with Single-Pulse Excitation, in *IEEE Transactions on Electromagnetic Compatibility*, vol. EMC-25, no. 3, pp. 358–366, Aug. 1983, H. M. Shen (Harvard U.), R. W. P. King (Harvard U.) and T. T. Wu (Harvard U.), DOI: https://doi.org/10.1109/TEMC.1983.304100.

H. M. Shen, Member, *IEEE*, R. W. P. King, Life Fellow, *IEEE*, and T. T. Wu

Abstract: The properties of the parallel-plate Harvard EMP simulator (HES) are studied in the time domain under single-pulse excitation. The behavior of the HES is observed through three windows: the voltage at the driving point, the current and charge densities on the surfaces of the top plate, and the electric field between the plates in the simulator. In order to eliminate the systematic interference which can be quite severe, a new technique of "dual measurements" has been developed and more accurate data have been obtained. The measurements have been interpreted with the help of knowledge acquired from earlier continuous-wave investigations and from studies with the similar rhombic EMP simulator.

44) An Experimental Investigation of the Rhombic EMP Simulator Under Pulse Excitation, in *IEEE Transactions on Electromagnetic Compatibility*, vol. EMC-25, no. 1, pp. 40–46, Feb. 1983, H.-M. Shen (Harvard U.), R. W. P. King (Harvard U.) and T. T. Wu (Harvard U.), DOI: https://doi.org/10.1109/TEMC.1983.304080.

Hao-Ming Shen, Ronold W. P. King, Fellow, *IEEE*, and Tai T. Wu

Abstract: The rhombic simulator excited by a pulse generator is investigated experimentally. The voltage pulse at the driving point of the simulator and the associated current pulses that travel along the conducting tubes that form the rhombus have been measured with different loads. The complex sequence of electromagnetic pulses (including the incident pulse, the reflected pulses from the load, and other reflection points) have been measured on the metal ground plane. The origin (localization) of each pulse has been carried out. From the Fourier transformation of the waveforms in the time domain, the spectrum of the electromagnetic pulse in the frequency domain is obtained. It is shown that the characteristics of the simulator with the two types of excitation (continuous wave and pulse) are consistent.

45) Theoretical Analysis of the Rhombic Simulator Under Pulse Excitation, in *IEEE Transactions on Electromagnetic Compatibility*, vol. EMC-25, no. 1, pp. 47–55, Feb. 1983, H.-M. Shen (Harvard U.), R. W. P. King (Harvard U.) and T. T. Wu (Harvard U.), DOI: https://doi.org/10.1109/TEMC.1983.304081.

Hao-Ming Shen, Ronold W. P. King, Fellow, *IEEE*, and Tai T. Wu

Abstract: An analysis is given of the electromagnetic pulse generated in the space between the wires of a rhombic simulator by pulses of current traveling along the wires. After the mechanism of pulse excitation is clarified with the help of measured results, the electromagnetic pulse is investigated in terms of a current pulse with known shape that travels along the rhombic wires. The calculated pulse sequences agree very well with the measured ones. The "subpulse" radiated by each small segment of the wires as the current pulse traverses it is also studied. The time delays of the sequence of subpulses play an important role in the formation of the complete electromagnetic pulse. The amplitudes of the subpulses are relatively less important.

46) The Dual-Measurement Procedure for Eliminating Systematic Interference, in *IEEE Transactions on Electromagnetic Compatibility*, vol. EMC-26, no. 1, pp. 14–18, Feb. 1984, H.-M. Shen (Harvard U.), R. W. P. King (Harvard U.) and T. T. Wu (Harvard U.), DOI: https://doi.org/10.1109/TEMC.1984.304172.

Hao-Ming Shen, Member, *IEEE*,
Ronold W. P. King, Fellow, *IEEE*, and Tai T. Wu

Abstract: A special "dual-measurement" procedure has been introduced for eliminating systematic interference in electromagnetic measurements. This interference has its origin in the same source as the measured signal and is usually picked up by the cables, chassis, and objects other than the sensor. The interference, when superimposed on the signal being measured, obscures the signal and degrades the accuracy of the measurement. The use of the dual-measurement procedure can notably reduce the systematic interference and greatly improve the clarity and accuracy of the measurement.

47) Electromagnetic missiles, *J. Appl. Phys.* **57**, 2370–2373 (1985), T. T. Wu (Harvard U.). DOI: https://doi.org/10.1063/1.335465; Citations: 105.

Gordon McKay Laboratory, Harvard University, Cambridge, Massachusetts 02138

(Received 16 November 1984; accepted for publication 14 December 1984)

Under transient excitation the energy transmitted by an antenna of finite size to a faraway receiver can decrease much more slowly than the usual R^{-2}. Instead, by a suitable choice of excitation, this quantity can decrease as slowly as one wishes, under the physical restriction that the total energy radiated by the antenna is finite. Such cases of slow decrease are referred to as electromagnetic missiles. Another way of stating the result is that the product of the following two quantities can approach zero as slowly as one wishes: (1) the energy transmitted per unit area of receiver, and (2) the total area of the receiver. Examples are constructed when either one, but not both, of these quantities remains finite at infinite distance.

Commentary by Dr. John Myers

His "electromagnetic missiles" put an end to widespread assumption that the Maxwell equations imply that the intensity of a propagating electromagnetic pulse must drop off at large distances as the inverse square of the distance.

48) Spreading of electromagnetic pulses, *Journal of Applied Physics* **58**: 2064–2065 (1985), T. T. Wu (Harvard U.) and Henry Lehmann (Hamburg U.), DOI: https://doi.org/10.1063/1.335964.

Tai Tsun Wu
Gordon McKay Laboratory, Harvard University, Cambridge, Massachusetts 02138
Harry Lehmann
II. Institut für Theoretische Physik der Universität Hamburg, Hamburg, Federal Republic of Germany

(Received 29 March 1985; accepted for publication 13 May 1985)

It is shown that for any finite-energy solution of the free-space Maxwell equations without sources, the energy contained in an infinite slab of finite thickness approaches zero as time tends to infinity.

Commentary by Dr. Dionisios Margetis

The theme of this work is motivated by and expands the subject of the previous paper (4). The authors prove that for any finite-energy solution of Maxwell's equations in free space (without sources), the electromagnetic energy contained in an infinite slab of finite thickness approaches zero in the long-time limit. This result has important implications. For example, in principle the electromagnetic energy in a slab can decrease to zero at an arbitrarily slow rate with increasing time.

49) Properties of lateral electromagnetic fields and their application, *Radio Science* **21**: 13–23 (1986), R. W. P. King (Harvard U.), Margaret Owens(Harvard U.), and T. T. Wu (Harvard U). DOI: https://doi.org/10.1029/RS021i001p00013.

Ronold W. P. King, Margaret Owens, and Tai Tsun Wu
Gordon McKay Laboratory, Harvard University,
Cambridge, Massachusetts

The formulas for the complete electromagnetic field in region 1 ($z \geq 0$, k_1) due to a horizontal electric dipole also located in region 1 near its plane boundary with region 2 ($z \leq 0$, k_2; $|k_2|^2 \ll |k_1|^2$) are reviewed and applied to determine the moment of a magnetic dipole that effectively generates the field in region 2. Use is made of this field to determine the locus of the Poynting vector, the maximum depth of penetration into region 2 of the lateral wave part of the field, and the fraction of power associated with the lateral wave in a specified range. Application is made to the use of horizontal antennas on the seafloor for geophysical exploration of the lithosphere.

50) Spherical lens as a launcher of electromagnetic missiles, *J. Appl. Phys.* **62**, 4036–4040 (1987), R. T. T. Wu (Harvard U.), W. P. King (Harvard U.), and Hao-Ming Shen (Harvard U). DOI: https://doi.org/10.1063/1.339115.

Tai Tsun Wu, Ronold W. P. King, and Hao-Ming Shen
Gordon McKay Laboratory, Harvard University, Cambridge, Massachusetts 02138

(Received 4 May 1987; accepted for publication 7 July 1987)

A uniform dielectric sphere, under transient excitation by a point source, is found to be capable of launching an electromagnetic missile, i.e., there is a direction where the decay of the total energy transmitted to a receiver is slower than r^{-2}. The conditions for the presence of an electromagnetic missile have a natural interpretation, which in turn leads to a classification of electromagnetic missiles according to the singularities of differentiable maps in two dimensions.

51) Generalized analysis of the spherical lens as a launcher of electromagnetic missiles, *J. Appl. Phys.* **63**, 5647–5653 (1988), Hao-Ming Shen (Harvard U.), T. T. Wu (Harvard U.), and R.W.P. King (Harvard U.). DOI: https://doi.org/10.1063/1.340347.

Hao-Ming Shen, Tai Tsun Wu, and Ronold W. P. King
Gordon McKay Laboratory, Harvard University, Cambridge, Massachusetts 02138

(Received 26 October 1987; accepted for publication 24 February 1988)

It is found that, like the directly transmitted radiation field previously investigated, an internally multiply reflected radiation field can also become an electromagnetic missile under certain different electromagnetic-missile conditions. The analysis is carried out in a general way, i.e., for the point source located either inside or outside the dielectric sphere. The electromagnetic-missile conditions are obtained for general multiply reflected radiation fields.

52) V-conical antenna, in *IEEE Transactions on Antennas and Propagation*, vol. 36, no. 11, pp. 1519–1525, Nov. 1988, Hao-Ming Shen (Harvard U.), R. W. P. King (Harvard U.) and Tai Tsun Wu (Harvard U.), DOI: https://doi.org/10.1109/8.9699.

Hao-Ming Shen, Senior Member, *IEEE*,
Ronold W.P. King, Life Fellow, *IEEE*, and Tai Tsun Wu

Abstract: A new angular antenna called the V-conical antenna was investigated. The characteristics of the antenna are frequency-independent, and the electromagnetic fields are pure spherical waves (transverse electromagnetic mode) even in the near region. The theoretical analysis makes use of conformal mapping. Closed-form solutions for the fields, currents, and characteristic parameters are obtained. The formulas are simple and precise. An experimental study of the antenna was conducted under pulse excitation. The shape of the measured electric-field pulses is the same as that of the source pulse from a picosecond pulse generator. The measured field patterns are very close to those predicted by the theory. Such a structure with a tapered resistive termination is an ideal directional pulse antenna or electromagnetic pulse simulator.

53) Curved electromagnetic missiles, *J. Appl. Phys.* **65**, 2604–2610 (1989), John M. Myers (Harvard U.), Hao-Ming Shen (Harvard U.), T. T. Wu (Harvard U), and Howard E. Brandt (Harry Diamong Laboratories), https://doi.org/10.1063/1.342790.

John M. Myers, Hao-Ming Shen, and Tai Tsun Wu
Gordon McKay Laboratory, Harvard University, Cambridge, Massachusetts 02138
Howard E. Brandt
Harry Diamond Laboratories, 2800 Powder Mill Road, Adelphi, Maryland 20783

(Received 6 September 1988; accepted for publication 29 November 1988)

An interesting topic has been the possible behavior of transient fields in the limit of great distances from their sources. Under the physical restriction that the total energy radiated is finite, it has been shown that the energy reaching a distant receiver can decrease with distance much more slowly than the usual r^{-2}. Such cases of slow decrease have been referred to as electromagnetic missiles. All of the wide variety of known missiles propagate in essentially straight lines. It is shown here that such a missile can follow a path that is strongly curved. An example of a curved electromagnetic missile is explicitly constructed and some of its simpler properties are discussed.

54) The properties of the electromagnetic missile, *J. Appl. Phys.* **66**, 4025–4034 (1989), Hao-Ming Shen (Harvard U.) and T. T. Wu (Harvard U). DOI: https://doi.org/10.1063/1.344011.

Hao-Ming Shen and Tai Tsun Wu
Gordon McKay Laboratory, Harvard University, Cambridge, Massachusetts 02138

(Received 1 March 1989; accepted for publication 6 July 1989)

The electromagnetic field from the current in a circular disk and its Poynting vector off the axis of propagation have been investigated both analytically and experimentally. These studies show that the electromagnetic (EM) missile has the following properties: (1) The waveform of the EM field when propagating remains similar in shape but diminishes in size. (2) The transverse distribution of energy around the axis is stable; i.e., when the longitudinal distance increases, the transverse pattern of the energy remains the same. This "plane-wave beam-like" property is different from continuous-wave (cw) radiation, in which the energy pattern is like that of a spherical wave. (3) The transverse energy pattern has a cusp on the axis. This property is also different from cw radiation. With cw, the radiation has a flat top on the axis of the main lobe.

55) Fun with pulses, *Physics World*, Vol. 3, 39–42 (1990), John M. Myers (Harvard U.), Hao-Ming Shen (Harvard U.), T. T. Wu (Harvard U.), and Howard E. Brandt (Harry Diamong Laboratories), DOI: https://doi.org/10.1088/2058-7058/3/11/28.

John M. Myers, Hao-Ming Shen, Tai Tsun Wu, and Howard Brandt
Gordon McKay Laboratory, Harvard University,
Cambridge, MA 02138, USA

In 1887, 26 years after Maxwell first published his seminal equations that define electromagnetic theory, Hertz produced the first controlled radio wave in the laboratory using an oscillating dipole. Since then, the use of electromagnetic waves has permeated our daily life. In the earlier applications the wave forms were almost sinusoidal, prominent examples being radio (AM and FM), television and radar. In each of these cases, there is a relatively narrow bandwidth around a carrier frequency.

56) The resonant circular array of electrically short elements, *J. Appl. Phys.* **68**, 431–439 (1990), G. Fikioris (Harvard U.), T. T. Wu (Harvard U.), and R. W. P. King (Harvard U.). https://doi.org/10.1063/1.346839.

George Fikioris, Ronold W. P. King, and Tai Tsun Wu
Gordon McKay Laboratory, Harvard University, Cambridge, Massachusetts 02138

(Received 5 February 1990; accepted for publication 3 April 1990)

Different resonant distributions of current in circular arrays with $N = 60$, 72, and 90 elements are investigated theoretically with the method of symmetrical components and a two-term representation of current in each element. The arrays studied have only one driven element. The remarkable results—which are displayed in a sequence of graphs—indicate that the relative currents in the elements can be distributed in triplets, triplets alternating with singlets, two independent sets of singlets, or one set of singlets. The driving-point impedance of this last distribution is a pure resistance, the Q of the 90-element array is over 30 000, and the half-power beamwidth of each of the 90 sharp peaks in the power pattern is 2°.

57) Backscattering of an electromagnetic missile by a metal cylinder of degree higher than two, *J. Appl. Phys.* **80**, 15–24 (1996), V. A. Houdzoumis (Harvard U.), T. T. Wu (Harvard U.), and John M. Myers (Harvard U.) DOI: https://doi.org/10.1063/1.362769.

Vassilios A. Houdzoumis, Tai T. Wu, and John M. Myers
Gordon McKay Laboratory, Harvard University, Cambridge, Massachusetts 02138-2901

(Received 10 August 1995; accepted for publication 20 March 1996)

The backscattering of an electromagnetic missile from a perfectly conducting curved obstacle is investigated. The obstacle is assumed to have zero curvature just at the point of reflection of the incident pulse. The asymptotic dependence of the backscattered energy is sought, as the distance separating the obstacle from the source of the incident pulse tends to infinity. The backscattered energy is found to depend on the rate at which the energy spectral density of the current pulse at the source decays with increasing frequency, as well as on the degree of flatness of the obstacle at the point of reflection. © *1996 American Institute of Physics.* [S0021-8979(96)00513-0]

58) Highly directive current distributions: General theory, *Phys. Rev. E* **58**, 2531 (1998), D. Margetis (Harvard U.), G. Fikioris (Air Force Research Laboratory), John M. Myers (Harvard U.), and T. T. Wu (Harvard U.) DOI: https://doi.org/10.1103/PhysRevE.58.2531.

Dionisios Margetis,[1] George Fikioris,[2] John M. Myers,[1] and Tai Tsun Wu[1]
[1]*Gordon McKay Laboratory, Harvard University, Cambridge, Massachusetts 02138-2901*
[2]*Air Force Research Laboratory, Sensors Directorate, Hanscom Air Force Base, Massachusetts 01731*
(Received 19 November 1997)

A theoretical scheme for studying the properties of localized, monochromatic, and highly directive classical current distributions in two and three dimensions is formulated and analyzed. For continuous current distributions, it is shown that maximizing the directivity D in the far field while constraining $C=N/T$, where N is the integral of the square of the magnitude of the current density and T is proportional to the total radiated power, leads to a Fredholm integral equation of the second kind for the optimum current. This equation is a useful analytical tool for studying currents that produce optimum directivities above the directivity of a uniform distribution. Various consequences of the present formulation are examined analytically for essentially arbitrary geometries of the current-carrying region. In particular, certain properties of the optimum directivity are derived and differences between the continuous and discrete cases are pointed out. When $C \rightarrow \infty$, the directivity tends to infinity monotonically, in accord with Oseen's "Einstein needle radiation."
[S1063-651X(98)02108-4]

PACS number(s): 41.20.Bt, 03.40.Kf

59) Electric field induced in cells in the human body when this is exposed to low-frequency electric fields, *Phys. Rev. E* **58**, 2363 (1998), R. W. P. King (Harvard U.), and T. T. Wu (Harvard U.), DOI: https://doi.org/10.1103/PhysRevE.58.2363.

Ronold W. P. King and Tai Tsun Wu
Gordon McKay Laboratory, Harvard University, Cambridge, Massachusetts 02138
(Received 11 March 1998; revised manuscript received 6 May 1998)

A detailed analysis is carried out of the electric field induced in a cell when the body is exposed to an incident axial electric field at 50–60 Hz. It is shown that the field in a spherical cell is effectively shielded by the membrane so that the induced field in its interior is negligibly small. It is also shown that the induced electric field in a cylindrical cell (which is long compared to its radius) is the same as the axial field outside the cell. In this case, the cell membrane has no shielding effect. [S1063-651X(98)09108-9]

PACS number(s): 87.50.−a, 87.10.+e

60) Calculated electric field induced in bundles of long cells in the human body when it is exposed to low-frequency electric fields. *Radiation Research* **153**(5):715–721 (2000), R. W. P. King (Harvard U.), and T. T. Wu (Harvard U.). https://doi.org/10.1667/0033-7587(2000)153[0715:CEFIIB]2.0.CO;2.

Ronold W. P. King[1], Tai Tsun Wu[1,*]
[1]*Gordon McKay Laboratory, Harvard University, Cambridge, Massachusetts 02138*
[*]Author to whom correspondence should be addressed.

King, R. W. P. and Wu, T. T., Calculated Electric Field Induced in Bundles of Long Cells in the Human Body When It Is Exposed to Low-Frequency Electric Fields.

A theoretical analysis is made of the electric field induced in the interior of long cylindrical cells arranged in bundles of parallel cells embedded in the saline fluid in the human body exposed to extremely low-frequency (ELF) and very low-frequency (VLF) electric fields parallel to the length of the cells. Such bundles characterize muscle. The analysis parallels in many respects the earlier study of isolated long cells (King and Wu, *Phys. Rev. E* **58**, 2363–2369, 1998) but has important added complications. It leads to a similar result: The field inside the membrane of each long cell is the same as the field just outside the membrane in the saline fluid; i.e., the membrane provides no shielding. While one might expect the presence of the neighboring cells to provide a shielding effect for each cell, the results of this study show quantitatively that this is not the case. In fact, the electric field in the interior of each of a bundle of long cells is seen to increase very slightly (of the order of 0.4%) when the cells in the bundle are moved close enough to be practically in contact.
(Received: 9 August 1999; Accepted: 1 November 1999; Published: 1 May 2000)

61) On the application of numerical methods to Hallen's equation *IEEE Transactions On Antennas and Propagation*, **49**: 383–392 (2001), G. Fikioris, and T. T. Wu. DOI: https://doi.org/10.1109/8.918612; Citations: 82.

George Fikioris, Member, *IEEE*, and Tai Tsun Wu

The so-called Hallen integral equation for the current on a finite linear antenna center-driven by a delta-function generator takes two forms depending on the choice of kernel. The two kernels are usually referred to as the exact and the approximate or reduced kernel. With the approximate kernel, the integral equation has no solution. Nevertheless, the same numerical method is often applied to both forms of the integral equation. In this paper, the behavior of the numerical solutions thus obtained is investigated, and the similarities and differences between the two numerical solutions are discussed. The numerical method is Galerkin's method with pulse functions. We first apply this method to the two corresponding forms of the integral equation for the current on a linear antenna of infinite length. In this case, the method yields an infinite Toeplitz system of algebraic equations in which the width of the pulse basis functions enters as a parameter. The infinite system is solved exactly for nonzero pulse width; the exact solution is then developed asymptotically for the case where the pulse width is small. When the asymptotic expressions for the case of the infinite antenna are used as a guide for the behavior of the solutions of the finite antenna, the latter problem is greatly facilitated. For the approximate kernel, the main results of this paper carry over to a certain numerical method applied to the corresponding equation of the Pocklington type.

Commentary by Dr. Dionisios Margetis

In antenna theory and design, it is often necessary to compute the current distributions numerically by use of appropriate models. For tubular dipole antennas and their direct variants, these models may involve a well-known integral equation, Hallen's equation, for the current. This paper addresses a key issue in this regard: the comparison of predictions of a typical numerical method, Galerkin's method with pulse functions, applied to the governing integral equation for a basic tubular dipole antenna under the exact or an approximate kernel. The authors first address analytically the case of the infinite antenna, and then use their results as a guide for the antenna of finite length. A far-reaching conclusion concerns the appearance and nature of large oscillations near the driving point in the imaginary part of the input admittance under an approximate kernel.

62) Exactly calculable field components of electric dipoles in planar boundary, *J. Math. Phys.* **42**, 713–745 (2001), D. Margetis (Harvard U.), and T. T. Wu (Harvard U.) DOI: https://doi.org/10.1063/1.1330731.

Dionisios Margetis[a)] and Tai Tsun Wu
Gordon McKay Laboratory, Harvard University, Cambridge, Massachusetts 02138-2901

(Received 19 January 2000; accepted for publication 13 October 2000)

The Sommerfeld integrals for the electromagnetic fields in the planar boundary between air and a homogeneous, isotropic medium, due to a horizontal and a vertical electric dipole each lying along the interface, are examined in detail. In the case of the horizontal dipole, the tangential electric field is given in terms of series that involve confluent hypergeometric functions, namely, the Fresnel and exponential integrals. A similar exposition is presented for the magnetic and vertical electric fields of the vertical dipole. When the index of refraction of the adjacent space is of a sufficiently large magnitude, the derived series converge rapidly and uniformly with the distance from the source. Specifically, their rates of convergence are shown to be independent of distance. It is pointed out that the corresponding formulas of King *et al*. are valid down to any distance close to the source, where they smoothly connect to known "quasi-static" approximations. [DOI: 10.1063/1.1330731]

63) Properties and applications of the large circular resonant dipole array, in *IEEE Transactions on Antennas and Propagation*, vol. 51, no. 1, pp. 103–109, Jan. 2003, R. W. P. King (Harvard U.), M. Owens (Harvard U.) and Tai Tsun Wu (Harvard U.), DOI: https://doi.org/10.1109/TAP.2003.809053.

Ronold W. P. King, Life Fellow, *IEEE*, Margaret Owens, and Tai Tsun Wu

Abstract: Large closed-curve antenna arrays have been a subject of research for many years and have been shown to have many interesting properties. The paper investigates some of the properties of such a dipole array when the closed curve is a circle. Recently, because of its unique horizontal field pattern, a 90-element circular array of this type has been proposed as a microwave beacon for the coastal navigation of ships and airplanes. In the design of these arrays, it is suggested that the array be rotated mechanically. The question arises: can the mechanical rotation be replaced by an electronic rotation? We show that electronic rotation is not possible for the 90-element circular array originally described, but is possible for a modified array. The subtle difference between these two arrays is clarified and a simple criterion is given for the general case. Also presented is the derivation of an asymptotic formula for the radiation pattern of a resonant circular array of *N* equal elements with only one element driven. Since the theory for such an array is complicated and involves numerous numerical difficulties, a simple asymptotic formula for the field pattern has advantages over other methods. The simple formula is shown to produce a vertical field pattern that is indistinguishable from its numerically calculated counterpart. Generalization to noncircular arrays is discussed briefly.

64) Electromagnetic Resonances of a Straight Wire, in *IEEE Transactions on Antennas and Propagation*, vol. 59, no. 1, pp. 129–134, Jan. 2011, John M. Myers (Harvard U.), Sheldon S. Sandler (Harvard U.), and T. T. Wu (Harvard U.), DOI: https://doi.org/10.1109/TAP.2010.2090479.

John M. Myers, Sheldon S. Sandler, Life Member, *IEEE*, and Tai Tsun Wu

Abstract: With an interest in finding wires and distinguishing them from other electrically conducting objects, we have looked for an electromagnetic "fingerprint" in terms of resonances of a straight wire of length $2h$ and radius a. The resonances of the wire are formulated using the theory of the linear antenna, leading to an integral equation for the current on the wire. Complex-valued resonant frequencies are defined as those for which the homogeneous integral equation for the current on the wire has non-zero solutions. By applying a variational technique we obtain approximate numerical solutions for the resonant frequencies and their widths. A table of the first five resonances is given for several ratios of wire half-length h to wire radius a. In a subsequent paper we propose to extend the method described here to deal with wires on an earth-air interface, for example as used to command the detonation of improvised explosive devices.

65) Electromagnetic resonances of a straight wire on an earth-air interface, *IEEE Transactions Antennas and Propagation*, **59**, 3154–3163 (2011). John M. Myers (Harvard U.), Sheldon S. Sandler (Harvard U.), and T. T. Wu (Harvard U.). DOI: https://doi.org/10.1109/TAP.2011.2161539.

John M. Myers; Sheldon S. Sandler; Tai Tsun Wu
Harvard School of Engineering and Applied Sciences,
Cambridge, Massachusetts

Using a variational method, we recently determined an electro-magnetic "signature" for characterizing a straight wire in free space. The signature consists of the first five resonant frequencies and their widths, more compactly expressed as the first five complex-valued resonant frequencies. Here we apply the variational method to the much more complicated case of determining the same signature for a straight wire or wire pair on a flat interface between a homogeneous earth and air. To calculate the resonances we obtain an integral equation for the current on a wire on the interface between two dielectric media. Complex-valued resonant frequencies are defined as those for which the homogeneous integral equation for the current in an equivalent thin strip on the interface has non-zero solutions. The variational method extracts good approximations to these complex-valued resonant frequencies, without having to solve the integral equation. A table of resonances is given for the case of a relative dielectric constant of the earth equal to 4 and for three values of the ratio of wire radius to wire half-length.

Commentary by Dr. Dionisios Margetis

This work addresses a fundamental aspect of the classic problem of recognizing buried objects by reading off their electromagnetic signatures. The authors consider a perfectly conducting strip of finite length and width located in the flat interface between air and homogeneous earth. First, they derive an approximation for the related "Green function" (including the interface) of this setting, which supplies the appropriate simplified kernel for the governing integral equation for the current on the strip. Subsequently, they solve the homogeneous integral equation for the current by a variational method, and compute the associated complex-valued resonant frequencies.

Chapter 7
Yang–Mills Theory & Monopoles
(23 papers)

66) Fermion-Fermion Scattering in a Yang–Mills Theory at High-Energy: Sixth Order Perturbation Theory, *Phys. Rev. D* **12** (1975), 3257, Barry M. McCoy (SUNY, Stony Brook), Tai Tsun Wu (Harvard U.), DOI: https://doi.org/10.1103/PhysRevD.12.3257.

Barry M. McCoy†
Institute for Theoretical Physics, State University of New York, Stony Brook, New York 11794

Tai Tsun Wu
Gordon McKay Laboratory, Harvard University, Cambridge, Massachusetts 02138
(Received 23 June 1975)

We study the high-energy behavior of the elastic scattering of two isospin-$\frac{1}{2}$ fermions interacting through the Yang-Mills field. The Higgs mechanism is invoked so that there is no infrared divergence. In the sixth order, the amplitude for isovector exchange is found to behave as $s \ln^2 s$ multiplied by a function of t, while the amplitude for isoscalar exchange behaves as $s \ln s$ multiplied by another function of t. These results are qualitatively different from the answers previously given in the literature. In particular, there is no contribution to the leading terms from large transverse momentum transfers; this is the reason why we obtain one less factor of $\ln s$ in sixth order.

67) Boson-Fermion and Boson-Boson Scattering in a Yang–Mills Theory at High-Energy: Sixth Order Perturbation Theory, *Phys. Rev. D* **13** (1976), 1076, Barry M. McCoy (SUNY, Stony Brook), Tai Tsun Wu (Harvard U.), DOI: https://doi.org/10.1103/PhysRevD.13.1076.

Barry M. McCoy†
Institute for Theoretical Physics, State University of New York at Stony Brook, Stony Brook, New York 11794

Tai Tsun Wu
Gordon McKay Laboratory, Harvard University, Cambridge, Massachusetts 02138
(Received 29 September 1975)

We extend our previous study of Yang-Mills fields by calculating the high-energy behavior of the boson-fermion and of the boson-boson amplitude in sixth-order perturbation theory. In the isovector and isoscalar channels of both these processes the behavior of the amplitude is the same as that found in fermion-fermion scattering.

68) Multiple Direct Exchange in a Yang–Mills Theory at High-Energy, *Phys. Rev. D* **14** (1976), 3482-3496, B. M. Mccoy (SUNY, Stony Brook), T. T. Wu (Harvard U.), DOI: https://doi.org/10.1103/PhysRevD.14.3482.

Barry M. McCoy[†]
Institute for Theoretical Physics, State University of New York, Stony Brook, New York 11794

Tai Tsun Wu
Gordon McKay Laboratory, Harvard University, Cambridge, Massachusetts 02138
(Received 10 February 1976)

For eighth and higher orders, we obtain the leading high-energy behavior of the sum of all one-layer Feynman diagrams in Yang-Mills theory. These are the contributions from Feynman diagrams where the two incident fast particles exchange directly Yang-Mills bosons that are much less energetic. The incident particles may be either bosons or fermions of arbitrary isospin, and the result is also generalized to include the case of the Higgs scalar. The scattering amplitudes in all these cases are closely related, and all behave as $s\ln^{n-1}s$ in the $2n+2$ order. Furthermore, in this leading order for $n \geq 2$, the exchanged isospins are always 0 and 2, no matter how high the isospins of the incident particles are.

69) Dirac's Monopole Without Strings: Classical Lagrangian Theory, *Phys. Rev. D* **14** (1976), 437–445, Tai Tsun Wu (Harvard U.), Chen Ning Yang (SUNY, Stony Brook), DOI: https://doi.org/10.1103/PhysRevD.14.437; Citations: 137.

Tai Tsun Wu*
Gordon McKay Laboratory, Harvard University, Cambridge, Massachusetts 02138

Chen Ning Yang[†]
Institute for Theoretical Physics, State University of New York, Stony Brook, New York 11794
(Received 16 March 1976)

The non-quantum-mechanical interaction of a Dirac magnetic monopole and a point charge through the electromagnetic field is studied. A classical action integral which is multiple-valued is found. Stability of this action integral against variations of the world lines of the point charge and the monopole, and against variations of the electromagnetic potentials, yields the correct Lorentz equations of motion of the particles and the Maxwell equations for the field. No strings are introduced in the formalism.

70) Dirac Monopole Without Strings: Monopole Harmonics, *Nucl.Phys. B* **107** (1976), 365–380, In Yang, C.N.: Selected Papers 1945–1980, 493–508. (*Nucl. Phys. B* **107** (1976) 365–380) and State Univ. New York Stony Brook - ITP-SB-76-05 (76,REC.FEB.) 30 P. (602076), Tai Tsun Wu (Harvard U.), Chen Ning Yang (SUNY, Stony Brook), DOI: https://doi.org/10.1016/0550-3213(76)90143-7; Citations: 771.

Tai Tsun WU *
Gordon McKay Laboratory, Harvard University, Cambridge, Massachusetts 02138

and

Chen Ning YANG **
Institute for Theoretical Physics, State University of New York,
Stony Brook, New York 11794

Received 17 February 1976

Using the ideas developed in a previous paper which are borrowed from the mathematics of fibre bundles, it is shown that the wave function ψ of a particle of charge Ze around a Dirac monopole of strength g should be regarded as a *section*. The section is without discontinuities. Thus the monopole *does not* possess strings of singularities in the field around it. The eigensections of the angular momentum operators are monopole harmonics which are explicitly exhibited.

71) Some Properties of Monopole Harmonics, *Phys. Rev. D* **16** (1977), 1018–1021, Tai Tsun Wu (Harvard U.), Chen Ning Yang (SUNY, Stony Brook), DOI: https://doi.org/10.1103/PhysRevD.16.1018; Citations: 159.

Tai Tsun Wu*
Gordon McKay Laboratory, Harvard University, Cambridge, Massachusetts 02138

Chen Ning Yang†
Institute for Theoretical Physics, State University of New York, Stony Brook, New York 11794
(Received 6 April 1977)

Some properties of monopole harmonics are derived.

72) Two-Dimensional Yang–Mills Theory in the Leading 1/N Expansion, *Phys. Lett. B* **71** (1977), 142–144, Tai Tsun Wu (CERN), DOI: https://doi.org/10.1016/0370-2693(77)90762-6; Citations: 78.

Tai Tsun WU [1]
CERN, Geneva, Switzerland

Received 30 August 1977

Recent controversies about the gauge invariance of the two-dimensional SU(N) Yang-Mills theory in the 't Hooft limit of large N are resolved. The fermion (quark) propagator is found explicitly, and is qualitatively different from those in the previous literature.

73) Failure of Bartels' Scheme of Applying Reggeon Field Theory to a Spontaneously Broken Yang–Mills Theory, *Phys. Lett. B* **71** (1977), 97–98, B. M. McCoy (SUNY, Stony Brook), Tai Tsun Wu (CERN), DOI: https://doi.org/10.1016/0370-2693(77)90749-3.

Barry M. McCOY
Institute for Theoretical Physics, State University of New York, Stony Brook, New York 11794, USA

and

Tai Tsun WU[1]
CERN, Geneva, Switzerland

Received 22 August 1977

The recently proposed scheme of Bartels of studying the high-energy behavior of Yang-Mills theory in terms of Reggeon field theory fails, because, in twelfth order or higher, not all transverse diagrams contribute to the leading power of ln *s*.

74) Interaction of Electrons, Magnetic Monopoles, and Photons 1, *Sci. Sin.* **21** (1978), 317–326, Also in *Yang, C.N.: Selected Papers 1945–1980*, 541–550.

Tu Tung-sheng (杜东生),
Wu Tai-tsun (吴大峻) and Yang Chen-ning (杨振宁)

Received February 18, 1978.

Abstract

Using a method developed for treating monopole harmonics, we present in this series of papers a second quantized theory of interacting electrons, magnetic monopoles, and photons. In this first paper we discuss a system of interacting electrons, monopoles, and the electromagnetic field. The electrons and monopoles are treated as spin 1/2 Dirac particles with the negative energy states not filled.

75) Two-Dimensional Yang–Mills Theory in the Leading 1/N Expansion, T. T. Wu (CERN), In Gervais, J. I. (Ed.), Jacob, M. (Ed.): *Non-linear and Collective Phenomena In Quantum Physics*, 297–305.

76) XV. Two-dimensional Yang–Mills Theory in the Leading 1/N Expansion, *Phys. Rept.* **49** (1979), 245–253, Tai Tsun Wu (CERN), DOI: https://doi.org/10.1016/0370-1573(79)90115-7.

Tai Tsun Wu
CERN, Geneva, Switzerland

Abstract:
For the two-dimensional Yang–Mills theory in the leading $1/N$ expansion, the question of gauge invariance for generalized axial gauges is reviewed. It is found that the commonly used principal-value prescription for the Yang–Mills propagator is ambiguous. While there are two natural interpretations of this principal-value prescription, neither one leads to gauge invariant results. If Wick rotation is used, then the Feynman rules can be reformulated in a gauge-invariant manner in terms of singular integrals of Calderon-Zygmund type. The non-linear integral equation in the light-cone gauge for the fermion propagator is solved analytically, and the resulting solution contains a branch cut. The disturbing properties of this branch cut are discussed.

77) Interaction of a Fermion With a Monopole. 1., *Nucl. Phys. B* **222** (1983), 411–418, Tai Tsun Wu (DESY), DOI: https://doi.org/10.1016/0550-3213(83)90542-4.

TAI TSUN WU[1]
Deutsches Elektronen-Synchrotron DESY, Hamburg, West Germany

Received 24 March 1983

It is proposed that some low-energy properties of the interaction of one or more fermions with a monopole are sensitive to, and hence can be used as probes for, extremely small distances. One possible example of such properties is studied in detail.

78) Monopole Chemistry, *Phys. Lett. B* **129** (1983), 209–212, T.W. Ruijgrok (Utrecht U.), J.A. Tjon (Utrecht U.), Tai Tsun Wu (DESY), DOI: https://doi.org/10.1016/0370-2693(83)90845-6.

Th.W. RUIJGROK, J.A. TJON
Instituut voor Theoretische Fysica, Utrecht, The Netherlands

and

Tai Tsun WU [1]
Deutsches Elektronen-Synchroton DESY, Hamburg, Germany

Received 20 May 1983

Possible chemical binding of the magnetic monopole is discussed. The interaction of a monopole with a hydrogen atom is investigated in the Born–Oppenheimer approximation with the electron described by the Dirac equation. The effective potential between the monopole and the proton is determined together with the resulting bound states.

79) Monopole Catalysis of Proton Decay, *Nucl. Phys. B* **232** (1984), 349–355, T. F. Walsh (DESY), P. Weisz (Hamburg U.), Tai Tsun Wu (DESY), DOI: https://doi.org/10.1016/0550-3213(84)90571-6.

T. F. WALSH

Deutsches Elektronen-Synchrotron DESY, Hamburg, Federal Republic of Germany

P. WEISZ†

II. Institut für Theoretische Physik der Universität Hamburg, Federal Republic of Germany

Tai Tsun WU*

Deutsches Elektronen-Synchrotron DESY, Hamburg, Federal Republic of Germany

Received 25 April 1983

The Rubakov effect (monopole catalysis of proton decay at strong-interaction cross sections) is absent for SU(5) grand unified monopoles, because of the non-existence of a zero-energy fermion-monopole bound state in this theory.

80) Monopole-Fermion and Dyon-Fermion Bound States. 1. General Properties and Numerical Results, *Nucl. Phys. B* **247** (1984), 421–449, Per Osland (DESY), Tai Tsun Wu (Harvard U.), DOI: https://doi.org/10.1016/0550-3213(84)90557-1.

Per OSLAND

Deutsches Elektronen-Synchrotron DESY, D-2000 Hamburg 52, Germany

Tai Tsun WU[1]

Gordon McKay Laboratory, Harvard University, Cambridge, Massachusetts 02138, USA

Received 6 June 1984

General properties of monopole-fermion and dyon-fermion bound states are investigated, for the case of Dirac monopoles and their dyon analogues. The attractive forces that lead to bound states are due to the interaction of a fermion anomalous magnetic moment with the monopole magnetic field and the additional Coulomb interaction in the dyon case. Detailed numerical results are presented.

81) Monopole-Fermion and Dyon-Fermion Bound States. 2. Weakly Bound States for the Lowest Angular Momentum, *Nucl. Phys. B* **247** (1984), 450–470, Per Osland (DESY), Tai Tsun Wu (Harvard U.), DOI: https://doi.org/10.1016/0550-3213(84)90558-3.

Per OSLAND

Deutsches Elektronen-Synchrotron DESY, D-2000 Hamburg 52, West Germany

Tai Tsun WU[1]

Gordon McKay Laboratory, Harvard University, Cambridge, Massachusetts 02138, USA

Received 6 June 1984

We investigate weakly bound dyon-fermion states of the lowest angular momentum. Both Coulomb attraction and Coulomb repulsion are studied. Binding energies are given by a transcendental equation which is solved explicitly in a number of limiting cases. Normalized wave functions are given in terms of Bessel functions and confluent hypergeometric functions.

82) Monopole-fermion and dyon-fermion bound states: (III). Monopole-fermion system with $j = |q| - 12$ and large $\kappa|q|$, *Nucl.Phys. B* **256** (1985), 13–31, Per Osland (DESY), Tai Tsun Wu (DESY and Harvard U.), DOI: https://doi.org/10.1016/0550-3213(85)90383-9.

Per OSLAND

Deutsches Elektronen-Synchrotron, DESY, D-2000 Hamburg 52, Federal Republic of Germany

Tai Tsun WU*

Deutsches Elektronen-Synchrotron, DESY, D-2000 Hamburg 52, Federal Republic of Germany and Gordon McKay Laboratory, Harvard University, Cambridge, Massachusetts 02138, USA

Received 8 January 1985

In the first part of the paper, we give analytic, approximate results for the Kazama-Yang monopole-fermion binding energies and wave functions, valid for large values of $A = \frac{1}{2}Z|eg|\kappa$, where κ is the extra magnetic moment. In the second part, more general results are obtained for the same problem that are valid when *either* A is large *or* the binding is weak. Numerical results for the binding energy are tabulated and compared.

83) Monopole-fermion and dyon-fermion bound states: (IV). Dyon-fermion system with $j = |q| - 12$ and large $\kappa|q|$, *Nucl. Phys. B* **256** (1985), 32–66, Per Osland (DESY and Harvard U.), Tai Tsun Wu (Harvard U.), DOI: https://doi.org/10.1016/0550-3213(85)90384-0.

Per OSLAND*

Deutsches Elektronen-Synchrotron DESY, D-2000 Hamburg 52, Federal Republic of Germany and Gordon McKay Laboratory, Harvard University, Cambridge, Massachusetts 02138, USA

Tai Tsun WU*

Gordon McKay Laboratory, Harvard University, Cambridge, Massachusetts 02138, USA

Received 8 January 1985

In the first part of the paper, we give analytic, approximate results for dyon-fermion binding energies and wave functions, valid for large values of $A = \frac{1}{2}Z|eg|\kappa$, where κ is the extra magnetic moment. In the second part, more general results are obtained for the same problem that are valid when *either* A is large *or* the binding is weak. Numerical results for the binding energy are tabulated and compared. The case of very strong binding is also discussed.

84) Monopole-Fermion and Dyon-Fermion Bound States. 5. Weakly Bound States for the Monopole-Fermion System, *Nucl. Phys. B* **256** (1985), 449–498, P. Osland (DESY and Harvard U.), C. L. Schultz (Harvard U.), T. T. Wu (Harvard U.), DOI: https://doi.org/10.1016/0550-3213(85)90404-3.

Per OSLAND*

Deutsches Elektronen-Synchrotron DESY, D-2000 Hamburg 52, Federal Republic of Germany and Gordon McKay Laboratory, Harvard University, Cambridge, Massachusetts 02138, USA

Cherie L. SCHULTZ*,** and Tai Tsun WU*

Gordon McKay Laboratory, Harvard University, Cambridge, Massachusetts 02138, USA

Received 18 February 1985

We present explicit, approximate, remarkably precise results for the Kazama–Yang hamiltonian, which describes a Dirac monopole interacting with a spin-$\frac{1}{2}$ fermion that has an extra magnetic moment. The results are valid for bound states of angular momentum $j \geqslant Z|eg| + \frac{1}{2}$, where the radial wave functions are determined by four coupled differential equations. These equations have been solved analytically for $M - E \ll M$, which is a limit of considerable practical interest. Binding energies and wave functions are given.

85) Generalized gauge transformations: SU(2) Yang–Mills case, *Phys. Rev. D* **54** (1996), 5302–5314, R. Gastmans (Leuven U.), Conrad Newton (Harvard U. and CERN), Tai Tsun Wu (Harvard U. and CERN), DOI: https://doi.org/10.1103/PhysRevD.54.5302.

R. Gastmans
Institute for Theoretical Physics, University of Leuven, B-3001 Leuven, Belgium

Conrad Newton and Tai Tsun Wu
Gordon McKay Laboratory, Harvard University, Cambridge, Massachusetts 02138
and Theory Division, CERN, CH-1211 Geneva 23, Switzerland
(Received 29 March 1996)

We consider generalized gauge transformations for SU(2) Yang-Mills theory involving products of field operators at different space-time points. The infinitesimal gauge transformations are explicitly constructed to third order in the coupling constant. Also the case of an SU(2) doublet in interaction with the Yang-Mills fields is worked out to third order. [S0556-2821(96)00820-X]

PACS number(s): 11.15.−q

86) Generalized gauge transformations: Pure Yang–Mills case, *Phys. Rev. D* **57** (1998), 1203–1224, R. Gastmans (Leuven U.), Tai Tsun Wu (Harvard U. and CERN), DOI: https://doi.org/10.1103/PhysRevD.57.1203.

R. Gastmans*
Institute for Theoretical Physics, University of Leuven, B-3001 Leuven, Belgium

Tai Tsun Wu
Gordon McKay Laboratory, Harvard University, Cambridge, Massachusetts 02138
and Theory Division, CERN, CH-1211 Geneva 23, Switzerland
(Received 20 August 1997; published 23 December 1997)

Gauge transformations with Dirac point splitting are systematically discussed for the case of a pure Yang-Mills theory. These generalized gauge transformations are based on two ingredients: a fixed four-vector, which defines the point splitting, and a weight function, which gives an average over the amount of point splitting and which provides a cutoff in momentum space in the direction of the point splitting four-vector. From the requirement that the group property must be satisfied, it is found, starting from a simple ansatz, that an infinitesimal generalized gauge transformation takes the form of an infinite series in the coupling constant. Using induction on the order of the coupling constant, it is shown that all higher-order terms indeed exist and that they can be expressed in terms of the lower-order formulas. That there are such generalized gauge transformations suggests the possibility of a Yang-Mills field theory with mitigated divergences. [S0556-2821(97)03524-8]

PACS number(s): 11.15.−q

87) Remarks on Yang–Mills theory, T. T. Wu (Harvard U. and CERN), Symposium on Symmetries and Reflections in Honor of C. N. Yang (1999), 199–204.

TAI TSUN WU*
Gordon McKay Laboratory, Harvard University, Cambridge, MA 02138, USA
and
Theory Division, CERN, CH-1211 Geneva 23, Switzerland

The Yang–Mills theory is the most important development in physics in the second half of this century. Some of the highlights are briefly summarized.

88) Yang–Mills theory for nonsemisimple groups, Jean Nuyts (UMH, Mons), Tai Tsun Wu (Harvard U. and CERN), DOI: https://doi.org/10.1103/PhysRevD.67.025014.

Jean Nuyts*
Université de Mons-Hainaut, 20 Place du Parc, 7000 Mons, Belgium

Tai Tsun Wu†
Gordon McKay Laboratory, Harvard University, Cambridge, Massachusetts 02138
and Theoretical Physics Division, CERN, CH-1211 Geneva 23, Switzerland
(Received 21 October 2002; published 27 January 2003)

For semisimple groups, possibly multiplied by U(1)'s, the number of Yang-Mills gauge fields is equal to the number of generators of the group. In this paper, it is shown that, for nonsemisimple groups, the number of Yang-Mills fields can be larger. These additional Yang-Mills fields are not irrelevant because they appear in the gauge transformations of the original Yang-Mills fields. Such nonsemisimple Yang-Mills theories may lead to physical consequences worth studying. The nonsemisimple group with only two generators that do not commute is studied in detail.

DOI: 10.1103/PhysRevD.67.025014 PACS number(s): 11.15.−q, 11.10.−z

Chapter 8
Mathematical & Theoretical Physics in High Energy Physics (131 papers)

89) On the perturbation of characteristic vibrations by dissipative forces. *Arch. Math* **5**, 175–181 (1954), T. T. Wu (University of Minnesota). DOI: https://doi.org/10.1007/BF01899335.

90) Domains of Definition for Feynman Integrals over Real Feynman Parameters, *Phys. Rev.* **123** (1961), 678–689, Tai Tsun Wu (Princeton, Inst. Advanced Study), DOI: https://doi.org/10.1103/PhysRev.123.678.

Tai Tsun Wu*
The Institute for Advanced Study, Princeton, New Jersey
(Received March 1, 1961)

Given a Feynman diagram, the corresponding integral over real Feynman parameters is meaningful and analytic in a certain domain in the space of the Lorentz invariants formed from the external momenta, each of which is on the mass shell. In the case where all the masses are equal, the intersection of these domains for all proper convergent diagrams is studied. For the cases of four and five external lines, the real intersections are explicitly found; for the case of six, seven, and eight external lines, procedures for finding the real intersections are given. A knowledge of the real intersection makes it possible to construct geometrically a subset of the complex intersection. Generalization to unequal external masses is briefly considered.

91) Schrödinger Scattering Amplitude. I. *J. Math. Phys.* **2**, 710–713 (1961), Alex Grossmann (Brandeis University) and Tai Tsun Wu (Harvard U.), DOI: https://doi.org/10.1063/1.1703760.

Alex Grossmann and Tai Tsun Wu†
Physics Department, Brandeis University, Waltham, Massachusetts, and Gordon McKay Laboratory, Harvard University, Cambridge, Massachusetts
(Received March 3, 1961)

The Schrödinger scattering amplitude for a fixed potential is studied as a function of the three components of the initial momentum, the three components of the final momentum, and the square root of the energy.

92) Perturbation Theory of Pion-Pion Interaction. 1. Renormalization, *Phys. Rev.* **125** (1962), 1436–1450, Tai Tsun Wu (Harvard U. and Princeton, Inst. Advanced Study), DOI: https://doi.org/10.1103/PhysRev.125.1436.

Tai Tsun Wu*
Gordon McKay Laboratory, Harvard University, Cambridge, Massachusetts and Institute for Advanced Study, Princeton, New Jersey
(Received October 9, 1961)

The problem of pion-pion scattering is studied on the basis of the model of a four-particle direct interaction without derivative coupling. Renormalization is carried out for this model with a detailed analysis of overlap insertions. To every finite order in the renormalized coupling constant, it is shown that the unitarity relation holds and that the Feynman integral representation is still valid, and hence renormalization has no effect on analytic properties.

93) Schrödinger Scattering Amplitude. III. *J. Math. Phys.* **3**, 684–689 (1962), Alex Grossmann (Institute for Advanced Study) and Tai Tsun Wu (Harvard U.), DOI: https://doi.org/10.1063/1.1724270.

Alex Grossmann
Institute for Advanced Study, Princeton, New Jersey
and
Tai Tsun Wu†
Gordon McKay Laboratory, Harvard University, Cambridge, Massachusetts
(Received February 9, 1962)

The methods of an earlier paper are used to obtain a domain of analyticity for the Schrödinger scattering amplitude minus the first Born term. The connection between the scattering integral equation and the Schrödinger equation is also studied.

94) Perturbation Theory of Scattering Amplitudes at High Energies, *Phys. Rev.* **130** (1963), 2566–2572, James D. Bjorken (Princeton, Inst. Advanced Study), Tai Tsun Wu (Princeton, Inst. Advanced Study), DOI: https://doi.org/10.1103/PhysRev.130.2566; Citations: 99.

James D. Bjorken* and Tai Tsun Wu†
Institute for Advanced Study, Princeton, New Jersey
(Received 13 February 1963)

The asymptotic behavior of Feynman-Dyson graphs for elastic scattering at high energies is studied by means of the Mellin transform. The leading terms of a set of graphs involving three-particle intermediate states are summed. Under certain conditions the asymptotic behavior of the sum is found to contain a factor $(\ln s)^{-3/2}$. This is incompatible with Regge behavior, but is consistent instead with a cut in the complex angular momentum plane.

95) Singular Potentials and Peratization. II. *J. Math. Phys.* **5**, 799–804 (1964), A. Pais (The Rockefeller Institute) and Tai Tsun Wu (Harvard U.), DOI: https://doi.org/10.1063/1.1704181.

A. PAIS
The Rockefeller Institute, New York, New York
and
TAI TSUN WU*
Gordon McKay Laboratory, Harvard University, Cambridge, Massachusetts
(Received 10 February 1964)

In a field theory with more than one unrenormalizable interaction (like the W theory with weak and electromagnetic coupling), the important problem arises of resumming multiple series of individually divergent terms. In order to get a first insight in the new questions which arise for multiple as compared to single series, we study the analogous question for a superposition of two singular potentials, using a family of exactly soluble cases. We ask whether one can expand resummed series for the zero-energy scattering amplitude in powers of one coupling constant with coefficients depending on the other. The answer depends both on the relative magnitude of the coupling constants and on the relative degree of singularity of the interactions. Depending on these two conditions one finds three regimes, one where a convergent power expansion holds, another where an asymptotic expansions obtains, and a third where it is impossible to expand in powers of either single constant separately. It is conjectured that a similar situation will be true in a field theory with leading power singularities only (and no logarithmic ones), if such a theory has meaning.

96) Scattering Formalism for Singular Potential Theory, *Phys. Rev.* **134** (1964), B1303–B1307, A. Pais (Rockefeller U.), Tai Tsun Wu (Harvard U.), DOI: https://doi.org/10.1103/PhysRev.134.B1303.

A. PAIS
The Rockefeller Institute, New York, New York
AND
TAI TSUN WU*
Gordon McKay Laboratory, Harvard University, Cambridge, Massachusetts
(Received 12 February 1964)

When using power-series expansions in the coupling constant(s) for the computation of scattering amplitudes in singular potential theory, it is necessary to introduce some cutoff in order to avoid divergences of the individual terms in this series. In earlier work this was done by cutting off the potential. It is here shown that the general scattering formalism actually contains an intrinsic limiting operation which makes it unnecessary to introduce a cutoff as an extraneous computational device. For regular potentials one tacitly and justifiedly exchanges this operation with another step in the derivations. It is not clear whether such a "built-in" cutoff procedure also exists in unrenormalizable field theory.

97) Scattering by the Singular Potential $-gr^{-4}\ln r$, *Phys. Rev.* **136**, B1176 (1964), Tai Tsun Wu (University of Wisconsin), DOI: https://doi.org/10.1103/PhysRev.136.B1176.

TAI TSUN WU*
Summer Institute of Theoretical Physics, University of Wisconsin, Madison, Wisconsin
(Received 1 July 1964)

In connection with higher order leptonic weak interactions, divergent series containing an infinite number of logarithmic factors are encountered. In order to have some understanding of series of this type, the problem of scattering by the singular potential $-gr^{-4}\ln r$ is examined for the special case of zero energy. This problem has the advantage that the correct answer can be obtained by other means. Generalization to the potential $-gr^{-\lambda}\ln r$ with $\lambda>3$ is also considered.

98) Some Speculations Concerning High-Energy Large Momentum Transfer Processes, *Phys. Rev.* **137** (1965), B708–B711, In

Yang, C.N.: Selected Papers 1945–1980, 365–368, and in *Lo, S. Y. (ed.): Geometrical pictures in hadronic collisions*, 183–186, T. T. Wu (Brookhaven), Chen-Ning Yang (Brookhaven), DOI: https://doi.org/10.1103/PhysRev.137.B708; Citations: 318.

TAI TSUN WU† AND C. N. YANG‡
Brookhaven National Laboratory, Upton, New York
(Received 23 September 1964)

It is speculated that the sharp decrease with increasing energy of differential cross sections at large angles is due to a mechanism independent of the method of excitation. Some consequences of such a possibility are discussed.

99) High-Energy Large-Momentum-Transfer Processes and Analyticity in Coordinate Space, *Phys. Rev.* **143**, 1117 (1966), Tai Tsun Wu (Harvard U.), DOI: https://doi.org/10.1103/PhysRev.143.1117.

TAI TSUN WU*
Gordon McKay Laboratory, Harvard University, Cambridge, Massachusetts
(Received 1 June 1965)

In nonrelativistic scattering theory, analytic potentials lead to exponentially decreasing differential cross sections for large momentum transfers. We attempt to adopt this connection for strong interactions. Since the differential cross section for proton-proton elastic scattering with large momentum transfers decreases exponentially with the momentum transfer, we postulate that the wave functions of strongly interacting particles are analytic functions of the coordinate variables. This assumption leads in a natural way to an exponential dependence on the transverse momentum transfer. On the basis of this, we speculate about various scattering processes involving a total of four strongly interacting particles, two incoming and two outgoing. The results are, in some cases, less specific than those of the previous consideration. Since it is natural to consider not only two-particle but also three-particle wave functions, the speculation is also extended to some processes where three particles are produced.

100) Born Approximation and Large-Momentum-Transfer Processes in Potential Scattering, *Phys. Rev.* **143**, 1110 (1966), Tai Tsun Wu (Harvard U.), DOI: https://doi.org/10.1103/PhysRev.143.1110.

TAI TSUN WU*
Gordon McKay Laboratory, Harvard University, Cambridge, Massachusetts
(Received 1 June 1965)

Several problems of nonrelativistic scattering by analytic potentials are studied, where both the energy and the momentum transfer are large. Specifically, we find asymptotically the reflection coefficients from the one-dimensional potentials $Ve^{x}(1+e^{x})^{-2}$, $Ve^{x/2}(1+e^{x})^{-1}$, and $V\exp(-x^2)$, and the scattering amplitudes for their three-dimensional generalizations $Ve^{r}(1+e^{r})^{-2}$, $Ve^{r/2}(1+e^{r})^{-1}$, and $V\exp(-r^2)$. The results are compared with these obtained from Born approximation, and it is found that the first Born approximation gives the correct answer asymptotically in the cases $Ve^{x/2}(1+e^{x})^{-1}$ and $Ve^{r/2}(1+e^{r})^{-1}$, but not in the other cases. Conjectures about more general cases are also given.

101) Asymptotic Form of the *S* Matrix for Large Angular Momentum in the Left Half-Plane, *Phys. Rev.* **144**, 1232 (1966), Hung Cheng (Harvard U.) and Tai Tsun Wu (Harvard U.), DOI: https://doi.org/10.1103/PhysRev.144.1232.

HUNG CHENG*
Department of Physics, Harvard University, Cambridge, Massachusetts

AND

TAI TSUN WU†
Gordon McKay Laboratory, Harvard University, Cambridge, Massachusetts
(Received 19 May 1965; revised manuscript received 8 December 1965)

Starting with the Schrödinger equation, we prove that for all energies, $S(\lambda,s)$ approaches $e^{2i\lambda\pi}$ as $|\lambda|$ becomes large in the direction $\frac{1}{2}\pi < \arg\lambda < \frac{3}{2}\pi$, for a class of potentials. These include the square-well potential, the cut-off Coulomb potential, a single Yukawa potential, and a superposition of Yukawa potentials of the form $\int_\mu^\infty (e^{-\mu' r}/r)\, e^{-\mu'}\, d\mu'$. The asymptotic forms of the Regge-pole parameters α_n and β_n are derived. We found that $\arg\lambda_n$ approaches $\frac{1}{2}\pi$ or $\frac{3}{2}\pi$ as $n\to\infty$, and β_n is proportional to $1+e^{2\pi i\alpha_n}$, which grows exponentially for the Regge poles in the lower half plane. The asymptotic forms for the Jost functions and the Y function are also given. A general proof for the asymptotic formula $S(\lambda,s)\to e^{2\pi i\lambda}$ as $|\lambda|\to\infty$, $\frac{1}{2}\pi < \arg\lambda < \frac{3}{2}\pi$, is also outlined.

102) Neutral Lepton Currents, *Phys. Rev.* **147**, 1033 (1966), Tai Tsun Wu (Harvard U.), DOI: https://doi.org/10.1103/PhysRev.147.1033.

TAI TSUN WU*
Gordon McKay Laboratory, Harvard University, Cambridge, Massachusetts
(Received 14 February 1966)

The question is raised whether there is a self-coupled neutral lepton current in weak interactions.

103) Theoretical Discussions Concerning η′–δ Degeneracy, *Phys. Rev. Lett.* **18** (1967) 10, 349, S. F. Tuan (Princeton, Inst. Advanced Study), Tai Tsun Wu (Rockefeller U.), DOI: https://doi.org/10.1103/PhysRevLett.18.349.

S. F. Tuan*†
Institute for Advanced Study, Princeton, New Jersey

and

Tai Tsun Wu‡
Rockefeller University, New York, New York
(Received 9 January 1967)

104) Analytic structure of energy levels in a field theory model, *Phys. Rev. Lett.* **21** (1968), 406–409, Carl M. Bender (Harvard U.), Tai Tsun Wu (Harvard U.), DOI: https://doi.org/10.1103/PhysRevLett.21.406; Citations: 150.

Carl M. Bender*
Department of Physics, Harvard University, Cambridge, Massachusetts
and
Tai Tsun Wu
Division of Engineering and Applied Physics, Harvard University, Cambridge, Massachusetts
(Received 17 May 1968)

We consider a model of $\lambda\varphi^4$ field theory in which perturbation theory diverges and analytically continue the energy levels into the complex λ plane. Using WKB technique, we determine that the energy levels have an infinite sequence of branch points (where level crossing occurs) with a limit point $a\lambda = 0$. Thus the origin is not an isolated singularity. The resolvent $(z-H)^{-1}$ has an infinite sequence of poles with a limit point at $\lambda = 0$.

Commentary by Carl M. Bender

It was known for many years that perturbation series are usually divergent. In the papers *Phys. Rev. Lett.* **21**, 406 (1968) and *Phys. Rev.* **184**, 1231 (1969), Bender and Wu have discover WHY perturbation series diverge: There are singularities, called Bender–Wu singularities, in the complex coupling-constant plane. They show that these singularities have a physical interpretation: The energy levels are all analytic continuations of one another, and the energy levels cross at these singularities. The energies, when considered as functions of complex coupling constant g, are all branches of a single energy function E(g); that is, the nth energy level is the value of E(g) on the nth sheet of a Riemann surface in complex coupling-constant space. Thus, Bender and Wu find an entirely new explanation of quantization; the discreteness of energy levels is due to the discreteness of the sheets in a Riemann surface.

105) Some Solutions of the classical Isotopic Gauge Field Equations, T. T. Wu (Harvard U.), Chen-Ning Yang (SUNY, Stony Brook), In *Yang, C.N.: Selected Papers 1945–1980*, 400–405 also in *H. Mark and S. Fernbach, Properties Of Matter Under Unusual Conditions*, New York 1969, 349–345.

T. T. WU*
Harvard University, Cambridge, Massachusetts
and
C. N. YANG
Institute for Theoretical Physics, State University of New York, Stony Brook, New York

106) Theoretical Considerations on Inelastic Electron Scattering at High Energies, *Phys. Rev.* **183**, 1324 (1969), Hung Cheng (MIT) and Tai Tsun Wu (Harvard U.), DOI: https://doi.org/10.1103/PhysRev.183.1324; Citations: 54.

HUNG CHENG*
Department of Mathematics, Massachusetts Institute of Technology, Cambridge, Massachusetts 02139
AND
TAI TSUN WU
Gordon McKay Laboratory, Harvard University, Cambridge, Massachusetts 02138
(Received 24 March 1969)

Physical arguments are presented for the conjecture that, compared with photoproduction and hadronic production, the difficulty in making large transverse momentum transfers at high energies is much less in electroproduction processes when the virtual photon is far off the mass shell. Model calculations are carried out to verify this result. In particular, in the c.m. systems of the outgoing hadrons, the angular distribution of the vector meson produced by the reactions $e^-+p \rightarrow e^-+p+\rho^0$ (or φ, or ω) is less peaked in the forward direction at high energies when the virtual photon is further off the mass shell. This and other possible experimental tests are discussed. This reduction of the forward peaks also makes electroproduction processes a fertile ground for testing statistical theories of high-energy collisions.

107) Form-factors and impact factors in high-energy scattering, *Phys. Rev.* **184** (1969), 1868–1872, Hung Cheng (MIT, LNS), Tai Tsun Wu (Harvard U.), DOI: https://doi.org/10.1103/PhysRev.184.1868.

HUNG CHENG*
Department of Mathematics, Massachusetts Institute of Technology, Cambridge, Massachusetts 02139
AND
TAI TSUN WU
Gordon McKay Laboratory, Harvard University, Cambridge, Massachusetts 02138
(Received 21 April 1969)

In this paper, we study the general properties of the impact factor. In particular, the fourth-order impact actor for the electron is shown to exist and is explicitly worked out. This implies that the sum of the cor-esponding sixth-order amplitudes for processes like electron-electron scattering satisfies the impact-factor epresentation. We also find that, in both the second and the fourth order, the impact factor is proportional o the electromagnetic form factor. An interesting relation between the impact factor and the form factor is hen shown to hold generally. These arguments are next extended to scalar mesons and the same conclu-ions are reached. Considerations on the impact factor in higher orders are also made.

108) Impact factor and exponentiation in high-energy scattering processes, *Phys. Rev.* **186** (1969), 1611–1618, Hung Cheng (MIT, LNS), Tai Tsun Wu (Harvard U.), DOI: https://doi.org/10.1103/PhysRev.186.1611; Citations: 133.

HUNG CHENG*
Department of Mathematics, Massachusetts Institute of Technology, Cambridge, Massachusetts 02139

AND

TAI TSUN WU
Gordon McKay Laboratory, Harvard University, Cambridge, Massachusetts 02138
(Received 14 May 1969)

We consider here the exchange of an arbitrary number of photons between two fermions at high energies, and an exponentiation, similar to that in the Glauber approximation and the droplet model, is obtained. When this result is applied to the processes of electron-electron scattering, Delbrück scattering, and photon-photon scattering, the scattering amplitude in each case can again be cast in the impact-factor representation. Although some modifications from the lowest-order result are necessary to account for the multiphoton effects, the impact factors themselves are not changed at all. The limitation of this exponentiation is discussed in connection with more complicated exchanges.

109) Theory of high energy diffraction scattering, H. Cheng (MIT, LNS and Harvard U.), T.T. Wu (MIT, LNS and Harvard U.), In Stony Brook 1969, Proceedings, Conference On High Energy Collisions, New York 1969, 329–365.

110) Anharmonic oscillator, *Phys. Rev.* **184** (1969), 1231–1260, Carl M. Bender (Harvard U.), Tai Tsun Wu (Harvard U.), DOI: https://doi.org/10.1103/PhysRev.184.1231; Citations: 1039.

CARL M. BENDER*†
Department of Physics, Harvard University, Cambridge, Massachusetts 02138

AND

TAI TSUN WU
Division of Engineering and Applied Physics, Harvard University, Cambridge, Massachusetts 02138
(Received 4 February 1969)

We consider the anharmonic oscillator defined by the differential equation $(-d^2/dx^2+\frac{1}{4}x^2+\frac{1}{4}\lambda x^4)\Phi(x) = E(\lambda)\Phi(x)$ and the boundary condition $\lim_{x\to\pm\infty}\Phi(x)=0$. This model is interesting because the perturbation series for the ground-state energy diverges. To investigate the reason for this divergence, we analytically continue the energy levels of the Hamiltonian H into the complex λ plane. Using WKB techniques, we find that the energy levels as a function of λ, or more generally of λ^α, have an infinite number of branch points with a limit point at $\lambda=0$. Thus, the origin is not an isolated singularity. Level crossing occurs at each branch point. If we choose $\alpha=\frac{1}{3}$, the resolvent $(z-H)^{-1}$ has no branch cut. However, for all z it has an infinite sequence of poles which have a limit point at the origin. The anharmonic oscillator is of particular interest to field theoreticians because it is a model of $\lambda\varphi^4$ field theory in one-dimensional space-time. The unusual and unexpected properties exhibited by this model may give some indication of the analytic structure of a more realistic field theory.

Commentary by Carl M. Bender

It was known for many years that perturbation series are usually divergent. In the papers *Phys. Rev. Lett.* **21**, 406 (1968) and *Phys. Rev.* **184**, 1231 (1969), Bender and Wu have discover WHY perturbation series diverge: There are singularities, called Bender–Wu singularities, in the complex coupling-constant plane. They show that these singularities have a physical interpretation: The energy levels are all analytic continuations of one another, and the energy levels cross at these singularities. The energies, when considered as functions of complex coupling constant g, are all branches of a single energy function E(g); that is, the nth energy level is the value of E(g) on the nth sheet of a Riemann surface in complex coupling-constant space. Thus, Bender and Wu find an entirely new explanation of quantization; the discreteness of energy levels is due to the discreteness of the sheets in a Riemann surface.

111) Physical picture for high-energy diffraction scattering, Phys. *Rev. Lett.* **23** (1969), 670–673, Hung Cheng (MIT, LNS), Tai Tsun Wu (Harvard U.), DOI: https://doi.org/10.1103/PhysRevLett.23.670.

Hung Cheng
Department of Mathematics, Massachusetts Institute of Technology, Cambridge, Massachusetts 02139

and

Tai Tsun Wu
Gordon McKay Laboratory, Harvard University, Cambridge, Massachusetts 02138
(Received 22 July 1969)

Recent theoretical results on high-energy scattering are extended. We give explicit rules to obtain the high-energy behavior of the sums of large classes of Feynman diagrams. This consideration suggests, for high-energy diffraction scattering, a physical picture which is simple and natural. We emphasize that this physical picture has the virtue of yielding correctly all the high-energy results of quantum electrodynamics.

112) High energy behavior in quantum electrodynamics, T. T. Wu (Harvard U.), In Madison 1969, Proceedings, Regge Cut Conference, Madison 1969, 69–82.

113) High-energy collision processes in quantum electrodynamics. i, *Phys. Rev.* **182** (1969), 1852–1867, Hung Cheng (MIT, LNS), Tai Tsun Wu (Harvard U.), DOI: https://doi.org/10.1103/PhysRev.182.1852; Citations: 120.

HUNG CHENG*
Department of Mathematics, Massachusetts Institute of Technology, Cambridge, Massachusetts 02139
AND
TAI TSUN WU
Gordon McKay Laboratory, Harvard University, Cambridge, Massachusetts 02138
(Received 4 February 1969)

We have made a systematic study of all two-body elastic scattering amplitudes in quantum electrodynamics at high energies. In particular, we have calculated the high-energy behavior of the following processes: (1) Delbrück scattering, (2) electron Compton scattering, (3) photon-photon scattering, (4) electron-electron scattering, (5) electron-positron scattering, and (6) electron-proton scattering. The processes (1) and (2) are calculated up to the sixth order in the coupling constant e, the process (3) up to the eighth order, and the processes (4), (5), and (6) up to the fourth order. Our calculations show that all of these amplitudes are proportional to s, the square of the center-of-mass energy, as s becomes large. In other words, we have found that, to these orders, $\lim_{s\to\infty} d\sigma/dt$ exists and is nonzero for all $t\neq 0$, where $-t$ is the square of the momentum transfer. Furthermore, we found it meaningful to assign a factor (we call it the impact factor) to each particle. More precisely, for the high-energy scattering of $a+b\to a+b$, the imaginary coefficient of s for the scattering amplitude is proportional to $\int d\mathbf{q}_\perp[(\mathbf{q}_\perp+\mathbf{r}_1)^2]^{-1}[(\mathbf{q}_\perp-\mathbf{r}_1)^2]^{-1}\mathcal{I}^a(\mathbf{r}_1,\mathbf{q}_\perp)\mathcal{I}^b(\mathbf{r}_1,\mathbf{q}_\perp)$, where $2\mathbf{r}_1$ is the momentum transfer, and $\mathcal{I}^a(\mathbf{r}_1,\mathbf{q}_\perp)$ and $\mathcal{I}^b(\mathbf{r}_1,\mathbf{q}_\perp)$ are the impact factors of particles a and b, respectively. The integration is over the two-dimensional transverse momentum of the virtual photons. The important point is that $\mathcal{I}^a$ ($\mathcal{I}^b$) does *not* depend on what particle b (a) is. We have explicitly found the impact factors for the photon (up to e^4) and for the electron, the positron, and the proton (up to e^2). In the case of Delbrück scattering, we have also taken care of all higher-order diagrams with an arbitrary number of photons exchanged between the virtual pair and the proton or nucleus. The coefficient of s in this case can be expressed as the integral of the above-mentioned product $\mathcal{I}^a\mathcal{I}^b$ times modified photon propagators. The impact factor therefore appears to express an intrinsic property of a particle. Our result is consistent with neither the most straightforward interpretation of the Regge-pole model nor that of the droplet model. These inconsistencies are closely related to the nonplanar nature of the diagrams under consideration. Our results on Delbrück scattering are also qualitatively different from those of Bethe and Rohrlich based on the impact-parameter approximation.

114) High-energy collision processes in quantum electrodynamics. ii, *Phys. Rev.* **182** (1969), 1868–1872, Hung Cheng (MIT, LNS), Tai Tsun Wu (Harvard U.), DOI: https://doi.org/10.1103/PhysRev.182.1868; Citations: 55.

HUNG CHENG*
Department of Mathematics, Massachusetts Institute of Technology, Cambridge, Massachusetts 02139
AND
TAI TSUN WU
Gordon McKay Laboratory, Harvard University, Cambridge, Massachusetts 02138
(Received 4 February 1969)

We consider here all the high-energy scattering processes $a+b\to a+b$ in quantum electrodynamics, where a and b can be fermions or antifermions. Aside from the photon pole terms, these high-energy amplitudes are shown to be conveniently expressed by the impact factors. The impact factors for the electron, the positron, and the static nucleus are explicitly given.

115) High-energy collision processes in quantum electrodynamics. iii, *Phys. Rev.* **182** (1969), 1873–1898, Hung Cheng (MIT, LNS), Tai Tsun Wu (Harvard U.), DOI: https://doi.org/10.1103/PhysRev.182.1873; Citations: 91.

Hung Cheng*
Department of Mathematics, Massachusetts Institute of Technology, Cambridge, Massachusetts 02139

AND

Tai Tsun Wu
Gordon McKay Laboratory, Harvard University, Cambridge, Massachusetts 02138
(Received 4 February 1969)

We obtain explicitly the asymptotic behavior of the matrix elements at high energies for Delbrück scattering, i.e., the elastic scattering of a photon by a static Coulomb field via a virtual electron-positron pair. This is the simplest nontrivial two-body elastic scattering process in quantum electrodynamics besides those discussed in the preceding paper. The considerations are limited to the lowest order, Z^2e^6, for the matrix elements, and it is found that, to this order, $\lim_{\omega\to\infty} d\sigma/dt$ exists and is nonzero for any fixed positive momentum transfer. This limiting value is expressed in terms of integrals, which are evaluated numerically. The behavior of these integrals is also studied in detail for the cases where the momentum transfer is either much larger or much smaller than the mass of the electron. Moreover, the scattered photon is significantly polarized in the scattering plane. None of the present results agree with the earlier ones of Bethe and Rohrlich using the impact-parameter approximation.

116) High-energy collision processes in quantum electrodynamics. iv, *Phys. Rev.* **182** (1969), 1899–1906, Hung Cheng (MIT, LNS), Tai Tsun Wu (Harvard U.), DOI: https://doi.org/10.1103/PhysRev.182.1899; Citations: 74.

Hung Cheng*
Department of Mathematics, Massachusetts Institute of Technology, Cambridge, Massachusetts 02139

AND

Tai Tsun Wu
Gordon McKay Laboratory, Harvard University, Cambridge, Massachusetts 02138
(Received 4 February 1969)

The rather complicated asymptotic behavior at high energies found in the preceding paper for the Delbrück-scattering matrix element is recast in a simpler form. This is accomplished by introducing a suitable form factor in the transverse plane. This factor, called the impact factor of the photon, also appears in other processes. Moreover, it is verified that the transverse momentum transfer is small in Delbrück scattering.

117) High-energy elastic scattering in quantum electrodynamics, *Phys. Rev. Lett.* **22** (1969), 666, Hung Cheng (MIT), Tai Tsun Wu (Harvard U.), DOI: https://doi.org/10.1103/PhysRevLett.22.666; Citations: 180.

Hung Cheng*
Department of Mathematics, Massachusetts Institute of Technology, Cambridge, Massachusetts

and

Tai Tsun Wu†
Gordon McKay Laboratory, Harvard University, Cambridge, Massachusetts
(Received 3 February 1969)

We have examined all two-body, elastic-scattering amplitudes in quantum electrodynamics. Aside from the photon-pole contribution, the amplitude for any elastic process is imaginary and proportional to s in the limit $s \to \infty$ with t held finite. The coefficients of s, to the lowest nonvanishing orders of e, can be expressed simply in terms of "impact factors." The results contradict both the Regge-pole model and the droplet model in their most straightforward interpretations.

118) Longitudinal Momentum Distribution of Pionization Products, *Phys. Rev. Lett.* **23**, 1311 (1969), Hung Cheng (MIT) and Tai Tsun Wu (Harvard U.), DOI: https://doi.org/10.1103/PhysRevLett.23.1311, Citations: 55.

Hung Cheng†
Department of Mathematics, Massachusetts Institute of Technology, Cambridge, Massachusetts 02139

and

Tai Tsun Wu
Gordon McKay Laboratory, Harvard University, Cambridge, Massachusetts 02138
(Received 6 October 1969)

We obtain, on the basis of general principles only, a quantitative formula for the distribution of pionization products in ultrahigh-energy scattering as a function of the longitudinal momentum. Field theory is then used as a model to show the existence of the pionization process, and it also serves as an example in which the formula is satisfied. A highly feasible storage-ring experiment is proposed to test this predicted distribution and hence the underlying principles.

119) High-energy Delbrück scattering close to the forward direction, *Phys. Rev. D* **2** (1970), 2444–2457, Hung Cheng (MIT, LNS and Harvard U.), T. T. Wu (MIT, LNS and Harvard U.), DOI: https://doi.org/10.1103/PhysRevD.2.2444.

HUNG CHENG†
Department of Mathematics, Massachusetts Institute of Technology, Cambridge, Massachusetts 02139

AND

TAI TSUN WU
Gordon McKay Laboratory, Harvard University, Cambridge, Massachusetts 02138
(Received 19 May 1970)

Delbrück scattering is the elastic scattering of a photon by a static Coulomb field via electron-positron pair creation. At high energies, there are two natural scales for the momentum transfer Δ, namely, m and m^2/ω, where m is the mass of the electron, and ω is the photon energy. When Δ is much larger than the smaller scale m^2/ω, the impact factor representation holds at high energies. The impact picture is here extended to give also the high-energy behavior of the Delbrück scattering amplitude when Δ is comparable to m^2/ω. The result can be expressed in terms of generalized hypergeometric functions, which reproduces the known result in the forward direction when Δ is set equal to zero, and also joins smoothly to the impact factor representation when Δ is much larger than m^2/ω. In the present analysis, the fine-structure constant α is assumed to be small, but not $Z\alpha$. In other words, all terms of the form $\alpha(Z\alpha)^n$ in the amplitude are taken into account. It is also shown that the result for $\Delta \ll m$ is independent of the mass of the target, and hence is in particular applicable to Compton scattering by an electron.

120) High-energy behavior near threshold — massive quantum electrodynamics, *Phys. Rev. D* **2** (1970), 2276–2284, Hung Cheng (MIT, LNS and Harvard U.), T. T. Wu (MIT, LNS and Harvard U.), DOI: https://doi.org/10.1103/PhysRevD.2.2276.

HUNG CHENG†
Department of Mathematics, Massachusetts Institute of Technology, Cambridge, Massachusetts 02139

AND

TAI TSUN WU
Gordon McKay Laboratory, Harvard University, Cambridge, Massachusetts 02138
(Received 20 April 1970)

In this paper we study all scattering amplitudes in quantum electrodynamics in the limit $s \to \infty$, with t near or at the two-photon threshold. In this limit, the power of s for a diffractive amplitude with a two-photon cut is found to be promoted from 1 to $\frac{3}{2}$. The fourth-order electron-electron scattering amplitude is first studied by the method of Feynman parameters. The method of impact diagrams is next generalized to handle the more complicated cases. We then apply the new method explicitly to the lowest-order diffractive amplitude for electron-electron scattering, Compton scattering, and photon-photon scattering. A general case $a+b \to a'+b'$ is also discussed. We find that the scattering amplitude is now factorized and the existence of a Regge pole is suggested. This is then verified by a study of the tower diagrams. Thus the leading singularity in the J plane, while being a pair of branch points for $t \leq 0$, is a moving Regge pole located to the right of $J=\frac{3}{2}$ as t is near $4\lambda^2$. The Gribov paradox is thereby automatically resolved.

121) High-Energy Behavior near Threshold: ϕ^3 Theory, *Phys. Rev. D* **2** (1970), 2285–2297, Hung Cheng (MIT, LNS), T. T. Wu (Harvard U.), DOI: https://doi.org/10.1103/PhysRevD.2.2285.

HUNG CHENG†
Department of Mathematics, Massachusetts Institute of Technology, Cambridge, Massachusetts 02139

AND

TAI TSUN WU
Gordon McKay Laboratory, Harvard University, Cambridge, Massachusetts 02138
(Received 20 April 1970)

In this paper we study the two-body elastic scattering process in φ^3 theory, in the limit $s \to \infty$ with t equal to or near the threshold value 4, where the mass of the scalar particle is taken to be unity. For the scattering amplitude corresponding to any ladder-type diagram, we find that there is a promotion for the power of s by $\frac{1}{2}$. For example, the scattering amplitude for the ladder diagram of $n+1$ rungs is promoted from $O(s^{-1}\ln^n s)$ away from threshold to $O(s^{-1/2}\ln^{n-1}s)$ at threshold. This means that, for small coupling constants, the leading Regge pole is promoted from the neighborhood of $l=-1$ away from threshold to the neighborhood of $l=-\frac{1}{2}$ at threshold. There are, in addition, infinitely many Regge trajectories approaching $l=-\frac{1}{2}$ as t approaches 4. The scattering amplitude for the n-rung diagram is explicitly given in the limit $s \to \infty$, with t at or near the threshold 4, and various scales for $t-4$ are pointed out. The two-rung and three-rung diagrams are especially studied in detail.

122) High-energy behavior near threshold — potential scattering, *Phys. Rev. D* **2** (1970), 2298–2305, Hung Cheng (MIT, LNS), T. T. Wu (Harvard U.), DOI: https://doi.org/10.1103/PhysRevD.2.2298.

HUNG CHENG†
Department of Mathematics, Massachusetts Institute of Technology, Cambridge, Massachusetts 02139

AND

TAI TSUN WU
Gordon McKay Laboratory, Harvard University, Cambridge, Massachusetts 02138
(Received 20 April 1970)

In this paper we study the promotion phenomenon in potential scattering. In Sec. 3 we give the general formulas for the scattering amplitude at threshold, and show that the asymptotic form of the scattering amplitude as the momentum transfer squared approaches infinity is determined not only by the right-hand-plane Regge poles $[\alpha(0)>-\frac{1}{2}]$, but also by the left-hand-plane zeros of $Y(\lambda)$. In Sec. 3 we study the leading Regge pole $\alpha(k_t^2)$ over the whole region $-\infty<k_t^2<0$ for a weak and attractive Yukawa potential, and find that it is indeed promoted from -1 to $-\frac{1}{2}$ within a small neighborhood of the threshold. We also obtain the leading zero of $Y(\lambda)$ if the potential is repulsive. Together with the results in Sec. 2, we have, therefore, the explicit asymptotic form of the scattering amplitude. In Sec. 4 we study a general nth Born term and find that its leading term is promoted from $s^{-1}(\ln s)^{n-1}$ to $s^{-1/2}(\ln s)^{n-2}$ at threshold. Summing over these leading terms, we obtain results in complete agreement with those in Secs. 2 and 3, both for attractive and for repulsive potentials, off or at threshold.

123) Impact picture — theory of high-energy scattering, T. T. Wu (Harvard U.), H. Cheng (MIT, LNS), In *Kiev Inst. Theor. Phys. A N*, Itf-70-99, Kiev 1970, 59–96.

124) What is the Pomeranchuk trajectory?, *Phys. Rev. Lett.* **24** (1970), 759–761, Hung Cheng (MIT, LNS), Tai Tsun Wu (Harvard U.), DOI: https://doi.org/10.1103/PhysRevLett.24.759.

Hung Cheng†
Department of Mathematics, Massachusetts Institute of Technology, Cambridge, Massachusetts 02139

and

Tai Tsun Wu
Gordon McKay Laboratory, Harvard University, Cambridge, Massachusetts 02138
(Received 16 February 1970)

We calculate the high-energy amplitude in quantum electrodynamics and scalar electrodynamics, and find that the Pomeranchon is a moving Regge pole near $J=\frac{3}{2}$ when t is equal to or near the two-particle threshold, and is a fixed branch point at $J=1$ for $t \leq 0$. This is due to the promotion from 1 to $\frac{3}{2}$ for the power of s of the tower-diagram amplitudes at threshold. The Gribov paradox is thereby automatically resolved.

125) Limit of Cross-Sections at Infinite Energy, *Phys. Rev. Lett.* **24** (1970), 1456–1460, In *Lo, S. Y. (ed.): Geometrical pictures in hadronic collisions*, 277–281, Hung Cheng (MIT, LNS), T. T. Wu (Harvard U.), DOI: https://doi.org/10.1103/PhysRevLett.24.1456; Citations: 322.

Hung Cheng†
Department of Mathematics, Massachusetts Institute of Technology, Cambridge, Massachusetts 02139

and

Tai Tsun Wu
Gordon McKay Laboratory, Harvard University, Cambridge, Massachusetts 02138
(Received 15 May 1970)

At infinite energy, we predict: (1) σ_{tot} approaches infinity; (2) the ratio of the real part to the imaginary part of the forward elastic amplitude approaches zero; (3) σ_{el}/σ_{tot} approaches $\frac{1}{2}$; (4) the width of diffraction peak approaches zero; its product with σ_{tot} is a constant. We give theoretical evidence based on massive quantum electrodynamics as well as experimental evidence in support of these predictions, and a physical picture for high-energy scattering.

126) Diffraction scattering for inelastic processes, *Phys. Rev. D* **1** (1970), 459–466, Hung Cheng (MIT, LNS), Tai Tsun Wu (Harvard U.), DOI: https://doi.org/10.1103/PhysRevD.1.459.

HUNG CHENG†
Department of Mathematics, Massachusetts Institute of Technology, Cambridge, Massachusetts 02139

AND

TAI TSUN WU
Gordon McKay Laboratory, Harvard University, Cambridge, Massachusetts 02138
(Received 24 June 1969)

Previous treatments on high-energy scattering in quantum electrodynamics are extended to diffractive inelastic processes. We explore the model in which the diffractive excitation proceeds through the exchange of vector mesons. Six inelastic impact factors are explicitly calculated in the lowest order: (1) scalar to scalar, (2) pseudoscalar to pseudoscalar, (3) vector to vector, (4) axial vector to axial vector, (5) scalar to axial vector, and (6) pseudoscalar to axial vector. For all six of these impact factors, the masses of the incoming and outgoing particles may be different; and for the last two impact factors, the spins and the parities of the incoming and outgoing particles may be different. The nonvanishing of the impact factors demonstrates that mass, spin, and parity can change during a diffractive process. We also conclude that C (charge conjugation quantum number), S (strangeness), I (isotopic spin), B (baryon number), and G parity, etc., must remain the same during a diffractive process. Some general properties of the impact factor are also discussed.

127) Elastic scattering processes in scalar electrodynamics at high energies. i, *Phys. Rev. D* **1** (1970), 467–473, Hung Cheng (MIT, LNS), Tai Tsun Wu (Harvard U.), DOI: https://doi.org/10.1103/PhysRevD.1.467.

HUNG CHENG†
Department of Mathematics, Massachusetts Institute of Technology, Cambridge, Massachusetts 02139

AND

TAI TSUN WU
Gordon McKay Laboratory, Harvard University, Cambridge, Massachusetts 02138
(Received 25 August 1969)

In this paper we study all elastic scattering amplitudes in scalar electrodynamics at high energies. The impact factors for the scalar meson and for the vector meson in this theory are explicitly calculated in the lowest order. We also show that the elastic amplitudes still satisfy an impact-factor representation after all vector-meson exchanges are summed, and an exponentiation, similar to that in quantum electrodynamics, is obtained.

128) Theory of high-energy diffraction scattering. I, *Phys. Rev. D* **1** (1970), 1069–1082, Hung Cheng (MIT, LNS), Tai Tsun Wu (Harvard U.), DOI: https://doi.org/10.1103/PhysRevD.1.1069.

HUNG CHENG†
Department of Mathematics, Massachusetts Institute of Technology, Cambridge, Massachusetts 02139

AND

TAI TSUN WU
Gordon McKay Laboratory, Harvard University, Cambridge, Massachusetts 02138
(Received 14 August 1969)

Recent theoretical results on high-energy scattering are extended. It is found that the form of the matrix elements in the high-energy limit can be readily understood in terms of energy denominators in pre-Feynman perturbation theory. On this basis, we give explicit rules to obtain the high-energy behavior of the sums of large classes of Feynman diagrams, for both the scattering of two extremely relativitistic particles and that of one such particle by an external static potential. Application of these rules yields immediately all the previous results on high-energy elastic scattering, and also leads to a deeper understanding of exponentiation. These considerations suggest a simple and natural physical picture for high-energy elastic scattering and, by trivial extensions, also for diffraction and some inelastic scattering processes. We emphasize that this physical picture has the virtue of correctly yielding all the high-energy results of quantum electrodynamics.

129) Theory of high-energy diffraction scattering. ii, *Phys. Rev. D* **1** (1970), 1083–1091, Hung Chen (MIT), Tai Tsun Wu (Harvard U.), DOI: https://doi.org/10.1103/PhysRevD.1.1083.

HUNG CHENG†
Department of Mathematics, Massachusetts Institute of Technology, Cambridge, Massachusetts 02139

AND

TAI TSUN WU
Gordon McKay Laboratory, Harvard University, Cambridge, Massachusetts 02138
(Received 28 August 1969)

In this paper we extend the treatment in Paper I of this series by studying high-energy scattering processes from the viewpoint of the projectile system, i.e., the system in which the incident particle is at rest. In particular, we give two equivalent sets of rules for obtaining the high-energy scattering amplitudes from this standpoint, one for the momentum space and the other for the position space. It is found that both sets of rules give directly the exact expression of the impact factor, and no more approximations need to be made even for the numerators of the integrals involved. We conclude that the impact factor is best studied in the projectile system, while the potential is most conveniently taken care of in the laboratory system, and that they can be separately treated. In the case of high-energy collision of two particles, each impact factor is best studied in its own projectile system. These two sets of rules are shown to be equivalent to those in the laboratory system. In the Appendix we also study high-energy scattering processes in the position space from the viewpoint of the laboratory system, and explicit rules are given.

130) Photon-photon scattering close to the forward direction, *Phys. Rev. D* **1** (1970), 3414–3415, Hung Cheng (MIT, LNS), Tai Tsun Wu (Harvard U.), DOI: https://doi.org/10.1103/PhysRevD.1.3414.

HUNG CHENG†
Department of Mathematics, Massachusetts Institute of Technology, Cambridge, Massachusetts 02139

AND

TAI TSUN WU
Gordon McKay Laboratory, Harvard University, Cambridge, Massachusetts 02138
(Received 16 February 1970)

The scattering amplitude for photon-photon elastic scattering in quantum electrodynamics is explicitly obtained for the case where the energy is very large while the momentum transfer is small compared with the rest energy of the electron (divided by c). This amplitude has no infrared divergence and, at high energies, the fourth- and sixth-order contributions are negligible. The explicit answer for the eighth order is remarkably simple. In terms of helicity states, it is found that the amplitudes $RR \to RR$ and $RL \to RL$ are the same, while the amplitude $RR \to LL$ is different but of the same order of magnitude. By the optical theorem, the total cross section for $\gamma+\gamma \to e^-+e^-+e^++e^+$ is found to be 6.5 μb at high energies.

131) Impact Pictures and Regge Poles, Hung Cheng (MIT), Tai Tsun Wu (Harvard U.), PRINT-70-1743 (MIT) (1970).

132) The impact factor representation, Hung Cheng (MIT), Tai Tsun Wu (Harvard U.), PRINT-70-2601 (1970).

133) High-energy production processes in quantum electrodynamics, *Phys. Rev. D* **1** (1970), 456–458, Hung Cheng (MIT, LNS), Tai Tsun Wu (Harvard U.), DOI: https://doi.org/10.1103/PhysRevD.1.456.

HUNG CHENG†
Department of Mathematics, Massachusetts Institute of Technology, Cambridge, Massachusetts 02139
AND
TAI TSUN WU
Gordon McKay Laboratory, Harvard University, Cambridge, Massachusetts 02138
(Received 8 September 1969)

Production amplitudes in quantum electrodynamics are studied in the high-energy limit. In particular, we calculate the impact factors for pair production and for bremsstrahlung explicitly. Numerous production amplitudes are then obtained in terms of these two impact factors. The method we use is the one of impact diagrams, which is applicable to all Feynman diagrams at high energies.

134) Logarithmic factors in the high-energy behavior of quantum electrodynamics, *Phys. Rev. D* **1** (1970), 2775–2794, Hung Cheng (MIT, LNS), Tai Tsun Wu (Harvard U.), DOI: https://doi.org/10.1103/PhysRevD.1.2775; Citations: 64.

HUNG CHENG†
Department of Mathematics, Massachusetts Institute of Technology, Cambridge, Massachusetts 02139
AND
TAI TSUN WU
Gordon McKay Laboratory, Harvard University, Cambridge, Massachusetts 02138
(Received 23 January 1970)

The perturbation series for electron-electron elastic scattering in quantum electrodynamics is studied in the limit of high energies. For this matrix element, in addition to the previously known terms which are proportional to s (the square of the c.m. energy) and hence lead to a constant total cross section at high energies, there are found terms of the orders of magnitudes $s \ln s$, $s\ (\ln s)^2$, $s(\ln s)^3$, etc. For $n=1, 2, 3, \ldots$, the coefficient of $s(\ln s)^n$ is a power series in the fine-structure constant α, where the leading term is proportional to $\alpha^{2(n+1)}$ and is due to Feynman diagrams with n closed electron loops. Physically, through the optical theorem, the presence of these terms is intimately related to the production of low-energy electron-positron pairs in high-energy electron-electron scattering, but is independent of whether the spin-1 particle is a photon with zero mass or massive neutral vector meson. These leading terms of order $\alpha^{2(n+1)}$ are explicitly found for all n, and are all imaginary, representing absorption. The procedure of summing the leading term is carried out, and the result demonstrates dramatically the importance of unitarity in the direct, or s, channel for high-energy processes. Generalization to two-body diffraction processes $a+b\to a'+b'$ is immediate.

135) High-energy elastic scattering close to the forward direction in quantum electrodynamics, Hung Cheng (MIT), Tai Tsun Wu (Harvard U.), PRINT-70-971 (MIT) (1970).

136) Quantum electrodynamics at high-energies, Hung Cheng (SLAC), Tai Tsun Wu (Harvard U.), Proc. 5th Int. Symp. on Electron and Photon Interactions at High Energies, Aug 1971, Ithaca. pp. 147–62.

137) Quantum electrodynamics at high energies, H. Cheng (MIT, LNS and Harvard U.), T. T. Wu (MIT, LNS and Harvard U.), In Ithaca 1971, Proceedings, Symposium On Electron An Photon Interactions, Cornell Univ, Ithaca 1972, 147–161.

138) Impact picture and the eikonal approximation, *Phys. Lett. B* **34** (1971), 647–650, Hung Cheng (DESY), T. T. Wu (DESY), DOI: https://doi.org/10.1016/0370-2693(71)90162-6.

H. CHENG‡ and T. T. WU‡*
Deutsches Elektronen-Synchrotron DESY, Hamburg, Germany

Received 19 February 1971

A detailed study of the ϕ^3-theory shows that the impact picture is correct, while the eikonal approximation is not valid. We discuss a relatively simple experimental test, at the CERN Intersecting Storage Ring, of the impact picture.

139) Theory of Pionization, *Phys. Rev. D* **3** (1971), 2195–2209, Hung Cheng (MIT, LNS and Harvard U.), T. T. Wu (MIT, LNS and Harvard U.), DOI: https://doi.org/10.1103/PhysRevD.3.2195.

HUNG CHENG†
Department of Mathematics, Massachusetts Institute of Technology, Cambridge, Massachusetts 02139

AND

TAI TSUN WU
Gordon McKay Laboratory, Harvard University, Cambridge, Massachusetts 02138
(Received 28 August 1970)

In this paper we make a detailed study of the phenomenon of pionization. We first concentrate on processes in which the two incident particles are identical, e.g., proton-proton scattering. By assuming that the c.m. system has no special significance in high-energy scattering or, more precisely, by assuming that the distribution of pionization products has a forward-backward symmetry not only in the c.m. system but also in a class of coordinate systems called the C systems (this assumption will be called the *symmetry hypothesis*), we can get rather precise information about the dependence of pionization distribution on the longitudinal momenta. In particular, the one-particle distribution function is $f(\mathbf{p}_\perp)d^3p/E$. These results are then explicitly verified in the model of quantum electrodynamics. Next, the considerations are extended to processes in which the two incident particles are different. For such processes, we propose the *limiting hypothesis:* The distribution of the pionization products is independent of the momenta of either of the incident particles in the limit the incident momenta are both infinite. With this hypothesis we can again obtain quantitative information on the dependence of the pionization distribution on the longitudinal momenta. When applied to the special case of identical particles, the limiting hypothesis is equivalent to the symmetry hypothesis. For the more general case where the two incident particles are not identical, it follows from the limiting hypothesis that the one-particle distribution function is again $f(\mathbf{p}_\perp)d^3p/E$, but the results on the multiparticle distribution functions are slightly weaker than those for identical particles. The limiting hypothesis is then shown to be valid in various field-theory models. The process of photon-fermion scattering is studied in particular detail. We also investigate the process of bremsstrahlung and show that it gives no pionization products. Finally, the above results are examined in the context of scalar electrodynamics, and all of the qualitative features are found to be the same.

140) High-energy behavior of a spin flip amplitude, *Phys. Rev. D* **3** (1971), 2377–2393, Hung Cheng (MIT, LNS and Harvard U. and Hamburg U.), T. T. Wu (Harvard U., DESY, and HH Eidelstedt.), DOI: https://doi.org/10.1103/PhysRevD.3.2377.

Hung Cheng†‡
Department of Mathematics, Massachusetts Institute of Technology, Cambridge, Massachusetts 02139‖
and
Deutsches Elektronen-Synchrotron, DESY, Hamburg, Germany

AND

Tai Tsun Wu‡
Gordon McKay Laboratory, Harvard University, Cambridge, Massachusetts 02138‖
and
Deutsches Elektronon-Synchrotron, DESY, Hamburg, Germany
and
HH Eidelstedt, Hinschstrasse 25, Germany
(Received 12 November 1970)

The perturbation series for the spin-flip amplitude, where two units of spin are exchanged, of the scattering of two vector mesons through fermion–fermion–vector-meson coupling is studied in the limit of high energies. Similar to the case of elastic scattering (without spin flip) in quantum electrodynamics, there are, for this spin-flip amplitude, terms in the perturbation series of the orders of magnitudes s, $s\ \ln s$, $s(\ln s)^2$, $s(\ln s)^3$, etc. For each n, the leading term in the coefficient of $s(\ln s)^n$ is due to Feynman diagrams with $n+2$ closed fermion loops. These leading terms are found explicitly for the exactly forward direction, and are then summed. Because these terms alternate in sign, at high energies this sum is smaller than any one of the leading terms in the series. This sum is studied in detail for the case where massless photons are exchanged, and is found to have a simple factorization property. Some of the results are extended to the more general case of exchanging massive neutral vector mesons. The high-energy behavior of this spin-flip amplitude is qualitatively different from that of the spin-nonflip amplitude studied previously.

141) High-energy scattering of a fermion with anomalous magnetic moment — nonexponentiation, *Phys. Rev. D* **3** (1971), 2394–2396, Hung Cheng (MIT, LNS and Harvard U. and Hamburg U.), T. T. Wu (Harvard U. and DESY, Hamburg U.), DOI: https://doi.org/10.1103/PhysRevD.3.2394.

Hung Cheng†‡
Department of Mathematics, Massachusetts Institute of Technology, Cambridge, Massachusetts 02139§
and
Deutsches Elektronen-Synchrotron DESY, Hamburg, Germany

AND

Tai Tsun Wu‡
Gordon McKay Laboratory, Harvard University, Cambridge, Massachusetts 02138§
and
Deutsches Elektronen-Synchrotron DESY, Hamburg, Germany
(Received 18 January 1971)

Although it is well known that the scattering amplitude for a Dirac particle in a potential takes a simple-exponentiation form at high energies, we show here that this is no longer the case if the Dirac particle has an anomalous magnetic moment as in the case of the proton. This serves as an example of our conclusion that simple exponentiation holds only if a particle can be treated as a point with no internal structure or internal degrees of freedom. This means that simple exponentiation is unlikely to occur for high-energy amplitudes of hadron-hadron scattering.

142) High-energy potential scattering. II, *Phys. Rev. D* **3** (1971), 2397–2410, Hung Cheng (MIT, LNS and Harvard U. and Hamburg U.), T. T. Wu (Harvard U. and DESY, Hamburg U.), DOI: https://doi.org/10.1103/PhysRevD.3.2397.

HUNG CHENG†‡
Department of Mathematics, Massachusetts Institute of Technology, Cambridge, Massachusetts 02139§
and
Deutsches Elektronen-Synchrotron DESY, Hamburg, Germany

AND

TAI TSUN WU‡
Gordon McKay Laboratory, Harvard University, Cambridge, Massachusetts 02138§
and
Deutsches Elektronen-Synchrotron DESY, Hamburg, Germany
(Received 28 December 1970)

In view of the advances of knowledge, both experimental and theoretical, about hadronic scattering processes at high energies, the potential-theory model studied before is found to exhibit a number of realistic features. Because partial differential equations can be dealt with in a much more straightforward manner than operator equations, this potential-theory model is useful in understanding some features of high-energy processes. The problem of exponentiation, i.e., the possible appearance of an exponential factor, is studied in detail. It is found that the simple-exponentiation form, found recently in field theories within certain approximations, does not hold in general when the scattering particles have a finite size or internal degrees of freedom. This result in turn is applied to the field-theoretical cases to understand further the nature of the approximations involved.

143) Large order behavior of Perturbation theory, *Phys. Rev. Lett.* **27** (1971), 461, Carl M. Bender (MIT), Tai Tsun WU (Harvard U. and DESY), DOI: https://doi.org/10.1103/PhysRevLett.27.461; Citations: 268.

Carl M. Bender*
Department of Mathematics, Massachusetts Institute of Technology, Cambridge, Massachusetts 02139

and

Tai Tsun Wu†
Gordon McKay Laboratory, Harvard University, Cambridge, Massachusetts 02138, ‡ and Deutsches Elektronen-Synchrotron, Hamburg, Germany
(Received 28 June 1971)

We examine the large-order behavior of perturbation theory for the anharmonic oscillator, a simple quantum-field-theory model. New analytical techniques are exhibited and used to derive formulas giving the precise rate of divergence of perturbation theory for all energy levels of the x^{2N} oscillator. We compute higher-order corrections to these formulas for the x^4 oscillator with and without Wick ordering.

Commentary by Carl M. Bender

Until the papers *Phys. Rev. Lett.* **27**, 461 (1971), *Phys. Rev. D* **7**, 1620 (1973), and *Phys. Rev. D* **8**, 3346 (1973), were written, the usual approach in perturbation theory was to calculate the first few terms in a perturbation expansion. In these papers it is shown how to find the asymptotic behavior of the *n*th term in a perturbation expansion in the limit of large *n*. The techniques used are called

Bender–Wu theory. Using Bender–Wu theory one can find the exact rate of divergence of a perturbation series. This information is crucial in understanding how to use a summation technique to approximate the function represented by the perturbation series.

144) High-Energy Scattering in φ^3 Theory and the Breakdown of Eikonal Approximation. II, *Phys. Rev. D* **6** (1972), 1693–1701, Hung Cheng (MIT, LNS and Harvard U. and DESY), T. T. Wu (MIT, LNS and Harvard U. and DESY), DOI: https://doi.org/10.1103/PhysRevD.6.1693.

Hung Cheng†‡
Department of Mathematics, Massachusetts Institute of Technology, Cambridge, Massachusetts 02139

and

Tai Tsun Wu‡
Gordon McKay Laboratory, Harvard University, Cambridge, Massachusetts 02138§
and Deutsches Elektronen-Synchrotron (DESY), Hamburg, Germany
(Received 4 June 1971)

In this paper we study, in ϕ^3 theory, the high-energy amplitudes for the exchange of two or more ladders. We obtain these amplitudes in closed forms to all orders of the coupling constant. The conclusions are as follows: (i) All of the scattering amplitudes satisfy the impact-factor representation. (ii) Except for the leading term (in the coupling constant) the scattering amplitude of two-ladder exchange is not in the form dictated by the eikonal approximation. (iii) The scattering amplitudes for the exchange of N ladders, $N \geq 3$, are never in the form dictated by the eikonal approximation – even for the leading term. (When the coupling constant approaches zero, all of them are of the order of s^{-3} instead of s^{-2N+1}, as dictated by the eikonal approximation.) Thus the eikonal approximation is not valid in ϕ^3 theory. The amplitude for the exchange of n scalar mesons, $n > 4$, is also given. Contrary to the popular notion, it is not of the order of s^{-n+1} when $n > 4$. Summing over such amplitudes does not lead to the exponentiation form commonly conceded in the past.

145) Scattering of a charged vector meson in a static field at high energies, *Phys. Rev. D* **5** (1972), 445–456, Hung Cheng (MIT, LNS and Harvard U. and DESY), T. T. Wu (Harvard U. and DESY), DOI: https://doi.org/10.1103/PhysRevD.5.445.

Hung Cheng†‡
Department of Mathematics, Massachusetts Institute of Technology, Cambridge, Massachusetts 02139

and

Tai Tsun Wu‡
Gordon McKay Laboratory, Harvard University, Cambridge, Massachusetts 02138§
and Deutsches Elektronen-Synchrotron DESY, Hamburg, Germany
(Received 14 June 1971)

We study here the scattering of a charged vector meson (without the anomalous magnetic moment) in a static field. The scattering amplitude for such a process is obtained explicitly in the case of a central field. We find that the simple-exponentiation form of the eikonal approximation does not occur.

146) Coulomb Scattering of a Charged Vector Meson with an Electric Dipole Moment, *Phys. Rev. D* **5** (1972), 3247–3252, Hung Cheng (MIT), Tai Tsun Wu (Harvard U., Phys. Dept. and DESY), DOI: https://doi.org/10.1103/PhysRevD.5.3247.

Hung Cheng†‡
Department of Mathematics, Massachusetts Institute of Technology, Cambridge, Massachusetts 02139

and

Tai Tsun Wu‡
Gordon McKay Laboratory, Harvard University, Cambridge, Massachusetts 02138§
and Deutsches Elektronen-Synchrotron, DESY, Hamburg, Germany
(Received 7 February 1972)

We generalize, by including an electric dipole moment that leads to parity nonconservation, our previous investigation of the scattering of a charged vector meson by a static Yukawa potential, or, in particular, a static Coulomb field. The existence of a solution that satisfies Meixner's corner condition requires the following inequality between the anomalous magnetic moment κ and the electric dipole moment w: $\kappa > 1 + w^2$. In the special case where the electric dipole moment is absent, this condition reduces to the one obtained previously.

147) Anharmonic oscillator. II: A Study of perturbation theory in large order, *Phys. Rev. D* **7** (1973), 1620–1636, In Le Guillou, J. C. (ed.), Zinn-Justin, J. (ed.): Large-order behaviour of perturbation theory 41–57, Carl M. Bender (MIT, LNS), T. T. Wu (Harvard U.), DOI: https://doi.org/10.1103/PhysRevD.7.1620; Citations: 526.

Carl M. Bender*
Department of Mathematics, Massachusetts Institute of Technology, Cambridge, Massachusetts 02139

and

Tai Tsun Wu†
Gordon McKay Laboratory, Harvard University, Cambridge, Massachusetts 02138
(Received 21 August 1972)

This paper is concerned with the nature of perturbation theory in very high order. Specifically, we study the Rayleigh-Schrödinger expansion of the energy eigenvalues of the anharmonic oscillator. We have developed two independent mathematical techniques (WKB analysis and difference-equation methods) for determining the large-n behavior of A_n^K, the nth Rayleigh-Schrödinger coefficient for the Kth energy level. We are not concerned here with placing bounds on the growth of A_n^K as n, the order of perturbation theory, gets large. Rather, we consider the more delicate problem of determining the precise asymptotic behavior of A_n^K as $n \to \infty$ for both the Wick-ordered and non-Wick-ordered oscillators. Our results are in exact agreement with numerical fits obtained from computer studies of the anharmonic oscillator to order 150 in perturbation theory.

Commentary by Carl M. Bender

Until the papers *Phys. Rev. Lett.* **27**, 461 (1971), *Phys. Rev. D* **7**, 1620 (1973), and *Phys. Rev. D* **8**, 3346 (1973), were written, the usual approach in perturbation theory was to calculate the first few terms in a perturbation expansion. In these papers it is shown how to find the asymptotic behavior of the *n*th term in a perturbation expansion in the limit of large *n*. The techniques used are called Bender–Wu theory. Using Bender–Wu theory one can find the exact rate of divergence of a perturbation series. This information is crucial in understanding how to use a summation technique to approximate the function represented by the perturbation series.

148) Coupled anharmonic oscillators. I. Equal-Mass Case, *Phys. Rev. D* **8** (1973), 3346–3378, Tom Banks (MIT), Carl M. Bender (MIT), Tai Tsun Wu (Harvard U.), DOI: https://doi.org/10.1103/PhysRevD.8.3346; Citations: 230.

Thomas Banks*
Department of Physics, Massachusetts Institute of Technology, Cambridge, Massachusetts 02139

Carl M. Bender†
Department of Mathematics, Massachusetts Institute of Technology, Cambridge, Massachusetts 02139

Tai Tsun Wu‡
Gordon McKay Laboratory, Harvard University, Cambridge, Massachusetts 02138
(Received 7 May 1973)

This is the first in a series of papers on the large-order behavior of perturbation theory for coupled anharmonic oscillators. We exploit previously published dispersion techniques to convert the calculation of perturbation theory in large order into a barrier-penetration problem. We then introduce new semiclassical methods for describing tunneling through nonspherically symmetric, N-dimensional potentials. To illustrate our new methods, we calculate the large-order behavior of perturbation theory for a simple system of two equal-mass oscillators with quartic coupling. Our predictions are in complete agreement with computer calculations. We then extend our results to oscillators with x^{2N} coupling, N-oscillator systems, and some infinite-oscillator systems.

Commentary by Carl M. Bender

Until the papers *Phys. Rev. Lett.* **27**, 461 (1971), *Phys. Rev. D* **7**, 1620 (1973), and *Phys. Rev. D* **8**, 3346 (1973), were written, the usual approach in perturbation theory was to calculate the first few terms in a perturbation expansion. In these papers it is shown how to find the asymptotic behavior of the *n*th term in a perturbation expansion in the limit of large *n*. The techniques used are called Bender–Wu theory. Using Bender–Wu theory one can find the exact rate of divergence of a perturbation series. This information is crucial in understanding how to use a summation technique to approximate the function representated by the perturbation series.

149) Classical SU(3) gauge field equations, *J. Math. Phys.* **15** (1974), 53–54, A. C. T. Wu (Michigan U.), T. T. Wu (Harvard U.), DOI: https://doi.org/10.1063/1.1666502.

Alfred C. T. Wu

Department of Physics, University of Michigan, Ann Arbor, Michigan 48104

Tai Tsun Wu

Gordon McKay Laboratory, Harvard University, Cambridge, Massachusetts 02138
(Received 14 August 1973)

Admissible forms of the static solutions to the $SU(3)$ gauge field equation are examined. It is shown that by a proper choice of the form of solutions which extricate the $SU(3)$ indices, the set of nonlinear *partial* differential equations is reducible to nonlinear *ordinary* differential equations for the radial functions.

150) Theory of hadron "bags" with scattering, *Phys. Rev. D* **9** (1974), 3495–3500, Tai Tsun Wu (Harvard U.), B. M. Mccoy (Harvard U.), Cheng Hung (MIT, LNS), DOI: https://doi.org/10.1103/PhysRevD.9.3495.

Tai Tsun Wu* and Barry M. McCoy†
Gordon McKay Laboratory, Harvard University, Cambridge, Massachusetts 02138

Hung Cheng‡
Department of Mathematics, Massachusetts Institute of Technology, Cambridge, Massachusetts 02139
(Received 25 March 1974)

We derive the boundary conditions satisfied by a boson field in the theory of hadron "bags." The scattering problem, the fission problem, and the fusion problem in this theory of one spatial dimension are discussed.

151) Mandelstam Diagrams and Bose Statistics, Barry M. McCoy (SUNY, Stony Brook), Tai Tsun Wu (Fermilab), ITP-SB-74-44 (1974).

152) Possible Quantum Number Assignment for J and psi-prime (3695), H. T. Nieh (SUNY, Stony Brook), Tai Tsun Wu (Harvard U.), Chen Ning Yang (SUNY, Stony Brook), ITP-SB-74-46 (1974).

153) Mandelstam Diagrams Are Not Enough to study the high-energy behavior of φ^3 theory, *Phys. Rev. D* **12** (1975), 546, Barry M. McCoy (SUNY, Stony Brook), Tai Tsun Wu (Fermilab), DOI: https://doi.org/10.1103/PhysRevD.12.546.

Barry M. McCoy†
Institute for Theoretical Physics, State University of New York, Stony Brook, New York 11794
Tai Tsun Wu‡
Fermi National Accelerator Laboratory, Batavia, Illinois 60510§
(Received 18 November 1974)

Viewed in the crossed channel, the ϕ^3 Mandelstam diagrams for the Reggeon-particle cut represent repeated interaction between two of the three exchanged particles. A much larger class of Feynman diagrams is obtained by allowing all three exchanged particles to participate in the interaction, as required by Bose statistics. It is found that, in the limit of high energies, all diagrams of the same order in this larger class contribute comparably. In particular, the contributions from the Mandelstam and non-Mandelstam diagrams are of the same order of magnitude. This larger class of diagrams is summed in terms of an integral equation. It is found from this integral equation that if the momentum transfer is not too large, then the leading singularity in the angular momentum plane is a pole. As the momentum transfer increases this pole eventually disappears into a cut. These considerations are generalized, in a nontrivial way, to the case of the Reggeon-Reggeon cut. The signature factor plays a most important role, and some of the contributing diagrams show unexpected high-energy behavior. The profound effect of these additional diagrams on the program of Reggeon calculus is discussed.

154) Remarks on e^+e^- Annihilation Into Hadrons in Quantum Field Theory, *Phys. Rev. D* **12** (1975), 302, Hung Cheng (MIT), Tai Tsun Wu (Fermilab), DOI: https://doi.org/10.1103/PhysRevD.12.302.

Hung Cheng*
Department of Mathematics, Massachusetts Institute of Technology, Cambridge, Massachusetts 02139

Tai Tsun Wu†
Fermi National Accelerator Laboratory, Batavia, Illinois 60510‡
(Received 18 December 1974)

We show that the absorptive effect of particle production on e^+e^- annihilation reactions is to reduce the cross sections and give a constant value to $R = \sigma(e^+e^- \rightarrow \text{hadrons})/\sigma(e^+e^- \rightarrow \mu^+\mu^-)$. In the approximation of summing leading terms, this effect simply gives a factor s^{-a} to all cross sections.

155) Quantum Number Exchanges at Extremely High-Energies, *Phys. Rev. Lett.* **35** (1975), 1190–1192, B. M. Mccoy (SUNY, Stony Brook), T. T Wu (Harvard U.), DOI: https://doi.org/10.1103/PhysRevLett.35.1190.

Barry M. McCoy†
Institute for Theoretical Physics, State University of New York at Stony Brook, Stony Brook, New York 11794

and

Tai Tsun Wu
Gordon McKay Laboratory, Harvard University, Cambridge, Massachusetts 02138
(Received 25 August 1975)

We study systematically the high-energy behavior of backward Compton scattering and pair annihilation in massive quantum electrodynamics. For both of these processes (where a fermion is exchanged), we find, in the complex angular-momentum plane, a fixed cut with a simple pole coming out at a physical value of momentum transfer. These results are used to discuss possible behaviors, at extremely high energies, of hadronic processes with the exchange of quantum numbers.

156) Some Remarks About Unquantized non-Abelian Gauge Fields, *Phys. Rev. D* **12** (1975), 3843–3844, and New York State Univ Stony Brook - ITP-SB-75-23 (75,REC.OCT) 6p, T. T Wu (Harvard U.), Chen-Ning Yang (SUNY, Stony Brook), DOI: https://doi.org/10.1103/PhysRevD.12.3843; Citations: 173.

Tai Tsun Wu*
Gordon McKay Laboratory, Harvard University, Cambridge, Massachusetts 02138

Chen Ning Yang†
Institute for Theoretical Physics, State University of New York, Stony Brook, New York 11794
(Received 8 September 1975)

For unquantized non-Abelian gauge fields, remarks are made on the question of whether the field strengths determine the gauge field, on the dual transformations, and on complex gauge fields.

157) Comment on High-Energy Behavior of Nonabelian Gauge Theories, *Phys. Rev. Lett.* **35** (1975), 604–605, B. M. Mccoy (SUNY, Stony Brook), T. T Wu (Harvard U.), DOI: https://doi.org/10.1103/PhysRevLett.35.604.

Barry M. McCoy*
Institute for Theoretical Physics, State University of New York, Stony Brook, New York 11794

and

Tai Tsun Wu†
Gordon McKay Laboratory, Harvard University, Cambridge, Massachusetts 02138
(Received 18 June 1975)

The high-energy behavior of the sixth-order fermion-fermion scattering amplitude in the Yang-Mills theory is recalculated, and found to be qualitatively different from that given previously.

158) Concept of Nonintegrable Phase Factors and Global Formulation of Gauge Fields, *Phys. Rev. D* **12** (1975), 3845–3857, Also in Yang, C.N.: Selected Papers 1945–1980, 460–472, Tai Tsun Wu (Harvard U.), Chen Ning Yang (SUNY, Stony Brook), DOI: https://doi.org/10.1103/PhysRevD.12.3845; Citations: 1059.

Tai Tsun Wu*
Gordon McKay Laboratory, Harvard University, Cambridge, Massachusetts 02138

Chen Ning Yang†
Institute for Theoretical Physics, State University of New York, Stony Brook, New York 11794
(Received 8 September 1975)

Through an examination of the Bohm-Aharonov experiment an intrinsic and complete description of electromagnetism in a space-time region is formulated in terms of a nonintegrable phase factor. This concept, in its global ramifications, is studied through an examination of Dirac's magnetic monopole field. Generalizations to non-Abelian groups are carried out, and result in identification with the mathematical concept of connections on principal fiber bundles.

159) High-Energy Behavior of Gauge Theories, Tai Tsun Wu (Harvard U.), 18th International Conference on High-Energy Physics (1976), A5.23–26.

160) Some Inequalities in the Eikonal Approximation, Alexander W. Chao (SLAC), Tai Tsun Wu (CERN), Chen Ning Yang (CERN), CERN-TH-2224, (1976).

161) Theory of Fermion Exchange in Massive Quantum Electrodynamics at High-Energy. VI. Summation of Diagrams, *Phys. Rev. D* **13** (1976), 508–512, B. M. McCoy (SUNY, Stony Brook), Tai Tsun Wu (Harvard U.), DOI: https://doi.org/10.1103/PhysRevD.13.508.

Barry M. McCoy[†]
The Institute for Theoretical Physics, State University of New York at Stony Brook, Stony Brook, New York 11794

Tai Tsun Wu
Gordon McKay Laboratory, Harvard University, Cambridge, Massachusetts 02138
(Received 4 August 1975)

We conclude our study of fermion exchange in massive quantum electrodynamics at high energy by summing the leading contributions which have been computed in perturbation theory in the four preceding papers.

162) Theory of Fermion Exchange in Massive Quantum Electrodynamics at High-Energy. V. Twelfth and Higher Order Perturbation Theory, *Phys. Rev. D* **13** (1976), 484–507, B. M. McCoy (SUNY, Stony Brook), Tai Tsun Wu (Harvard U.), DOI: https://doi.org/10.1103/PhysRevD.13.484.

Barry M. McCoy[†]
The Institute for Theoretical Physics, State University of New York at Stony Brook, Stony Brook, New York 11794

Tai Tsun Wu
Gordon McKay Laboratory, Harvard University, Cambridge, Massachusetts 02138
(Received 4 August 1975)

We extend to twelfth order our study of the high-energy behavior of Compton scattering near the backward direction and of pair annihilation in massive quantum electrodynamics. We then generalize our considerations by introducing a matrix method of calculating higher-order diagrams and present the result of the high-energy expansion of an arbitrarily high order of perturbation theory.

163) Theory of Fermion Exchange in Massive Quantum Electrodynamics at High Energy IV. Tenth-Order Perturbation Theory, *Phys. Rev. D* **13**, 424 (1976), B. M. McCoy (SUNY, Stony Brook), Tai Tsun Wu (Harvard U.), DOI: https://doi.org/10.1103/PhysRevD.13.424.

Barry M. McCoy†
The Institute for Theoretical Physics, State University of New York at Stony Brook, Stony Brook, New York 11794

Tai Tsun Wu
Gordon McKay Laboratory, Harvard University, Cambridge, Massachusetts 02138
(Received 4 August 1975)

The detailed calculations of the preceding two papers are continued to tenth order. There are 41 Feynman diagrams which must be considered. These diagrams fall into 3 categories: (1) fifteen which contribute to the leading real part, (2) six which individually contribute to the leading imaginary part that cancel in pairs in the sum, and (3) twenty which contribute to the leading imaginary part which do not cancel identically in the sum.

164) Theory of Fermion Exchange in Massive Quantum Electrodynamics at High-Energy. III. Eighth-Order Perturbation Theory, *Phys. Rev. D* **13** (1976), 395–423, B. M. McCoy (SUNY, Stony Brook), Tai Tsun Wu (Harvard U.), DOI: https://doi.org/10.1103/PhysRevD.13.395.

Barry M. McCoy†
Institute for Theoretical Physics, State University of New York at Stony Brook, Stony Brook, New York 11794

Tai Tsun Wu
Gordon McKay Laboratory, Harvard University, Cambridge, Massachusetts 02138
(Received 4 August 1975)

We extend the detailed calculations of the preceding paper to eighth order. We find 7 diagrams are needed to evaluate the leading real part and an additional 5 are needed for the complete evaluation of the leading imaginary part.

165) Theory of Fermion Exchange in Massive Quantum Electrodynamics at High-Energy. II. Sixth-Order Perturbation Theory, *Phys. Rev. D* **13** (1976), 379–394, B. M. McCoy (SUNY, Stony Brook), Tai Tsun Wu (Harvard U.), DOI: https://doi.org/10.1103/PhysRevD.13.379.

Barry M. McCoy†
The Institute for Theoretical Physics, State University of New York at Stony Brook, Stony Brook, New York 11794

Tai Tsun Wu
Gordon McKay Laboratory, Harvard University, Cambridge, Massachusetts 02138
(Received 4 August 1975)

In this paper we present the details of our sixth-order perturbation theory calculation of the high-energy behavior of Compton scattering near the backward direction and pair annihilation. For each of these matrix elements, a formula is obtained not only for the leading term, which is real, but also for the leading imaginary part. Momentum-space techniques are used throughout. In particular we impose a cutoff in the transverse momentum and then compute the behavior as s (the square of the center-of-mass energy)$\rightarrow\infty$. With this procedure there are 3 Feynman diagrams which contibute to the leading real part. These 3 diagrams each give a factor of $\ln^2 s$ and an integral over transverse-momentum space. Each of these 3 transverse integrals separately diverge as $k_{max}\rightarrow\infty$ but when added together the $k_{max}\rightarrow\infty$ limit may be taken and the result agrees with the result previously obtained by Feynman parameters. To complete the evaluation of the leading imaginary part a fourth diagram is needed.

166) Theory of Fermion Exchange in Massive Quantum Electro-dynamics at High-Energy. I., *Phys. Rev. D* **13** (1976), 369–378, B. M. McCoy (SUNY, Stony Brook), Tai Tsun Wu (Harvard U.), DOI: https://doi.org/10.1103/PhysRevD.13.369.

Barry M. McCoy†
The Institute for Theoretical Physics, State University of New York at Stony Brook, Stony Brook, New York 11794

Tai Tsun Wu
Gordon McKay Laboratory, Harvard University, Cambridge, Massachusetts 02138
(Received 4 August 1975)

We consider the high-energy behavior of Compton scattering near the backward direction and of pair annihilation into two photons in massive quantum electrodynamics. We study the limit where s, the center-of-mass energy squared, becomes large and t, the square of the momentum transfer, is fixed and spacelike. In each order of perturbation theory we calculate the leading terms in the $s \to \infty$ expansion, and then these leading terms are summed. In the positive-signature channel this leads to a Regge pole. In the negative-signature channel the amplitude is dominated by a pole when $-t$ is sufficiently small. However, as $-t$ increases the pole moves to the left and eventually the amplitude is dominated by the fixed Mandelstam cut. In this paper we present the results of our calculations, summarize the most important features of the calculations, and discuss the physics of the result. The details of the calculations are presented in the 5 succeeding papers.

167) A Static Sourceless Gauge Field, *Phys. Rev. D* **13** (1976), 3233, Tai Tsun Wu (Harvard U.), Chen Ning Yang (SUNY, Stony Brook), DOI: https://doi.org/10.1103/PhysRevD.13.3233.

Tai Tsun Wu*
Gordon McKay Laboratory, Harvard University, Cambridge, Massachusetts 02138

Chen Ning Yang†
Institute for Theoretical Physics, State University of New York, Stony Brook, New York 11794
(Received 23 January 1976)

A static sourceless gauge field, coupled only to itself, is exhibited for the group SL(2, C). Its field strength is $O(r^{-2})$ as $r \to \infty$. It is also a static sourceless gauge field for SO(3, 1), the orthochronous Lorentz group without inversion. A conjecture is made that such solution does not exist for compact groups.

168) Statistical Analysis of Feynman Diagrams, *Phys. Rev. Lett.* **37** (1976), 117–120, Also in *Le Guillou, J. C. (ed.), Zinn-Justin, J. (ed.): Large-order behaviour of perturbation theory* 58–61, Carl M. Bender (MIT, LNS), Tai Tsun Wu (Harvard U.), DOI: https://doi.org/10.1103/PhysRevLett.37.117; Citations: 61.

Carl M. Bender*
Department of Mathematics, Massachusetts Institute of Technoloy, Cambridge, Massachusetts 02139

and

Tai Tsun Wu†
Gordon McKay Laboratory, Harvard University, Cambridge, Massachusetts 02138
(Received 5 May 1976)

The methods of statistical mechanics are used to evaluate sums of many-vertex Feynman graphs in a simple quantum-field-theory model.

Commentary by Carl M. Bender

In quantum field theory the *n*th term in a perturbation expansion is the sum of all Feynman diagrams having *n* vertices (that is, *n* powers of the coupling constant). Using powerful analytical techniques Bender and Wu show how Feynman diagrams behave when they have many vertices. The answer is surprising. One might think that when *n* is large, the most important diagrams have special topological characteristics; for example, one might think that a special class of diagrams (such as ladder graphs, for example) would dominate when there are many vertices. In fact, in the limit of large *n* the most important contribution comes from average graphs that are utterly featureless. This paper shows how to perform the sum of such graphs to find the precise large-order behavior of perturbation theory.

169) Formation of Maxwellian Tails, *Phys. Rev. Lett.* **36**, 1107 (1976), Max Krook and Tai Tsun Wu, DOI: https://doi.org/10.1103/PhysRevLett.36.1107; Citations: 192.

Max Krook and Tai Tsun Wu
Division of Engineering and Applied Physics, Harvard University, Cambridge, Massachusetts 02138
(Received 9 February 1976)

Using two models, we study the relaxation to a Maxwell distribution in the context of classical kinetic theory. For the first model, an exact solution of the nonlinear Boltzmann equation is derived. For the second model, an asymptotic solution exhibits the remarkable feature of a transient tail population sometimes much larger than the equilibrium Maxwell distribution. This phenomenon may be of importance for calculating rates of fast chemical reactions and for controlled thermonuclear fusion.

170) Exact Solution of Boltzmann Equations for Multicomponent Systems, *Phys. Rev. Lett.* **38** (1977), 991–993, Max Krook (Harvard U.) and Tai Tsun Wu (Harvard U.), DOI: https://doi.org/10.1103/PhysRevLett.38.991.

Max Krook and Tai Tsun Wu
Division of Engineering and Applied Physics, Harvard University, Cambridge, Massachusetts 02138
(Received 24 November 1976)

Exact solutions of similarity type are obtained for Boltzmann equations describing the temporal relaxation to equilibrium of an N-component gas with elastic collision cross sections of special form.

171) Exact solutions of the Boltzmann equation, *Phys. Fluids* **20**, 1589–1595 (1977), Max Krook (Harvard U.) and Tai Tsun Wu (Harvard U.), DOI: https://doi.org/10.1063/1.861780; Citation: 197.

Max Krook[a] and Tai Tsun Wu

Division of Engineering and Applied Physics, Harvard University, Cambridge, Massachusetts 02138
(Received 9 March 1977)

The nonlinear Boltzmann equation for the relaxation to equilibrium of a homogeneous one-component gas, is considered for a class of collision models. The models are characterized by elastic cross sections inversely proportional to the relative speed, but with arbitrary dependence on center-of-mass scattering angle. The Boltzmann equation is solved exactly for a particular family of physically interesting initial distributions. The distribution functions are of the similarity form and consist of the product of a Maxwell function with "time-dependent temperature" and a linear function of v^2.

172) Introduction to Gauge Theory, Tai Tsun Wu (CERN), In Bonn 1977, Proceedings, Differential Geometrical Methods In *Mathematical Physics.Ii.*, Berlin 1977, 161–169, DOI: https://doi.org/10.1007/BFb0063669.

173) Perturbation Theory for Two-dimensional Gauge Models, *Nucl. Phys. B* **143** (1978), 503–520, Ion Olimpiu Stamatescu (CERN), Tai Tsun Wu (CERN), DOI: https://doi.org/10.1016/0550-3213(78)90066-4.

Ion Olimpiu STAMATESCU * and Tai Tsun WU **

CERN, Geneva, Switzerland

Received 10 July 1978

Euclidean symmetric integration, previously proposed for the SU(N) Yang-Mills theory in the limit of large N, is used to study the propagators in the Schwinger model and the massive vector-meson model in two dimensions. The result of summing the perturbation series agrees with the exact solution in each case. Therefore, perturbation theory is here capable of dealing with non-analytic behaviour in the coupling constant.

174) Numerical aspects of the approach to a Maxwellian distribution, *Phys. Rev. A* **19** (1979), 883–888, John Tjon (Instituut voor Theoretische Fysica der Rijksuniversiteit, Utrecht, The Netherlands) and Tai Tsun Wu (Harvard U.), DOI: https://doi.org/10.1103/PhysRevA.19.883; Citations: 67.

John Tjon and Tai Tsun Wu*

Instituut voor Theoretische Fysica der Rijksuniversiteit, Utrecht, The Netherlands
(Received 13 September 1978)

It has been conjectured, for the Boltzmann equation, that an arbitrary initial state tends first to relax towards a state characterized by the similarity solution. We present here a simple model where the possible validity of the statement can be studied numerically. The results of extensive numerical computation for the first time give support for the validity of this conjecture.

175) Nonperturbative quantum field theory, B. M. Mccoy (SUNY, Stony Brook), T. T. Wu (Harvard U.), *Sci. Sin.* **22** (1979), 1021–1032.

BARRY M. MCCOY
(*Institute for Theoretical Physics, State University of New York*)
AND TAI TSUN WU (吴大峻)
(*Gordon McKay Laboratory, Harvard University*)

Received May 25, 1979.

ABSTRACT

176) Lattice Renormalization of Non-Perturbative Quantum Field Theory, in *Bifurcation Phenomena in Mathematical Physics*, p. 69 (1979), B. M. Mccoy (SUNY, Stony Brook), T. T. Wu (Max Planck Institute, Munchen and Harvard U.), edited by C. Bardos and D. Bessis. DOI: https://doi.org/10.1007/978-94-009-9004-3_5.

Barry M. Mc Coy

Institute for Theoretical Physics, State University of New York, Stony Brook, New York 11794, U.S.A.*)

and

Tai Tsun Wu
Max-Planck-Institut für Physik und Astrophysik, München, Fed. Rep. of Germany ; Gordon McKay Laboratory, Harvard University, Cambridge, Massachusetts 02138, U.S.A.**)

ABSTRACT : We discuss lattice renormalization as a basis for understanding non-perturbative effects in quantum field theory. We first study two solvable examples of quantum field theory in two dimensions where lattice renormalization may be explicitly carried out, namely the Ising field theory and the Thirring model. The Ising theory demonstrates that, with the same Lagrangian, two different relativistic quantum field theories can be constructed by using different lattice renormalizations. The Thirring model without a bare mass exhibits the remarkable phenomenon of dynamic generation of mass accompanied by dynamic determination of the coupling constant. We then turn to four dimensions and, after contrasting Euclidean with Minkowski lattice renormalization, discuss their application to non-perturbative phenomena in the Wess-Zumino model, scalar electrodynamics, and extended non-Abelian gauge theory.

* Work supported in part by the United States National Science Foundation under grant No. DMR-7707863A01.

** Permanent address.
Work supported in part by the United States Department of Energy under contract No.EY-76-S-02-3227.

177) Infra-red asymptotic behavior of one fermion green's function in a scalar model with isospin, *Phys. Lett. B* **85** (1979), 395–398, Victor N. Popov (CERN), Tai Tsun Wu (CERN), DOI: https://doi.org/10.1016/0370-2693(79)91281-4.

Victor N. POPOV [1] and Tai Tsun WU [2]
CERN, Geneva, Switzerland

Received 18 May 1979

In a theory where massive fermions interact with a massless scalar field of isospin 1, the behaviour of the one-fermion Green's function is found to differ from the free Green's function by a factor $(1 - (2g^2/\pi^2)\ln m|x - y|)^{-3/8}$, in the limit of large separation $|x - y|$.

178) On a Possible Structure in the Proton Elastic Form-factor, *Z. Phys. C* **5** (1980), 159, C. Bourrely (Marseille, CPT), Jacques Soffer (CERN and Marseille, CPT), Tai Tsun Wu (Harvard U.), DOI: https://doi.org/10.1007/BF01576198.

C. Bourrely
Centre de Physique Théorique, CNRS Marseille, Luminy, Case 907, F-13288 Marseille Cedex 2, France

J. Soffer
CERN Geneva and Centre de Physique Théorique, CNRS Marseille, Luminy, Case 907, F-13288 Marseille Cedex 2, France

Tai Tsun Wu
Harvard University, Cambridge, MA 02138, USA

Received 11 January 1980

Abstract. We investigate in the proton form factor the possibility to have a zero in the space-like region. As a consequence the e–p differential cross-section would have a dip and a second maximum which are accessible to experiment. If the existence of such structure would be revealed it would be very important for understanding the composite nature of the proton.

179) Collective Effects of Induced Behavior, Regina Graber (Munich, Max Planck Inst.), Ion Olimpiu Stamatescu (Wuppertal U.), Tai Tsun Wu (Harvard U.), WU B 80-22 (1980).

180) A completely solvable model of the nonlinear boltzmann equation, T. W. Ruijgrok (Utrecht U.), T. T. Wu (DESY and Harvard U.), Preprint - RUIJGROK, T. W. (81,REC.AUG.) 28p (1981).

181) Four-dimensional lattice gauge theory, B. M. Mccoy (SUNY, Stony Brook), T. T. Wu (Harvard U.), In Marseille 1981, Proceedings, Theoretical Aspects Of Quantum Chromodynamics, 307–319.

182) A Class of Exactly Solvable Three-body Quantum Mechanical Problems and the Universal Low-energy Behavior, *Phys. Lett. A* **83** (1981), 105–109, S. Albeverio (Provence U. and Bielefeld U. and Marseille, CPT and Harvard U. and Oslo U. and Ruhr U., Bochum), R. Hoegh-Krohn (Provence U. and Bielefeld U. and Marseille, CPT and Harvard U. and Oslo U. and Ruhr U., Bochum), T. T. Wu (Provence U. and Bielefeld U. and Marseille, CPT and Harvard U. and Oslo U. and Ruhr U., Bochum), DOI: https://doi.org/10.1016/0375-9601(81)90507-7; Citations: 74.

Sergio ALBEVERIO, Raphael HØEGH-KROHN and Tai Tsun WU[1]
Université d'Aix-Marseille II, Faculté des Sciences de Luminy, Marseille, France
Fakultät für Mathematik, Universität Bielefeld, Bielefeld, Fed. Rep. Germany
Centre de Physique Théorique II, CNRS, Marseille, France
Gordon McKay Laboratory, Harvard University, Cambridge, MA, USA
Matematisk Institutt, Universitetet i Oslo, Oslo, Norway
Mathematisches Institut, Ruhr-Universität Bochum, Bochum, Fed. Rep. Germany

Received 24 February 1981

Three-body systems with two-body point interactions are studied. These systems are the universal low energy limits of three-body problems with short-range two-body forces. Hence if there are infinitely many spherically symmetric three-body bound states with energies E_n then $\lim_{n\to\infty} E_n/E_{n+1} = e^{2\pi/\sigma}$, where σ is explicitly computed.

183) Nonperturbative quantum field theory, Barry M. McCoy (SUNY, Stony Brook), Tai Tsun Wu (Harvard U.), ITP-SB-79-37 (1981).

184) Helicity Amplitudes for Massless QED, *Phys. Lett. B* **105** (1981), 215, P. De Causmaecker (Marseille, CPT and Leuven U. and Harvard U. and Provence U.), R. Gastmans (Marseille, CPT and Leuven U. and Harvard U. and Provence U.), W. Troost (Marseille, CPT and Leuven U. and Harvard U. and Provence U.), Tai Tsun Wu (Marseille, CPT and Leuven U. and Harvard U. and Provence U.), DOI: https://doi.org/10.1016/0370-2693(81)91025-X; Citations: 145.

P. De CAUSMAECKER [a,1], R. GASTMANS [a,2], W. TROOST [a,3] and Tai Tsun WU [b,4]
[a] *Instituut voor Theoretische Fysica, University of Leuven, B-3030 Leuven, Belgium*
[b] *Centre de Physique Théorique, CNRS, Université de Provence, 13288 Marseille Cedex 9, France*

Received 2 July 1981

The introduction of helicity states, for both fermions and photons, is shown to lead to simple expressions for QED amplitudes, in which an arbitrary number of photons are radiated, provided that the fermion masses can be neglected. As an example, the results are given for the process $e^+e^- \to 4\gamma$.

185) Large-Order perturbation theory. *Int. J. Quantum Chem.* **21**: 105–118 (1982), T. T. Wu (Harvard U.), DOI: https://doi.org/10.1002/qua.560210107.

Abstract: The original motivation for studying the asymptotic behavior of the coefficients of perturbation series came from quantum field theory. An overview is given of some of the attempts to understand quantum field theory beyond finite-order perturbation series. At least in the case of the Thirring model and probably in general, the full content of a relativistic quantum field theory cannot be recovered from its perturbation series. This difficulty, however, does not occur in quantum mechanics, and the anharmonic oscillator is used to illustrate the methods used in large-order perturbation theory. Two completely different methods are discussed, the first one using the WKB approximation, and a second one involving the statistical analysis of Feynman diagrams. The first one is well developed and gives detailed information about the desired asymptotic behavior, while the second one is still in its infancy and gives instead information about the distribution of vertices of the Feynman diagrams.

186) Multiple Bremsstrahlung in Gauge Theories at High-Energies (I). General Formalism for Quantum Electrodynamics, *Nucl. Phys. B* **206** (1982), 53–60, P. De Causmaecker (Leuven U.), R. Gastmans (Leuven U.), W. Troost (Leuven U.), Tai Tsun Wu (DESY and Harvard U.), DOI: https://doi.org/10.1016/0550-3213(82)90488-6; Citations: 281.

P DE CAUSMAECKER[1], R GASTMANS[2] and W TROOST[1]
Instituut voor Theoretische Fysica, University of Leuven, B-3030 Leuven, Belgium

Tai Tsun WU[4]
Deutsches Elektronen-Synchrotron DESY, Hamburg, Germany
and
Gordon McKay Laboratory, Harvard University, Cambridge, Mass, USA

Received 9 February 1982

Multiple bremsstrahlung is studied on the level of tree diagrams for gauge theories At high energies and in most of the kinematic region, the fermion mass can be neglected In this case, it is natural to introduce helicity states for both fermions and gauge particles Our general formalism is given in detail for quantum electrodynamics In particular, it is expedient to use photon polarization vectors which depend on the fermion helicities In this way, extensive cancellations between Feynman diagrams are accomplished automatically

187) Impact Picture in the Coulomb Interference Region at High-energies, *Phys. Lett. B* **196** (1987), 237, *Phys. Lett. B* 198 (1987), 591 (erratum), C. Bourrely (Marseille, CPT), Jacques Soffer (Marseille, CPT), Tai Tsun Wu (CERN and Harvard U.), DOI: https://doi.org/10.1016/0370-2693(87)90611-3.

C. BOURRELY and J. SOFFER
Centre de Physique Théorique, CNRS, F-13288 Marseille Cedex 9, France

and

Tai Tsun WU[1]
CERN, CH-1211 Geneva 23, Switzerland
and Gordon McKay Laboratory, Harvard University, Cambridge, MA 02138, USA

Received 24 June 1987

We present the predictions on $p\bar{p}$ elastic differential cross sections in the Coulomb interference region for energies 0.546 to 2.0 TeV on the basis of the impact picture.

188) Quantum approach to beamstrahlung, *Phys. Lett. B* **197** (1987), 253, CERN Geneva - TH. 4761 (87,REC.AUG.) 13p, M. Jacob (CERN), T. T. Wu (CERN and Harvard U.), DOI: https://doi.org/10.1016/0370-2693(87)90377-7.

M. JACOB
CERN, CH-1211 Geneva 23, Switzerland

and

Tai Tsun WU [1]
CERN, CH-1211 Geneva 23, Switzerland
and Gordon McKay Laboratory, Harvard University, Cambridge, MA 62138, USA

Received 20 July 1987

Bremsstrahlung energy losses by electrons and positrons in very high energy linear colliders are discussed in terms of Feynman diagrams. At least in some cases, it is possible to separate different contributions in the quantum radiation of the impinging particles. In view of this, we here focus on the radiation before beam crossing. This is in addition to the radiation during beam crossing.

189) Quantum Calculation of Beamstrahlung: The Spinless Case, *Nucl. Phys. B* **303** (1988), 373–388, M. Jacob (CERN), Tai Tsun Wu (CERN and Harvard U.), DOI: https://doi.org/10.1016/0550-3213(88)90388-4.

M. JACOB

CERN, Geneva, Switzerland

Tai Tsun WU*

CERN, Geneva, Switzerland and Gordon McKay Laboratory, Harvard University, Cambridge, Massachusetts, USA

Received 14 October 1987

Bremsstrahlung energy losses by electrons and positrons in very high energy linear colliders are calculated in terms of Feynman graphs. A follow-up to a previous paper "Quantum approach to beamstrahlung", which focuses upon radiation before bunch-crossing, this paper presents a calculation of the fractional energy loss of an electron as it radiates inside the positron bunch which it traverses. The calculation applies to the special conditions set by machines in the TeV range.

190) Semiclassical Analysis of the Weak Coupling Limit of SU(2) Lattice Gauge Theory: The Subspace of Constant Fields, *Phys. Rev. D* **37** (1988), 2307, Jochen Bartels (Hamburg U.), T. T. Wu (Harvard U.), DOI: https://doi.org/10.1103/PhysRevD.37.2307.

J. Bartels
II. Institut für Theoretische Physik, Universität Hamburg, Hamburg, West Germany

T. T. Wu
CERN, Geneva, Switzerland and Gordon McKay Laboratory, Harvard University, Cambridge, Massachusetts 02138
(Received 9 November 1987)

This paper contains the first part of a systematic semiclassical analysis of the weak-coupling limit of lattice gauge theories, using the Hamiltonian formulation. The model consists of an N^3 cubic lattice of pure SU(2) Yang-Mills theory, and in this first part we limit ourselves to the subspace of constant field configurations. We investigate the flow of classical trajectories, with a particular emphasis on the existence and location of caustics. There the ground-state wave function is expected to peak. It is found that regions densely filled with caustics are very close to the origin, i.e., in the domain of weak field configurations. This strongly supports the expectation that caustics are essential for quantities of physical interest.

191) Quantum Calculation of Beamstrahlung: The Dirac Case, *Nucl. Phys. B* **303** (1988), 389-406, M. Jacob (CERN), Tai Tsun Wu (CERN and Harvard U.), DOI: https://doi.org/10.1016/0550-3213(88)90389-6.

M. JACOB
CERN, Geneva, Switzerland

Tai Tsun WU*
CERN, Geneva, Switzerland and Gordon McKay Laboratory, Harvard University, Cambridge, Massachusetts, USA

Received 4 December 1987

Bremsstrahlung energy losses by electrons and positrons in very high energy linear colliders are calculated in terms of Feynman graphs. A follow-up to two previous papers, "Quantum approach to beamstrahlung" and "Quantum calculation of beamstrahlung: The spinless case", this paper deals with the calculation of radiative energy losses taking spin effects into account. The calculation applies to the special conditions set by machines in the TeV and multi-TeV ranges, when the disruption parameter remains small.

192) $N = 4$ supersymmetric quantum mechanics and its large order behavior, *Phys. Lett. B* **240** (1990), 401–403, J. J. M. Verbaarschot (SUNY, Stony Brook), Peter C. West (Geneva U. and King's Coll. London), T. T. Wu (Harvard U.), DOI: https://doi.org/10.1016/0370-2693(90)91119-V.

J.J.M. VERBAARSCHOT
Department of Physics, SUNY, Stony Brook, NY 11794, USA

P. WEST
Department of Physics, University of Geneva, CH-1211 Geneva 4, Switzerland and Mathematics Department, King's College, Strand, London WC2R 2LS, UK [1]

and

T.T. WU
Physics Department, Harvard University, Cambridge, MA 02138, USA

Received 9 February 1990

A model of $N=4$ supersymmetric quantum mechanics is shown to have a perturbative expansion for which the large order behaviour is $-(3/2)^n n!(9/\pi+5/n+...)$. In the space of the coupling constants the supersymmetric case is found to be the dividing point between two different types of large order behaviour. In this way the large order behaviour of supersymmetric theories is special.

193) Semiclassical analysis of the weak coupling limit of SU(2) lattice gauge theory: the extreme infrared region, *Phys. Rev. D* **42** (1990), 1233–1245, Jochen Bartels (Hamburg U.), Bernd Raabe (Hamburg U.), Tai Tsun Wu (Harvard U.), DOI: https://doi.org/10.1103/PhysRevD.42.1233.

Jochen Bartels and Bernd Raabe
II. Institut für Theoretische Physik, Universität Hamburg, Hamburg, West Germany

Tai Tsun Wu
Gordon McKay Laboratory, Harvard University, Cambridge, Massachusetts 02138
(Received 23 March 1990)

This paper contains the second part of our systematic investigation of the weak-coupling limit of SU(2) lattice gauge theory, using the semiclassical approximation. We study the flow of solutions close to the subspace of spatially constant fields ("extreme infrared") and establish the existence of a dense cobweb of caustics in the region of very small fields. The origin of these caustics are oscillations of neighboring classical solutions around each other.

194) Completely solvable models of the nonlinear Boltzmann equation I. Case of three velocities, *J. Math. Phys.* **31** (1990): 3003–3029, Per Osland (Bergen U.) and Tai Tsun Wu (Harvard U.), DOI: https://doi.org/10.1063/1.528954.

Per Osland
Department of Physics, University of Bergen, Allégt. 55, N-5007 Bergen, Norway

Tai Tsun Wu
Gordon McKay Laboratory, Harvard University, Cambridge, Massachusetts 02138

(Received 29 May 1990; accepted for publication 8 August 1990)

In one space and one time dimension, a class of models of the nonlinear Boltzmann equation is presented that is exactly solvable for all initial conditions. The models have three velocity components and the following desirable properties: (a) conservation of the number of particles; (b) energy conservation; (c) nonlinearity; (d) positivity of distribution functions; and (e) unique equilibrium state (for any given density), which is approached as $t \to \infty$. These models are very rich in structure, and some of their simple properties are studied.

195) The Asymptotics of phi**3 ladder diagrams in six-dimensions: High-energy and large momentum transfer, *Nucl. Phys. B* **365** (1991), 404–430, R. Gastmans (Leuven U.), W. Troost (Leuven U.), Tai Tsun Wu (Harvard U.), DOI: https://doi.org/10.1016/S0550-3213(05)80027-6.

R. GASTMANS* and W. TROOST**

Institute for Theoretical Physics, University of Leuven, B-3001 Leuven, Belgium

Tai Tsun WU***

Gordon McKay Laboratory, Harvard University, Cambridge, MA 02138, USA

Received 8 March 1991

The first three leading terms of the asymptotic form of the scattering amplitude for the n-rung ϕ^3 ladder diagram in six dimensions are derived in the high-energy, large-momentum transfer limit. The behavior of these terms is qualitatively different from that of the same diagram in four dimensions.

196) Macroscopic lack of localization in quantum radiative processes, *Phys. Lett. A* **162** (1992), 304–308, and CERN Geneva - TH. 6176 (91/07,rec.Nov.) 6 p, M. Jacob (CERN), Tai Tsun Wu (CERN and Harvard U.), DOI: https://doi.org/10.1016/0375-9601(92)90021-D.

M. Jacob [a] and Tai Tsun Wu [a,b]

[a] *Theoretical Physics Division, CERN, CH-1211 Geneva 23, Switzerland*
[b] *Gordon McKay Laboratory, Harvard University, Cambridge, MA, USA*

Received 19 November 1991; accepted for publication 16 December 1991
Communicated by J.P. Vigier

We use results on beamstrahlung and pair production in bunch collisions to illustrate the macroscopic lack of localization of the corresponding quantum processes. Localization is usually associated with a coherence length. However, for some relatively important contributions to the production rate, the concept of inside and outside a bunch of particles loses its meaning. The effect should be clearly observable with electron–positron colliders in the multi-TeV range.

197) Point splitting regularization for gauge theories: Quantum electrodynamics, *Z. Phys. C* **61** (1994), 441–448, Conrad Newton (Harvard U. and CERN), Per Osland (Bergen U.), Tai Tsun Wu (Harvard U. and CERN), DOI: https://doi.org/10.1007/BF01413183.

Conrad Newton[1,2,★], Per Osland[3,★★], Tai Tsun Wu[1,2,★]

[1] Gordon McKay Laboratory, Harvard University Cambridge, MA 02138, USA
[2] Theoretical Physics Division, CERN, CH-1211 Geneva 23, Switzerland
[3] Department of Physics, University of Bergen, Allégt. 55 N-5007 Bergen, Norway

Received 28 September 1993

Abstract. As a method of regularization, point splitting has played an essential role in the recent theoretical determination of the masses of the Higgs boson and the top quark. It is the purpose of this paper to put this point-splitting regularization on a firm basis. The result turns out to be extremely simple: replace the usual vertex factor $-ie\gamma_\mu$ in quantum electrodynamics by

$$-ie\left(\gamma_\mu - \frac{\not{p}}{p\cdot\delta}\delta_\mu\right),$$

where p is the momentum of the photon line, and δ_μ is the distance for point splitting. No additional vertices are needed.

198) Behavior of total and elastic cross-sections at very high-energies, Tai Tsun Wu (Harvard U. and CERN), Quantum Infrared Phys.1994:0291-302 (1994).

199) Lattice Gauge Theory in Four-dimensions: Decoupling Points and a Dual Symmetric Perturbation Theory, *Nucl. Phys. B* **190** (1981), 519–541, Barry M. McCoy (SUNY, Stony Brook), Tai Tsun Wu (Harvard U.), DOI: https://doi.org/10.1016/0550-3213(81)90446-6.

Barry M. McCOY

Institute for Theoretical Physics, State University of New York at Stony Brook, Stony Brook, New York 11794, USA

Tai Tsun WU

Gordon McKay Laboratory, Harvard University, Cambridge, Mass. 02138, USA

Received 3 April 1981

For a general compact gauge group, the anisotropic lattice gauge theory is studied where the interaction energies $E_{\mu\nu}$ for the six different elementary plaquettes are independent. By considering the limit $E_{xt} \to \infty$ and $E_{yz} \to 0$, an exact relation is found between the four-dimensional lattice gauge theories and the corresponding two-dimensional systems. For the Z_2 lattice gauge theory, the corresponding system is the two-dimensional Ising model. In this case, a dual symmetric perturbation theory about such special points is constructed with coupling of the two-dimensional subsystems through both order and disorder variables. We discuss the requirements of Lorentz invariance and the implications of having the critical temperatures of the subsystems the same or different.

200) Exact relation between four-dimensional lattice gauge theories and two-dimensional spin systems, *Phys. Lett. B* **108** (1982), 213–216, Barry M. McCoy (SUNY, Stony Brook), Tai Tsun Wu (Harvard U.), DOI: https://doi.org/10.1016/0370-2693(82)91178-9.

Barry M. McCOY

Institute for Theoretical Physics, State University of New York at Stony Brook, Stony Brook, NY 11794, USA and Research Institute for Mathematical Sciences, Kyoto University, Japan

and

Tai Tsun WU

Gordon McKay Laboratory, Harvard University, Cambridge, MA 02138, USA

Received 10 April 1981
Revised manuscript received 24 July 1981

We consider a four-dimensional lattice gauge theory where the interaction energies $E_{\mu\nu}$ of the six different orientations of elementary plaquettes are taken to be independent. In the limit $E_{xt} \to \infty$ and $E_{yz} \to 0$ we show that the gauge system decouples into corresponding two-dimensional spin systems.

201) Classical Models of Confinement, *Nucl. Phys. B* **237** (1984), 205–225, Harry Lehmann (Hamburg U.), Tai Tsun Wu (DESY), DOI: https://doi.org/10.1016/0550-3213(84)90156-1.

Harry LEHMANN

II. Institut für Theoretische Physik der Universität Hamburg, Fed. Rep. Germany

Tai Tsun WU[1]

Deutsches Elektronen-Synchrotron DESY, Hamburg, Fed. Rep. Germany

Received 19 September 1983

We investigate the structure of classical models of confinement by analytic methods. The models considered involve just one abelian gauge field and can be viewed as non-linear versions of Maxwell's theory. We study in detail the case of two opposite static point charges. A scaling law is found for small separations of the charges, while a perturbation treatment is developed for large separations. The results obtained for large separations include an exact evaluation of the first correction term to the linear potential and a classification of the asymptotic field configurations.

202) Fermi Pseudopotential in Higher Dimensions, *J. Math. Phys.* **25** (1984), 1742, Alexander Grossmann (Marseille, CPT), Tai Tsun Wu (DESY and Harvard U.), DOI: https://doi.org/10.1063/1.526337.

Alexander Grossmann
Centre de Physique Théorique II, CNRS, Marseille, France

Tai Tsun Wu
Deutsches Elektronen-Synchrotron DESY, Hamburg, Germany, and Gordon McKay Laboratory, Harvard University, Cambridge, Massachusetts 02138

(Received 16 June 1983; accepted for publication 9 December 1983)

The Fermi pseudopotential is generalized from three to five dimensions, and the case of an infinite, uniform, equidistant, linear chain of such pseudopotentials is studied in detail. Similar to the three-dimensional case, zero-width resonances are also present in five dimensions. While this generalization is natural and can be carried through formally when the strength is negative, there are basic changes in the underlying structure. These results in five dimensions also apply in four dimensions.

PACS numbers: 02.30.Tb, 12.90. + b

203) Classical models of confinement II, *Commun. Math. Phys.* **97** (1985), 161, Harry Lehmann (Hamburg U.), Tai Tsun Wu (Harvard U.), DOI: https://doi.org/10.1007/BF01206184.

Harry Lehmann* and Tai Tsun Wu**
1 II. Institut für Theoretische Physik der Universität Hamburg, D-2000 Hamburg, Federal Republic of Germany
2 Deutsches Elektronen-Synchrotron DESY, D-2000 Hamburg, Federal Republic of Germany, and Gordon McKay Laboratory, Harvard University, Cambridge, MA 02138, USA

Dedicated to the memory of Kurt Symanzik

Abstract. We generalize the class of Abelian models of paper I which lead to a linear potential between opposite charges. The electric field E is here taken as a power series in the electric displacement D raised to the σth power. We solve to first order the case of two opposite static point charges asymptotically for large separations.

204) Classical models of confinement with moving charges, *Nucl. Phys. B* **271** (1986), 166–174, Harry Lehmann (Hamburg U., Inst. Theor. Phys. II), Tai Tsun Wu (Harvard U.), DOI: https://doi.org/10.1016/S0550-3213(86)80007-4.

Harry LEHMANN[1]

II. Institut für Theoretische Physik der Universität Hamburg, Hamburg, Federal Republic of Germany

Tai Tsun WU[1†]

Gordon McKay Laboratory, Harvard University, Cambridge, MA 02138, USA

Received 21 August 1985

We investigate a class of abelian models which lead to a linear potential between opposite charges. For uniformly moving charges with opposite velocities, the electromagnetic fields and the confining domains are obtained asymptotically for large separations.

205) New techniques and results in gauge theory calculations, CALKUL Collaboration • R. Gastmans (Leiden U. and Leuven U. and DESY and Harvard U.) *et al.*, In *Erice 1983, Proceedings, Electroweak Effects At High Energies*, 599–609 (1987).

206) High-energy, large momentum transfer processes: The N rung ladder diagram of phi**3 theory, *J. Math. Phys.* **32** (1991), 1619–1637, C. Newton (Harvard U.), T. T. Wu (Harvard U.), DOI: https://doi.org/10.1063/1.529275.

Conrad Newton and Tai Tsun Wu
Gordon McKay Laboratory, Harvard University, Cambridge, Massachusetts 02138

(Received 3 January 1991; accepted for publication 12 February 1991)

The asymptotic form of the amplitude for the N-rung ladder diagram in φ^3 theory is calculated in the limit $s \gg |t| \gg m^2$. The result is proportional to $s^{-1}|t|^{-N+1}$ times a homogeneous polynomial of degree $2N-2$ in $\ln s$ and $\ln|t|$. It is not an analytic function of s and t, although the first $N+1$ derivatives are continuous in the physical region. Points of nonanalyticity occur when $\ln|t| = \gamma \ln s$ for $\gamma = \frac{1}{2}, \frac{1}{3}, \ldots, 1/(N-2)$. The notion of reduced Symanzik functions is introduced. With the aid of these reduced Symanzik functions, a generating function is derived for the leading logarithms of all ladder diagrams.

207) Stability of three unit - charge systems, *Phys. Rev. A* **46** (1992), 3697–3703, Andre Martin (CERN), Jean-Marc Richard (LPSC, Grenoble and CERN), Tai Tsun Wu (Harvard U. and CERN), DOI: https://doi.org/10.1103/PhysRevA.46.3697.

André Martin
CERN, Theory Division, CH 1211 Genève 23, Switzerland

Jean-Marc Richard
Institut des Sciences Nucléaires, Université Joseph Fourier, Grenoble, France
and CERN, Theory Division, CH 1211 Genève 23, Switzerland

Tai Tsun Wu
Gordon McKay Laboratory, Harvard University, Cambridge, Massachusetts 02138
and CERN, Theory Division, CH 1211 Genève 23, Switzerland
(Received 20 April 1992)

We discuss the binding of three unit charges $q_i = \pm 1, \mp 1, \mp 1$, with various constituent masses m_i. It is found that the most convenient variables are the normalized inverse masses defined by $\alpha_i = m_i^{-1}/(m_1^{-1}+m_2^{-1}+m_3^{-1})$. The region where this three-body system is stable against dissociation into a neutral atom and an isolated charge is restricted to a narrow band that is characterized by small values of $|\alpha_2 - \alpha_3|$ and includes the well-known cases H_2^+, $e^+e^-e^-$, and H^-. Some properties of this region of stability are presented. A method of proof of instability of e^-pe^+ is proposed.

PACS number(s): 34.20.Cf, 36.10.−k, 31.10.+z, 31.20.Di

208) Optimized lower bounds in the three-body problem, *Nucl. Phys. B* **393** (1993), 111–125, Jean-Louis Basdevant (Orsay, IPN and Paris U., VI-VII), Andre Martin (CERN), Jean-Marc Richard (LPSC, Grenoble), Tai Tsun Wu (Harvard U. and CERN), DOI: https://doi.org/10.1016/0550-3213(93)90240-P.

Jean-Louis Basdevant
Division de Physique Théorique [1], Institut de Physique Nucléaire, F-91406 Orsay Cedex, France
and
LPTPE, Université Pierre et Marie Curie, Paris, France

André Martin
CERN, Theory Division, CH 1211 Geneva 23, Switzerland

Jean-Marc Richard
Institut des Sciences Nucléaires-CNRS-IN2P3, Université Joseph Fourier, F-38026 Grenoble Cedex, France

Tai Tsun Wu [2]
Gordon McKay Laboratory, Harvard University, Cambridge, MA 02138, USA
and
CERN, Theory Division, CH 1211 Geneva 23, Switzerland

Received 2 July 1992
Accepted for publication 25 September 1992

We derive new lower bounds on three-body ground-state energies in terms of two-body binding energies. For three-body systems with different masses, we obtain a significant improvement as compared to previous results. In particular, the present method leads to exact results in the case of harmonic interactions, as opposed to what was found previously. It sheds light on the mechanism by which such bounds are obtained and appears to be extendable to four and more particles. The numerical results displayed show up to be very accurate estimates of the exact values, and can be used as checks in actual numerical calculations. Possible applications or generalizations are briefly outlined.

209) Generalized gauge transformation: Abelian case, *Phys. Lett. B* **341** (1994), 84–89, R. Gastmans (Leuven U.), Conrad Newton (Harvard U. and CERN), Tai Tsun Wu (Harvard U. and CERN), DOI: https://doi.org/10.1016/0370-2693(94)01259-8.

R. Gastmans [a,1], Conrad Newton [b,2], Tai Tsun Wu [b,2]
[a] *Institute for Theoretical Physics, University of Leuven, B-3001 Leuven, Belgium*
[b] *Gordon McKay Laboratory, Harvard University, Cambridge, MA 02138, USA and Theory Division, CERN, CH-1211 Geneva 23, Switzerland*

Received 14 September 1993; revised manuscript received 16 September 1994
Editor: R. Gatto

Abstract

A gauge transformation in quantum electrodynamics involves the product of field operators at the same space-time point and hence does not have a well-defined meaning. One way to avoid this difficulty is to generalize the gauge transformation by using different space-time points in the spirit of Dirac's point splitting. It is found that such a generalization indeed exists and the resulting generalized infinitesimal gauge transformation takes the form of an infinite series in the coupling constant.

210) Stability of systems of three arbitrary charges. I. General properties, *Phys. Rev. A* **52** (1995), 2557–2563, Andre Martin (CERN), Jean Marc Richard (LPSC, Grenoble and ECT, Trento), Tai Tsun Wu (Harvard U. and CERN), DOI: https://doi.org/10.1103/PhysRevA.52.2557.

André Martin
CERN, Theory Division, CH-1211 Genève 23, Switzerland

Jean-Marc Richard
Institut des Sciences Nucléaires CNRS-IN2P3, Université Joseph Fourier, Grenoble Cédex, France and European Centre for Theoretical Studies in Nuclear Physics and Related Areas, University of Trento, Trento, Italy

Tai Tsun Wu
Gordon McKay Laboratory, Harvard University, Cambridge, Massachussets 02138 and CERN, Theory Division, CH-1211 Genève 23, Switzerland
(Received 31 May 1994; revised manuscript received 1 March 1995)

The stability of systems consisting of a negative charge $-q_1$ and two positive charges q_2 and q_3 is discussed as a function of the values of the charges q_i and of the constituent masses m_i. We give general results and list some open questions of interest, which hopefully will stimulate further investigations.

PACS number(s): 03.65.Ge, 31.15.−p, 36.10.−k

211) Universality of low-energy scattering in (2+1)-dimensions, Khosrow Chadan (Orsay, LPT), N. N. Khuri (Rockefeller U.), Andre Martin (CERN and Annecy, LAPTH and Rockefeller U.), Tai Tsun Wu (Harvard U. and CERN), DOI: https://doi.org/10.1103/PhysRevD.58.025014.

Khosrow Chadan
Laboratoire de Physique Théorique et Hautes Energies, Université Paris-Sud, F-91405 Orsay, France

N. N. Khuri
Department of Physics, The Rockefeller University, New York, New York 10021

André Martin
TH Division, CERN, Geneva, Switzerland; Laboratoire de Physique Théorique ENSLAPP F-74941 Annecy-le-Vieux, France; and Department of Physics, The Rockefeller University, New York, New York 10021

Tai Tsun Wu
Gordon McKay Laboratory, Harvard University, Cambridge, Massachusetts 02138-2901 and TH Division, CERN, Geneva, Switzerland
(Received 22 January 1998; published 24 June 1998)

For any relativistic quantum field theory in 2+1 dimensions, with no zero mass particles, and satisfying the standard axioms, we establish a remarkable low-energy theorem. The S-wave phase shift, $\delta_0(k)$, k being the c.m. momentum, vanishes as either $\delta_0 \to c/\ln(k/m)$or $\delta_0 \to O(k^2)$ as $k \to 0$. The constant c is universal and $c = \pi/2$. This result follows only from the rigorously established analyticity and unitarity properties for 2-particle scattering. This kind of universality was first noted in non-relativistic potential scattering, albeit with an incomplete proof which missed, among other things, an exceptional class of potentials where $\delta_0(k)$ is $O(k^2)$ near $k=0$. We treat the potential scattering case with full generality and rigor, and explicitly define the exceptional class. Finally, we look at perturbation theory in ϕ_3^4 and study its relation to our non-perturbative result. The remarkable fact here is that in n-th order the perturbative amplitude diverges like $(\ln k)^n$ as $k \to 0$, while the full amplitude vanishes as $(\ln k)^{-1}$. We show how these two facts can be reconciled.
[S0556-2821(98)02714-3]

PACS number(s): 11.10.Jj, 11.10.Cd, 11.10.Kk, 11.80.−m

212) Optimized Lower Bound for Four-Body Hamiltonians. In: *Few-Body Systems* **24**.1 (May 1998), pp. 39–54, A. Benslama (Universite de Constantine), A. Metatla (Universite de Constantine), A. Bachkhaznadji (Universite de Constantine), S. R. Zouzou (Universite de Constantine), A. Krikeb (University Calude Bernard, CNRS-IN2P3), J.-L. Basdevant (University of Paris Orsay), J.-M. Richard (University Joseph Fourier, CNRS-IN2P3), and T. T. Wu (Harvard University). DOI: https://doi.org/10.1007/s006010050075.

A. Benslama[1], A. Metatla[1], A. Bachkhaznadji[1], S. R. Zouzou[1,*], A. Krikeb[2], J.-L. Basdevant[3], J.-M. Richard[4], and T. T. Wu[5,**]

[1] Département de Physique Théorique, Institut de Physique, Université de Constantine, Route d'Ain El Bey, 25000 Constantine, Algeria

[2] Institut de Physique Nucléaire, Université Claude Bernard, CNRS-IN2P3, 43 Boulevard du 11 Novembre 1918, F-69622 Villeurbanne Cedex, France

[3] Division de Physique Théorique[†], Institut de Physique Nucléaire, Université Paris-Sud, F-91406 Orsay, and Département de Physique, École Polytechnique, F-91128 Palaiseau, France

[4] Institut des Sciences Nucléaires, Université Joseph Fourier, CNRS-IN2P3, 53 Avenue des Martyrs, F-38026 Grenoble Cedex, France

[5] Gordon McKay Laboratory, Harvard University, Cambridge, MA 02138, USA, and CERN, Theory Division, CH-1211 Genève 23, Switzerland

Abstract. Generalizing a method elaborated for three-body systems, we derive a new lower bound on four-body ground-state energies in terms of two-body binding energies in the unequal-mass case. For simple power-law potentials, this bound is compared to variational calculations and is shown to be very close to the exact result. In particular, it gives the exact answer for harmonic interactions.

213) Stability Domain of Systems of Three Arbitrary Charges. In: *Few-Body Systems* **29**.4 (Dec. 2000), pp. 237–257, A. Krikeb (University Calude Bernard, CNRS-IN2P3), A. Martin (CERN), J.-M. Richard (University Joseph Fourier, CNRS-IN2P3), and T.-T. Wu (Harvard University), DOI: https://doi.org/10.1007/s006010070003.

A. Krikeb[1], A. Martin[2], J.-M. Richard[3], and T.-T. Wu[4,**]

[1] Institut de Physique Nucléaire de Lyon, CNRS-IN2P3, Université Claude Bernard, F-69622 Villeurbanne, France
[2] CERN, Theory Division, CH-1211 Genève 23, Switzerland, and LAPP, B.P. 110, F-74941 Annecy Le Vieux, France
[3] Institut des Sciences Nucléaires, CNRS-IN2P3, Université Joseph Fourier, 53, Avenue des Martyrs, F-38026 Grenoble, France
[4] CERN, Theory Division, CH-1211 Genève 23, Switzerland, and Gordon McKay Laboratory, Harvard University, Cambridge, MA, USA

Abstract. We present results on the stability of quantum systems consisting of a negative charge $-q_1$ with mass m_1 and two positive charges q_2 and q_3, with masses m_2 and m_3, respectively. We show that, for given masses m_i, each instability domain is convex in the plane of the variables $(q_1/q_2, q_1/q_3)$. A new proof is given of the instability of muonic ions (α, p, μ^-). We then study stability in some critical regimes where $q_3 \ll q_2$: Stability is sometimes restricted to large values of some mass ratios; the behaviour of the stability frontier is established to leading order in q_3/q_2. Finally we present some conjectures about the shape of the stability domain, both for given masses and varying charges, and for given charges and varying masses.

214) Mikheyev–Smirnov–Wolfenstein effect for linear electron density, Harry Lehmann (Hamburg U.), Per Osland (Bergen U.), Tai Tsun Wu (Harvard U. and CERN), DOI: https://doi.org/10.1007/s002200100394.

Harry Lehmann[1,*], Per Osland[2,], Tai Tsun Wu[3,4,***]**

[1] II. Institut für Theoretische Physik der Universität Hamburg, Hamburg, Germany
[2] Department of Physics, University of Bergen, Allégaten 55, 5007 Bergen, Norway
[3] Gordon McKay Laboratory, Harvard University, Cambridge, MA 02138, USA
[4] Theoretical Physics Division, CERN, 1211 Geneva 23, Switzerland

Received: 3 April 2000 / Accepted: 10 April 2000

Abstract: When the electron density is a linear function of distance, it is known that the MSW equations for two neutrino species can be solved in terms of known functions. It is shown here that more generally, for any number of neutrino species, these MSW equations can be solved exactly in terms of single integrals. While these integrals cannot be expressed in terms of known functions, some of their simple properties are obtained. Application to the solar neutrino problem is briefly discussed.

215) Bound states in n dimensions (especially $n = 1$ and $n = 2$), N. N. Khuri (Rockefeller U.), Andre Martin (CERN and Annecy, LAPTH), Tai Tsun Wu (Harvard U. and CERN), DOI: https://doi.org/10.1007/s006010200003.

N. N. Khuri[1], A. Martin[2,3], and T.-T. Wu[2,4]

[1] Department of Physics, The Rockefeller University, New York, NY 10021, USA
[2] Theoretical Physics Division, CERN, CH-1211 Geneva 23, Switzerland
[3] Laboratoire de Physique Théorique ENSLAPP, F-74941 Annecy-le-Vieux, France
[4] Gordon McKay Laboratory, Harvard University, Cambridge, MA 02138-2901, USA

Abstract. We stress that in contradiction with what happens in space dimensions $n \geq 3$, there is no strict bound on the number of bound states with the same structure as the semi-classical estimate for a large coupling constant. We give, in two dimensions, examples of weak potentials with one or infinitely many bound states. We derive bounds for one and two dimensions which have the "right" coupling-constant behaviour for large coupling.

216) Theory and application of Fermi pseudo-potential in one dimension, *J. Math. Phys.* **43**, 5949–5976 (2002), Tai Tsun Wu (Harvard U.); Ming Lun Yu (Applied Materials, Inc), DOI: https://doi.org/10.1063/1.1519940.

Tai Tsun Wu[a)]
Gordon McKay Laboratory, Harvard University, Cambridge, Massachusetts 02138, and Theoretical Physics Division, CERN, Geneva, Switzerland

Ming Lun Yu[b)]
41019 Pajaro Drive, Fremont, California 94539

(Received 17 May 2002; accepted 30 August 2002)

The theory of interaction at one point is developed for the one-dimensional Schrödinger equation. In analog with the three-dimensional case, the resulting interaction is referred to as the Fermi pseudo-potential. The dominant feature of this one-dimensional problem comes from the fact that the real line becomes disconnected when one point is removed. The general interaction at one point is found to be the sum of three terms, the well-known delta-function potential and two Fermi pseudo-potentials, one odd under space reflection and the other even. The odd one gives the proper interpretation for the $\delta'(x)$ potential, while the even one is unexpected and more interesting. Among the many applications of these Fermi pseudo-potentials, the simplest one is described. It consists of a superposition of the delta-function potential and the even pseudo-potential applied to two-channel scattering. This simplest application leads to a model of the quantum memory, an essential component of any quantum computer.
[DOI: 10.1063/1.1519940]

217) Bound states in one and two spatial dimensions, K. Chadan (Orsay, LPT), N. N. Khuri (Rockefeller U.), A. Martin (CERN and Annecy, LAPP), Tai Tsun Wu (Harvard U., Phys. Dept. and CERN), DOI: https://doi.org/10.1063/1.1532538.

K. Chadan
Laboratoire de Physique Théorique, Unité Mixte de Recherche UMR 8627-CNRS, Université de Paris XI, Bâtiment 210, 91405 Orsay Cedex, France

N. N. Khuri
Department of Physics, The Rockefeller University, New York, New York 10021

A. Martin[a)]
TH Division, CERN, Geneva, Switzerland and Laboratoire de Physique Théorique, F-74941 Annecy-le-Vieux, France

Tai Tsun Wu
Gordon McKay Laboratory, Harvard University, Cambridge, Massachusetts 02138 and TH Division, CERN, Geneva, Switzerland

(Received 25 September 2002; accepted 18 October 2002)

In this article we study the number of bound states for potentials in one and two spatial dimensions. We first show that in addition to the well-known fact that an arbitrarily weak attractive potential has a bound state, it is easy to construct examples where weak potentials have an infinite number of bound states. These examples have potentials which decrease at infinity faster than expected. Using somewhat stronger conditions, we derive explicit bounds on the number of bound states in one dimension, using known results for the three-dimensional zero angular momentum. A change of variables which allows us to go from the one-dimensional case to that of two dimensions results in a bound for the zero angular momentum case. Finally, we obtain a bound on the total number of bound states in two dimensions, first for the radial case and then, under stronger conditions, for the noncentral case. © *2003 American Institute of Physics.* [DOI: 10.1063/1.1532538]

218) Bound states in two spatial dimensions in the noncentral case, Andre Martin (CERN and Annecy, LAPP), Tai Tsun Wu (Harvard U. and CERN), DOI: https://doi.org/10.1063/1.1639956.

André Martin[a)]
Theoretical Physics Division, CERN, CH-1211 Geneva 23, Switzerland and Laboratoire de Physique Théorique LAPP, F-74941 Annecy-le-Vieux, France

Tai Tsun Wu
Gordon McKay Laboratory, Harvard University, Cambridge, Massachusetts 02138-2901 and Theoretical Physics Division, CERN, CH-1211 Geneva 23, Switzerland

(Received 17 October 2003; accepted 10 November 2003)

We derive a bound on the total number of negative energy bound states in a potential in two spatial dimensions by using an adaptation of the Schwinger method to derive the Birman–Schwinger bound in three dimensions. Specifically, counting the number of bound states in a potential gV for $g=1$ is replaced by counting the number of g_i's for which zero energy bound states exist, and then the kernel of the integral equation for the zero-energy wave functon is symmetrized. One of the keys of the solution is the replacement of an inhomogeneous integral equation by a homogeneous integral equation. © *2004 American Institute of Physics.*
[DOI: 10.1063/1.1639956]

219) Universality of low-energy scattering in (2+1)-dimensions: The Non-symmetric case, N. N. Khuri (Rockefeller U.), Andrc Martin (Rockefeller U. and CERN), Pierre C. Sabatier (Montpellier U.), Tai Tsun Wu (Harvard U. and CERN), DOI: https://doi.org/10.1063/1.1843274.

N. N. Khuri
Department of Physics, The Rockefeller University, New York, New York 10021
André Martin
Department of Physics, The Rockefeller University, New York, New York 10021 and Department of Physics, CERN, Theory Division, Geneva, Switzerland
Pierre C. Sabatier
Laboratoire de Physique Mathématique, Université de Montpellier II, Sciences et Techniques du Languedoc, F-34095, Montpellier, France
Tai Tsun Wu
Gordon McKay Laboratory, Harvard University, Cambridge, Massachusetts 02138 and Department of Physics, CERN, Theory Division, Geneva, Switzerland

(Received 23 September 2004; accepted 24 September 2004; published online 4 February 2005)

For a very large class of potentials, $V(\vec{x}), \vec{x} \in R^2$, we prove the universality of the low-energy scattering amplitude, $f(\vec{k}', \vec{k})$. The result is $f = \sqrt{\pi/2}(1/\log k) + o(1)/[\log(1/k)]$. The only exceptions occur if V happens to have a zero-energy bound state. Our new result includes as a special subclass the case of rotationally symmetric potentials, $V(|\vec{x}|)$.
[DOI: 10.1063/1.1843274]

220) Low-energy potential scattering in two and three dimensions, N. N. Khuri (Rockefeller U.), Andre Martin (CERN), Jean-Marc Richard (LPSC, Grenoble), Tai Tsun Wu (Harvard U. and CERN), DOI: https://doi.org/10.1063/1.3167803.

N. N. Khuri,[1,a)] André Martin,[2,b)] J.-M. Richard,[3,c)] and Tai Tsun Wu[4,2,d)]
[1]*Department of Physics, The Rockefeller University, New York, New York 10021, USA*
[2]*TH Division, CERNack, CH—1211 Geneva 23, Switzerland*
[3]*Laboratoire de Physique Subatomique et Cosmologie, IN2P3-CNRS, Université Joseph Fourier, INPG, 53, Avenue des Martyrs, 38026 Grenoble, France*
[4]*Gordon McKay Laboratory, Harvard University, Cambridge, Massachusetts 02138-2901, USA*

(Received 16 February 2009; accepted 9 June 2009; published online 20 July 2009)

Conditions are established for the existence of a scattering length and an effective range in the low-energy expansion of the S-wave phase shift of a central potential in two and three dimensions. The behavior of the phase shift as a function of the momentum is also derived for longer-range power-law potentials which do not fulfill these conditions.
[DOI: 10.1063/1.3167803]

Chapter 9
Quantum Field Theory and Statistical Physics (40 papers)

221) Ground State of a Bose System of Hard Spheres, *Phys. Rev.* **115** (1959), 1390–1404, Tai Tsun Wu (Princeton, Inst. Advanced Study), DOI: https://doi.org/10.1103/PhysRev.115.1390; Citations: 260.

TAI TSUN WU*
The Institute for Advanced Study, Princeton, New Jersey, and the Bell Telephone Laboratories, Murray Hill, New Jersey
(Received April 3, 1959)

It is shown that the pseudopotential method can be extended to yield further terms in the low-density expansion of the ground-state energy of a system of Boltzmann or Bose particles with hard-sphere interaction. Two terms beyond the known result are found, and the expansion is no longer a power series in $(a^3\rho)^{\frac{1}{2}}$. Other related properties of the system are discussed.

222) Some Nonequilibrium Properties of a Bose System of Hard Spheres at Extremely Low Temperature, Tai Tsun Wu (Harvard U & University of California), *J. Math. Phys.* 1 January 1961; **2** (1): 105–123. https://doi.org/10.1063/1.1724205.

TAI TSUN WU
Gordon McKay Laboratory, Harvard University, Cambridge, Massachusetts, and Lawrence Radiation Laboratory, University of California, Berkeley, California
(Received July 6, 1960)

The pseudopotential method is used to study a special type of flow for a Bose system of hard spheres. In the first-order approximation, the wave function of the entire system is assumed to be the product of identical single-particle wave functions, which in general are time-dependent. Such a flow is necessarily irrotational, and the single-particle wave function satisfies a Schrödinger equation with a nonlinear self-coupling term. On the basis of this equation of motion, the following properties of the Bose system are discussed: the effect of the rigid wall, the moment of inertia, the compressional wave, and a type of "vortex filament." In the second-order approximation, the wave function of the system is expressed in terms of two functions such that one of them describes the single-particle state suitable for most of the particles while the other one describes the pair excitations. The much more complicated equations of motion are found, but in this approximation the flow is no longer strictly irrotational. The compressional waves are also studied in the second-order approximation.

223) Properties of Normal Thresholds in Perturbation Theory, *Phys. Rev.* **123**, 689 (1961), Tai Tsun Wu (The Institute for Advanced Study, Princeton), DOI: https://doi.org/10.1103/PhysRev.123.689.

TAI TSUN WU*
The Institute for Advanced Study, Princeton, New Jersey
(Received March 1, 1961)

Making use of the relation between a Feynman diagram and the corresponding electric circuit, several properties of the normal thresholds are established.

224) Perturbation Theory of Pion-Pion Interaction. II. Two-Pion Approximation, *Phys. Rev.* **126**, 2219 (1962), Tai Tsun Wu (Harvard U.), DOI: https://doi.org/10.1103/PhysRev.126.2219.

TAI TSUN WU*
Gordon McKay Laboratory, Harvard University, Cambridge, Massachusetts
(Received February 2, 1962)

On the basis of the model of a four-particle direct interaction without derivative coupling, relations are obtained between the various renormalized quantities related to pion-pion interaction. By considering only two-pion intermediate states but preserving crossing symmetry and unitarity up to the production threshold, closed systems of nonlinear integral equations are obtained from these relations to describe approximately the low-energy pion-pion scattering.

225) Effect of Higher Order Weak Interactions on the Production Cross Section of Intermediate Bosons by Neutrinos, *Phys. Rev. Lett.* **11**, 395 (1963), Tai Tsun Wu (Institute for Advanced Study, Princeton), DOI: https://doi.org/10.1103/PhysRevLett.11.395.

Tai Tsun Wu*
Institute for Advanced Study, Princeton, New Jersey
(Received 5 September 1963)

226) Theory of Toeplitz Determinants and the Spin Correlations of the Two-Dimensional Ising Model. I, *Phys. Rev.* **149**, 380–401 (1966), Tai Tsun Wu (Harvard U.), DOI: https://doi.org/10.1103/PhysRev.149.380; Citations: 254.

TAI TSUN WU*
Gordon McKay Laboratory, Harvard University, Cambridge, Massachusetts
(Received 11 April 1966)

We study in detail the asymptotic behavior, for large separations, of the correlation between two spins in the case of the two-dimensional Ising model without magnetic field. In the limit of infinite separation, this correlation is equal to the square of the spontaneous magnetization per spin. This paper is devoted to answering the question how, for fixed temperature, the correlation approaches this limiting value. The investigation is restricted to the situation where the two spins under consideration lie on the same row of the two-dimensional lattice. There are three distinct cases where the fixed temperature is above, below, or equal to the critical temperature or the Curie temperature. In the first two cases, the limiting value is approached exponentially as a function of the separation, while at the critical temperature the correlation behaves asymptotically as the inverse fourth root of the separation. In this paper, we evaluate exactly the first four terms of the asymptotic expansion in the first case, the first three terms in the second case, and the coefficients of the terms proportional to the $-1/4$ and the $-9/4$ powers of the separation in the third case. All these results are obtained by first expressing the correlation as a Toeplitz determinant, and the method used is capable of giving in principle the entire asymptotic expansion for large classes of such determinants, when the size of the determinant approaches infinity. The explicit results for the two-dimensional Ising model also serve as an example where the prescription of summing the leading terms, or the most divergent terms, fails.

Commentary by Barry McCoy

This is the first of a long series of papers published in 1966–1981 on the correlation functions of the two-dimensional Ising model.

This series of papers makes the Ising model the best studied of all solvable models in statistical mechanics. In this first paper Wu invents a method using Wiener Hopf sum equations to compute the long distance asymptotic behavior of the correlation of two spins in the same row for the three cases of $T < T_c$, $T = T_c$ and $T > T_c$.

227) Theory of Toeplitz Determinants and the Spin Correlations of the Two-Dimensional Ising Model. II, *Phys. Rev.* **155** (1967), 438–452, Barry M. McCoy (Harvard U.), Tai Tsun Wu (Brookhaven), DOI: https://doi.org/10.1103/PhysRev.155.438; Citations: 53.

BARRY M. MCCOY*
Lyman Laboratory, Harvard University, Cambridge, Massachusetts
AND
TAI TSUN WU†
Brookhaven National Laboratory, Upton, New York
(Received 18 October 1966)

We study in detail the asymptotic behavior, for large separations, of the correlation between two spins in the same row of the two-dimensional Ising model with the particular value of a purely imaginary magnetic field $H = \frac{1}{2}i\pi kT$. We derive the partition function and the magnetization per spin first proposed by Lee and Yang and find that the correlation function approaches its limiting value exponentially. In contrast to the case of no magnetic field, the approach is shown to be oscillatory rather than monotonic. The first three terms of the asymptotic behavior are explicitly given. The correlation function is obtained by expressing it as a block Toeplitz determinant whose behavior for large separation is studied by solving a pair of coupled Wiener-Hopf sum equations.

Commentary by Barry McCoy

This paper is the beginning of my 15 year collaboration with Tai and is the first part of my PhD thesis. In 1952 C.N. Yang and T.D. Lee computed the free energy of the Ising model in the presence of an imaginary magnetic field $H/kT = \pi i/2$ and conjectured an expression for the magnetization. We proved this conjecture and in addition computed the long distance behavior of the correlations by extending the Wiener–Hopf methods of Tai Tsun Wu's paper *Phys. Rev.* **149**, 380 (1966) from the scalar to a 2×2 matrix case. This is the first case of an explicit solution of a matrix Wiener–Hopf sum equation.

228) Theory of Toeplitz Determinants and the Spin Correlations of the Two-Dimensional Ising Model. III, *Phys. Rev.* **164** (1967), 719–735, Hung Cheng (MIT, LNS), Tai Tsun Wu (Harvard U.), DOI: https://doi.org/10.1103/PhysRev.164.719; Citations: 90.

HUNG CHENG*
Massachusetts Institute of Technology, Cambridge, Massachusetts
AND
TAI TSUN WU†
Rockefeller University, New York, New York
(Received 6 March 1967)

We study the asymptotic behavior, for large separations, of the spin-spin correlation function $\langle\sigma_{0,0}\sigma_{M,N}\rangle$ in the two-dimensional Ising model, where the two spins are not necessarily on the same row. Besides the limiting value for infinite separation, which is the square of the spontaneous magnetization, we evaluate the two leading terms in the asymptotic expression in each of the two cases $T<T_c$ and $T>T_c$. It is found that the nearest singularity of the generating function for the correlation is quite simple in the case $T>T_c$, but much more complicated for $T<T_c$. In an Appendix, we also give exactly in a very simple form the correlation $\langle\sigma_{0,0}\sigma_{N,N}\rangle$ for symmetrical Ising lattice at the critical temperature T_c.

Commentary by Barry McCoy

This paper with H. Cheng extends the computation of Tai Tsun Wu's paper *Phys. Rev.* **149**, 380 (1966) from the row correlation to the correlation of two spins at arbitrary positions on the lattice and demonstrates that at $T = T_c$ the long range correlations are constant on ellipses.

229) Theory of Toeplitz Determinants and the Spin Correlations of the Two-Dimensional Ising Model IV, *Phys. Rev.* **162**, 436 (1967), Barry M. McCoy (Harvard U.), Tai Tsun Wu (Harvard U.), DOI: https://doi.org/10.1103/PhysRev.162.436; Citations: 197.

BARRY M. McCOY*
Lyman Laboratory, Harvard University, Cambridge, Massachusetts
AND
TAI TSUN WU†
AND
Gordon McKay Laboratory, Harvard University, Cambridge, Massachusetts‡ and Rockefeller University, New York, New York
(Received 5 April 1967)

We consider the rectangular Ising model on a half-plane of infinite extent and study some of the consequences connected with the presence of the boundary. Only the spins on the boundary row are allowed to interact with a magnetic field $\mathfrak{H}$. The method of Pfaffians is employed to obtain exact expressions for the partition function. It is found that the free energy is the sum of two terms, one of which is independent of $\mathfrak{H}$ and proportional to the total number of lattice sites, while the other depends on $\mathfrak{H}$ and is proportional to the number of lattice sites on the boundary. This separation makes it possible to define various thermodynamic quantities associated with the boundary. In particular, the boundary magnetization is shown to be discontinuous, in the ferromagnetic case, at zero magnetic field for temperatures below the bulk critical temperature T_c. This discontinuity, which is the spontaneous boundary magnetization, goes to zero as $(1-T/T_c)^{1/2}$ as $T\to T_c-$. For $T=T_c$, the discontinuity is of course absent, and the boundary magnetization behaves as $-\mathfrak{H}\ln\mathfrak{H}$ for small $\mathfrak{H}$. The boundary susceptibility at zero magnetic field in the ferromagnetic case exhibits a logarithmic singularity at $T=T_c$, both above and below transition. An interesting feature is that the ferromagnetic boundary magnetization, although discontinuous for $T<T_c$, may be analytically continued beyond the point $\mathfrak{H}=0$. We interpret this as a hystersis phenomenon which we study in detail by computing the probability distribution function for the average boundary spin. The correlation function for two spins, both on the boundary row, is also obtained exactly and its asymptotic behavior is given. Finally, we derive an expression for the magnetization in any row and explicitly evaluate it for the second row, i.e., the row next to the boundary.

Commentary by Barry McCoy

This paper is the second part of my thesis. We consider the Ising model on a half plane with a magnetic field Hb on the boundary and compute the correlation functions near the boundary as a function of the boundary field. We found that the magnetization on the *j*th row in from the boundary is not discontinuous at $H_b = 0$ but can be analytically continued from $H_b > 0$ to $H_b < 0$.

This is an example of hysteresis.

230) Theory of Toeplitz Determinants and the Spin Correlations of the Two-Dimensional Ising Model. V, *Phys. Rev.* **174** (1968), 546–559, Barry M. McCoy (YITP, Stony Brook), Tai Tsun Wu (Harvard U.), DOI: https://doi.org/10.1103/PhysRev.174.546.

BARRY M. MCCOY
Institute for Theoretical Physics, State University of New York, Stony Brook, New York 11790
AND
TAI TSUN WU*
Gordon McKay Laboratory, Harvard University, Cambridge, Massachusetts 02138
(Received 8 April 1968)

In a previous paper of this series, we studied $\mathfrak{P}(\bar{\sigma})$, the probability that the average boundary spin for a half-plane of Ising spins is $\bar{\sigma}$. We extend our study of such functions by generalizing the previous derivation to any magnetic system and using them to relate the magnetic behavior at the critical isotherm to the spin-spin correlations at the critical temperature. We also clarify the meaning of the secondary maxima previously found in $\mathfrak{P}(\bar{\sigma})$ by a more accurate calculation and by examining the magnetization on interior rows of the half-plane. These considerations show that $\mathfrak{P}(\bar{\sigma})$ is more properly to be considered as the sum of two separate spin probability functions. The various ways in which the thermodynamic limit may be taken and the effects that these different limiting procedures have on $\mathfrak{P}(\bar{\sigma})$ are discussed in detail. The influence of boundary conditions on spin probability functions is studied by reversing the sign of one column of horizontal bonds. We compute the additional free energy resulting from the misfit bonds and use this to show that $\mathfrak{P}(\bar{\sigma})$ is no longer bimodal but is rectangular for $|\bar{\sigma}|$ less than the boundary spontaneous magnetization. Finally, we present in graphical form numerical integrations of the boundary magnetization and susceptibility.

Commentary by Barry McCoy

The hysteresis found in Barry McCoy and Tai Tsun Wu's paper *Phys. Rev.* **162**, 436 (1967) is examined in further detail and it is numerically shown how the analytically continued magnetization in the *j*th row connects with the bulk magnetization at $j \to \infty$.

231) Hydrogen-Bonded Crystals and the Anisotropic Heisenberg Chain, *Il*

Nuovo Cimento **56**, 311 (1968), Barry M. McCoy (Harvard U.), Tai Tsun Wu (Harvard U.), DOI: https://doi.org/10.1007/BF02710156; Citations: 52.

B. M. McCoy (*)

Lyman Laboratory, Harvard University - Cambridge, Mass. (**)

Tai Tsun Wu (***)

Gordon McKay Laboratory, Harvard University, Cambridge, Mass.

(ricevuto il 13 Gennaio 1968)

Summary. — We show that the transfer matrix for Yang's two-dimensional model of a hydrogen-bonded crystal in the presence of an electric field commutes with a generalized form of the Hamiltonian of the anisotropic Heisenberg chain. Thus the connection between the two problems initially found for the ground-state eigenvector is extended to all the eigenvectors.

Commentary by Barry McCoy

By 1968 it was known that the recently solved six vertex model and quantum XXZ spin chain could both be solved by use of Bethe's ansatz. We explained this by demonstrating that the transfer matrix of the six vertex model commuted with the Hamiltonian of the XXZ chain. This has the curious feature that the six vertex transfer matrix has one more parameter than does the XXZ spin chain. Baxter subsequently explained this by exhibiting a parametrization the six vertex model which has a free parameter such that the transfer matrices with two different values of this parameter commute with each other and that there is a special value of the parameter the transfer matrix specializes to the Hamiltonian of the XXZ chain. This leads to the Yang–Baxter equation.

232) Random Impurities as the Cause of Smooth Specific Heats Near the Critical Temperature, *Phys. Rev. Lett.* **21**, 549 (1968), Barry M. McCoy (YITP, Stony Brook), Tai Tsun Wu (Harvard U.), DOI: https://doi.org/10.1103/PhysRevLett.21.549; Citations: 85.

Barry M. McCoy
Institute for Theoretical Physics, State University of New York, Stony Brook, New York 11790

and

Tai Tsun Wu*
Gordon McKay Laboratory, Harvard University, Cambridge, Massachusetts 02138
(Received 1 July 1968)

We present a modification of the two-dimensional Ising model which incorporates random impurities. The specific heat of this model is infinitely differentiable even at the critical temperature where it possesses an essential singularity. We find this specific heat to be in perfect quantitative agreement with the smooth peak recently observed by van der Hoeven, Teaney, and Moruzzi for $T \gtrsim T_c$ in the specific heat of EuS.

Commentary by Barry McCoy

This is a letter announcing the results of Barry McCoy and Tai Tsun Wu's paper *Phys. Rev.* **176**, 631 (1968) and a comparison with experimental data.

233) Theory of a Two-Dimensional Ising Model with Random Impurities .I. Thermodynamics, *Phys. Rev.* **176**, 631 (1968), Barry M. McCoy (YITP, Stony Brook), Tai Tsun Wu (Harvard U.), DOI: https://doi.org/10.1103/PhysRev.176.631; Citations: 373.

Barry M. McCoy
Institute for Theoretical Physics, State University of New York at Stony Brook, Stony Brook, New York 11790
AND
Tai Tsun Wu*
Gordon McKay Laboratory, Harvard University, Cambridge, Massachusetts 02138
(Received 9 July 1968)

Recent experiments demonstrate that at the Curie temperature the specific heat may be a smooth function of the temperature. We propose that this effect can be due to random impurities and substantiate our proposal by a study of an Ising model containing such impurities. We modify the usual rectangular lattice by allowing each row of vertical bonds to vary randomly from row to row with a prescribed probability function. In the case that this probability is a particular distribution with a narrow width, we find that the logarithmic singularity of Onsager's lattice is smoothed out into a function which at T_c is infinitely differentiable but not analytic. This function is expressible in terms of an integral involving Bessel functions and is computed numerically.

Commentary by Barry McCoy

This paper modifies the translationally invariant two dimensional Ising model by allowing the couplings between neighboring rows to be randomly chosen from a probability distribution. This models a system which has quenched random impurities which vary randomly from row to row. Even though the local row to row coupling constants vary from row to row the specific heat is shown to be a probability one object. The specific heat of this quenched randomly layered Ising model smooths out the logarithmic singularity at T_c of the specific heat of the homogeneous Ising into an infinitely differentiable essential singularity. The temperature range where this smoothing occurs is determined from the standard deviation of the probability distribution of the random bonds.

234) Theory of a Two-Dimensional Ising Model with Random Impurities II, Spin Correlation Functions, *Phys. Rev.* **188**, 982 (1969), Barry M. McCoy (YITP, Stony Brook), Tai Tsun Wu (Harvard U.), DOI: https://doi.org/10.1103/PhysRev.188.982; Citations: 120.

Barry M. McCoy
Institute for Theoretical Physics, State University of New York, Stony Brook, New York 11790

AND

Tai Tsun Wu
Gordon McKay Laboratory, Harvard University, Cambridge, Massachusetts 02138
(Received 2 June 1969)

We continue our investigation of an Ising model with immobile random impurities by studying the spin-spin correlation functions. These correlations are not probability-1 objects and have a probability distribution. When the random bonds have the particular distribution function studied in the first paper of this series, we demonstrate that the average value and the second moment of the temperature derivatives of these correlations are infinitely differentiable but fail to be analytic at T_c, the temperature at which the observable specific heat fails to be analytic. When $T<T_c$, we consider $S_\infty(l)=\lim_{m\to\infty}\langle\sigma_{0,0}\sigma_{l,m}\rangle$. This limit is not independent of l. In the special case that the random bonds are symmetrically distributed about the lth row, the geometric mean of $S_\infty(l)$ is computed and shown to vanish exponentially rapidly when $T\to T_c-$. We contrast this with a lower bound that shows that the spontaneous magnetization can vanish no more rapidly than T_c-T, and present a description of how the local magnetization $S_\infty(l)^{\frac{1}{2}}$ behaves as $T\to T_c-$.

Commentary by Barry McCoy

The correlations of two spins in the same row of the quenched randomly layered Ising model are studied. It is shown that these correlations are not probability one objects but instead have a probability distribution.

235) Zero-Field Susceptibility of the Two-Dimensional Ising Model near T_c, *Phys. Rev. Lett.* **31** (1973), 1409–1411, Eytan Barouch (MIT), Barry M. McCoy (Harvard U.), Tai Tsun Wu (Harvard U.), DOI: https://doi.org/10.1103/PhysRevLett.31.1409; Citations: 221.

Eytan Barouch*
Department of Mathematics, Massachusetts Institute of Technology, Cambridge, Massachusetts 02139

and

Barry M. McCoy† and Tai Tsun Wu
Gordon McKay Laboratory, Harvard University, Cambridge, Massachusetts 02138
(Received 9 October 1973)

We compute exactly the four coefficients $C_{0\pm}$ and $C_{1\pm}$ in the expansion
$\beta^{-1}\chi = C_{0\pm}|1 - T_c/T|^{-7/4} + C_{1\pm}|1 - T_c/T|^{-3/4} + O(1)$
for the susceptibility of the rectangular two-dimensional Ising model.

Commentary by Barry McCoy

This letter announces the result, which is derived in Barry McCoy, Tai Tsun Wu, Craig A. Tracy, and Eytan Barouch's paper *Phys. Rev. B* **13**, 316 (1976) that the two spin correlation function in the scaling region can be exactly expressed in terms of Painlevé function of the third kind. Using this exact result the singular behavior of the magnetic susceptibility at T_c is exactly computed.

236) Spin spin correlation functions for the two-dimensional Ising model: Exact theory in the scaling region, *Phys. Rev. B* **13** (1976), 316–374, Tai Tsun Wu (Harvard U.), Barry M. McCoy (SUNY, Stony Brook), Craig A. Tracy (Rochester U.), Eytan Barouch (Clarkson U.), DOI: https://doi.org/10.1103/PhysRevB.13.316; Citations: 568.

Tai Tsun Wu
Division of Engineering and Applied Physics, Harvard University, Cambridge, Massachusetts 02138

Barry M. McCoy*
Institute for Theoretical Physics, State University of New York, Stony Brook, New York 11794

Craig A. Tracy†§
Institute for Fundamental Studies, Department of Physics and Astronomy, University of Rochester, Rochester, New York 14627

Eytan Barouch‡
Department of Mathematics, Clarkson College of Technology, Potsdam, New York 13676
(Received 13 May 1975)

We compute exactly the spin-spin correlation functions $\langle\sigma_{0,0}\sigma_{M,N}\rangle$ for the two-dimensional Ising model on a square lattice in zero magnetic field for $T > T_c$ and $T < T_c$. We then analyze the correlation functions in the scaling limit $T \to T_c$, $M^2+N^2 \to \infty$ such that $(M^2+N^2)^{1/2}(T-T_c)$ is fixed. In this scaling limit $\langle\sigma_{0,0}\sigma_{M,N}\rangle = R^{-1/4}F_\pm(t) + R^{-5/4}F_{1\pm}(t) + o(R^{-5/4})$, where t is the scaling variable R/ξ and $F_\pm(t)$ and $F_{1\pm}(t)$ are the scaling functions (ξ is the correlation length). We derive exact expressions for these scaling functions, in terms of a Painlevé function of the third kind and analyze both the small- and large-t behavior. A table of values for $F_\pm(t)$ (good to ten significant digits) is also given. As an application we compute the coefficients $C_{0\pm}$ and $C_{1\pm}$ in the expansion $k_BT\chi(T) = C_{0\pm}|1 - T_c/T|^{-7/4} + C_{1\pm}|1 - T_c/T|^{-3/4} + O(1)$ of the zero-field susceptibility $\chi(T)$ as $T \to T_c^\pm$.

Commentary by Barry McCoy

This paper derives the Painlevé III result announced in Eytan Barouch, Barry McCoy, and Tai Tsun Wu's paper *Phys. Rev. Lett.* **31**, 1409–1411 (1973).

The first part in this paper extends the leading asymptotic behavior of the Ising two spin correlation derived in Tai Tsun Wu's paper *Phys. Rev.* **164**, 719 (1967) with Cheng to an all orders exact computation. From this all orders computation on the lattice an all orders expression in the scaling limit was derived.

In 1965 John Myers made the remarkable discovery that the scattering of electromagnetic radiation from a finite strip can be expressed in terms of a Painlevé III function. Myers was a student of Wu and Wu recognized that in the scaling limit near T_c the essential methods of Myers could be adapted to the Ising model.

This introduction of Painlevé functions into physics led to the development of the mathematics of isomonodromic deformation theory. In particular M. Jimbo and T. Miwa showed that on the lattice (not in the scaling limit) the correlation of two spins on the diagonal are given in terms of a Painlevé VI function. There is much, however, which to this day is not understood. In particular there should be an isomonodromic deformation expression for the spin correlations in general position off of the diagonal which remains to be discovered. In 1985 Myers found a matrix generalization of Painlevé III which describes the scattering of radiation from a broken corner and this may suggest a further extension to the Ising model.

237) High-Energy Behavior of Two-Body Elastic Amplitudes in Gauge Field Theories, *Stud. Nat. Sci.* **11** (1976), 287–300, B. M. Mccoy (SUNY, Stony Brook), T. T Wu (Harvard U.), DOI: https://doi.org/10.1007/978-1-4684-2925-1_14.

Barry M. McCoy*
Institute for Theoretical Physics
State University of New York at Stony Brook
Stony Brook, New York 11794
and
Tai Tsun Wu† (Presented by Tai Tsun Wu)
Gordon McKay Laboratory
Harvard University
Cambridge, Massachusetts 02138

In this talk, we discuss the high-energy behavior of two-body elastic amplitudes in gauge field theories. We shall be concerned only with the limit:

$$s \to \infty, \tag{1}$$

with t fixed at a physical value.

238) Remarks on the Rotational Invariance of Two-Dimensional Ising Model Correlation Functions, *Phys. Lett. A* **57** (1976), 111, E. Barouch (Clarkson U.), B. M. Mccoy (SUNY, Stony Brook), C. A. Tracy (SUNY, Stony Brook), T. T Wu (Harvard U.), DOI: https://doi.org/10.1016/0375-9601(76)90179-1.

E. BAROUCH
Department of Mathematics, Clarkson College of Technology, Potsdam, New York 13676, USA

B.M. McCOY, C.A. TRACY
The Institute for Theoretical Physics, State University of New York at Stony Brook, Stony Brook, New York 11794, USA

and

T.T. WU
Gordon McKay Laboratory, Harvard University, Cambridge, Massachusetts 02138, USA

Received 21 April 1976

We refute a contention of Bariev concerning the proof of the rotational invariance of Ising model correlation functions.

240) Painlevé Functions of the Third Kind, *J. Math. Phys.* **18** (1977), 1058, Barry M. McCoy (SUNY, Stony Brook), Craig A. Tracy

(SUNY, Stony Brook), Tai Tsun Wu (Harvard U.), DOI: https://doi.org/10.1063/1.523367; Citations: 224.

Barry M. McCoy and Craig A. Tracy

The Institute for Theoretical Physics, State University of New York at Stony Brook, Stony Brook, New York 11794

Tai Tsun Wu

Gordon McKay Laboratory, Harvard University, Cambridge, Massachusetts 02138
(Received 4 October 1976)

We explicitly construct the one-parameter family of solutions, $\eta(\theta;\nu,\lambda)$, that remain bounded as $\theta \to \infty$ along the positive real θ axis for the Painlevé equation of third kind

$$ww'' = (w')^2 - \theta^{-1}ww' + 2\nu\theta^{-1}(w^3 - w) + w^4 - 1,$$

where, as $\theta \to \infty$, $\eta \sim 1 - \lambda\Gamma(\nu + 1/2)2^{-2\nu}\theta^{-\nu-1/2}e^{-2\theta}$. We further construct a representation for $\psi(t;\nu,\lambda) = -\ln[\eta(t/2;\nu,\lambda)]$, where $\psi(t;\nu,\lambda)$ satisfies the differential equation

$$\psi'' + t^{-1}\psi' = (1/2)\sinh(2\psi) + 2\nu t^{-1}\sinh(\psi).$$

The small-θ behavior of $\eta(\theta;\nu,\lambda)$ is described for $|\lambda| < \pi^{-1}$ by

$$\eta(\theta;\nu,\lambda) \sim 2^{\sigma}B\theta^{\sigma}.$$

The parameters σ and B are given as explicit functions of λ and ν. Finally an identity involving the Painlevé transcendent $\eta(\theta;\nu,\lambda)$ is proved. These results for the special case $\nu = 0$ and $\lambda = \pi^{-1}$ make rigorous the analysis of the scaling limit of the spin–spin correlation function of the two-dimensional Ising model.

Commentary by Barry McCoy

This paper gives a different mathematical derivation of the Painlevé III representation of the Ising scaling function based in the series expansion of Barry McCoy, Tai Tsun Wu, Craig A. Tracy, and Eytan Barouch's paper *Phys. Rev. B* **13**, 316 (1976) which does not use Myers methods. We note that the Painlevé III equation is a second order differential equation and we show how to determine the correct boundary conditions for the Ising model by computing a connection formula which computes the behavior at infinity in terms of the initial conditions at zero. This is one of the first such connection functions derived for Painlevé functions.

240) Spin-Spin Correlation Functions for the Two-Dimensional Ising Model, in *Statistical Mechanics and Statistical Methods in Theory and Application*, Barry M. McCoy (SUNY, Stony Brook), Craig A. Tracy (SUNY, Stony Brook), Tai Tsun Wu (Harvard U.), edited by U. Landman (Plenum, N.Y., 1977), p. 83, DOI: https://doi.org/10.1007/978-1-4613-4166-6_5.

BARRY M. McCOY[‡] AND CRAIG A. TRACY

Institute for Theoretical Physics
State University of New York at Stony Brook
Stony Brook, L.I., New York

and

TAI TSUN WU

Division of Engineering and Applied Physics
Harvard University
Cambridge, Massachusetts

We give a simplified derivation of an exact expression for the spin-spin correlation function of the two-dimensional Ising model suitable for studying large spin separation for $T<T_c$.

241) Connection Between the KdV Equation and the Two-Dimensional Ising Model, *Phys. Lett. A* **61** (1977), 283–284, Barry M. McCoy (SUNY, Stony Brook), Craig A. Tracy (SUNY, Stony Brook), Tai Tsun Wu (Harvard U.), DOI: https://doi.org/10.1016/0375-9601(77)90613-2.

Barry M. McCOY, Craig A. TRACY
Institute for Theoretical Physics, State University of New York at Stony Brook, Stony Brook, L.I., New York 11794, USA

and

Tai Tsun WU
Gordan McKay Laboratory, Harvard University, Cambridge, Massachusetts 02138, USA

Received 24 March 1977

We construct the one-parameter family of solutions to $d^2w/dz^2 = zw + 2w^3$ that tend to zero for $z \to +\infty$ by specializing an equation previously solved in connection with the two-dimensional Ising model. These solutions are intimately related to the KdV equation.

242) Two-Dimensional Ising Model as an Exactly Solvable Relativistic Quantum Field Theory: Explicit Formulas for *n* Point Functions, *Phys. Rev. Lett.* **38** (1977), 793–796, Barry M. McCoy (SUNY, Stony Brook), Craig A. Tracy (SUNY, Stony Brook), Tai Tsun Wu (Harvard U.), DOI: https://doi.org/10.1103/PhysRevLett.38.793; Citations: 128.

Barry M. McCoy and Craig A. Tracy
Institute for Theoretical Physics, State University of New York, Stony Brook, New York 11794
and
Tai Tsun Wu
Gordon McKay Laboratory, Harvard University, Cambridge, Massachusetts 02138
(Received 28 February 1977)

For both $T > T_c$ and $T < T_c$, we report exact expressions for the n-spin correlation function of the two-dimensional Ising model suitable for studying large separations between spins. In particular, the scaling limit of these correlation functions can be shown to exist and yields n-point Schwinger functions of some relativistic quantum field theories.

Commentary by Barry McCoy

In this paper the Ising model is interpreted as a relativistic quantum field theory and the series expansions of the two point function found in Barry McCoy, Tai Tsun Wu, Craig A. Tracy, and Eytan Barouch's paper *Phys. Rev. B* **13**, 316 (1976) are extended to the general n-point function.

243) Two-dimensional ising model near T_c: approximation for small magnetic field, *Phys. Rev. B* **18**, 4886 (1978), Barry M. McCoy (SUNY, Stony Brook), Tai Tsun Wu (Utrecht U.), DOI: https://doi.org/10.1103/PhysRevB.18.4886.

Barry M. McCoy
Institute for Theoretical Physics, State University of New York, Stony Brook, New York 11794
Tai Tsun Wu*
Institute for Theoretical Physics, Utrecht, The Netherlands
(Received 3 May 1978)

The study of the equation of state of the two-dimensional Ising model is initiated by use of the recently calculated n-point spin-correlation functions. In the scaling region where $T \to T_c$ and $h = B|1 - T_c/T|^{-15/8} (H/kT_c)$ is of order 1 (B is a lattice-dependent constant) the free energy below T_c is of the form $f(T,h) - f(T,0) = \text{const}(1 - T_c/T)^2 \sum_{n=1}^{\infty} (1/n!) C_n^- h^n$ where the lattice-dependent constant is chosen so that $C_1^- = 1$. The constant C_2^- has been studied previously. In this paper we study C_3^- and C_4^- and find as a first approximation $C_3^-/3! \sim -11/72\pi^2$ and $C_4^-/4! \sim 5/189\pi + 59/162\pi^3$. These are in close agreement with the low-temperature-series values of -0.01538 and 0.0195.

244) Two-dimensional Ising Field Theory for $T < T_c$: String Structure of the Three Point Function, *Phys. Rev. D* **18** (1978), 1243, Barry M. McCoy (SUNY, Stony Brook), Tai Tsun Wu (Utrecht U.), DOI: https://doi.org/10.1103/PhysRevD.18.1243.

Barry M. McCoy
Institute for Theoretical Physics, State University of New York, Stony Brook, New York 11794

Tai Tsun Wu*
Institute for Theoretical Physics, Utrecht, The Netherlands
(Received 14 March 1978)

We use the recently derived expressions for the n-point functions of the two-dimensional Ising field theory for $T < T_c$ to study the three-point function when the separations between the spins is large. We find that the functional form is different depending on whether or not the three spins are collinear. When the three spins are collinear the asymptotic behavior of the connected three-point function is shown to be independent of the location of the middle spin. We interpret these results in terms of a string of correlation.

245) Two-dimensional Ising Field Theory for $T < T_c$: Green's Functions Strings in n-point Functions, *Phys. Rev. D* **18** (1978), 1253, Barry M. McCoy (SUNY, Stony Brook), Tai Tsun Wu (Utrecht U.), DOI: https://doi.org/10.1103/PhysRevD.18.1253.

Barry M. McCoy
Institute for Theoretical Physics, State University of New York, Stony Brook, New York 11794

Tai Tsun Wu*
Institute for Theoretical Physics, Utrecht, The Netherlands
(Received 27 March 1978)

We continue our investigation of the large-separation behavior of the n-point connected Green's functions of the two-dimensional Ising field theory for $T < T_c$ by studying the nearly collinear limit for $n \geq 4$. The results extend the description of the string structure found in the three-point function.

246) Two-dimensional Ising Field Theory in a Magnetic Field: Breakup of the Cut in the Two Point Function, *Phys. Rev. D* **18** (1978), 1259, Barry M. McCoy (SUNY, Stony Brook), Tai Tsun Wu (Utrecht U.), DOI: https://doi.org/10.1103/PhysRevD.18.1259; Citations: 174.

Barry M. McCoy
Institute for Theoretical Physics, State University of New York, Stony Brook, New York 11794

Tai Tsun Wu*
Institute for Theoretical Physics, Utrecht, The Netherlands
(Received 12 April 1978)

We demonstrate that the cut which is present as the leading singularity in the two-point function of the Ising field theory for $T < T_c$ and $H = 0$ breaks up into a sequence of poles for $H \neq 0$. Both the positions and the residues of the low-lying poles are calculated.

Commentary by Barry McCoy

This paper gives the first exact computation illustrating the phenomenon of confinement in quantum field theory.

For $T < T_c$ the Ising two point function has the structure of a 2 particle state instead of the one particle state spectrum it has for $T > T_c$. We use the technical results of Barry McCoy and Tai Tsun Wu's papers: *Phys. Rev. Lett.* **38**, 793 (1977), *Phys. Rev. D* **18**, 1243 (1978) and *Phys. Rev. D* **18**, 1253 (1978) to examine the two point function for $T < T_c$ in the presence of a small magnetic field. We found that the two point function breaks up into an infinite number of single particle excitations. This leads to the scenario that on a path in the T, H plane connecting $T < T_c$, $H = 0$ to $T > T_c$, $H = 0$ the excitation spectrum consists of a finite number of particles which changes as the path is traversed. This is interpreted as saying that the two particle states at $T < T_c$, $H = 0$ are confined for $H > 0$ to produce a series of one particle "bound" states. At $T > T_c$, $H = 0$ the confinement is so strong that only one particle survives. In 1989 Zamalodchikov showed that at $T = T_c$, $H > 0$ there are 8 particles in the spectrum; three of which are below the 2 particle threshold. The study of this scenario of confinement is still under active investigation.

247) Dynamic Mass Generation and the Thirring Model, *Phys. Lett. B* **87** (1979), 50–52, Barry M. McCoy (SUNY, Stony Brook), Tai Tsun Wu (Munich, Max Planck Inst. and Harvard U.), DOI: https://doi.org/10.1016/0370-2693(79)90015-7.

Barry M McCOY and Tai Tsun WU
Max-Planck-Institut fur Physik und Astrophysik, Munich, Fed Rep Germany,
Institute for Theoretical Physics, State University of New York, Stony Brook, NY 11794, USA
and Gordon McKay Laboratory, Harvard University, Cambridge, MA 02138, USA

Received 15 August 1979

A massive phase is shown to exist for the Thirring model with no bare mass, when the coupling constant assumes a critical value

Commentary by Barry McCoy

This paper discusses the the relation of the XXZ model with the massless Thirring model and uses this relation to demonstrate the phenomenon of dynamical mass generation at the value of the coupling constant where the massless Thirring model becomes unstable.

248) Quantum Field Theory and Statistical Mechanics, Barry M. McCoy (SUNY, Stony Brook), Tai Tsun Wu (Harvard U.), in Proceedings of the 1980 Guangzhou Conference on Theoretical Particle Physics, p. 68, Science Press, Beijing, Distributed by van Nostrand Reinhold, N.Y.

249) Nonlinear Partial Difference Equations for the Two-dimensional Ising Model, *Phys. Rev. Lett.* **45** (1980), 675–678, Barry M. McCoy (SUNY, Stony Brook), Tai Tsun Wu (Harvard U.), DOI: https://doi.org/10.1103/PhysRevLett.45.675; Citations: 61.

Barry M. McCoy
The Institute for Theoretical Physics, State University of New York, Stony Brook, New York 11794

and

Tai Tsun Wu
Gordon McKay Laboratory, Harvard University, Cambridge, Massachusetts 02138
(Received 30 April 1980)

The two-point function of the two-dimensional Ising model at arbitrary temperature is expressed in terms of the solution of a nonlinear partial difference equation. From this difference equation the known results for the two-point function of the Ising field theory may be regained as a special case.

PACS numbers: 05.50.+q

Commentary by Barry McCoy

This is the first of the series of Barry McCoy and Tai Tsun Wu's papers: *Phys. Rev. Lett.* **45**, 675 (1980), *Phys. Lett.* **80A**, 159 (1980), *Nucl. Phys. B* **180**, [FS2], 89 (1981), *Phys. Rev. Lett.* **46**, 757 (1981), and Statistical Mechanics of Quarks and Hadrons, H. Satz, ed., North Holland, **176** (1981) which derive bilinear partial difference equations for the Ising correlation functions. Barry McCoy and Tai Tsun Wu's papers: *Phys. Rev. Lett.* **45**, 675 (1980), *Phys. Lett.* **80A**, 159 (1980), and *Nucl. Phys. B* **180**, [FS2], 89 (1981)) consider 2 point functions and Barry McCoy and Tai Tsun Wu's paper *Phys. Rev. Lett.* **46**, 757 (1981) extends these results to the general n point functions. Using these equations all two point correlations can be generated from the two nearest neighbor correlations.

250) Some Properties of the Nonlinear Partial Difference Equation for the Ising Model Correlation Function, *Phys. Lett. A* **80**, 159

(1980), Barry M. McCoy (Marseille, CPT and Harvard U. and SUNY, Stony Brook), Tai Tsun Wu (Marseille, CPT and Harvard U. and SUNY, Stony Brook), DOI: https://doi.org/10.1016/0375-9601(80)90211-X.

Barry M. McCOY [1] and Tai Tsun WU [2]
Faculté des Sciences, Luminy, Université d'Aix-Marseille II, France
Centre de Physique Théorique, Section II, CNRS, Luminy, France
Gordon McKay Laboratory, Harvard University, Cambridge, MA, USA
Institute for Theoretical Physics, State University of New York, Stony Brook, NY, USA

Received 1 August 1980

Some basic properties are obtained for the non-linear partial difference equation underlying the spin–spin correlation function of the square two-dimensional Ising model. In particular, a convergent iterative solution is given, for arbitrary boundary conditions, and a version of maximum modulus theorem is derived.

251) Nonlinear Partial Difference Equations for the Two Spin Correlation Function of the Two-dimensional Ising Model, *Nucl. Phys. B* **180** (1981), 89–115, Barry M. McCoy (Marseille, CPT and Harvard U. and SUNY, Stony Brook), Tai Tsun Wu (Marseille, CPT and Harvard U. and SUNY, Stony Brook), DOI: https://doi.org/10.1016/0550-3213(81)90156-5.

Barry M. McCOY[1] and Tai Tsun WU[2]

Université d'Aix-Marseille II, Faculté des Sciences de Luminy, Centre de Physique Théorique, Section II, CNRS, Marseille, France
Gordon McKay Laboratory, Harvard University, Cambridge, Massachusetts, USA
and
Institute for Theoretical Physics, State University of New York, Stony Brook, New York, USA

Received 11 August 1980

For all temperatures, the two-point function of the two-dimensional Ising model is shown to be expressible in terms of the solution of a non-linear partial difference equation on the lattice. From this difference equation, the known results for the two-point function of the Ising field theory are regained by taking the scaling limit.

252) Non-Linear Partial Difference Equations for the Green's Functions of the Two-Dimensional Ising Model, Statistical Mechanics of Quarks and Hadrons, H. Satz, ed., North Holland, **176** (1981), Barry M. McCoy (SUNY, Stony Brook) and Tai Tsun Wu (Harvard U.).

253) Normalization sum for the Bethe's hypothesis wave functions of the Heisenberg-Ising chain, *Phys. Rev. D* **23** (1981) 2, 417, Michel Gaudin (Saclay), Barry M. McCoy (SUNY, Stony Brook), Tai Tsun Wu (Harvard U.), DOI: https://doi.org/10.1103/PhysRevD.23.417; Citations: 95.

Michel Gaudin
Centre d'Etudes Nucleaire de Saclay, Division de la Physique, Service de Physique Théorique, B.P. No 2-91190 Gif-sur-Yvette, France

Barry M. McCoy
Institute for Theoretical Physics, State University of New York at Stony Brook, Stony Brook, New York 11794

Tai Tsun Wu
Gordon McKay Laboratory, Harvard University, Cambridge, Massachusetts 02138
(Received 13 August 1980)

We calculate the normalization sum for wave functions of the one-dimensional Heisenberg-Ising model of L sites with N overturned spins.

Commentary by Barry McCoy

In order to compute correlation functions in the XXZ model it is necessary to compute scalar products of two Bethe's ansatz wave functions. This paper begins this process by computing the scalar product of a Bethe's ansatz wave function with itself. This begins a very fruitful line of research for XXZ correlations which continues to the present day.

254) Ising model correlation functions: difference equations and applications to gauge theory., B. M. Mccoy (SUNY, Stony Brook), T. T. Wu (Harvard U.), In Kyoto 1981, Proceedings, Non-linear Integrable Systems — Classical Theory and Quantum Theory, 121–134.

255) Ising Field Theory: Quadratic Difference Equations for the n-Point Green's Functions on the Lattice, *Phys. Rev. Lett.* **46** (1981), 757, Barry M. McCoy (SUNY, Stony Brook), Jacques H. H. Perk (SUNY, Stony Brook), Tai Tsun Wu (Harvard U.), DOI: https://doi.org/10.1103/PhysRevLett.46.757; Citations: 58.

Barry M. McCoy and Jacques H. H. Perk
The Institute for Theoretical Physics, State University of New York, Stony Brook, New York 11794

and

Tai Tsun Wu
Gordon McKay Laboratory, Harvard University, Cambridge, Massachusetts 02138
(Received 26 November 1980)

For the two-dimensional Ising model at arbitrary temperature, a system of quadratic difference equations is presented, involving the correlations of n order variables and that of $n-2$ order and two disorder variables. With suitable boundary conditions these equations specify the correlations.

PACS numbers: 05.50.+q

256) On Bose-Einstein condensation, Tai Tsun Wu (Harvard U. and CERN), CERN-TH-96-248 (1996).

257) Bose-Einstein condensation in an external potential. 2, Tai Tsun Wu (Harvard U. and CERN), CERN-TH-96-250 (1996).

258) Bose-Einstein condensation in an external potential. 1, Tai Tsun Wu (Harvard U. and CERN), CERN-TH-96-249 (1996).

259) Bose-Einstein condensation in an external potential at zero temperature: General theory, *Phys. Rev. A* **58**, 1465 (1998), Tai Tsun Wu (Harvard U.), DOI: https://doi.org/10.1103/PhysRevA.58.1465.

Tai Tsun Wu
Gordon McKay Laboratory, Harvard University, Cambridge, Massachusetts 02138-2901
(Received 13 December 1996; revised manuscript received 28 January 1998)

Bose-Einstein condensation is described in terms of the condensate wave function and the pair-excitation function, the latter being responsible for the existence of phonons. This minimal description in terms of these two functions is generalized to the case with an external potential. For a dilute gas with short-range pairwise repulsive interaction and at very low temperatures when the Bose-Einstein condensation is nearly complete, a partial differential equation is obtained for the condensate wave function and an integro-differential equation for the pair excitation. Experimentally, the external potential is used to trap the atoms, i.e., to keep them together. Since the trap is of macroscopic dimensions, the resulting external potential is often slowly varying. In these cases and when the condensate is in the lowest state, the partial differential equation for the condensate wave function and the integro-differential equation for the pair excitation are solved approximately for the case of a time-independent trap. [S1050-2947(98)04208-5]

PACS number(s): 03.75.Fi, 32.80.Pj, 05.20.−y

260) Fermion-boson symmetry and quantum field theory, Sau Lan Wu (Wisconsin U., Madison), Tai Tsun Wu (Harvard U.), Chen Zhou (Wisconsin U., Madison), e-Print: 1812.01988 [physics.gen-ph] (2018).

The application of fermion-boson symmetry to the standard model leads to the following: first, there are three generations of scalar quarks and scalar leptons in addition to the known quarks and leptons, and, secondly, the divergences in the perturbation series for the standard model are reduced. In the light of experimental data from LEP, Tevatron Collider, and LHC, some consequences of these two statements taken together are discussed.

A series of experiments are proposed to search for the scalar quarks and scalar leptons at the Large Hadron Collider. The first step in this search is to look for new fermions by analyzing events with a pair of oppositely changed leptons both with large transverse momenta. The scalar quarks and the scalar leptons are then searched for through their decays into these new fermions plus a known quark or lepton.

Chapter 10
Particle Physics Phenomenology and Higgs Physics (132 papers)

261) Intermediate Bosons: Pair Production Cross Section by Photons, *Phys. Rev. Lett.* **11** (1963), 390–394, Alfred C. T. Wu (MIT and Michigan U.), Tai Tsun Wu (Princeton, Inst. Advanced Study), Kurt Fuchel (Brookhaven), DOI: https://doi.org/10.1103/PhysRevLett.11.390.

Alfred C. T. Wu
Laboratory for Nuclear Sciences, Massachusetts Institute of Technology, Cambridge, Massachusetts*
and Harrison M. Randall Laboratory of Physics, University of Michigan, Ann Arbor, Michigan†

and

Tai Tsun Wu‡
The Institute for Advanced Study, Princeton, New Jersey

and

Kurt Fuchel‖
Applied Mathematics Department, Brookhaven National Laboratory, Upton, New York
(Received 15 July 1963)

262) Uncrossed Ladder Graphs in the Feinberg-Pais Theory of Leptonic Weak Interactions, *Phys. Rev.* **133**, B778 (1964), Yih Pwu (University of Pennsylvania) and Tai Tsun Wu (Princeton, Inst. Advanced Study), DOI: https://doi.org/10.1103/PhysRev.133.B778.

Yih Pwu†
University of Pennsylvania, Philadelphia, Pennsylvania

AND

Tai Tsun Wu*
Institute for Advanced Study, Princeton, New Jersey
(Received 23 July 1963)

The Feinberg-Pais theory of summing the uncrossed ladder graphs is re-examined. A single regularization of the W propagator is used throughout. Instead of the iterative procedure used previously, the leading terms of the individual graphs are summed exactly. The result of Feinberg and Pais for energies much below 300 BeV is obtained for all energies in the limit of infinite regulator mass.

263) Higher Order Leptonic Weak Interactions, *Phys. Rev.* **133**, B1299 (1964), Yih Pwu (University of Pensylvania) & Tai Tsun Wu (Princeton, Inst. Advanced Study), DOI: https://doi.org/10.1103/PhysRev.133.B1299.

YIH PWU*
University of Pennsylvania, Philadelphia, Pennsylvania

AND

TAI TSUN WU†
Institute for Advanced Study, Princeton, New Jersey
(Received 8 October 1963)

Higher order leptonic weak interactions as mediated by a charged intermediate boson are studied on the basis of summing the leading terms in perturbation theory. An important question is whether it is consistent to assume that such a procedure is meaningful and leads to finite results. This is indeed the case for the uncrossed ladder graphs, as shown by Feinberg and Pais. Here an attempt is made to study all higher order graphs. The regularization procedure adopted is the ξ-limiting formalism of Lee and Yang, and the leading terms taken into account are those which, for a given power of the coupling constant, contain the highest power of ξ^{-1} and also the highest possible power of $\ln \xi$ consistent with this maximum power of ξ^{-1}. The sum is determined by a comparison with the electrodynamics of vector mesons on the assumption that it is meaningful to sum the leading terms in both theories. It is then found that, for four-fermion allowed processes at energies much below 300 BeV, no inconsistencies are found and the conventional g^2 should be replaced by $\frac{4}{5}g^2$.

264) Phenomenological Analysis of Violation of CP Invariance in Decay of K0 and anti-K0, *Phys. Rev. Lett.* **13** (1964), 380–385, In Yang, C.N.: Selected Papers 1945–1980, 359–364, and in *Wolfenstein, L. (ed.): CP violation* 36–41, T. T. Wu (Brookhaven), Chen-Ning Yang (Brookhaven), DOI: https://doi.org/10.1103/PhysRevLett.13.380; Citations: 305.

Tai Tsun Wu* and C. N. Yang‡
Brookhaven National Laboratory, Upton, New York
(Received 18 August 1964)

1. It was recently discovered[1] that the long-lived component $K_L{}^0$ of K^0-$\bar{K}^0$ decays into the $\pi^+\pi^-$ mode. Now if CP invariance holds, the $CP = +1$ and $CP = -1$ components of K^0-$\bar{K}^0$ decay independently. The $\pi^+\pi^-$ mode in the S-wave state has $CP = 1$. Hence either the short-lived component $K_S{}^0$, or $K_L{}^0$, does not decay into $\pi^+ + \pi^-$, in contradiction to the new discovery.

Accepting the experimental result of reference 1, one is thus forced to the conclusion that CP invariance is violated in K^0-$\bar{K}^0$ decay, as explicitly stated in reference 1. Notice that this conclusion is independent of the details of the Weisskopf-Wigner formulation[2] of decay amplitudes, as applied to the K_0-$\bar{K}_0$ case by Lee, Oehme, and Yang,[3] whose notation we shall follow.[4] (In particular, small corrections to the exponential decay rule of the formalism cannot alter the conclusion that CP invariance is violated.)

In the present note we shall analyze the decay properties of K^0-$\bar{K}^0$, mostly from the phenomenological viewpoint. Possible further experiments will be discussed for their theoretical significance.

We shall assume CPT invariance, the validity of the Weisskopf-Wigner formulation,[2,3] and that for the strong and electromagnetic interactions, separate C, P, and T invariance hold.

265) Some Considerations on the intermediate-boson hypothesis and CP Non-conservation, *Phys. Rev. Lett.* **20** (1968) 26, 1546, S. Pakvasa (Hawaii U.), Tai-Tsun Wu (Harvard U.), S.F. Tuan (Hawaii U.), DOI: https://doi.org/10.1103/PhysRevLett.20.1546.

S. Pakvasa and S. F. Tuan*
University of Hawaii, Honolulu, Hawaii

and

Tai Tsun Wu
Harvard University, Cambridge, Massachusetts
(Received 20 December 1967)

The consequences of the assumptions that (1) the intermediate boson W has pair strong interactions and (2) $W^{\pm}=a^{\pm}$ particles of Lee (thus leading to CP nonconservation) are studied and found to lead to a consistent picture.

266) Vanishing of total cross sections at infinite energy, *Phys. Rev. Lett.* **22** (1969), 1405–1408, Hung Cheng (MIT, LNS), Tai Tsun Wu (Harvard U.), DOI: https://doi.org/10.1103/PhysRevLett.22.1405.

Hung Cheng*
Department of Mathematics, Massachusetts Institute of Technology, Cambridge, Massachusetts 02139

and

Tai Tsun Wu†
Gordon McKay Laboratory, Harvard University, Cambridge, Massachusetts 02138
(Received 1 May 1969)

We have examined the scattering amplitude of two scalar particles in scalar electrodynamics for arbitrary order of perturbation, and found uncanceled terms proportional to $s(\ln s)^n$, for n any positive integer. This suggests a branch point or an essential singularity at $J=1$, and that total cross sections vanish at infinite energy as some negative power of $\ln s$. This form fits very well the sum of the measured π^+p and π^-p cross sections between 8 and 22 BeV/c.

267) Inelastic electron scattering at high energies, *Phys. Rev. Lett.* **22** (1969), 1409–1411, Hung Cheng (MIT, LNS), Tai Tsun Wu (Harvard U.), DOI: https://doi.org/10.1103/PhysRevLett.22.1409.

Hung Cheng*
Department of Mathematics, Massachusetts Institute of Technology, Cambridge, Massachusetts 02139

and

Tai Tsun Wu
Gordon McKay Laboratory, Harvard University, Cambridge, Massachusetts 02138
(Received 19 March 1969)

Field-theoretic model calculations for inelastic electron scattering show that (1) scaling laws must be weakly violated; (2) there is no evidence for $\sigma_T/\sigma_S=0$ or ∞; and (3) the angular spread in production amplitudes must increase as $|Q^2|$ increases and is roughly proportional to $|Q^2|$ for large $|Q^2|$, where Q is the four-momentum of the virtual photon. Some experimental tests are suggested.

268) Polarization effects in high-energy elastic proton proton scattering, *Phys. Rev. Lett.* **23** (1969), 351–353, Hung Cheng (MIT, LNS), Tai Tsun Wu (Harvard U.), DOI: https://doi.org/10.1103/PhysRevLett.23.351.

Hung Cheng
Department of Mathematics, Massachusetts Institute of Technology, Cambridge, Massachusetts 02139

and

Tai Tsun Wu
Gordon McKay Laboratory, Harvard University, Cambridge, Massachusetts 02138
(Received 12 May 1969)

We propose to scatter unpolarized protons from a proton target polarized in three mutual directions and measure the polarization of the recoil proton in the same direction as the target polarization. The directions of polarization are perpendicular to the scattering plane, in the direction of the recoil proton, and orthogonal to the above two. The results of this measurement can be used to differentiate Regge-pole theory from the impact-factor representation.

269) Cross sections for two-pair production at infinite energy, *Phys. Rev. D* **2** (1970), 2103–2105, Hung Cheng (MIT, LNS), Tai Tsun Wu (Harvard U.), DOI: https://doi.org/10.1103/PhysRevD.2.2103.

Hung Cheng†
Department of Mathematics, Massachusetts Institute of Technology, Cambridge, Massachusetts 02139

AND

Tai Tsun Wu
Gordon McKay Laboratory, Harvard University, Cambridge, Massachusetts 02138
(Received 25 May 1970)

We give here the total cross sections for the following three high-energy processes: (i) $\gamma+\gamma\rightarrow e^++e^++e^-+e^-$, (ii) $\gamma+\gamma\rightarrow\pi^++\pi^++\pi^-+\pi^-$, and (iii) $\gamma+\gamma\rightarrow\pi^++\pi^-+e^++e^-$. Up to the eighth order, all of them approach constants in the high-energy limit. These constants are 6.5, 0.23×10^{-5}, and 0.26×10^{-3} μb, respectively, for (i), (ii), and (iii). The corresponding photon-photon elastic scattering amplitudes in the forward direction are also given. The creation of μ pairs is also briefly discussed.

270) Proposal to Set up a Tagged Photon Facility and to Survey Photon Induced Reactions in 50–300 BeV Region, B. Margolis (McGill U.), P. M. Patel (McGill U.), W. Ross (McGill U.), D. G. Stairs (McGill U.), L. Bird (Carleton U.) *et al.*, FERMILAB-PROPOSAL-0144 (1971).

271) Considerations on measuring gamma gamma → pi+ pi− with colliding beams, *Phys. Lett. B* **36** (1971), 241–245, Hung Cheng (MIT, LNS and Harvard U. and DESY), T. T. Wu (MIT, LNS and Harvard U. and DESY), DOI: https://doi.org/10.1016/0370-2693(71)90078-5.

H. CHENG ** ***
Department of Mathematics, Massachusetts Institute of Technology, Cambridge, Massachusetts 02139, *USA and Deutsches Elektronen-Synchrotron DESY, Hamburg, Germany*

and

T. T. WU ***
Gordon McKay Laboratory, Harvard University, Cambridge, Massachusetts 02138, *USA and Deutsches Elektronen-Synchrotron DESY, Hamburg, Germany*

Received 26 July 1971

Differential cross sections for pionization in electron-electron (or electron-positron) colliding beams are largest when the produced pions have nearly equal but opposite transverse momenta. Measurement in this case can determine the two amplitudes and relative phase for $\gamma + \gamma \to \pi^+ + \pi^-$.

272) Diffractive inelastic processes at extremely high energies, *Phys. Lett. B* **36** (1971), 357–360, Hung Cheng (MIT, LNS), T. T. Wu (Harvard U. and DESY), DOI: https://doi.org/10.1016/0370-2693(71)90725-8.

H. CHENG **‡
Department of Mathematics, Massachusetts Institute of Technology, Cambridge, Massachusetts 02139, *USA*

and

T. T. WU ‡
Gordon McKay Laboratory, Harvard University Cambridge, Massachusetts 02138, *USA* †
and
Deutsches Elektronen-Synchrotron DESY, Hamburg, Germany

Received 9 July 1971

At infinite energy, we predict that for all diffractive processes, initiated by either hadrons or photons, the integrated cross section approaches infinity and dips appear in the angular distribution. We also give results for pionization and discuss the theoretical basis.

273) Production of pion pairs in electron colliding beams, *Nucl. Phys. B* **32** (1971), 461–485, Hung Cheng (MIT, LNS), T. T. Wu (Harvard U. and DESY), DOI: https://doi.org/10.1016/0550-3213(71)90487-1.

Hung CHENG ** †
Department of Mathematics, Massachusetts Institute of Technology
Cambridge, Massachusetts 02139, USA

Tai Tsun WU†
Gordon McKay Laboratory, Harvard University
Cambridge, Massachusetts 02138, USA††
and
Deutsches Elektronen-Synchrotron DESY, Hamburg, Germany

Received 15 June 1971

Abstract: Pionization in electron-electron (or electron-positron) colliding beams is studied in detail for the dominant case where the produced pions have nearly equal but opposite transverse momenta. Accurate measurements in this case, without detecting the outgoing electrons, can be used to determine all amplitudes and phases for $\gamma\gamma \to \pi^+\pi^-$. The results are obtained directly from Feynman diagrams and thus inaccuracies due to Weizsäcker-Williams approximation are avoided. All the formulas are equally applicable to the production of any pair of spinless bosons in electron-electron (or electron-positron) colliding beams.

274) Impact picture and scattering of deuterons. i, *Phys. Rev. D* **6** (1972), 2637–2646, Hung Cheng (MIT, LNS), T. T. Wu (Harvard U.), DOI: https://doi.org/10.1103/PhysRevD.6.2637.

Hung Cheng†
Department of Mathematics, Massachusetts Institute of Technology, Cambridge, Massachusetts 02139

and

Tai Tsun Wu
Gordon McKay Laboratory, Harvard University, Cambridge, Massachusetts 02138
(Received 7 February 1972)

In this paper we apply the theory of the impact picture to the process of scattering of deuterons. We show that the impact-factor representation plays the same role in a complete relativistic scattering process as the Glauber form does in the scattering of a particle on a static deuteron. Therefore, as we use the impact-factor representation for the deuteron scattering amplitude, we include automatically the recoil effects of the deuteron, which must be taken into account when the recoil momentum of the deuteron is not much smaller than the deuteron mass. The main difference between the impact-factor representation and the Glauber form is that in the impact-factor representation, the internal wave function of the outgoing particle is Lorentz-contracted. This also accounts for the difference we found earlier between our results in quantum electrodynamics and the corresponding ones in the droplet model. To obtain the Lorentz-contracted wave function explicitly, we give the Lorentz transformation equations relating the laboratory system and the recoil system, defined to be the system in which the outgoing deuteron is at rest.

275) Electron-positron annihilation into hadrons at high energies, *Phys. Lett. B* **41** (1972), 375–378, Hung Cheng (MIT, LNS), T. T. Wu (Harvard U. and DESY), DOI: https://doi.org/10.1016/0370-2693(72)90599-0.

H. CHENG*
Department of Mathematics, Massachusetts Institute of Technology, Cambridge, Massachusetts 02139, USA

and

T.T. WU**
*Gordon McKay Laboratory, Harvard University ***, Cambridge, Massachusetts 02138, USA, and Deutsches Elektronen-Synchrotron DESY, Hamburg, Germany*

Received 1 August 1972

In the limit of extremely high energies with the transverse momentum of each hadron fixed, the reaction $e^+ + e^- \to$ hadrons proceeds by the annihilation of this pair into two virtual photons which turn into two fireballs of hadrons with negligible final-state interaction.

276) Anomalous magnetic moment of the charged intermediate vector boson, *Phys. Rev. D* **5** (1972), 2978–2998, Hung Cheng (MIT, LNS), T. T. Wu (Harvard U. and DESY), DOI: https://doi.org/10.1103/PhysRevD.5.2978.

Hung Cheng†‡
Department of Mathematics, Massachusetts Institute of Technology, Cambridge, Massachusetts 02139

and

Tai Tsun Wu‡
Gordon McKay Laboratory, Harvard University, Cambridge, Massachusetts§ 02138 and Deutsches Elektronen-Synchrotron, DESY, Hamburg, Germany
(Received 12 January 1972)

Assuming that the intermediate vector boson for weak interactions has no strong interactions, we reinvestigate the Corben-Schwinger problem of the scattering of a charged vector meson with anomalous magnetic moment by an external static Yukawa potential, or in particular a static Coulomb field. Meixner's corner condition of finite integrated energy in every bounded region is applied in the neighborhood of the singularity of the Yukawa potential, and it is found that the Corben-Schwinger problem, although nonrenormalizable, possesses a finite, unique solution if and only if the anomalous magnetic moment is larger than unity. It is therefore a characteristic of nonrenormalizable theories that there may be lower bounds to coupling constants. Experimental determination of the anomalous magnetic moment of the intermediate vector boson is thus of great importance. Furthermore, because certain differential equations happen to be solvable in terms of Bessel functions, an explicit perturbation expansion for weak coupling can be given for this scattering process.

277) High-energy delbrueck scattering from nuclei, *Phys. Rev. D* **5** (1972), 3077–3087, Hung Cheng (MIT, LNS), T. T. Wu (Harvard U. and DESY), DOI: https://doi.org/10.1103/PhysRevD.5.3077.

Hung Cheng†‡
Department of Mathematics, Massachusetts Institute of Technology, Cambridge, Massachusetts 02139

and

Tai Tsun Wu‡
Gordon McKay Laboratory, Harvard University, Cambridge, Massachusetts 02138
and Deutsches Elektronen-Synchrotron (DESY), Hamburg, Germany
(Received 12 January 1972)

We study the elastic scattering of a high-energy photon from a heavy nucleus, considered to be a static Coulomb field due to the charge Ze. Exchange of an arbitrarily large number of photons with the nucleus is taken into account, and the effect of this multiphoton exchange is found to be very large, particularly for momentum transfers which are large compared with the rest mass of the electron. In addition, an interesting theoretical problem in this connection is formulated but unfortunately not solved.

278) High-energy scattering in phi-to-the-third theory and the breakdown of the eikonal approximation. i, *Phys. Rev. D* **5** (1972), 3170–3189, Hung Cheng (MIT, LNS and and DESY), T. T. Wu (Harvard U. and DESY), DOI: https://doi.org/10.1103/PhysRevD.5.3170.

Hung Cheng†
Department of Mathematics, Massachusetts Institute of Technology, Cambridge, Massachusetts 02139§
and Deutsches Elektronen-Synchrotron DESY, Hamburg, Germany

and

Tai Tsun Wu‡
Gordon McKay Laboratory, Harvard University, Cambridge, Massachusetts 02138§
and Deutsches Elektronen-Synchrotron, DESY, Hamburg, Germany
(Received 22 March 1971)

In order to test the general validity of the impact picture which was extracted from quantum electrodynamics at high energies, we make a systematic study of high-energy scattering amplitudes in ϕ^3 theory. The method we use is the one developed in our treatment of quantum electrodynamics, and it enables us to obtain in closed forms the asymptotic amplitudes for multiple exchanges of Regge poles and scalar particles. Our results are nonperturbative and are valid to all orders of the coupling constant. We show in various examples that (i) the longitudinal momentum of the incident particle is always distributed in positive fractions by the particles created; (ii) an exchanged particle, either composite (Regge pole) or elementary, can carry only transverse momentum; (iii) the scattering amplitude satisfies the impact factor representation; (iv) except in the leading order of the coupling constant, the simple-exponentiation form of the eikonal approximation does not appear. The eikonal approximation fails in ϕ^3 theory because of particle creation: More than one particle may share the incident longitudinal momentum and it is impossible to single out a particular particle to be associated with the eikonal path. The above conclusions are reached by a study of the one-ladder amplitude and the amplitude for the double exchange of a ladder and a scalar particle. An integral equation for the Regge parameters of a ladder was derived and many properties of the Regge parameters were given. In particular, in the Wick-Cutkosky model, the Regge pole at zero invariant mass $\alpha(0) = 1$ when the coupling constant g is equal to $4\sqrt{6}\pi m$, where m is the mass of the scalar meson. Thus the physical picture of an expanding black disk in quantum electrodynamics is also realized in the Wick-Cutkosky model as long as $g > 4\sqrt{6}\pi m$. The scattering amplitude for the double exchange of a ladder and a scalar particle is given explicitly in terms of the Regge parameters of the ladder diagrams. Throughout our investigation of ϕ^3 theory at high energies, there is exact correspondence with the results in quantum electrodynamics at every step. This reaffirms our belief of the general validity of the impact picture.

279) Regge poles for large coupling constants. i, *Phys. Rev. D* **5** (1972), 3189–3191, Hung Cheng (MIT, LNS and DESY), T. T. Wu (Harvard U. and DESY), DOI: https://doi.org/10.1103/PhysRevD.5.3189.

Hung Cheng†‡
Department of Mathematics, Massachusetts Institute of Technology, Cambridge, Massachusetts 02139§ and Deutsches Elektronen-Synchrotron, DESY, Hamburg, Germany

and

Tai Tsun Wu‡
Gordon McKay Laboratory, Harvard University, Cambridge, Massachusetts 02138§ and Deutsches Elektronen-Synchrotron, DESY, Hamburg, Germany
(Received 3 May 1971)

We study the problem of scattering by an attractive Yukawa potential. In particular, the positions of Regge poles are found approximately when the coupling constant is large.

280) Regge poles for large coupling constants. ii, *Phys. Rev. D* **5** (1972), 3192–3201, Hung Cheng (MIT, LNS), T. T. Wu (Harvard U. and DESY), DOI: https://doi.org/10.1103/PhysRevD.5.3192.

Hung Cheng†‡
Department of Mathematics, Massachusetts Institute of Technology, Cambridge, Massachusetts 02139

and

Tai Tsun Wu‡
Gordon McKay Laboratory, Harvard University, Cambridge, Massachusetts 02138‖ and Deutsches Elektronen-Synchrotron DESY, Hamburg, Germany
(Received 15 June 1971)

Since we define the Pomeranchon to be the leading singularity of the "potential," and find that a high-energy particle resembles an expanding black disk if this singularity is located at $J>1$, we study as a model the generation of such a singularity by the ladder diagrams in ϕ^3 theory. We find that, when the coupling constant is sufficiently large, there are Regge poles located at $J>1$. Thus the Pomeranchon in this model is a Regge pole with $\alpha(0)>1$. This conclusion is reached by solving the Bethe-Salpeter equation to determine, in the strong coupling limit, the positions of the Regge (or Toller) poles at $t=0$. We find a family of approximately equally spaced Toller poles. The position of the leading singularity, but not the spacing, depends critically on whether the exchanged particle is massless or massive. Some numerical results are presented.

281) Stage I to Proposal #144: A Simple Spectrometer to Measure Photoproduction of Vector Mesons and Pions at High Energy, B. Margolis (McGill U.), P. M. Patel (McGill U.), W. Ross (McGill U.), D. G. Stairs (McGill U.), L. Bird (Carleton U.) *et al.*, DOI: https://doi.org/10.2172/990370 (1972).

282) Impact picture of pp, $\bar{p}p$, $\pi^{\pm}p$, and $K^{\pm}p$ elastic scattering from 20 to 5000 GeV, *Phys. Lett. B* **44** (1973), 97–101, Hung Cheng (MIT, LNS), J. K. Walker (Fermilab), T. T. Wu (Harvard U.), DOI: https://doi.org/10.1016/0370-2693(73)90311-0; Citations: 121.

H. CHENG*
Department of Mathematics, Massachusetts Institute of Technology, Cambridge, Mass. 02139, USA

J.K. WALKER**
Physics Department, National Accelerator Laboratory, Batavia, Ill. 60510, USA

and

T.T. WU***
Gordon McKay Laboratory, Harvard University, Cambridge, Mass. 02138, USA

Received 28 February 1973

We give a simple model of hadron scattering from 20 to 5000 GeV which is based on the impact picture derived from quantum field theory. Numerical results, consistent with existing data are given and in addition we predict the general behavior of high energy hadron-hadron elastic scattering.

283) Ratio of real to imaginary parts for *pp*, *pp*, $\pi^{\pm}p$, and $K^{\pm}p$ forward elastic scattering amplitudes from 50 to 5000 GeV, *Phys. Lett. B* **44** (1973), 283–285, Hung Cheng (MIT), James K. Walker (Fermilab), Tai Tsun Wu (Harvard U.), DOI: https://doi.org/10.1016/0370-2693(73)90227-X.

H. CHENG[☆]
Department of Mathematics, Massachusetts Institute of Technology, Cambridge, Massachusetts 02139, USA

J.K. WALKER[☆☆]
Physics Department, National Accelerator Laboratory, Batavia, Illinois 60510, USA

T.T. WU[☆☆☆]
Gordon McKay Laboratory, Harvard University, Cambridge, Massachusetts 02138, USA

Received 8 March 1973

Without introducing any additional parameter or assumption, we use our previous model for hadron scattering based on the impact picture to predict the ratio of the real to imaginary parts of various forward elastic scattering processes at high energies.

284) Beyond the Rising Total Cross Sections. *AIP Conf. Proc.* 1 June 1973; **15** (1): 54–97, Tai Tsun Wu (Harvard U.) and Hung Cheng (Harvard U.), DOI: https://doi.org/10.1063/1.2947187.

Tai Tsun Wu*
Gordon McKay Laboratory, Harvard University
Cambridge, Massachusetts 02138

Hung Cheng†
Department of Mathematics,
Massachusetts Institute of Technology,
Cambridge, Massachusetts 02138

ABSTRACT

The theoretical reasoning that leads to the prediction of, among others, total cross section rising like $(\ell n\ s)^2$ is reviewed. Briefly, relativistic kinematics dictate the dependence on longitudinal momentum of particles produced in the pionization region, and leads to a factor of $\ell n\ s$ in the integrated cross section. Unitarity in the t-channel then amplifies this $\ell n\ s$ to be a power of s, divided possibly by a power of $\ell n\ s$. This power of s is a realization of the strongly absorptive potential with a coupling constant increasing with energy, and thus the Froissart bound for total cross sections is saturated. Many other results, concerning in particular the phases of the scattering amplitudes in the forward direction and differential cross sections for both elastic and inelastic diffractive scattering, are obtained at the same time.

In order to make quantitative comparison with experimental data, past and future, we describe a simple model of hadron scattering. When applied to the elastic scattering of pp, $\bar{p}p$, $\pi^{\pm}p$, and $K^{\pm}p$, this model describes reasonably well existing data beyond 20GeV. From the resulting parameters, quantitative predictions about these processes are given in detail up to the highest energy that may become available from accelerators in this century, namely 1.5TeV on 1.5TeV. These considerations are also extended to inelastic diffractive scattering without spin-parity exchange.

A number of problems related to increasing total cross sections, some unsolved, are discussed.

(Presented by Tai Tsun Wu)

285) Implication of the impact picture on multiplicity, inclusive distribution, and production processes, *Phys. Lett. B* **45** (1973), 367–370, Hung Cheng (MIT, LNS), T. T. Wu (Harvard U.), DOI: https://doi.org/10.1016/0370-2693(73)90057-9.

H. CHENG*

Department of Mathematics, Massachusetts Institute of Technology, Cambridge, Mass. 02139, USA

and

T.T. WU**

Gordon McKay Laboratory, Harvard University, Cambridge, Mass. 02138, USA

Received 17 April 1973
Revised manuscript received 30 May 1973

We predict that both the average multiplicity and the inclusive distribution for pionization products increase like a positive power of s times a power of $\ln s$. Thus scaling is strongly violated. An experiment involving a fragmentation product is proposed.

286) Beyond the Rising Total Cross-Sections, *AIP Conf. Proc.* **15** (1973), 54–97, Tai Tsun Wu (Harvard U.), Hung Cheng (MIT), DOI: https://doi.org/10.1063/1.2947187.

Tai Tsun Wu*
Gordon McKay Laboratory, Harvard University
Cambridge, Massachusetts 02138

Hung Cheng†
Department of Mathematics,
Massachusetts Institute of Technology,
Cambridge, Massachusetts 02138

ABSTRACT

The theoretical reasoning that leads to the prediction of, among others, total cross section rising like $(\ell n\ s)^2$ is reviewed. Briefly, relativistic kinematics dictate the dependence on longitudinal momentum of particles produced in the pionization region, and leads to a factor of $\ell n\ s$ in the integrated cross section. Unitarity in the t-channel then amplifies this $\ell n\ s$ to be a power of s, divided possibly by a power of $\ell n\ s$. This power of s is a realization of the strongly absorptive potential with a coupling constant increasing with energy, and thus the Froissart bound for total cross sections is saturated. Many other results, concerning in particular the phases of the scattering amplitudes in the forward direction and differential cross sections for both elastic and inelastic diffractive scattering, are obtained at the same time.

In order to make quantitative comparison with experimental data, past and future, we describe a simple model of hadron scattering. When applied to the elastic scattering of pp, $\bar{p}p$, $\pi^{\pm}p$, and $K^{\pm}p$, this model describes reasonably well existing data beyond 20GeV. From the resulting parameters, quantitative predictions about these processes are given in detail up to the highest energy that may become available from accelerators in this century, namely 1.5TeV on 1.5TeV. These considerations are also extended to inelastic diffractive scattering without spin-parity exchange.

A number of problems related to increasing total cross sections, some unsolved, are discussed.

(Presented by Tai Tsun Wu)

287) Impact Picture and Diffractive Dissociation, *Phys. Rev. D* **9** (1974), 749–751, Hung Cheng (MIT, LNS), James K. Walker (Fermilab), Tai Tsun Wu (Harvard U.), DOI: https://doi.org/10.1103/PhysRevD.9.749.

Hung Cheng*
Department of Mathematics, Massachusetts Institute of Technology, Cambridge, Massachusetts 02

James K. Walker†
Physics Department, National Accelerator Laboratory, Batavia, Illinois 60510

Tai Tsun Wu‡
Gordon McKay Laboratory, Harvard University, Cambridge, Massachusetts 02138
(Received 1 August 1973)

Based on the impact picture, we give quantitative predictions on the following three reactions: (a) $p+p\to p+N^*(1470)$, (ii) $\pi+p\to\pi+N^*(1470)$, and (iii) $K+p\to K+N^*(1470)$. Factorization is predicted to be approximately valid except near the dip in reaction (i).

288) Impact Picture Predictions for K^+p Scattering at NAL, Hung Cheng (MIT), James K. Walker (Fermilab), Tai Tsun Wu (Harvard U.), NAL-PUB-74-052-THY-EXP, NAL-PUB-74-52-THY-EXP, FERMILAB-PUB-74-052-T-E (1974).

289) Three-particle Regge pole φ^3 theory, *Phys. Rev. D* **12** (1975), 578, Barry M. McCoy (SUNY, Stony Brook), Tai Tsun Wu (Fermilab), DOI: https://doi.org/10.1103/PhysRevD.12.578.

Barry M. McCoy†
Institute for Theoretical Physics, State University of New York, Stony Brook, New York 11794

Tai Tsun Wu‡
Fermi National Accelerator Laboratory, Batavia, Illinois 60510§
(Received 18 November 1974)

We prove that the location $-2+\alpha_3(\vec{\Delta})$ of the three-particle Regge pole in ϕ^3 theory obeys $\alpha_3(0) > 2.2292714924\alpha(0)$, where $-2+\alpha(0)$ is the position of the Reggeon-particle cut.

290) Possible Interactions of the *J* Particle, *Phys. Rev. Lett.* **34** (1975), 49–52, H. T. Nieh (SUNY, Stony Brook), Tai Tsun Wu (Harvard U.), Chen Ning Yang (SUNY, Stony Brook), DOI: https://doi.org/10.1103/PhysRevLett.34.49.

H. T. Nieh
Institute for Theoretical Physics, State University of New York, Stony Brook, New York 11794

and

Tai Tsun Wu
Gordon McKay Laboratory, Harvard University, Cambridge, Massachusetts 02138

and

Chen Ning Yang
Institute for Theoretical Physics, State University of New York, Stony Brook, New York 11794
(Received 25 November 1974)

We discuss some possible interaction schemes for the newly discovered particle J and their experimental implications, as well as the possible existence of two J^0's like the K_S-K_L case. Of particular interest is the case where the J particle has strong interactions with the hadrons. In this case J can be produced by associated production in hadron-hadron collisions and also singly in relative abundance in ep and μp collisions.

291) Impact Picture and the Ratio of Elastic to Total Cross-Section at ISR Energies, *Phys. Rev. D* **11** (1975), 68, Hung Cheng (MIT), J. K. Walker (Fermilab), T. T. Wu (Harvard U.), DOI: https://doi.org/10.1103/PhysRevD.11.68.

Hung Cheng*
Department of Mathematics, Massachusetts Institute of Technology, Cambridge, Massachusetts 02139

James K. Walker†
Physics Department, Fermi National Accelerator Laboratory, Batavia, Illinois 60510

Tai Tsun Wu‡
Gordon McKay Laboratory, Harvard University, Cambridge, Massachusetts 02138
(Received 24 June 1974)

We show that the impact picture of the Pomeron together with f^0 and ω exchange account for the shrinkage (antishrinkage) of the elastic forward peak for proton-proton (antiproton-proton) scattering for energies greater than 20 GeV. We predict that the ratio of the elastic to total pp cross section rises by 7% in the energy range available at the CERN ISR.

292) Speculations on Quark Observation, *Phys. Lett. B* **72** (1977), 219–223, B. M. McCoy (SUNY, Stony Brook), T. T. Wu (CERN), DOI: https://doi.org/10.1016/0370-2693(77)90706-7.

Barry M. McCOY‡
Institute for Theoretical Physics, State University of New York, Stony Brook, NY, USA

and

Tai Tsun WU‡
CERN, Geneva, Switzerland

Received 25 October 1977

Results of non-perturbative calculations in several models of relativistic quantum field theory lead to the concept of a stable particle whose mass is not well defined. Assuming that quarks are such particles, we propose several possible methods of quark observation.

293) On the Absence of a Second DIP in High-Energy p p Elastic Scattering, C. Bourrely (Marseille, CPT), Jacques Soffer (Marseille, CPT), Tai Tsun Wu (Utrecht U.), CPT-77-P-966, CNRS-CPT-77-P-966 (1977).

294) Impact Picture for Polarization and Rotation Parameter in High-energy Proton Proton Scattering, *Phys. Lett. B* **76** (1978), 481–483, C. Bourrely (Marseille, CPT), Jacques Soffer (Marseille, CPT), Tai Tsun Wu (CERN), DOI: https://doi.org/10.1016/0370-2693(78)90911-5.

C. BOURRELY and J. SOFFER
Centre de Physique Théorique, CNRS, Marseille, France

and

Tai Tsun WU[1]
CERN, Geneva, Switzerland

Received 17 March 1978

We propose to use rigid rotation as an approximation to the hadronic matter current inside a proton. This approximation leads to predictions of the polarization and the R parameter in proton–proton elastic scattering. The angular velocity of rigid rotation may be related to the isoscalar magnetic moment of the nucleon.

295) Impact picture and internal matter motion for elastic scattering, C. Bourrely (Marseille, CPT), Jacques Soffer (Marseille, CPT), Tai Tsun Wu (Utrecht U.), Marseille Colloq.1978:II.41.

296) Indeterminate mass particles, *Phys. Rept.* **49** (1979), 193–197, Barry M. McCoy (SUNY, Stony Brook), Tai Tsun Wu (CERN), DOI: https://doi.org/10.1016/0370-1573(79)90108-X.

Barry M. McCOY*

Institute for Theoretical Physics, State University of New York, Stony Brook, N.Y. 11794, U.S.A.

and

Tai Tsun WU**

CERN, Geneva, Switzerland

***Abstract*:**

We discuss a number of properties of stable particles whose masses are not well defined. Some of these properties can be used for their possible experimental detection.

297) A New Impact Picture for Low and High-Energy Proton Proton Elastic Scattering, *Phys. Rev. D* **19** (1979), 3249, C. Bourrely (Marseille, CPT), Jacques Soffer (Marseille, CPT), Tai Tsun Wu (Utrecht U.), DOI: https://doi.org/10.1103/PhysRevD.19.3249; Citations: 93.

C. Bourrely and J. Soffer
*Centre de Phsique Théorique, Centre National de la Recherche Scientifique, Marseille, France**

Tai Tsun Wu†
Instituut voor Theoretische Fysica, Rijksuniversiteit, Utrecht, The Netherlands
(Received 24 May 1978)

We improve significantly the impact picture that was used several years ago to predict the increase of total and integrated differential cross sections at high energies. The major improvements consist of the following: (1) The dependence of the Pomeron term on the momentum transfer is taken from a modified version of the relation between matter distribution and charge distribution, (2) Regge backgrounds are properly taken into account, and (3) a simple nontrivial form is used for the hadronic-matter current in the proton. For proton-proton elastic scattering, the phenomenological differential cross section is in good agreement with the experimental data in the laboratory momentum range of 14 to 2000 GeV/c, and is predicted for ISABELLE energy. Because of the third improvement, predictions are obtained for both polarization and R parameters for proton-proton elastic scattering.

298) Symmetric treatment of ultrarelativistic $e^+ e^-$ scattering, Frits A. Berends (Leiden U.), R. Gastmans (Leuven U.), Tai Tsun Wu (Marseille, CPT and Harvard U.), KUL-TF-79-022 (1979).

299) Theoretical cross-section of vector boson pair production from photons, *Phys. Rev. D* **21** (1980), 701–705, A. C. T. Wu (Michigan U.), T. T. Wu (Harvard U.), DOI: https://doi.org/10.1103/PhysRevD.21.701.

Alfred C. T. Wu
Department of Physics, University of Michigan, Ann Arbor, Michigan 48109

Tai Tsun Wu
Gordon McKay Laboratory, Harvard University, Cambridge, Massachusetts 02138
(Received 3 July 1979)

We record here a brief analytic account of the calculation of the electromagnetic production cross section of charged-vector-meson pairs. The differential cross section is given in closed form. The total cross section is expressed as a two-dimensional integral to be evaluated numerically.

300) Simplicity of bremsstrahlung cross-sections, D. Danckaert (Leuven U.), P. De Causmaecker (Leuven U.), R. Gastmans (Leuven U.), W. Tro, In Cambridge 1981, Proceedings, Asymptotic Realms Of Physics, 139–146.

301) A class of potentials with extremely narrow resonances. 1. case with discrete rotational symmetry, Alex Grossmann (Marseille, CPT), Tai Tsun Wu (Harvard U.), CPT-81/PE-1291 (1981), DOI: https://doi.org/10.2172/5195270.

Abstract: We consider a quantum-mechanical particle in three dimensions, subject to point interactions (Fermi pseudopotentials) placed on the vertices of a regular plane polygon. If N, the number of vertices, tends to infinity and the distance between two consecutive interaction centers tends to a constant, we show that our system has resonances that tend exponentially fast to the real axis. We discuss several conjectures on the generality of this result and stress its relevance as a simplified model for the Yagi-Uda antenna array of classical electromagnetism.

302) Single Bremsstrahlung Processes in Gauge Theories, *Phys. Lett. B* **103** (1981), 124–128, Frits A. Berends (Leiden U.), R. Kleiss (Leiden U.), P. De Causmaecker (Leuven U.), R. Gastmans (Leuven U.), Tai Tsun Wu (Harvard U.), DOI: https://doi.org/10.1016/0370-2693(81)90685-7; Citations: 391.

F.A. BERENDS and R. KLEISS
Instituut-Lorentz, Leiden, The Netherlands

P. De CAUSMAECKER[1] and R. GASTMANS[2]
Instituut voor Theoretische Fysika, University of Leuven, B-3030 Leuven, Belgium

and

Tai Tsun WU[3]
Gordon McKay Laboratory, Harvard University, Cambridge, MA 02138, USA

Received 14 April 1981

We find that, for QED and for SU(N) gauge theories, single hard bremsstrahlung cross sections are remarkably simple in the ultrarelativistic limit. They can be written as a product of two factors: one factor is connected with the lowest order, elastic process, and the second one is related to the infrared factor which describes soft quantum emission. We present explicit formulae for various processes.

303) Asymptotic behavior at very high-energies for hadron-hadron collisions, *Phys. Rev. D* **26** (1982), 328, T. T. Chou (Georgia U.), Tai Tsun Wu (Harvard U.), Chen Ning Yang (SUNY, Stony Brook), DOI: https://doi.org/10.1103/PhysRevD.26.328.

T. T. Chou
Department of Physics, University of Georgia, Athens, Georgia 30602

Tai Tsun Wu
Gordon McKay Laboratory, Harvard University, Cambridge, Massachusetts 02138

Chen Ning Yang
Institute for Theoretical Physics, State University of New York, Stony Brook, New York 11794
(Received 19 May 1981)

Asymptotic expressions for dip positions and other experimental parameters in hadron-hadron elastic and diffraction dissociation processes are computed for very large total cross section σ_T. The only assumptions are the geometrical picture, and the Yukawa-type shape of the tail of the hadron distribution.

304) Parity Violation in Elastic N N Scattering at High-energies, *J. Phys. G* **8** (1982), L93, P. Chiappetta (Marseille, CPT), Jacques Soffer (Marseille, CPT), Tai Tsun Wu (Provence U. and Harvard U.), DOI: https://doi.org/10.1088/0305-4616/8/7/002.

P Chiappetta†§, J Soffer† and Tai Tsun Wu‡||

† Centre de Physique Théorique, CNRS, F-13288 Marseille Cedex 9, France
‡ Université de Provence and Gordon McKay Laboratory, Harvard University, Cambridge, MA 02138, USA

Received 14 April 1982

Abstract. We calculate the asymmetry in nucleon–nucleon elastic scattering at high energies arising from heavy boson exchange in the standard Weinberg–Salam model. By neglecting the strong interaction enhancement effects, our result for p–n scattering is one order of magnitude higher than for p–p scattering. Enhancement effects increase the p–p asymmetry, which may become comparable with the p–n asymmetry. Our asymmetries, however, are too small if compared with the total cross section asymmetry measured at Argonne at 6 GeV/*c* on a water target.

305) Impact Picture Description of High-energy Elastic Proton Proton Polarization, *Phys. Rev. D* **26** (1982), 1781, C. Bourrely (LBL, Berkeley), H. A. Neal (SUNY, Stony Brook), H. O. Ogren (Indiana U.), Jacques Soffer (Marseille, CPT), Tai Tsun Wu (Harvard U.), DOI: https://doi.org/10.1103/PhysRevD.26.1781.

C. Bourrely*
Lawrence Berkeley Laboratory, Berkeley, California 94720

H. A. Neal
Physics Department, State University of New York, Stony Brook, New York 11794

H. A. Ogren
Physics Department, Indiana University, Bloomington, Indiana 47405

J. Soffer
Centre de Physique Théorique, Centre National de la Recherche Scientifique, 13288 Marseille, France

Tai Tsun Wu
Gordon McKay Laboratory, Harvard University, Cambridge, Massachusetts 02138
(Received 8 December 1981)

A model of rotating matter inside the proton was recently proposed to explain low- and high-energy proton-proton elastic scattering. Predictions made for the polarization at high energy and recent experiments at CERN and Fermilab support this concept of matter current inside a hadron.

306) Multiple Bremsstrahlung in Gauge Theories at High-Energies (II). Single Bremsstrahlung, *Nucl. Phys. B* **206** (1982), 61–89, Frits A. Berends (Leiden U.), R. Kleiss (Leiden U.), P. De Causmaecker (Leuven U.), R. Gastmans (Leuven U.), W. Troost (Leuven U.), Tai Tsun Wu (DESY and Harvard U.), DOI: https://doi.org/10.1016/0550-3213(82)90489-8; Citations: 231.

F A BERENDS and R KLEISS[1]

Instituut-Lorentz, Leiden, The Netherlands

P DE CAUSMAECKER[2], R GASTMANS[3] and W TROOST[4]

Instituut voor Theoretische Fysica, Universiteit of Leuven, B-3030 Leuven, Belgium

Tai Tsun WU[5]

Deutsches Elektronen-Synchrotron DESY, Hamburg, Germany, and Gordon McKay Laboratory, Harvard University, Cambridge, Mass., USA

Received 9 February 1982

We calculate the helicity amplitudes for the QED processes $e^+e^- \to \gamma\gamma\gamma$, $\mu^+\mu^-\gamma$, and $e^+e^-\gamma$, in the limit of vanishing fermion masses This is done by introducing explicit polarization vectors for the radiated photons With the same technique, we also calculate these processes taking into account Z^0 exchange Finally, we present the unpolarized cross sections in which mass terms are retained to describe the radiation of photons in directions parallel to those of the fermions

307) Four-jet production in e^+e^- annihilation, *Phys. Lett. B* **114** (1982), 203–207, D. Danckaert (Leuven U.), P. De Causmaecker (Leuven U.), R. Gastmans (Leuven U.), W. Troost (Leuven U.), Tai Tsun Wu (Harvard U.), DOI: https://doi.org/10.1016/0370-2693(82)90147-2; Citations: 58.

D. DANCKAERT, P. De CAUSMAECKER [1], R. GASTMANS [2], W. TROOST [3]
Institute of Theoretical Physics, University of Leuven, B3030 Leuven, Belgium

and

Tai Tsun WU [4]
Gordon McKay Laboratory, Harvard University, Cambridge, MA 02138, USA

Received 25 February 1982

For $e^+e^- \to q\bar{q}gg$ and $e^+e^- \to q\bar{q}q\bar{q}$, we find simple expressions for the helicity amplitudes, from which the cross section for $e^+e^- \to$ four jets is readily obtained.

308) Radiative Processes in Gauge Theories, CALKUL Collaboration R. Gastmans (Leiden U. and Leuven U. and Harvard U.) *et al.*, In Kazimierz 1982, Proceedings, Elementary Particle Physics, 247–259.

309) Comparison of impact-picture predictions with data from the CERN $p\bar{p}$ collider, *Phys. Lett. B* **121** (1983), 284–286, C. Bourrely (Marseille, CPT and Queen Mary, U. of London), Jacques Soffer (Marseille, CPT), Tai Tsun Wu (DESY), DOI: https://doi.org/10.1016/0370-2693(83)90806-7.

C. BOURRELY, J. SOFFER and Tai Tsun WU [1]
Centre de Physique Théorique, Centre National de la Recherche Scientifique, Marseille, France
Deutsches Elektronen-Synchrotron DESY, Hamburg, Germany
and Physics Department Westfield College, London, England

Received 16 November 1982

Recent data on the total cross section and the elastic slope parameters at small momentum transfers from the UA1 and UA4 Collaborations at the CERN $p\bar{p}$ collider are compared with the impact-picture predictions.

310) High-energy Pion — Proton Elastic Scattering and the Pion Form-factor, *Phys. Lett. B* **132** (1983), 191–194, C. Bourrely (Marseille, CPT and Queen Mary, U. of London), P. Chiappetta (Marseille, CPT), Jacques Soffer (Marseille, CPT), Tai Tsun Wu (DESY), DOI: https://doi.org/10.1016/0370-2693(83)90249-6.

C. BOURRELY [1], P. CHIAPPETTA [2], J. SOFFER and Tai Tsun WU [3]
Centre de Physique Théorique, CNRS – Luminy, Case 907, F-13288 Marseille Cedex 9, France

Received 16 June 1983

Recent data on π p elastic scattering at 200 GeV/*c* from Fermilab allow a further application of the Chou–Yang relation between scattering amplitudes and form factors. Using a carefully selected parametrization of the elastic amplitude, we find that the pion form factor has a *zero* for $|t|$ between 15 and 20 $(\mathrm{GeV})^2$. In order to locate this zero accurately, more experimental data at large momentum transfers are needed.

311) Multiple bremsstrahlung in gauge theories, CALKUL Collaboration, Frits A. Berends (Leiden U.) *et al.*, In Trieste 1983, Proceedings, Radiative Corrections In Su(2)l X U(1), 103–121.

312) Remarks About the Counting of Anti-neutrino Neutrino Generations Using the $e^+e^- \to \gamma Z^0$ Process, *Z. Phys. C* **23** (1984) 295–300 and Hamburg Desy - DESY 83-090 (83,REC.SEP.) 17p, Jochen Bartels (Hamburg U.), A. Fridman (DESY), Tai Tsun Wu (DESY), Andreas S. Schwarz (Hamburg U.), DOI: https://doi.org/10.1007/BF01546198.

J. Bartels[1], A. Fridman[2*], A. Schwarz[3**] and Tai Tsun Wu[2***]
1 II. Institut für Theoretische Physik der Universität D-2000 Hamburg, Federal Republic of Germany
2 Deutsches Elektronen-Synchrotron DESY, D-2000 Hamburg, Federal Republic of Germany
3 I. Institut für Experimentalphysik der Universität D-2000 Hamburg, Federal Republic of Germany

Received 28 November 1983

Abstract. In the present note we further explore the possibility of measuring the number of $\bar{\nu}\nu$ generations by means of $Z^0 \to \bar{\nu}\nu$ decays. We investigate in particular the case where the radiative transitions $e^+e^- \to \gamma Z^0$ are studied on the Z^0 resonance itself. The various background reactions are discussed as well as the statistical significance that can be obtained for an additional $\bar{\nu}\nu$ generation.

313) Multiple Bremsstrahlung in Gauge Theories at High-energies: 3 Finite Mass Effects in Collinear Photon Bremsstrahlung, *Nucl. Phys. B* **239** (1984), 382–394, CALKUL Collaboration Frits A. Berends (Leiden U.) *et al.*, DOI: https://doi.org/10.1016/0550-3213(84)90254-2; Citations: 84.

CALKUL Collaboration

F.A. BERENDS[1], P. DE CAUSMAECKER[2,**], R. GASTMANS[2,†], R. KLEISS[1],
W. TROOST[2,‡] and TAI TSUN WU[3,¶]

[1] *Instituut-Lorentz, Leiden, The Netherlands*
[2] *Instituut voor Theoretische Fysica, University of Leuven, B-3030 Leuven, Belgium*
[3] *Deutsches Electronen-Synchrotron DESY, Hamburg, Germany and Gordon McKay Laboratory, Harvard University, Cambridge MA 02138, USA*

Received 28 December 1983

We present a method for calculating the various spin amplitudes for QED processes in which an arbitrary number of photons is radiated in directions nearly parallel to the fermion directions. This is accomplished by introducing explicit polarization vectors for the photons and by working in the high energy limit, where finite mass effects are treated in leading order.

314) Multiple Bremsstrahlung in Gauge Theories at High-Energies. 4. The Process $e^+ e^- \to \gamma\gamma\gamma\gamma$, *Nucl. Phys. B* **239** (1984), 395–409, CALKUL Collaboration Frits A. Berends (Leiden U.) *et al.*, DOI: https://doi.org/10.1016/0550-3213(84)90255-4; Citations: 82.

CALKUL Collaboration

F.A. BERENDS[1], P. DE CAUSMAECEKER[2,**], R. GASTMANS[2,†], R. KLEISS[1],
W. TROOST[2,‡], and TAI TSUN WU[3,¶]

[1] *Instituut-Lorentz, Leiden, The Netherlands*
[2] *Instituut voor Theoretische Fysica, University of Leuven, B-3030 Leuven, Belgium*
[3] *Deutsches Electronen-Synchrotron DESY, Hamburg, Germany and Gordon McKay Laboratory, Harvard University, Cambridge, MA 02138, USA*

Received 2 January 1984

We calculate the helicity amplitudes and the cross section for the process $e^+e^- \to 4\gamma$ in the high-energy limit. The resulting expressions are presented in a form which allows an easy numerical evaluation. They are valid for the kinematical configurations where at most two photons are emitted in directions nearly parallel to the lepton directions.

315) FERMI Pseudopotentials and Resonances in Arrays, T.T. Wu (Harvard U.), In: Albeverio, S., Ferreira, L. S., Streit, L. (eds.) Resonances — Models and Phenomena. Lecture Notes in Physics, vol. 211. Springer, Berlin, Heidelberg (1984), DOI: https://doi.org/10.1007/3-540-13880-3_80.

Tai Tsun Wu*

Gordon McKay Laboratory, Harvard University
Cambridge, Massachusetts 02138, U.S.A.

Abstract

The method of Fermi pseudopotential for the Schrödinger equation is generalized to the case of the Maxwell's equations. Infinite, linear, uniform arrays of such pseudopotentials are analyzed, and infinitely narrow resonances are found. A list is given of related problems that are not yet solved but are guaranteed to be solvable.

316) Impact Picture Expectations for Very High-Energy Elastic *pp* and *p* Scattering, *Nucl. Phys. B* **247** (1984), 15–28, C. Bourrely (CERN), Jacques Soffer (Marseille, CPT), T.T. Wu (Harvard U.), DOI: https://doi.org/10.1016/0550-3213(84)90369-9; Citations: 112.

C. BOURRELY

CERN, Geneva, Switzerland and Centre de Physique Théorique, CNRS, Marseille, France

J. SOFFER

Centre de Physique Théorique, CNRS, Marseille, France

and

Tai Tsun WU

Harvard University, Cambridge, Massachusetts, USA

Received 7 May 1984

We review the predictions of the impact-picture for the many features which characterize elastic nucleon–nucleon scattering at very high energy. The comparison with recent CERN data up to the SPS collider energy shows the consistency of this theoretical approach and we give some predictions for the next generation of collider machines with the highest energy $\sqrt{s} = 40$ TeV

317) *pp* and *p* Elastic Scattering at $\sqrt{s} = 0.55$ to 40 TeV, *Phys. Rev. Lett.* **54** (1985), 757–759, Claude Bourrely (Marseille, CPT), Jacques Soffer (Marseille, CPT), Tai Tsun Wu (Harvard U.), DOI: https://doi.org/10.1103/PhysRevLett.54.757; Citations: 52.

Claude Bourrely and Jacques Soffer
Centre de Physique Théorique, Centre National de la Recherche Scientifique, Marseille, France

and

Tai Tsun Wu
Harvard University, Cambridge, Massachusetts 02138
(Received 29 October 1984)

The recent data at $\sqrt{s} = 0.546$ TeV from the CERN $p\bar{p}$ collider confirm once more the predictions of the impact picture, which is based on quantum field theory. These predictions on pp and $p\bar{p}$ total and elastic differential cross sections are extended to higher energies up to $\sqrt{s} = 40$ TeV. They are of direct interest for future experiments at CERN, Fermilab, and the superconducting super collider.

PACS numbers: 13.85.Dz, 12.40.Pp

318) Monopole-Fermion and Dyon-Fermion Bound States: 6. Weakly Bound States for the Dyon-Fermion System, *Nucl. Phys. B* **261** (1985), 687–730, Per Osland (Harvard U.), Tai Tsun Wu (Harvard U.), DOI: https://doi.org/10.1016/0550-3213(85)90594-2.

Per OSLAND and Tai Tsun WU

Gordon McKay Laboratory, Harvard University, Cambridge, Massachusetts 02138, USA

Received 13 June 1985

We present analytic, approximate results for dyon-fermion binding energies and the corresponding bound-state wave functions for angular momentum $j \gg Z|eg| + \frac{1}{2}$ The results are valid in the limit of weak binding, $M - E \ll M$, and for small dyon charges, $Z_d Ze^2 \ll 1$, with Z_d and Z the dyon and fermion charges, respectively With Coulomb attraction, $Z_d Z < 0$, the bound states can be classified into two types with opposite radial components of the fermion spin, provided that there is no near degeneracy between the two types Only one of these two types remains bound in the monopole limit, $Z_d \to 0$

319) Multiple Bremsstrahlung in Gauge Theories at High-energies. 5. The Process $e^+e^- \to \mu^+\mu^- \gamma\gamma$, *Nucl. Phys. B* **264** (1986), 243–264 and Leuven Cathol. Univ. - KUL-TF-84-007 (84,REC.JUL.85) 33p, CALKUL Collaboration • Frits A. Berends (Leiden U.) *et al.*, DOI: https://doi.org/10.1016/0550-3213(86)90481-5; Citations: 65.

F.A. BERENDS[1], P. DE CAUSMAECKER[2,*], R. GASTMANS[2,**], R. KLEISS[1],
W. TROOST[2,†] and T.T. WU[3,††]

[1] *Instituut-Lorentz, Leiden, The Netherlands*
[2] *Instituut voor Theoretische Fysica, University of Leuven, B-3030 Leuven, Belgium*
[3] *Gordon McKay Laboratory, Harvard University, Cambridge, MA 02138, USA*

Received 1 July 1985

We calculate the helicity amplitudes and the cross section for the process $e^+e^- \to \mu^+\mu^-\gamma\gamma$ in the high-energy limit. The resulting expressions are presented in a form which allows an easy numerical evaluation. They are also valid for the kinematical configuration where one or two photons are emitted in directions nearly parallel to any lepton direction.

320) Multiple Bremsstrahlung in Gauge Theories at High-energies: 6. The Process The process $e^+e^- \to e^+e^- \gamma\gamma$, *Nucl. Phys. B* **264** (1986), 265–276, CALKUL Collaboration • Frits A. Berends (Leiden U.) *et al.*, DOI: https://doi.org/10.1016/0550-3213(86)90482-7; Citations: 61.

CALKUL Collaboration

F.A. BERENDS[(a)], P. De CAUSMAECKER[(b),(i)], R. GASTMANS[(b),(ii)], R. KLEISS[(a)],
W. TROOST[(b),(iii)] and T.T. WU[(c),(iv)]

[(a)] *Instituut-Lorentz, Leiden, The Netherlands*
[(b)] *Instituut voor Theoretische Fysica, University of Leuven, B-3030 Leuven, Belgium*
[(c)] *Gordon McKay Laboratory, Harvard University, Cambridge, MA 02138, USA*

Received 1 July 1985

We calculate the helicity amplitudes and the cross section for the process $e^+e^- \to e^+e^-\gamma\gamma$ in the high-energy limit. The resulting expressions are presented in a form which allows an easy numerical evaluation. We also present the cross section for configurations where one or two photons are emitted nearly parallel to an incoming or outgoing lepton.

321) A viable axion model, *Phys. Lett. B* **172** (1986), 435–440, R. D. Peccei (DESY), Tai Tsun Wu (DESY), T. Yanagida (DESY), DOI: https://doi.org/10.1016/0370-2693(86)90284-4; Citations: 93.

R.D. PECCEI, Tai Tsun WU [1,2] and T. YANAGIDA [3]

Deutsches Elektronen Synchrotron DESY, D-2000 Hamburg, Fed. Rep. Germany

Received 4 February 1986; revised manuscript received 10 March 1986

It is discussed whether an axion-like excitation can be the source for the monoenergetic positrons observed at GSI. Although a direct extension of the original Peccei–Quinn model is experimentally ruled out, it is possible to construct an alternative model which avoids all previous axion bounds, involving quarkonia decays, K decays, nuclear decays and beam-dump experiments. The model predicts, at some level, the possibility of flavor-changing interactions involving charmed quarks and suggests an appealing regularity for the quark and lepton masses. The expectations of the model for resonant e^+e^- scattering are briefly discussed.

322) *pp* elastic scattering at Fermilab, C. Bourrely (Marseille, CPT), Jacques Soffer (Marseille, CPT), Tai Tsun Wu (Harvard U.), CPT-86-P-1947, CERN-TH-4643-87 (1986).

323) Caustics in a Simple SU(2) Lattice Gauge Theory Model, *Z. Phys. C* **33** (1987), 583, Jochen Bartels (Hamburg U.), Tai Tsun Wu (DESY and Harvard U.), DOI: https://doi.org/10.1007/BF01548272.

J. Bartels [1] and T.T. Wu [2]*

[1] II. Institut für Theoretische Physik, Universität, D-2000 Hamburg, Federal Republic of Germany
[2] Deutsches Elektronen-Synchrotron, D-2000 Hamburg, Federal Republic of Germany, and Gordon McKay Laboratory, Harvard University, Cambridge, MA 02138, USA

Received 29 October 1986

Abstract. This paper describes the beginning of an attempt to study the weak coupling limit of lattice gauge theories, using the Hamiltonian formulation and the semiclassical approximation. The topics are caustics and the behavior of the ground state wave function in the vicinity of caustics. For two very simple $SU(2)$ models (one plaquette, two plaquettes) we demonstrate the existence of caustics, determine their locations and study the peaking behavior of the ground state wave function on them in the limit $g^2 \to 0$.

324) High-energy, large-momentum-transfer processes: Ladder diagrams in ϕ^3 theory (I), *Nucl. Phys. B* **288** (1987), 77–94, Per Osland (Harvard U.), Tai Tsun Wu (DESY and Harvard U.), DOI: https://doi.org/10.1016/0550-3213(87)90209-4.

Per OSLAND*

Gordon McKay Laboratory, Harvard University, Cambridge, Massachusetts 02138, USA

Tai Tsun WU*

Deutsches Elektronen-Synchrotron DESY, D-2000 Hamburg 52, Federal Republic of Germany, and Gordon McKay Laboratory, Harvard University, Cambridge, Massachusetts 02138, USA

Received 23 September 1986

Relativistic quantum field theories may give us useful guidance to understanding high-energy, large-momentum-transfer processes, where the center-of-mass energy is much larger than the transverse momentum transfers, which are in turn much larger than the masses of the participating particles. With this possibility in mind, we study the ladder diagrams in φ^3 theory. In this paper, some of the necessary techniques are developed and applied to the simplest cases of the fourth- and sixth-order ladder diagrams.

325) Cross-Sections for Gluon + Gluon → Heavy Quarkonium + Gluon, *Phys. Lett. B* **184** (1987), 257–260, R. Gastmans (Leuven U.), W. Troost (Leuven U.), Tai Tsun Wu (CERN and Harvard U.), DOI: https://doi.org/10.1016/0370-2693(87)90578-8; Citations: 64.

R. GASTMANS [1], W. TROOST [2]

Institute for Theoretical Physics, University of Leuven, B-3030 Leuven, Belgium

and

Tai Tsun WU [3]

CERN, CH-1211 Geneva, Switzerland,
and Gordon McKay Laboratory, Harvard University, Cambridge, MA 02138, USA

Received 13 November 1986

Using the helicity amplitude formalism, we derive simple cross section formulae for the production of heavy quarkonia through the process $g + g \to (^{2S+1}L_J) + g$ when $L = $ S or P. Within the framework of perturbative QCD, these subprocesses are important for the hadroproduction of heavy quarkonia.

326) High-energy, Large Momentum Transfer Processes: Ladder Diagrams in φ^3 Theory. 2., *Nucl. Phys. B* **288** (1987), 95–130, P. Osland (Harvard U.), T. T. Wu (CERN and Harvard U.), DOI: https://doi.org/10.1016/0550-3213(87)90210-0.

Per OSLAND*

Gordon McKay Laboratory, Harvard University, Cambridge, Massachusetts 02138, USA

Tai Tsun WU*

CERN, Geneva, Switzerland, and Gordon McKay Laboratory, Harvard University, Cambridge, Massachusetts 02138, USA

Received 1 December 1986

The scattering amplitude for the four-rung ladder diagram in φ^3 theory is evaluated at high energies and for large momentum transfers. The result takes the form of $s^{-1}|t|^{-3}$ multiplied by a homogeneous sixth-order polynomial in $\ln s$ and $\ln|t|$. The novel and unexpected feature is that this polynomial is different depending on whether $\ln|t|$ is larger or less than $\frac{1}{2}\ln s$. Thus the asymptotic formula is not analytic at $\ln|t| = \frac{1}{2}\ln s$, although the first five derivatives are continuous.

327) Production of Heavy Quarkonia From Gluons, *Nucl. Phys. B* **291** (1987), 731, San Miniato Hvy.Flavors 1987:0259, R. Gastmans (Leuven U.), W. Troost (Leuven U.), Tai Tsun Wu (CERN and Harvard U.), DOI: https://doi.org/10.1016/0550-3213(87)90493-7; Citations: 103.

R. GASTMANS[1] and W. TROOST[2]

Institute for Theoretical Physics, University of Leuven, B-3030 Leuven, Belgium

Tai Tsun WU[3]

CERN, Geneva, Switzerland
and Gordon McKay Laboratory, Harvard University, Cambridge, MA 01238, USA

Received 11 February 1987

By introducing, in a covariant way, explicit polarization vectors for the gluons, we calculate the various helicity amplitudes for the processes $g + g \to ({}^{2S+1}L_J) + g$, with $L = S$ or P. In this way, we obtain simple formulae for the cross sections of these processes, which, in the framework of the perturbative QCD, are related to the hadroproduction of heavy quarkonia.

328) Complex moments of parton distributions and new sum rules, *Phys. Rev. Lett.* **59** (1987), 2009, C. Bourrely (Marseille, CPT), Jacques Soffer (Marseille, CPT), Tai Tsun Wu (CERN and Harvard U.), DOI: https://doi.org/10.1103/PhysRevLett.59.2009.

C. Bourrely and J. Soffer
Centre de Physique Théorique du Centre National de la Recherche Scientifique, Faculté des Sciences Luminy, 13288 Marseille Cédex 9, France

and

Tai Tsun Wu
CERN, Geneva, Switzerland, and
Gordon McKay Laboratory, Harvard University, Cambridge, Massachusetts 02138
(Received 25 June 1987)

The Q^2 evolution of quark and gluon distributions predicted by perturbative QCD is reexamined by consideration of the complex moments of these distributions. This analytic continuation leads us to derive several interesting new sum rules which depend on the number of flavors.

PACS numbers: 12.38.Bx

329) Relativistic theory of the decay of upsilon into Higgs boson + photon, *Phys. Rev. D* **38** (1988), 164, Goran Faldt (Uppsala U.), Per Osland (Harvard U.), Tai Tsun Wu (Harvard U.), DOI: https://doi.org/10.1103/PhysRevD.38.164.

Göran Fäldt
Gustaf Werners Institut, Box 531, S-751 21 Uppsala, Sweden

Per Osland
Gordon McKay Laboratory, Harvard University, Cambridge, Massachusetts 02138
*and Fysisk Institutt, Allégaten 55, Universitetet i Bergen, N-5007 Bergen, Norway**

Tai Tsun Wu
Gordon McKay Laboratory, Harvard University, Cambridge, Massachusetts 02138
(Received 6 July 1987)

We present a technique for calculating the decay rate for quarkonium→Higgs boson+photon. It is applied to upsilon decay into a Higgs boson of mass in the range 7–9 GeV. Relativistic effects are accounted for by employing Bethe-Salpeter equations to determine the $\Upsilon b\bar{b}$ and $Hb\bar{b}$ vertex functions. The decay rate is controlled by two amplitudes that interfere destructively. It is found that the ratio $\Gamma(\Upsilon \to H\gamma)/\Gamma(\Upsilon \to \mu^+\mu^-)$ is substantially smaller than that of nonrelativistic calculations.

330) Impact picture predictions for *pp* and *pp* elastic scattering at CERN collider, FNAL collider, LHC and SSC., *Z. Phys. C* **37** (1988), 369–375, *Z. Phys. C* **53** (1992) 538. (*Z. Phys. C* **37** (1988) 369), C. Bourrely (Marseille, CPT), Jacques Soffer (Marseille, CPT), T. T. Wu (CERN and Harvard U.), DOI: https://doi.org/10.1007/BF01578130.

C. Bourrely[1], J. Soffer[1], Tai Tsun Wu[2]**
[1] Centre de Physique Théorique*, CNRS, Luminy, Case 907, F-13288 Marseille Cedex 9, France
[2] CERN CH-1211 Geneva 23, Switzerland, and Gordon McKay Laboratory, Harvard University, Cambridge, MA 02138, USA

Received 4 September 1987

Abstract. We present in detail the predictions on $\bar{p}p$ and pp total and elastic differential cross-sections for energies 0.546 to 40 TeV obtained on the basis of the impact picture, including the Coulomb contribution.

331) Beamstrahlung With Fluctuating Charge Density, *Nucl. Phys. B* **314** (1989), 334–346, M. Jacob (CERN), Tai Tsun Wu (CERN and Harvard U.), DOI: https://doi.org/10.1016/0550-3213(89)90155-7.

M. JACOB

CERN, Geneva

Tai Tsun WU*

CERN, Geneva and Gordon McKay Laboratory, Harvard University, USA

Received 31 August 1988

The calculation of bremsstrahlung during bunch-bunch collisions in the quantum regime, relevant to electron-positron colliders in the multi-TeV range, is extended to take into account fluctuations in bunch density.

332) Beamstrahlung for Longitudinally Nonuniform Bunch, *Nucl. Phys. B* **318** (1989), 53–74, M. Jacob (CERN), Tai Tsun Wu (CERN and Harvard U.), DOI: https://doi.org/10.1016/0550-3213(89)90047-3.

M. JACOB

CERN, Geneva, Switzerland

Tai Tsun WU*

CERN, Geneva, Switzerland,
and
Gordon McKay Laboratory, Harvard University, Cambridge, MA, USA

Received 28 October 1988

Bremsstrahlung energy loss by electrons and positrons in very high energy e^+e^- linear colliders are calculated systematically in terms of Feynman diagrams for the case of longitudinally non-uniform bunches. The radiation rate is given as a sum of three terms, which include not only the known leading contributions but also the end and non-uniformity effects.

333) Beamstrahlung in the deep quantum regime, Part. World 1, 1 (1989) pp. 20–23, M. Jacob (CERN), T. T. Wu (CERN and Harvard U.).

M. Jacob and Tai Tsun Wu

Abstract

Electron–positron linear colliders working in the multi-TeV range cannot achieve the very high luminosities which are required, without bunch densities which imply an abundant bremsstrahlung. We discuss this radiation in the deep quantum regime, where such machines are likely to operate. Bringing radiation losses to a tolerable level is an interesting challenge to the design of such linear colliders.

334) Beamstrahlung in High-energy Electron Positron Linear Colliders With Nonuniform Bunches, *Phys. Lett. B* **216** (1989), 442–446, M. Jacob (CERN and Harvard U.), T. T. Wu (CERN and Harvard U.), DOI: https://doi.org/10.1016/0370-2693(89)91147-7.

M. JACOB [a] and Tai Tsun WU [a,b,1]

[a] *CERN, CH-1211 Geneva 23, Switzerland*

[b] *Gordon McKay Laboratory, Harvard University, Cambridge, MA 02138, USA*

Received 20 October 1988

A general method for the calculation of beamstrahlung radiation in bunch–bunch collisions in the case of electron–positron linear colliders in the TeV range is presented. It applies to bunches with varying longitudinal densities and takes edge effects into account. An extensive use of the Mellin transform is made in a Feynman-graph approach.

335) Pair Production in Bunch Crossing, *Phys. Lett. B* **221** (1989), 203–208, M. Jacob (CERN), Tai Tsun Wu (CERN and Harvard U.), DOI: https://doi.org/10.1016/0370-2693(89)91499-8.

M. JACOB

CERN, CH-1211 Geneva 23, Switzerland

and

Tai Tsun WU [1]

CERN, CH-1211 Geneva 23, Switzerland

and Gordon McKay Laboratory, Harvard University, Cambridge, MA 02138, USA

Received 31 January 1989

Electron–positron colliders operating in the multi-TeV range, at the required very high luminosity, will have to contend with an intense beamstrahlung. Serious problems will originate from the energetic electrons and positrons which the radiated photons will produce before they leave the bunch. We calculate the pair production rate by a photon traversing a bunch, focusing here on the contribution proportional to the bunch length, in the deep quantum regime. The probability for pair production is of the same order as that for beamstrahlung. It is considerable.

336) Coherent Pair Production in Linear Colliders, *Nucl. Phys. B* **327** (1989), 285–306, M. Jacob (CERN), Tai Tsun Wu (CERN and Harvard U.), DOI: https://doi.org/10.1016/0550-3213(89)90271-X.

M. JACOB

CERN, Geneva, Switzerland

TAI TSUN WU*

CERN, Geneva, Switzerland
and
Gordon McKay Laboratory, Harvard University, Cambridge, MA, USA

Received 19 May 1989

The production of an electron–positron pair by a photon impinging on a high-density bunch of electrons is calculated in the deep quantum regime, typically found in linear colliders in the multi-TeV range. We obtain separately a leading contribution proportional to the bunch length and logarithmic edge effects. Pair production is found to be considerable.

337) The Decay $Z \to \pi^0\gamma$, *Phys. Lett. B* **232** (1989), 529–532, M. Jacob (CERN), Tai Tsun Wu (CERN and Harvard U.), DOI: https://doi.org/10.1016/0370-2693(89)90454-1.

M. JACOB [a] and Tai Tsun WU [a,b,1]

[a] *CERN, CH-1211 Geneva 23, Switzerland*
[b] *Gordon McKay Laboratory, Harvard University, MA 02138, USA*

Received 15 September 1989

The rate for Z decay into $\pi^0\gamma$ is calculated using the pion dominance of the divergence of the axial vector current, which is intimately related to the chiral anomaly. The branching ratio can be sizeable. The relatively frequent observation of such spectacular events would give a test of PCAC in a new regime. This $Z \to \pi^0\gamma$ is the first example of a new class of Z and W decays.

338) On the Existence of Bound States for a Massive Spin 1 Particle and a Magnetic Monopole, *Phys. Rev. D* **42** (1990), 665–689, Haakon A. Olsen (Norwegian U. Sci. Tech.), Per Osland (Bergen U.), Tai Tsun Wu (Harvard U.), DOI: https://doi.org/10.1103/PhysRevD.42.665.

Haakon A. Olsen
Institute of Physics, University of Trondheim, AVH, N-7055 Dragvoll, Norway

Per Osland
Department of Physics, University of Bergen, Allégaten 55, N-5007 Bergen, Norway

Tai Tsun Wu
Gordon McKay Laboratory, Harvard University, Cambridge, Massachusetts 02138
(Received 18 December 1989)

We analyze the interaction of a magnetic monopole with a massive spin-one particle. The necessary angular momentum algebra, involving vector monopole harmonics, is developed. Physically acceptable states must lead to Hamiltonians whose differences are integrable. While an extra magnetic moment can lead to attraction at large distances, no value of the magnetic moment leads simultaneously to a physically acceptable behavior of the wave function at the origin. For no value of the magnetic moment do bound states exist, unless an additional short-range interaction is introduced.

339) Proton proton and proton — anti-proton elastic scattering at high-energies: Theory, phenomenology, and experiment, T. T. Wu (Harvard U.), In Norman 1990, Proceedings, Beyond the standard model 2, 314–324,

340) Large order behavior of the supersymmetric anharmonic oscillator, *Phys. Rev. D* **42** (1990), 1276–1284, J.J.M. Verbaarschot (SUNY, Stony Brook), Peter C. West (Geneva U. and King's Coll. London), Tai Tsun Wu (CERN and Harvard U.), DOI: https://doi.org/10.1103/PhysRevD.42.1276.

J. J. M. Verbaarschot
Department of Physics, State University of New York at Stony Brook, Stony Brook, New York 11794

P. West
Department of Physics, University of Geneva, CH-1211 Geneva 4, Switzerland
*and Department of Mathematics, King's College, Strand, London WC2R 2LS, United Kingdom**

Tai Tsun Wu
Theory Division, CERN, CH-1211 Geneva 23, Switzerland
*and Gordon McKay Laboratory, Harvard University, Cambridge, Massachusetts 02138**
(Received 15 March 1990)

The large-order behavior of two supersymmetric extensions of the anharmonic oscillator with two supersymmetries is found by explicit calculations. This behavior is also studied for deviations of the model away from supersymmetry. The supersymmetric point is a dividing point in that the behavior at the supersymmetric point and away from it in one direction is of the form $a_0 a^n \Gamma(n+\alpha)$, while in the opposite direction it is of the form $b_0 a^n \Gamma(n+\beta)\sin[(n+\gamma)2\pi/\tau]$.

341) Experimental confirmation of the impact picture predictions for elastic scattering at FNAL collider, *Phys. Lett. B* **252** (1990),

287–289, Claude Bourrely (Marseille, CPT), Jacques Soffer (Marseille, CPT), Tai Tsun Wu (CERN and Harvard U.), DOI: https://doi.org/10.1016/0370-2693(90)90874-6.

Claude Bourrely, Jacques Soffer
Centre de Physique Théorique, CNRS Luminy, Case 907, F-13288 Marseille Cedex 9, France

and

Tai Tsun Wu [1]
CERN, CH-1211, Geneva 23, Switzerland
and Gordon McKay Laboratory, Harvard University, Cambridge, MA 02138, USA

Received 12 September 1990

Theoretical predictions on the basis of the impact picture for $\bar{p}p$ elastic scattering are compared with recent experimental results from FNAL at the center-of-mass energy of 1.8 TeV. At this energy, the theoretical predictions for σ_{tot} and σ_{el}/σ_{tot} are 74.8 mb and 0.230 while the experimental results are 72.1 ± 3.3 mb and 0.23 ± 0.02, respectively. The data also confirm the predictions for the elastic differential cross section.

342) Real part of the anti-p p elastic scattering amplitude at s**(1/2) = 546-GeV, *Mod. Phys. Lett. A* **6** (1991), 2973–2975, *Mod. Phys. Lett. A* **7** (1992), 457 (erratum), Claude Bourrely (Marseille, CPT), Jacques Soffer (Marseille, CPT), Tai Tsun Wu (CERN and Harvard U.), DOI: https://doi.org/10.1142/S021773239100347X.

CLAUDE BOURRELY and JACQUES SOFFER
Centre de Physique Théorique, C.N.R.S. Luminy, Case 907,
13288 Marseille Cedex 9, France

and

TAI TSUN WU*
CERN, CH-1211, Geneva 23, Switzerland
and
Gordon McKay Laboratory, Harvard University, Cambridge, MA 02138, USA

Received 10 June 1991

In order to get an accurate determination of the real part of the $\bar{p}p$ elastic scattering amplitude at $\sqrt{s}$ = 546 GeV, the Genoa–Palaiseau–Roma–Valencia collaboration plans to take experimental data on $d\sigma/dt$ in the Coulomb interference region. We present detailed theoretical predictions for this cross-section. These predictions can be compared directly with the forthcoming data.

343) Beamstrahlung: A Status report with recent results, M. Jacob (CERN), Tai Tsun Wu (CERN and Harvard U.), *Conf. Proc. C* 910725V1

(1991), 60–65, Lepton/Photon Symp.1991:v.1:60–65, In *Geneva 1991, Proceedings, Joint International Lepton-Photon Symposium, vol. 1* 60–65 and CERN Geneva - TH. 6219 (91/08,rec.Dec.) 5 p.

344) Excess of tau$^+$ tau$^-$ V events in Z decays: A Naive look, *Phys. Lett. B* **271** (1991), 425–430, R. Gastmans (Leuven U.), W. Troost (Leuven U.), Tai Tsun Wu (CERN and Harvard U.), DOI: https://doi.org/10.1016/0370-2693(91)90112-4.

R. Gastmans [1,2], W. Troost [3,4]
Institute for Theoretical Physics, University of Leuven, B-3001 Leuven, Belgium

and

Tai Tsun Wu [5]
Theoretical Physics Division, CERN, CH-1211 Geneva 23, Switzerland
and Gordon McKay Laboratory, Harvard University, Cambridge, MA 02138, USA

Received 6 September 1991

The ALEPH Collaboration has analyzed the decays $Z\to \ell^+\ell^-V$ ($V=e^+e^-$, $\mu^+\mu^-$, $\pi^+\pi^-$), where ℓ is a lepton ($\ell=e$, μ, τ). A possible excess of events is observed when ℓ is the τ lepton. We find that the observed kinematic distributions can be reproduced by assuming that (i) a new vector boson S couples vectorially to the τ, and (ii) the vector boson S decays into a V (through virtual Z exchange) and a pair of neutral scalar particles ξ.

345) Multi-photon radiation in quantum beamstrahlung, *Nucl. Phys. B* **371** (1992), 59–78, and CERN Geneva - TH. 6050 (91/03,rec.Jun.) 22 p, M. Jacob (CERN), Tai Tsun Wu (CERN and Harvard U.), DOI: https://doi.org/10.1016/0550-3213(92)90229-5.

M. Jacob
Theory Division, CERN, CH-1211 Geneva 23, Switzerland

Tai Tsun Wu *
Theory Division, CERN, CH-1211 Geneva 23, Switzerland
and
Gordon McKay Laboratory, Harvard University, Cambridge, MA, USA

Received 10 May 1991
(Revised 9 October 1991)
Accepted for publication 18 October 1991

A detailed analysis of the dominant piece of the two-photon yield in electron–positron bunch collisions at high energy is used to obtain a factorization property which gives directly the leading contribution to all orders in the parameter relevant to beamstrahlung in the deep quantum regime. This provides a simplified probabilistic approach to the global radiation yield in beamstrahlung.

346) Radiated energy loss due to beamstrahlung in high-energy electron — positron linear colliders, *Z. Phys. C* **53** (1992), 479–484, M. Jacob (CERN), Tai Tsun Wu (CERN and Harvard U.), G. Zobernig (Wisconsin U., Madison), DOI: https://doi.org/10.1007/BF01625908.

M. Jacob,[1] Tai Tsun Wu,[2,a] G. Zobernig[3,b]

[1] Theoretical Physics Division, CERN, CH-1211 Geneva 23, Switzerland
[2] Theoretical Physics Division, CERN, CH-1211 Geneva 23, Switzerland and Gordon McKay Laboratory, Harvard University, Cambridge, MA 02138, USA
[3] Department of Physics, University of Wisconsin, Madison, WI 53706, USA

Received 2 August 1991

Abstract. Using the recently discovered property of independent photon emissions at fixed impact distance, an integro-differential equation is set up to give the radiated energy due to beamstrahlung to all orders in the expansion parameter $\alpha L_b/\ell_c$. Here L_b is the bunch length and ℓ_c is the coherence length. This integro-differential equation is solved for both the Klein–Gordon case and the physically important Dirac case.

347) Parameters in the electroweak theory: 2. Quadratic divergences, *Z. Phys. C* **55** (1992), 585–592, Per Osland (Bergen U.), Tai Tsun Wu (CERN and Harvard U.), DOI: https://doi.org/10.1007/BF01561295.

Per Osland[1,*] and Tai Tsun Wu[2,**]

[1] Department of Physics, University of Bergen, Allégt. 55, N-5007 Bergen, Norway
[2] CERN, CH-1211 Geneva 23, Switzerland, and Gordon McKay Laboratory, Harvard University, Cambridge, MA 02138, USA

Received 18 February 1992

Abstract. One of the major differences between electroweak theory and the better understood quantum electrodynamics is the appearance of quadratic divergences. We discuss this issue using point-splitting regularization. It is shown that these quadratic divergences vanish if a certain mass relation is satisfied. The relation is linear in the squares of all the masses of the theory.

348) Parameters in the electroweak theory: 3. masses of the top quark and the Higgs boson, *Z. Phys. C* **55** (1992), 593–604, Per Osland (Bergen U.), Tai Tsun Wu (CERN and Harvard U.), DOI: https://doi.org/10.1007/BF01561296.

Per Osland[1,*] and Tai Tsun Wu[2,**]

[1] Department of Physics, University of Bergen, Allégt. 55, N-5007 Bergen, Norway
[2] CERN, CH-1211 Geneva 23, Switzerland and Gordon McKay Laboratory, Harvard University, Cambridge, MA 02138, USA

Received 18 February 1992

Abstract. From the *eeH* coupling in the electroweak theory, it is found that the one-loop correction is finite (i.e., not logarithmically divergent) when a certain quadratic mass relation is satisfied. When this relation is used together with the one given previously, and the lepton masses are neglected compared with those of the W and the Z, it is found that

$m_H^2 = 9m_Z^2 - 6m_W^2$.

349) Parameters in the electroweak theory: 1. Point splitting as regularization, *Z. Phys. C* **55** (1992), 569–584, Per Osland (Bergen U.), Tai Tsun Wu (CERN and Harvard U.), DOI: https://doi.org/10.1007/BF01561294.

Per Osland[1,*] and Tai Tsun Wu[2,**]

[1] Department of Physics, University of Bergen, Allégt. 55, N-5007 Bergen, Norway
[2] CERN, CH-1211 Geneva 23, Switzerland and Gordon McKay Laboratory, Harvard University, Cambridge, MA 02138, USA

Received 18 February 1992

Abstract. This series of papers deals with the problem of finding relations between the parameters of the Standard Model of Glashow, Weinberg and Salam. In this first paper, the method of point splitting, suitably modified, is proposed as a useful regularization. When applied to quantum electrodynamics, this point-split regularization is indeed found to have desirable properties. More specifically, to one loop, the photon is shown to remain massless and the photon-photon scattering amplitude gauge invariant. In the following papers, this point-split regularization is applied to the electroweak theory. In the second paper, a first mass relation is obtained from the quadratic divergences. In the third paper, a second mass relation is found from the Higgs interactions. From these two relations, the masses of the Higgs boson and the top quark are determined theoretically for the Standard Model as $m_H = 190$ GeV and $m_t = 120$ GeV.

350) Theoretical determination of the masses of the Higgs boson and the top quark in the standard model, Per Osland (Bergen U.), Tai Tsun Wu (Harvard U.), BERGEN-1992-02, HARVARD-DAS-92-1 (1992).

351) Predictions for the masses of the Higgs boson and the top quark, P. Osland (Bergen U.), T. T. Wu (Harvard U. and CERN), Trieste Workshop on the Search for New Elementary Particles: Status and Prospects (1992).

352) Masses of the Higgs boson and the top quark in the Standard Model from absence of divergences, *Phys. Lett. B* **291** (1992), 315–319, Per Osland (Bergen U. and CERN), Tai Tsun Wu (CERN and Harvard U.), DOI: https://doi.org/10.1016/0370-2693(92)91051-A.

Per Osland [a,b,1] **and Tai Tsun Wu** [b,c,2]

[a] *Department of Physics, University of Bergen, Allégt. 55, N-5007 Bergen, Norway*
[b] *Theoretical Physics Division, CERN, CH-1211 Geneva 23, Switzerland*
[c] *Gordon McKay Laboratory, Harvard University, Cambridge, MA 02138, USA*

Received 9 July 1992

By purely theoretical arguments involving the absence of certain divergences in the standard model, the masses of the Higgs and the top quark have been found to be related to those of the W and the Z by $m_H^2 = 9m_Z^2 - 6m_W^2$, $m_t^2 = \frac{3}{2}m_Z^2 - m_W^2$, where terms of order m_b^2 have been neglected. The numerical values are $m_H = 190$ GeV and $m_t = 120$ GeV. This value of m_t is within the range favored by existing phenomenology. The Higgs of this mass has a width of 1 GeV, with the major branching ratios of 77%, 22% and 1% into W^+W^-, ZZ and $b\bar{b}$, and is quite easy to detect at SSC and LHC even without enhanced luminosity.

353) Transverse distribution effects in beamstrahlung, *Z. Phys. C* **58** (1993), 279–284, M. Jacob (CERN), Tai Tsun Wu (CERN and Harvard U.), DOI: https://doi.org/10.1007/BF01560346.

M. Jacob[1], Tai Tsun Wu[1,2,★]

[1] Theory Division, CERN, CH-1211 Geneva 23, Switzerland
[2] Gordon McKay Laboratory, Harvard University, Cambridge, MA 02138, USA

Received 23 November 1992

Abstract. Beamstrahlung is studied for bunches with densities that are transversely non-uniform. By treating an interesting special case analytically, one elicits the consequences of the new stationary phase conditions. Focusing effects are found to be relatively unimportant on the global radiated rate. Results are compared with the previously known case of uniform cylindrical bunches.

354) A New formulation of lattice gauge theory with fermions, Ion Olimpiu Stamatescu (Heidelberg, FEST and Heidelberg U. and CERN), Tai Tsun Wu (Harvard U. and CERN), CERN-TH-6631-92 (1993).

355) Comparison of theoretical predictions from the impact picture with the recent ua4 data, *Phys. Lett. B* **315** (1993), 195–197, Claude Bourrely (Marseille, CPT), Jacques Soffer (Marseille, CPT), Tai Tsun Wu (CERN and Harvard U.), DOI: https://doi.org/10.1016/0370-2693(93)90179-L.

Claude Bourrely, Jacques Soffer
Centre de Physique Théorique, CNRS Luminy, Case 907, F-13288 Marseille Cedex 9, France

and

Tai Tsun Wu [1]
CERN, CH-1211 Geneva, Switzerland
and Gordon McKay Laboratory, Harvard University, Cambridge, MA 02138, USA

Received 13 July 1993
Editor: R. Gatto

The theoretical predictions of the proton–antiproton elastic scattering cross section is compared with the recent UA4 data in the Coulomb interference region at the center-of-mass energy of 541 GeV. The agreement is excellent. A plot of the ratio of the experimental data to the theoretical prediction shows an oscillation of about (2–3)%. Further experimental studies are needed to ascertain whether this oscillation is actually present.

356) Expanding protons seen by HERA, Claude Bourrely (Marseille, CPT), *Phys. Lett. B* **339** (1994) 322–324, Jacques Soffer (Marseille, CPT), Tai Tsun Wu (CERN and Harvard U.), DOI: https://doi.org/10.1016/0370-2693(94)90627-0.

Claude Bourrely, Jacques Soffer, Tai Tsun Wu [1,2]
Centre de Physique Théorique - CNRS - Luminy, Case 907, F-13288 Marseille Cedex 9, France [3]

Received 15 August 1994
Editor: R. Gatto

Abstract

We show that the rising total photoproduction cross section recently observed at HERA is entirely consistent with the impact-picture in the theory of expanding protons which was proposed more than twenty years ago. More accurate data, which is expected to be forthcoming soon, will give this prediction a stringent test.

357) Mass relations in the two Higgs doublet model from the absence of quadratic divergences, *Z. Phys. C* **62** (1994), 253–264, Conrad Newton (Harvard U. and CERN), Tai Tsun Wu (Harvard U. and CERN), DOI: https://doi.org/10.1007/BF01560241.

Conrad Newton, Tai Tsun Wu
Gordon McKay Laboratory, Harvard University Cambridge, MA 02138, USA and Theory Division, CERN CH-1211 Geneva 23, Switzerland

Received 20 September 1993

Abstract. By setting the quadratic divergences to zero, four mass relations are obtained for the Standard Model generalized to two Higgs doublets. These four mass relations are obtained most simply in terms of the original fields in the Lagrangian, before spontaneous $SU(2) \times U(1)$ symmetry breaking is applied. Unlike the case of the Standard Model, the Higgs tadpoles of the two Higgs doublet theory do not furnish a complete set of mass relations, giving only three of the four.

358) Can the Higgs boson be detected at the Fermilab Tevatron collider?, *Z. Phys. C* **61** (1994), 421–424, Conrad Newton (Harvard U. and CERN), Per Osland (Bergen U.), Tai Tsun Wu (Harvard U. and CERN), DOI: https://doi.org/10.1007/BF01413180.

Conrad Newton[1,2,*] Per Osland[3,**] Tai Tsun Wu[1,2,*]

[1] Gordon McKay Laboratory, Harvard University, Cambridge, MA 02138, USA
[2] Theoretical Physics Division, CERN, CH-1211 Geneva 23, Switzerland
[3] Department of Physics, University of Bergen, Allégt. 55, N-5007 Bergen, Norway

Received 3 September 1993

Abstract. By theoretical arguments involving the absence of certain divergences in the Standard Model, the masses of the Higgs boson and the top quark have been found to be $m_H = 190$ GeV and $m_t = 120$ GeV. Among the operating accelerators, only the Tevatron collider at Fermilab has sufficient energy to produce the Higgs boson and the top quark. The top quark is already being searched for at the Tevatron collider; it is the purpose of this paper to discuss possible ways of finding the Higgs boson there. The cross section for producing it at Fermilab (1.8 TeV) is about 0.15 pb. With an upgraded luminosity of $\mathscr{L} = 10^{32}$ cm^{-2}s^{-1}, about 150 Higgs particles are produced each year.

359) Lattice fermion formulation with one-sided derivatives, *Nucl. Phys. B Proc. Suppl.* **42** (1995), 838–840. In *Bielefeld 1994, Lattice 94* 838–840, I.O. Stamatescu (Heidelberg, FEST and Heidelberg U.), T. T. Wu (Harvard U.), DOI: https://doi.org/10.1016/0920-5632(95)00397-R.

Ion-Olimpiu Stamatescu [a]and Tai Tsun Wu [b]

[a] FEST, Schmeilweg 5, D-69118 Heidelberg, Germany, and
Inst. Theor. Physik, Universität Heidelberg, D-69120 Heidelberg, Germany

[b] Gordon McKay Laboratory, Harvard University, Cambridge, MA 02138, U. S. A.

We discuss a lattice fermionic formulation based on one-sided differences. The formulation is non-hermitean, has no doubling and requires an average over directions akin to the point splitting regularization. We discuss the weak coupling continuum limit, the hopping parameter expansion and the possibility of numerical analyses.

360) Mass relations in the standard model: The Higgs boson and the top quark masses, Conrad Newton (Harvard U. and CERN), Per Osland (Bergen U.), Tai Tsun Wu (Harvard U. and CERN), Quantum Systems: New Trends and Methods (QS 94), BERGEN-1994-08 (1994).

361) Scattering by spheres at short wavelengths: From Maxwell to HERA, Tai Tsun Wu (CERN and Harvard U.), CERN-TH-95-238 (1995).

362) Quarter century of rising total cross-sections, 7th Rencontres de Blois on Frontiers in Strong Interactions and 6th International Conference on Elastic and Diffractive Scattering, Claude Bourrely (Marseille, CPT), Jacques Soffer (Marseille, CPT), Tai Tsun Wu (Harvard U.), CPT-95-P-3242 (1995).

363) A new variant of symmetry breaking for quark mass matrices, *Phys. Lett. B* **384** (1996), 249–254, Harry Lehmann (Hamburg U.), Conrad Newton (Harvard U. and CERN), Tai Tsun Wu (Harvard U. and CERN), DOI: https://doi.org/10.1016/0370-2693(96)00609-0.

Harry Lehmann [a], Conrad Newton [b,1], Tai Tsun Wu [b,1]

[a] *II. Institute für Theoretische Physik der Universität Hamburg, Hamburg, Germany*

[b] *Gordon McKay Laboratory, Harvard University, Cambridge, Massachusetts 02138, USA and Theoretical Physics Division, CERN, CH 1211 Geneva 23, Switzerland*

Received 19 April 1996
Editor: R. Gatto

Abstract

We propose a definite pattern for the breaking of a discrete flavor symmetry of the quark mass matrices which is caused by the masses of the first two generations of quarks. We discuss the consequences of this proposal for the KM matrix.

364) New singularities in nonrelativistic coupled channel scattering: 1. Second order, *Phys. Rev. D* **56** (1997), 6779–6784, N. N. Khuri

(Rockefeller U.), Tai Tsun Wu (Harvard U. and CERN), DOI: https://doi.org/10.1103/PhysRevD.56.6779.

N. N. Khuri
Department of Physics, The Rockefeller University, New York, New York 10021

Tai Tsun Wu
Gordon McKay Laboratory, Harvard University, Cambridge, Massachusetts 02138-2901
and Theoretical Physics Division, CERN, CH-1211 Geneva 23, Switzerland
(Received 7 May 1997)

New and surprising singularities are found in the forward scattering amplitude for nonrelativistic potential scattering with coupled channels. In the simplest case of two coupled channels, these singularities appear when the energy difference between the two channels is larger than the inverse range of the potential. They are similar to singularities recently discovered by one of us for potential scattering on $R^3 \otimes S^1$.
[S0556-2821(97)03923-4]

PACS number(s): 03.80.+r, 03.65.Nk

365) New singularities in nonrelativistic coupled channel scattering 2. Fourth order, *Phys. Rev. D* **56** (1997), 6785–6797, N. N. Khuri (Rockefeller U.), Tai Tsun Wu (Harvard U. and CERN), DOI: https://doi.org/10.1103/PhysRevD.56.6785.

N. N. Khuri
Department of Physics, The Rockefeller University, New York, New York 10021

Tai Tsun Wu
Gordon McKay Laboratory, Harvard University, Cambridge, Massachusetts 02138-2901
and Theoretical Physics Division, CERN, CH-1211 Geneva 23, Switzerland
(Received 7 May 1997)

We consider a two-channel nonrelativistic potential scattering problem, and study perturbation theory in fourth order for the forward amplitude. The main result is that the new singularity demonstrated in second order in the preceding paper I also occurs at the same point in fourth order. Its strength is again that of a pole.
[S0556-2821(97)04023-X]

PACS number(s): 03.80.+r, 03.65.Nk

366) $\gamma\gamma$ Total cross-section at high energies, *Mod. Phys. Lett.* **A15**:9–13 (2000), C. Bourrely (Marseille, CPT), Jacques Soffer (Marseille, CPT), Tai Tsun Wu (CERN and Harvard U.), DOI: https://doi.org/10.1142/S0217732300000037.

C. BOURRELY and J. SOFFER
Centre de Physique Théorique, CNRS, Luminy Case 907, F-13288 Marseille Cedex 09, France*

TAI TSUN WU
Gordon McKay Laboratory, Harvard University, Cambridge, MA 02138, USA

Received 3 October 1999

We show that the rising total cross-sections $\sigma(\gamma\gamma \to \text{hadrons})$ recently observed by the L3 and OPAL collaborations at LEP are fully consistent with the impact-picture for high-energy scattering. The impact picture is then used to predict this $\gamma\gamma$ total cross-section at higher energies, and confirm the universal increase of total cross-sections including those of pp, $\bar{p}p$ and γp.

367) Solar Mikheev–Smirnov–Wolfenstein effect with three generations of neutrinos, *Phys. Rev. D* **62**, 013008 (2000), Per Osland (Bergen U.), Tai Tsun Wu (Harvard U. and CERN), DOI: https://doi.org/10.1103/PhysRevD.62.013008.

Per Osland
Department of Physics, University of Bergen, Allégaten 55, N-5007 Bergen, Norway

Tai Tsun Wu
Gordon McKay Laboratory, Harvard University, Cambridge, Massachusetts 02138
and Theoretical Physics Division, CERN, CH-1211 Geneva 23, Switzerland
(Received 29 December 1999; published 7 June 2000)

Under the assumption that the density variation of the electrons can be approximated by an exponential function, the solar Mikheyev-Smirnov-Wolfenstein effect is treated for three generations of neutrinos. The generalized hypergeometric functions that result from the exact solution of this problem are studied in detail, and a method for their numerical evaluation is presented. This analysis plays a central role in the determination of neutrino masses, not only the differences of their squares, under the assumption of universal quark-lepton mixing.

PACS number(s): 14.60.Pq, 13.15.+g

368) Universal quark lepton mixing and determination of neutrino masses, Per Osland (Bergen U.), Tai Tsun Wu (Harvard U. and CERN), e-Print: hep-ph/0006185 [hep-ph] (2000).

Per Osland[a,1] and Tai Tsun Wu[b,2]

[a]*Department of Physics, University of Bergen, Allégaten 55, N-5007 Bergen, Norway*

[b]*Gordon McKay Laboratory, Harvard University, Cambridge, Massachusetts 02138, U.S.A.*
and
Theoretical Physics Division, CERN, CH-1211 Geneva 23, Switzerland

Abstract

If three right-handed neutrinos are added to the Standard Model, then, for the three known generations, there are six quarks and six leptons. It is then natural to assume that the symmetry considerations that have been applied to the quark matrices are also valid for the lepton mass matrices. Under this assumption, the solar and atmospheric neutrino data can be used to determine the individual neutrino masses. Three minima have been found, using the χ^2 fit, and, from these minima, it is determined that the mass of the lightest neutrino is 1.3×10^{-3} eV, that of the next heavier neutrino is 1.3×10^{-2} eV, while the mass of the heaviest neutrino is 3.4×10^{-2}, 5.8×10^{-2} or 9.4×10^{-2} eV.

CERN-TH/99-285
March 2000

369) Three neutrino MSW effect and the LNW mass matrix, Per Osland (Bergen U.), *Phys. Scr.* **2001**, 37 (2001), Tai Tsun Wu (Harvard U. and CERN), DOI: https://doi.org/10.1238/Physica.Topical.093a00037.

Per Osland[1] and Tai Tsun Wu[2]

[1]Department of Physics, University of Bergen, Allégaten 55, N-5007 Bergen, Norway
[2]Gordon McKay Laboratory, Harvard University, Cambridge, Massachusetts 02138, U.S.A. and Theoretical Physics Division, CERN, CH-1211 Geneva 23, Switzerland

Received March 4, 2001

PACS Ref: 14.60.Pq, 02.30.Gp

Abstract

We review recent work on analytical solutions to the MSW equations for three neutrino flavours, for exponential and linear potentials. An application to a particular mass matrix is also discussed. The three neutrino masses are determined, respectively, to be 0.001–0.004, and roughly 0.01 and 0.05 eV.

370) Impact-picture phenomenology for $\pi^{\pm}p$, $k^{\pm}p$ and pp, $\bar{p}p$ elastic scattering at high energies, *Eur. Phys. J. C* **28**, 97–105 (2003), Claude Bourrely (Marseille, CPT), Jacques Soffer (Marseille, CPT), Tai Tsun Wu (Harvard U. and CERN), DOI: https://doi.org/10.1140/epjc/s2003-01159-7; Citations: 63.

C. Bourrely[1], J. Soffer[1], T.T. Wu[2,3]

[1] Centre de Physique Théorique[a], CNRS-Luminy Case 907, 13288 Marseille Cedex 9, France
[2] Gordon McKay Laboratory, Harvard University Cambridge, MA 02138, USA
[3] Theoretical Physics Division, CERN, 1211 Geneva 23, Switzerland

Received: 21 October 2002 / Revised version: 31 January 2003 /
Published online: 7 March 2003 – © Springer-Verlag / Società Italiana di Fisica 2003

Abstract. We present an extension to $\pi^{\pm}p$ and $K^{\pm}p$ elastic scattering at high energies of the impact-picture phenomenology we first proposed more than twenty years ago, for pp and $\bar{p}p$ elastic scattering. We show, in particular, that the analytic form of the opacity function for the proton obtained previously is compatible with the experimental results on πp and Kp elastic scattering at high energies. It is proposed that $\pi^{\pm}$ and $K^{\pm}$ external beams be provided from CERN-LHC, so that their elastic scattering from protons can be studied at higher energies. Our phenomenology for pp and $\bar{p}p$ elastic scattering is updated by including new data and we give predictions for future experiments at BNL-RHIC and CERN-LHC.

371) Neutrino masses from universal fermion mixing, *Phys. Rev. D* **67**, 053005 (2003), R. Gastmans (Leuven U.), Per Osland (Bergen U.), Tai Tsun Wu (Harvard U. and CERN), DOI: https://doi.org/10.1103/PhysRevD.67.053005.

R. Gastmans
Institute for Theoretical Physics, Katholieke Universiteit Leuven, B-3001 Leuven, Belgium

Per Osland
Department of Physics, University of Bergen, Allégaten 55, N-5007 Bergen, Norway

Tai Tsun Wu
Gordon McKay Laboratory, Harvard University, Cambridge, Massachusetts 02138
and Theoretical Physics Division, CERN, CH-1211 Geneva 23, Switzerland
(Received 23 October 2002; published 28 March 2003)

If three right-handed neutrinos are added to the standard model, then, for the three known generations, there are six quarks and six leptons. It is then natural to assume that the symmetry considerations that have been applied to the quark matrices are also valid for the lepton mass matrices. Under this assumption, the solar and atmospheric neutrino data can be used to determine the individual neutrino masses. Using the χ^2 fit, it is found that the mass of the lightest neutrino is $(2–5)\times 10^{-3}$ eV, that of the next heavier neutrino is $(10–13)\times 10^{-3}$ eV, while the mass of the heaviest neutrino is $(52–54)\times 10^{-3}$ eV.

DOI: 10.1103/PhysRevD.67.053005 PACS number(s): 14.60.Pq, 12.15.Ff, 13.15.+g

372) Why the real part of the proton-proton forward scattering amplitude should be measured at the LHC, Gribov Memorial Volume, pp. 125–128 (2006), C. Bourrely (Marseille, CPT), N. N. Khuri (Rockefeller U.), Andre Martin (CERN), Jacques Soffer (Marseille, CPT), Tai Tsun Wu (CERN and Harvard U.), DOI: https://doi.org/10.1142/9789812773784_0014.

Abstract:
We stress the importance of testing dispersion relations at the LHC.
A violation might be a sign for new physics.

373) Impact picture of very high-energy hadron interactions (with numerical results), Hung Cheng (MIT), James K. Walker (Fermilab), Tai Tsun Wu (Harvard U. and DESY), PRINT-72-3477 (2006).

374) Impact picture for the analyzing power A(N) in very forward *pp* elastic scattering, *Phys. Rev. D* **76**, 053002 (2007), Claude Bourrely (Marseille, CPT), Jacques Soffer (Temple U.), Tai Tsun Wu (Harvard U. and CERN), DOI: https://doi.org/10.1103/PhysRevD.76.053002.

Claude Bourrely*
Centre de Physique Théorique, CNRS Luminy, Case 907 F-13288 Marseille Cedex 09, France

Jacques Soffer†
Department of Physics, Temple University, Philadelphia, Pennsylvania 19122-6082, USA

Tai Tsun Wu‡
Harvard University, Cambridge, Massachusetts 02138, USA,
and Theoretical Physics Division, CERN, 1211 Geneva 23, Switzerland
(Received 10 June 2007; published 13 September 2007)

In the framework of the impact picture we compute the analyzing power A_N for pp elastic scattering at high energy and in the very forward direction. We consider the full set of Coulomb amplitudes and show that the interference between the hadronic nonflip amplitude and the single-flip Coulomb amplitude is sufficient to obtain a good agreement with the present experimental data. This leads us to conclude that the single-flip hadronic amplitude is small in this low momentum transfer region and it strongly suggests that this process can be used as an absolute polarimeter at the Relativistic Heavy Ions Collider, at Brookhaven National Laboratory, working as a polarized pp collider.

DOI: 10.1103/PhysRevD.76.053002 PACS numbers: 13.40.−f, 13.85.Dz, 13.88.+e

375) Scattering and production at high energies, Conference in Honor of C.N. Yang's 85th Birthday: Statistical Physics, High Energy, Condensed Matter and Mathematical Physics (2008), Tai Tsun Wu (Harvard U. and CERN), DOI: https://doi.org/10.1142/9789812794185_0008.

TAI TSUN WU
Harvard University
Cambridge, MA 02138, U.S.A.
and
CERN, Geneva, Switzerland

A summary is given of some of the developments in the elastic scattering and production processes at high energies. First, Professor Yang's geometrical model of hadronic and nuclear collisions is reviewed. When there was some preliminary experimental evidence that the total cross section for proton–proton collision might not approach a constant at very high energies, theoretical studies were initiated to determine the high-energy asymptotic behavior in relativistic quantum gauge theory. With this starting point, the successive stages of development are:

I. Theoretical prediction of increasing total cross sections;
II. Development of phenomenology and quantitative predictions that were verified experimentally afterwards; and
III. Theory and phenomenology of production processes, especially that of the Higgs particle, at the Large Hadron Collider.

Since the Large Hadron Collider (LHC) is still being built, the last topic is at an early stage of development, and unexpected results may be forthcoming.

Keyword: Scattering, production, high energy, cross section, Yang–Mills non-Abelian gauge theory, phenomenology, Large Hadron Collider, Higgs particle.

376) Transverse Momentum Distributions in High-Energy Hadron Interactions, R. Gastmans (Leuven U.), Sau Lan Wu (Wisconsin U., Madison), Tai Tsun Wu (Harvard U. and CERN), CERN-PH-TH-2010-115, CERN-PH-TH-2010-115 (2010).

R. Gastmans[a1], Sau Lan Wu[b2], and Tai Tsun Wu[c]

[a] Instituut voor Theoretische Fysica, Katholieke Universiteit Leuven, Celestijnenlaan 200D, B-3001 Leuven, Belgium

[b] Department of Physics, University of Wisconsin, Madison WI 53706, USA

[c] Gordon McKay Laboratory, Harvard University, Cambridge MA 02138, USA,

and

Theory Division, CERN, CH-1211 Geneva 23, Switzerland

Abstract

From the recent theoretical result on the production of the Higgs boson at the Large Hadron Collider, it follows that the transverse momenta of the gluons in the protons are of the order of 1 GeV/c. Thus, a fraction of the Z's is expected to be produced with a transverse momentum of this order of magnitude and this phenomenon should be observable in the near future through the leptonic decay of the Z. If this fraction is found experimentally to be sizeable, then one possibility is the appearance of paired jets, i.e., two jets with the property that the vector sum of their transverse momenta is of the order of 1 GeV/c, even if these jets do not come from the decay of the Z. Such narrow transverse momenta imply that the quark and the gluon distribution functions appropriate for proton-proton and proton-antiproton interactions are significantly different from those for electron-proton interactions.

[1]Work supported in part by the FWO-Vlaanderen, project G.0235.05, and in part by the Federal Office for Scientific, Technical and Cultural Affairs through the 'Interuniversity Attraction Poles Programme – Belgian Science Policy' P6/11-P.

[2]Work supported in part by the United States Department of Energy Grant No. DE-FG02-95ER40896.

377) Transverse momentum distribution of the Z produced at the Large Hadron Collider and related phenomena, *Phys. Lett. B* **693** (2010), 452–455, R. Gastmans (Leuven U.), Sau Lan Wu (Wisconsin U., Madison), Tai Tsun Wu (Harvard U. and CERN), DOI: https://doi.org/10.1016/j.physletb.2010.08.075.

R. Gastmans [a,*,1], Sau Lan Wu [b,2], Tai Tsun Wu [c,d]

[a] *Instituut voor Theoretische Fysica, Katholieke Universiteit Leuven, Celestijnenlaan 200D, B-3001 Leuven, Belgium*
[b] *Department of Physics, University of Wisconsin, Madison, WI 53706, USA*
[c] *Gordon McKay Laboratory, Harvard University, Cambridge, MA 02138, USA*
[d] *Theory Division, CERN, CH-1211 Geneva 23, Switzerland*

ARTICLE INFO

Article history:
Received 1 July 2010
Received in revised form 20 August 2010
Accepted 31 August 2010
Available online 9 September 2010
Editor: L. Alvarez-Gaumé

Keywords:
High-energy physics
Proton–proton scattering
LHC physics
Transverse momentum distributions
Paired jets

ABSTRACT

From the recent theoretical result on the production of the Higgs boson at the Large Hadron Collider, it follows that other particles will also be produced with small transverse momentum, of the order of 1 GeV/c. The leptonic decay mode of the Z is especially suited for a first observation of this phenomenon. Other related effects, such as paired jets, are also discussed.

378) Production of isolated Higgs particle at the Large Hadron Collider, *Phys. Lett. B* **683** (2010), 354–357, R. Gastmans (Leuven U.), Sau Lan Wu (Wisconsin U., Madison), Tai Tsun Wu (Harvard U. and CERN), DOI: https://doi.org/10.1016/j.physletb.2009.12.004.

R. Gastmans [a,*,1], Sau Lan Wu [b,2], Tai Tsun Wu [c,d]

[a] Instituut voor Theoretische Fysica, Katholieke Universiteit Leuven, Celestijnenlaan 200D, B-3001 Leuven, Belgium
[b] Department of Physics, University of Wisconsin, Madison, WI 53706, USA
[c] Gordon McKay Laboratory, Harvard University, Cambridge, MA 02138, USA
[d] Theory Division, CERN, CH-1211, Geneva 23, Switzerland

ARTICLE INFO

Article history:
Received 9 August 2009
Accepted 3 December 2009
Available online 21 December 2009
Editor: G.F. Giudice

ABSTRACT

Theoretical predictions are presented for the process $p + p \to A + H + B$, where H is the Higgs particle and A (B) is a group of particles that mostly goes down one (the other) beam pipe. These predictions are obtained using results from relativistic quantum gauge field theory and the fact that, for the Large Hadron Collider (LHC), the center-of-mass energy $\gg$ Higgs mass $\gg$ proton mass. The average transverse momentum of the produced Higgs particle is of the order of 1 GeV/c. This process leads naturally to the concept of considering the LHC as a Pomeron Collider.

379) Determination of the forward slope in pp and elastic scattering up to LHC energy, Claude Bourrely (Marseille U., Luminy), Jacques Soffer (Temple U.), Tai Tsun Wu (Harvard U. and CERN), DOI: https://doi.org/10.1140/epjc/s10052-011-1601-x.

Claude Bourrely[1], **Jacques Soffer**[2,a], **Tai Tsun Wu**[3,4]

[1] Département de Physique, Faculté des Sciences de Luminy, Université de la Méditerranée-Aix-Marseille II, 13288 Marseille Cedex 09, France
[2] Physics Department, Temple University, Barton Hall, 1900 N. 13th Street, Philadelphia, PA 19122-6082, USA
[3] Harvard University, Cambridge, MA 02138, USA
[4] Theoretical Physics Division, CERN, 1211 Geneva 23, Switzerland

Received: 2 November 2010 / Revised: 21 February 2011 / Published online: 25 March 2011

Abstract In the analysis of experimental data on pp (or $\bar{p}p$) elastic differential cross section it is customary to define an average forward slope b in the form $\exp(-b|t|)$, where t is the momentum transfer. Taking as working example the results of experiments at Tevatron and SPS, we will show with the help of the impact picture approach that this simplifying assumption hides interesting information on the complex non-flip scattering amplitude, and that the slope b is not a constant. We investigate the variation of this slope parameter, including a model-independent way to extract this information from an accurate measurement of the elastic differential cross section. An extension of our results to the LHC energy domain is presented in view of future experiments.

380) An upper bound on the total inelastic cross-section as a function of the total cross-section, Tai Tsun Wu (Harvard U. and CERN), Andre Martin (CERN), Shasanka Mohan Roy (Tata Inst.), Virendra Singh (Tata Inst.), DOI: https://doi.org/10.1103/PhysRevD.84.025012.

Tai Tsun Wu,[1,2,*] André Martin,[3,†] Shasanka Mohan Roy,[4,‡] and Virendra Singh[5,§]
[1]Harvard University, Cambridge, Massachusetts, USA
[2]Theoretical Physics Division, CERN, Geneva, Switzerland
[3]Theoretical Physics Division, CERN, Geneva, Switzerland
[4]Homi Bhabha Centre for Science Education, TIFR, V. N. Purav Marg, Mankhurd, Mumbai, 400 088, India
[5]Tata Institute of Fundamental Research, Mumbai, 400005, India
(Received 10 November 2010; published 15 July 2011)

Recently, André Martin has proved a rigorous upper bound on the inelastic cross sectionσ_{inel} at high energy, which is one-fourth of the known Froissart-Martin-Lukaszuk upper bound on σ_{tot}. Here, we obtain an upper bound on σ_{inel} in terms of σ_{tot} and show that the Martin bound on σ_{inel} is improved significantly with this added information.

DOI: 10.1103/PhysRevD.84.025012 PACS numbers: 03.67.−a, 03.65.Ud, 42.50.−p

381) Higgs Decay into Two Photons, Revisited, R. Gastmans (Leuven U.), Sau Lan Wu (Wisconsin U., Madison), Tai Tsun Wu (Harvard U. and CERN), e-Print: 1108.5872 [hep-ph] (2011); Citations: 54.

Abstract

The one-loop calculation of the amplitude for the Higgs decay $H \to \gamma\gamma$ due to virtual W's in the unitary gauge is presented. As the Higgs does not directly couple to the massless photons, the one-loop amplitude is finite. The calculation is performed in a straightforward way, without encountering divergences. In particular, artifacts like dimensional regularization are avoided. This is achieved by judiciously routing the external momenta through the loop and by combining the integrands of the amplitudes before carrying out the integration over the loop momentum. The present result satisfies the decoupling theorem for infinite Higgs mass, and is thus different from the earlier results obtained in the $\xi = 1$ gauge using dimensional regularization. The difference between the results is traced to the use of dimensional regularization.

*Work supported in part by the FWO-Vlaanderen, project G.0651.11, and in part by the Federal Office for Scientific, Technical and Cultural Affairs through the 'Interuniversity Attraction Poles Programme – Belgian Science Policy' P6/11-P.
†Work supported in part by the United States Department of Energy Grant No. DE-FG02-95ER40896.

382) Higgs Decay H → $\gamma\gamma$ through a W Loop: Difficulty with Dimensional Regularization, R. Gastmans (Leuven U.), Sau Lan Wu (Wisconsin U., Madison), Tai Tsun Wu (Harvard U. and CERN), e-Print: 1108.5322 [hep-ph] (2011).

R. Gastmans[1,*], Sau Lan Wu[2,†], and Tai Tsun Wu[3],

[1] Instituut voor Theoretische Fysica, Katholieke Universiteit Leuven, Celestijnenlaan 200D, B-3001 Leuven, Belgium.

[2] Department of Physics, University of Wisconsin, Madison WI 53706, USA.

[3] Gordon McKay Laboratory, Harvard University, Cambridge MA 02138, USA,

and

Theory Division, CERN, CH-1211 Geneva 23, Switzerland.

Abstract

Since the photon has no mass, it does not couple directly to the Higgs particle. This implies that the one-loop correction to the decay $H \to \gamma\gamma$ is necessarily finite. Therefore, this correction should be calculable without introducing either regularization or ghosts. Such a calculation is carried out in this paper for the case of one W loop. The result obtained this way turns out *not* to agree with the previous, well-known one, and it is argued that the present result is to be preferred because it satisfies the decoupling theorem. The discrepancy can be traced to the use of dimensional regularization in the previous result.

*Work supported in part by the FWO-Vlaanderen, project G.0651.11, and in part by the Federal Office for Scientific, Technical and Cultural Affairs through the 'Interuniversity Attraction Poles Programme – Belgian Science Policy' P6/11-P.

†Work supported in part by the United States Department of Energy Grant No. DE-FG02-95ER40896.

383) Higgs production at the large hadron collider: Phenomenological model and theoretical predictions, *Nucl. Phys. B* **850** (2011), 53–95, R. Gastmans (Leuven U.), Sau Lan Wu (Wisconsin U., Madison), Tai Tsun Wu (Harvard U. and CERN), DOI: https://doi.org/10.1016/j.nuclphysb.2011.04.012.

R. Gastmans [a,*,1], Sau Lan Wu [b,2], Tai Tsun Wu [c,d]

[a] *Instituut voor Theoretische Fysica, Katholieke Universiteit Leuven, Celestijnenlaan 200D, B-3001 Leuven, Belgium*
[b] *Department of Physics, University of Wisconsin, Madison, WI 53706, USA*
[c] *Gordon McKay Laboratory, Harvard University, Cambridge, MA 02138, USA*
[d] *Theory Division, CERN, CH-1211 Geneva 23, Switzerland*

Received 23 September 2010; received in revised form 18 April 2011; accepted 19 April 2011

Available online 23 April 2011

Abstract

Using the results from relativistic gauge field theory, a phenomenological model is developed for the production of an isolated Higgs particle H. The specific process is $p + p \to A + H + B$, where the group of particles A (B) mostly goes down one (the other) beam pipe. The theory and the phenomenology apply when the center-of-mass energy $\sqrt{s} \gg M \gg m$, with M and m the masses of the Higgs particle and the proton respectively. Thus, there are two large parameters, namely $\sqrt{s}/M$ and M/m. That $M \gg m$ plays a central role. This phenomenological model is applied to the Large Hadron Collider (LHC) to predict the differential production cross sections. With high probability, this isolated Higgs particle is produced with a small transverse momentum of the order of 1 GeV/c. Because of this fact and the relatively small number of observed particles in such events, the method of data analysis is different from those developed so far for Higgs detection. These events can be described as due to Pomeron–Pomeron annihilation, and are 'clean' in the sense that those from TeV linear colliders are called 'clean'. The LHC, with its center-of-mass design energy of 14 TeV and its design luminosity of 10^{34} cm^{-2} s^{-1}, can function exceptionally well as a Pomeron collider.

384) High-energy asymptotic behavior of the Bourrely-Soffer-Wu model for elastic scattering, Claude Bourrely (Aix-Marseille III U.), John M. Myers (Harvard U.), Jacques Soffer (Temple U.), Tai Tsun Wu (Harvard U. and CERN), DOI: https://doi.org/10.1103/PhysRevD.85.096009.

Claude Bourrely*
Aix-Marseille Université, Département de Physique, Faculté des Sciences de Luminy, 13288 Marseille, Cedex 09, France

John M. Myers†
School of Engineering and Applied Sciences, Harvard University, Cambridge, Massachusetts 02138, USA

Jacques Soffer‡
Physics Department, Temple University, Barton Hall, 1900 N, 13th Street Philadelphia, Pennsylvania 19122-6082, USA

Tai Tsun Wu§
Harvard University, Cambridge, Massachusetts 02138, USA and Theoretical Physics Division, CERN, 1211 Geneva 23, Switzerland
(Received 7 February 2012; published 31 May 2012)

Some time ago, an accurate phenomenological approach, the BSW model, was developed for proton-proton and antiproton-proton elastic scattering cross sections at center-of-mass energies above 10 GeV. This model has been used to give successful theoretical predictions for these processes, at successive collider energies. The BSW model involves a combination of integrals that, while computable numerically at fairly high energies, require some mathematical analysis to reveal the high-energy asymptotic behavior. In this paper we present a high-energy asymptotic representation of the scattering amplitude at moderate momentum transfer, for the leading order in an expansion parameter closely related to the logarithm of the center-of-mass energy. The fact that the expansion parameter goes as the logarithm of the energy means that the asymptotic behavior is accurate only for energies greatly beyond any foreseeable experiment. However, we compare the asymptotic representation against the numerically calculated model for energies in a less extreme region of energy. The asymptotic representation is given by a simple formula which, in particular, exhibits the oscillations of the differential cross section with momentum transfer. We also compare the BSW asymptotic behavior with the Singh-Roy unitarity upper bound for the diffraction peak.

DOI: 10.1103/PhysRevD.85.096009 PACS numbers: 11.80.Fv, 13.85.Dz, 13.85.Lg, 25.40.Cm

385) Production processes at extremely high energies, *Phys. Lett. B* **720** (2013), 205–209, R. Gastmans (Leuven U.), Sau Lan Wu (Wisconsin U., Madison), Tai Tsun Wu (Harvard U. and CERN), DOI: https://doi.org/10.1016/j.physletb.2013.01.064.

R. Gastmans[a,*,1], Sau Lan Wu[b,2], Tai Tsun Wu[c,d]

[a] Instituut voor Theoretische Fysica, KU Leuven, B-3001 Leuven, Belgium
[b] Department of Physics, University of Wisconsin, Madison, WI 53706, USA
[c] Gordon McKay Laboratory, Harvard University, Cambridge, MA 02138, USA
[d] Theory Division, CERN, CH-1211 Geneva 23, Switzerland

ARTICLE INFO

Article history:
Received 20 December 2012
Accepted 28 January 2013
Available online 1 February 2013
Editor: G.F. Giudice

ABSTRACT

The production processes are identified that contribute to the rise of the total cross section in proton-proton scattering at extremely high energies, $s \to \infty$. At such energies, the scattering can be described by a black disk (completely absorptive) with a radius expanding logarithmically with energy surrounded by a gray fringe (partially absorptive). For the leading term of $(\ln s)^2$ in the increasing total cross section, the gray fringe is neglected, and geometrical optics is generalized to production processes. It is known that half of the rise in the total cross section is due to elastic scattering. The other half is found to originate from the production of jets with relatively small momenta in the center-of-mass system.

386) Scalar Quarks at the Large Hadron Collider, Jean-Marc Richard (Lyon U. and Lyon, IPN), Tai Tsun Wu (CERN and Harvard U.), e-Print: 1308.5379 [hep-ph] (2013).

387) Yang–Mills Gauge Theory and Higgs Particle, Published in: *Int. J. Mod. Phys. A* **30** (2015) 34, 1530065, Tai Tsun Wu (Harvard U.), Sau Lan Wu (Wisconsin U., Madison), DOI: https://doi.org/10.1142/9789814725569_0008.

Tai Tsun Wu

Gordon McKay Laboratory, Harvard University, Cambridge, MA 02138, USA

Sau Lan Wu†

Physics Department, University of Wisconsin-Madison, Madison, WI 53706, USA

Motivated by the experimental data on the Higgs particle from the ATLAS Collaboration and the CMS Collaboration at CERN, the standard model, which is a Yang–Mills non-Abelian gauge theory with the group $U(1) \times SU(2) \times SU(3)$, is augmented by scalar quarks and scalar leptons without changing the gauge group and without any additional Higgs particle. Thus there is fermion–boson symmetry between these new particles and the known quarks and leptons. In a simplest scenario, the cancellation of the quadratic divergences in this augmented standard model leads to a determination of the masses of all these scalar quarks and scalar leptons. All these masses are found to be less than $100\,\mathrm{GeV}/c^2$, and the right-handed scalar neutrinos are especially light. Alterative procedures are given with less reliance on the experimental data, leading to the same conclusions.

388) Do we understand near-forward elastic scattering up to TeV energies?, Claude Bourrely (Aix-Marseille U.), Jacques Soffer (Temple U.), Tai Tsun Wu (CERN and Harvard U.), DOI: https://doi.org/10.1142/S0217751X15420063.

Claude Bourrely
Département de Physique, Faculté des Sciences, Aix-Marseille Université, Campus Universite de Luminy, Marseille 13288, Cedex 09, France

Jacques Soffer
Department of Physics, Temple University, PA 19122-1801, USA
Jacques.soffer@gmail.com

Tai Tsun Wu
Gordon McKay Laboratory, Harvard University, Cambridge, MA 02138, USA
Theoretical Physics Division, CERN, 1211 Geneva 23, Switzerland

Received 31 July 2014
Published 26 February 2015

In 1970, on purely theoretical grounds, all total hadronic total cross-sections were predicted to increase without limit for higher and higher energies. This was contrary to the conventional belief at that time. In 1978, an accurate phenomenological model was formulated for the case of proton–proton and antiproton-proton interactions. The parameters for this model were slightly improved in 1984 using the additional available experimental data. Since then, for 30 years these parameters have not changed. This development, including especially the difficult task of formulating this phenomenological model and the comparison of the predictions of this model with later experimental results, is summarized.

Keywords: Elastic hadronic reactions; pomeron.

PACS numbers: 11.80.Fv, 13.85.Dz, 25.40.Cm

389) Impact picture for near-forward elastic scattering up to LHC energies, *AIP Conf. Proc.* **1654**, 050008 (2015), Jacques Soffer (Temple U.), Claude Bourrely (Aix-Marseille U.), Tai Tsun Wu (CERN and Harvard U.), DOI: https://doi.org/10.1063/1.4915981.

Jacques Soffer[*], Claude Bourrely[†] and Tai Tsun Wu[,‡]**

[*]*Physics Department, Temple University, 1835 N, 12 Street, Philadelphia, PA 19122-6082, USA*
[†]*Aix-Marseille Université, Département de Physique, Faculté des Sciences, site de Luminy, 13288 Marseille, Cedex 09, France*
[**]*Gordon McKay Laboratory, Harvard University, Cambridge, MA 02138, USA*
[‡]*Theoretical Physics Division, CERN, 1211 Geneva 23, Switzerland*

Abstract. We will recall the main features of an accurate phenomenological model to describe successfully near-forward elastic scattering in a wide energy range, including ISR, SPS and Tevatron colliders. A large step in energy domain is accomplished with the LHC collider, presently running, giving the opportunity to confront the new data with the predictions of our theoretical approach.

Keywords: elastic scattering, total cross section, pomeron, LHC
PACS: 11.80.Fv, 13.85.Dz, 13.85.Lz, 25.40.Cm

390) Higgs decay H→γγ: New theoretical results and possible experimental implications, *Int. J. Mod. Phys. A* **30** (2015) 32, 1550200, R. Gastmans (KU Leuven, FET), Sau Lan Wu (Wisconsin U., Madison), Tai Tsun Wu (CERN and Harvard U. (main)), DOI: https://doi.org/10.1142/S0217751X15502000.

R. Gastmans

Instituut voor Theoretische Fysica, Katholieke Universiteit Leuven, Celestijnenlaan 200D, B-3001 Leuven, Belgium

Sau Lan Wu

Physics Department, University of Wisconsin-Madison, Madison, WI 53706, USA

Tai Tsun Wu

Gordon McKay Laboratory, Harvard University, Cambridge, MA 02138, USA
Theory Division, CERN, CH-1211 Geneva 23, Switzerland

Received 16 October 2015
Accepted 19 October 2015
Published 16 November 2015

The contribution of one W loop to the decay of the Higgs particle into two gammas is recalculated. Since this matrix element is finite in the standard model, it should be calculated in a straightforward way without artifacts such as regularization or ghosts. The result of the present calculation differs from the previous one, giving a Higgs decay rate smaller by a factor of 2 in the mass range of physical interest. In view of the results from the ATLAS Collaboration and the CMS Collaboration, this leads to possible far-reaching experimental consequences.

Keywords: Higgs decay; LHC physics; standard model; quantum field theory.

PACS numbers: 11.16.Bt, 12.15.−y, 12.15.Ji, 14.80.Bn

391) Positivity of the real part of the forward scattering amplitude, André Martin (CERN), Tai Tsun Wu (Harvard U.), DOI: https://doi.org/10.1103/PhysRevD.97.014011.

André Martin*

CERN, 1211 Geneva, Switzerland

Tai Tsun Wu

Gordon McKay Laboratory, Harvard University, 02138 Cambridge, Massachusetts, USA

(Received 5 September 2017; published 23 January 2018; corrected 26 January 2018)

We prove the general theorem that the real part of the crossing even forward two-body scattering amplitude is positive at sufficiently high energies if, above a certain energy, the total cross section increases monotonically to infinity at infinite energy.

DOI: 10.1103/PhysRevD.97.014011

392) Alex Grossmann, Scattering Amplitude, Fermi Pseudopotential, and Particle Physics, Springer International Publishing, Book: "Theoretical Physics, Wavelets, Analysis, Genomics: An Indisciplinary Tribute to Alex Grossmann", Tai Tsun Wu (Harvard U.), DOI: https://doi.org/10.1007/978-3-030-45847-8_8 (2023).

Abstract:
Alex Grossmann and I knew each other for more than sixty years. This article describes some of our collaborations, including those in physics and engineering.

Chapter 11
Concept of the Basic Standard Model and a Relation between the Three Gauge Coupling Constants (5 papers)

393) Augmented standard model and the simplest scenario, *Int. J. Mod. Phys. A* **30** (2015) 32, 1550201, Tai Tsun Wu (Harvard U.), Sau Lan Wu (Wisconsin U., Madison), DOI: https://doi.org/10.1142/S0217751X15502012.

Tai Tsun Wu

Gordon McKay Laboratory, Harvard University, Cambridge, MA 02138, USA

Sau Lan Wu

Physics Department, University of Wisconsin-Madison, Madison, WI 53706, USA

Received 16 October 2015
Accepted 19 October 2015
Published 16 November 2015

The experimental discovery of the Higgs particle in 2012 by the ATLAS Collaboration and the CMS Collaboration at CERN ushers in a new era of particle physics. On the basis of these data, scalar quarks and scalar leptons are added to each generation of quarks and leptons. The resulting augmented standard model has fermion–boson symmetry for each of three generations, but only one Higgs doublet giving masses to all the elementary particles.

A specific special case, the simplest scenario, is studied in detail. In this case, there are twenty six quadratic divergences, and all these divergences are cancelled provided that one single relation between the masses is satisfied. This mass relation contains a great deal of information, and in particular determines the masses of all the right-handed scalar quarks and scalar leptons, while gives relations for the masses of the left-handed ones.

An alternative procedure is also given with a different starting point and less reliance on the experimental data. The result is of course the same.

Keywords: Higgs decay; LHC physics; standard model; quantum field theory.

PACS numbers: 11.16.Bt, 12.15.−y, 12.15.Ji, 14.80.Bn

394) Failure of the Feynman R_1 gauge for the standard model: An explicit example, *Int. J. Mod. Phys. A* **31** (2016) 04n05, 1650028, Tai Tsun Wu (Harvard U.), Sau Lan Wu (Wisconsin U., Madison), DOI: https://doi.org/10.1142/S0217751X16500287.

Tai Tsun Wu

Gordon McKay Laboratory, Harvard University,
Cambridge, MA 02138, USA

Sau Lan Wu

Physics Department, University of Wisconsin-Madison,
Madison, WI 53706, USA

Received 15 January 2016
Accepted 22 January 2016
Published 2 February 2016

The decay of the Higgs particle into two photons through a W loop was calculated in a straightforward way four years ago on the basis of the standard model. This calculation was carried out in the unitary gauge. Nevertheless, all attempts to reproduce this correct answer using the Feynman rules in the R_1 gauge, or the more general R_ξ gauge, have failed. In this paper, a detailed analysis is carried out to compare the unitary gauge with the R_ξ gauge; through this comparison, the underlying reason is determined why the answer cannot be obtained using the Feynman rules in the R_ξ gauge.

This is the first example where the use of the Feynman rules in the R_ξ gauge leads to an incorrect answer, and this incorrect answer cannot be ruled out by any simple argument such as gauge invariance. It means that great care must be exercised in using Feynman rules.

Keywords: Higgs decay; LHC physics; standard model; quantum field theory.

PACS numbers: 11.16.Bt, 12.15.−y, 12.15.Ji, 14.80.Bn

395) Comparing the R_ξ gauge and the unitary gauge for the standard model: An example, *Nucl. Phys. B* **914** (2017), 421–445, Tai Tsun Wu (Harvard U.), Sau Lan Wu (Wisconsin U., Madison), DOI: https://doi.org/10.1016/j.nuclphysb.2016.11.007.

Tai Tsun Wu [a,*], Sau Lan Wu [b]

[a] Gordon McKay Laboratory, Harvard University, Cambridge, MA 02138, USA
[b] Physics Department, University of Wisconsin-Madison, Madison, WI 53706, USA

Received 23 September 2016; accepted 4 November 2016
Available online 16 November 2016
Editor: Hubert Saleur

Abstract

For gauge theory, the matrix element for any physical process is independent of the gauge used. However, since this is a formal statement, it does not guarantee this gauge independence in every case. An example is given here where, for a physical process in the standard model, the matrix elements calculated with two different gauge – the R_ξ gauge and the unitary gauge – are explicitly verified to be different. This is accomplished by subtracting one matrix element from the other. This non-zero difference turns out to have a subtle origin. Two simple operators are found not to commute with each other: in one gauge these two operations are carried out in one order, while in the other gauge these same two operations are carried out in the opposite order. Because of this result, a series of question are raised such that the answers to these question may lead to a deeper understanding of the Yang–Mills non-Abelian gauge theory in general and the standard model in particular.

396) Gauge invariance for a class of tree diagrams in the standard model, Tai Tsun Wu (Harvard U. (main)), Sau Lan Wu (Wisconsin U., Madison), CERN-TH-2018-198 (2018), e-Print: 1812.02740.

Tai Tsun Wu[a,*], Sau Lan Wu[b]

[a] Gordon McKay Laboratory, Harvard University, Cambridge, MA 02138, U. S. A.
[b] Physics Department, University of Wisconsin-Madison, Madison, WI 53706, U. S. A.

Abstract

For gauge theory, the matrix element for any physical process is independent of the gauge used. Since this is a formal statement and examples are known where gauge invariance is violated, for any specific process this gauge invariance needs to be checked by explicit calculation. In this paper, gauge invariance is found to hold for a large non-trivial class of processes described by tree diagrams in the standard model – tree diagrams with two external W bosons and any number of external Higgs bosons. This verification of gauge invariance is quite complicated, and is based on a direct study of the difference between different gauges through induction on the number of external Higgs bosons.

397) Concept of the basic standard model and a relation between the three gauge coupling constants, *Int. J. Mod. Phys. A* **38** (2023) 28, 2350128, Tai Tsun Wu (Wisconsin U., Madison), Sau Lan Wu (Wisconsin U., Madison), DOI: https://doi.org/10.1142/S0217751X23501282; Downloaded 13,132 times.

Tai Tsun Wu* and Sau Lan Wu

Department of Physics,
University of Wisconsin-Madison,
Madison, WI 53706, USA
**ttwu.harvard@gmail.com*

Received 17 August 2023
Accepted 21 August 2023
Published 9 November 2023

Among the known elementary particles, 15 are not massless. In the Lagrangian of the standard model, the Higgs particle couples to all 15 of them to give them their respective masses. In this paper, a new model is proposed, where in the Lagrangian, the Higgs particle couples only to the four heaviest known elementary particles t, H, Z and W, all with masses in the vicinity of $100\,\mathrm{GeV}/c^2$, but not to the other 11. In this model, the masses of these 11 elementary particles that do not couple directly to the Higgs particle are generated by radiative corrections. Because of the reduction of the number of terms in the Lagrangian, this model is referred to as the basic standard model. By studying how the bottom quark can acquire a mass, it is found that there must be a relation between the three gauge coupling constants of the gauge group $U(1) \times SU(2) \times SU(3)$. This relation between these coupling constants is the first prediction, on purely theoretical grounds without using any experimental input, of the basic standard model, and is studied in some detail. In particular, this new prediction is found to be in good agreement with experimental measurements.

Keywords: Quantum field theory; Yang–Mills non-Abelian gauge theory; standard model; quantum chromodynamics.

PACS numbers: 03.70.+k, 12.10.-g, 14.80.Bn, 12.28.-t

Commentary from Sau Lan Wu

Tai Tsun Wu completed his last publication one year before his death, at the age of 89. He was most excited about his proposal of the modification of the standard model, which he called the basic standard model. In this basic standard model, he was able to derive a relation between the three gauge coupling constants successfully. He was most proud of this work and he was hoping to be able to derive the mass of the bottom quark. He was amazed that the mass of the bottom quark is so different from the masses of W, Z, Higgs, and Top. Between the publication of this paper (November 2023) to the time soon after his death (July 19, 2024), this publication has been downloaded 12,921 times. The hope is that some bright young theorists will continue Tai Tsun Wu's proposed concept of the basic standard model to solve the mystery of the mass problem.

Chapter 12
Books (6 books)

1) The Scattering and Diffraction of Waves (Harvard University Press, 1959), DOI: https://doi.org/10.4159/harvard.9780674498389.

Ronold W. P. King
Gordon McKay Professor of Applied Physics
Harvard University
and
Tai Tsun Wu
Assistant Professor of Applied Physics
Harvard University

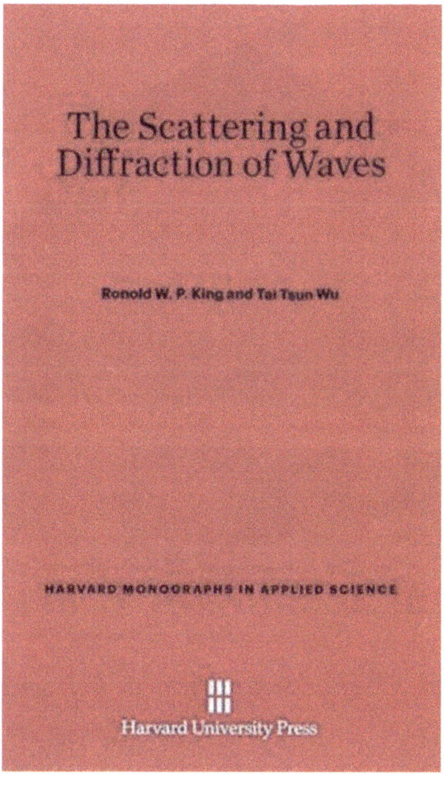

2) Antennas in Matter: Fundamentals, Theory, and Applications, M.I.T. Press, 1981.

Ronold W. P. King, Glenn S. Smith with Margaret Owens, Tai Tsun Wu

3) Expanding Protons: Scattering at High-Energies, Cambridge, USA: MIT-PR. (1987).

Hung Cheng and Tai Tsun Wu

This book presents an original theory, based on more than a decade of research, of the behavior of strongly interacting particles in high-energy collision processes. Using gauge field theories, the authors systematically study elastic as well as multiparticle production processes and diffractive scattering, arriving at the conclusion that the total cross section should rise, that is, the particles should "expand," as energies increase.

Cheng and Wu compare their results with experimental data, including the outcomes of a series of experiments conducted in the mid-1970s that dramatically confirmed their predictions of rising total cross section at high energies and the dependence of cross sections on the momenta of incoming and outgoing particles. The presentation is systematic and mathematically rigorous yet broad enough to be accessible to graduate students.

After a clear presentation of the basic elements of particle physics and the quantum theory of gauge fields, the authors take up fermion-fermion elastic scattering, high-energy scattering models, multiparticle production, elastic amplitude, the rising total

4) The Ubiquitous Photon: Helicity Methods for QED and QCD (International Series of Monographs on Physics), 1990.

R. Gastmans and Tai Tsun Wu

5) Lateral Electromagnetic Waves: Theory and Applications to Communications, Geophysical Exploration, and Remote Sensing (Springer-Verlag, 1992).

P. King, Margaret Owens, Tai Tsun Wu

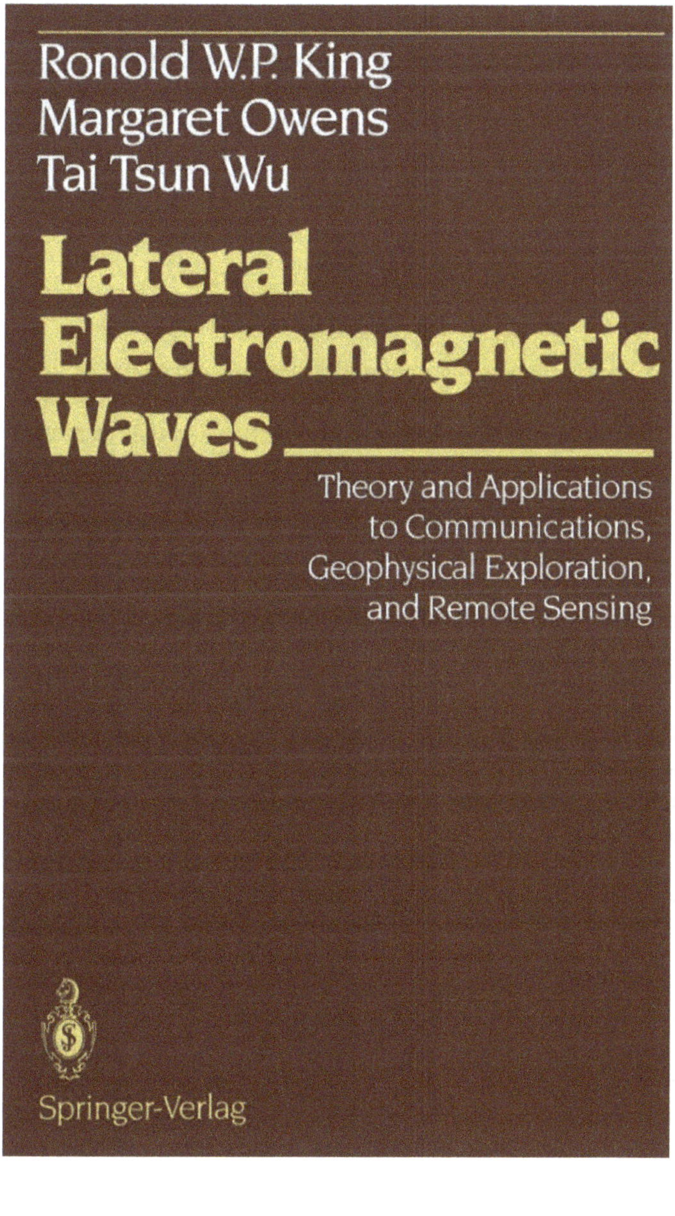

A lateral wave is the electromagnetic wave near the interface between two different planar media such as air and sea water. Although most antennas are located near an interface, the effect of an interface is often ignored or described as "ground clutter." In many situations, an understanding of the effect of the interface is essential. After the publication of the book Professor Wu extended the work in a number of directions such as non-planar interfaces.

6) The Two-Dimensional Ising Model, Harvard University Press, 1973. Second edition Dover, 2014.

Barry M. McCoy, Stony Brook University
Tai Tsun Wu, Harvard University

1. INTRODUCTION

Statistical mechanics is an old and venerable branch of physics which has received the attention of many physicists from the time of Gibbs. Since then it has developed in several directions so that at present it is possible to distinguish at least three different theoretical approaches to the subject: (1) the foundational approach, which is concerned with establishing the general properties of, and proving existence theorems for, statistical mechanical systems by rigorous mathematical means, (2) the phenomenological approach, which is concerned with correlating and quantitatively explaining the results of experiments by any available method, and (3) the model-building approach, which attempts to gain insight into practical situations by studying simple models in which at least some physically interesting quantities may be exactly computed. Each of these approaches has made such valuable contributions to our understanding of statistical mechanics that it is neither feasible nor desirable to separate them completely. However, because each approach has developed such a large body of literature, it is likewise not possible to give an adequate treatment of all of them in a single book. Therefore, while we will attempt to make this book self-contained by giving a brief discussion of the foundation of statistical mechanics, and while we will attempt to place the book in a somewhat broader context by making contact with the existing experimental situation, we will concentrate our efforts on the study of certain solvable models.

The number of exactly solvable problems in a field depends on the complexity of the subject. For example, there are innumerable solvable problems in classical mechanics, whereas, at the other extreme, very few problems in relativistic quantum field theory have ever been exactly solved. The scarcity of solvable models in both statistical mechanics and relativistic quantum field theory is due basically to the fact that, in a system with a very large number of particles, each particle may indirectly interact with an enormous number of others even if the fundamental interaction is two-body and of short range. However, it is the purpose of statistical mechanics to study systems with a large number of particles, and the phenomena of greatest interest are precisely those which are not present in the classical or quantum mechanics of a small number of particles. Therefore, a criterion for the usefulness of any model in statistical mechanics is its capability of giving us insight into the new phenomena characteristic of a large number of particles.

The most characteristic feature of statistical mechanical systems is the existence of phase transitions. Surely the most familiar phase transition is either the condensation of steam into water or the freezing of water into ice. Only slightly less familiar is the ferromagnetic phase transition that takes place at the Curie temperature, which, as an example, is roughly 1043°K for iron. Of the several existing models which exhibit a phase transition, the most famous is the Ising model. In three dimensions the model is so complicated that no exact computation has ever been made, while in one dimension the Ising model does not undergo a phase transition. However, it is one of the most beautiful discoveries of twentieth-century physics that in two dimensions the Ising model not only has a ferromagnetic phase transition but also has very many physical properties which may be exactly computed. Indeed, despite the restriction on dimensionality, the two-dimensional Ising model exhibits all of the phenomena peculiar to magnetic systems near the Curie temperature. For that reason, the two-dimensional Ising model forms the basis of almost all our theoretical understanding of the phase transition to the ferromagnetic state.

Chapter 13
Selected Publications of Tai Tsun Wu (11 papers)

The following papers have been selected for reprint in this memorial volume in recognition of their lasting importance and scientific significance. Many other valuable and influential works by Tai Tsun Wu are not included here but remain integral to his legacy.

1. Reprinted article 21, listed in Chapter 6:

The Cylindrical Antenna with Nonreflecting Resistive Loading, *IEEE Trans. Antennas and Propagation*, Vol. 13, 369–373 (1965), T. T. Wu (Gordon McKay Laboratory and Harvard University) and R. W. P. King (Gordon McKay Laboratory, Harvard University), DOI: https://doi.org/10.1109/TAP.1965.1138429; Citations: 373.

[Pages 286–290]

2. Reprinted article 70, listed in Chapter 7:

Dirac Monopole Without Strings: Monopole Harmonics, *Nucl.Phys. B* **107** (1976), 365–380, In Yang, C.N.: Selected Papers 1945–1980, 493–508. (*Nucl. Phys. B* **107** (1976) 365–380) and State Univ. New York Stony Brook - ITP-SB-76-05 (76,REC.FEB.) 30 P. (602076) (SEE BOOK INDEX), Tai Tsun Wu (Harvard U.), Chen Ning Yang (SUNY, Stony Brook), DOI: https://doi.org/10.1016/0550-3213(76)90143-7; Citations: 771.

Reprinted from The Lancet, 107, Tai Tsun Wu and Chen Ning Yang, Dirac Monopole Without Strings: Monopole Harmonics, 365–380, (1976), with permission from Elsevier.

[Pages 291–306]

3. Reprinted article 87, listed in Chapter 7:

Remarks on Yang–Mills theory, T. T. Wu (Harvard U. and CERN), Symposium on Symmetries and Reflections in Honor of C. N. Yang (1999), 199–204.

Reprinted with permission from Tai Tsun Wu, Symmetry and Modern Physics, pp. 199–204 (2003). World Scientific Publishing Company, https://doi.org/10.1142/9789812795083_0013.

[Pages 307–312]

4. Reprinted article 110, listed in Chapter 8:

Anharmonic oscillator, *Phys. Rev.* **184** (1969), 1231–1260, Carl M. Bender (Harvard U.), Tai Tsun Wu (Harvard U.), DOI: https://doi.org/10.1103/PhysRev.184.1231; Citations: 1039.

Reprinted with permission from Carl M. Bender and Tai Tsun Wu, The Physical Review, vol. 184, no. 5, 1231–1260 (1969). Copyright (1969) by the American Physical Society.

[Pages 313–342]

5. Reprinted article 125, listed in Chapter 8:

Limit of Cross-Sections at Infinite Energy, *Phys. Rev. Lett.* **24** (1970), 1456–1460, In *Lo, S. Y. (ed.): Geometrical pictures in hadronic collisions*, 277–281. (see Book Index), Hung Cheng (MIT, LNS), T. T. Wu (Harvard U.), DOI: https://doi.org/10.1103/PhysRevLett.24.1456; Citations: 322.

Reprinted with permission from Hung Cheng and Tai Tsun Wu, Physical Review Letters, vol. 24, no. 25, 1456–1460 (1970). Copyright (1970) by the American Physical Society.

[Pages 343–347]

6. Reprinted article 158, listed in Chapter 8:

Concept of Nonintegrable Phase Factors and Global Formulation of Gauge Fields, *Phys. Rev. D* **12** (1975), 3845–3857, Also in Yang, C.N.: Selected Papers 1945–1980, 460–472, Tai Tsun Wu (Harvard U.), Chen Ning Yang (SUNY, Stony Brook), DOI: https://doi.org/10.1103/PhysRevD.12.3845; Citations: 1059.

[Pages 348–360]

7. Reprinted article 233, listed in Chapter 9:

Theory of a Two-Dimensional Ising Model with Random Impurities I. Thermodynamics, *Phys. Rev.* **176**, 631 (1968), Barry M. McCoy (SUNY, Stony Brook), Tai Tsun Wu (Harvard U.); DOI: https://doi.org/10.1103/PhysRev.176.631; Citations: 373.

[Pages 361–373]

8. Reprinted article 236, listed in Chapter 9:

Spin spin correlation functions for the two-dimensional Ising model: Exact theory in the scaling region, *Phys. Rev. B* **13** (1976), 316–374, Tai Tsun Wu (Harvard U.), Barry M. McCoy (SUNY, Stony Brook), Craig A. Tracy (Rochester U.), Eytan Barouch (Clarkson U.), DOI: https://doi.org/10.1103/PhysRevB.13.316; Citations: 568.

[Pages 374–432]

9. Reprinted article 264, listed in Chapter 10:

Phenomenological Analysis of Violation of CP Invariance in Decay of K0 and anti-K0, *Phys. Rev. Lett.* **13** (1964), 380–385, In Yang, C.N.: Selected Papers 1945-1980, 359-364, and in *Wolfenstein, L. (ed.): CP violation* 36–41, T.T. Wu (Brookhaven), Chen-Ning Yang (Brookhaven), DOI: https://doi.org/10.1103/PhysRevLett.13.380; Citations: 305.

Reprinted with permission from Tai Tsun Wu and Chen-Ning Yang, Physical Review Letters, vol. 13, no. 12, 380–385 (1964). Copyright (1964) by the American Physical Society.

[Pages 433–438]

10. Reprinted article 302, listed in Chapter 10:

Single Bremsstrahlung Processes in Gauge Theories, *Phys. Lett. B* **103** (1981), 124–128, Frits A. Berends (Leiden U.), R. Kleiss (Leiden U.), P. De Causmaecker (Leuven U.), R. Gastmans (Leuven U.), Tai Tsun Wu (Harvard U.), DOI: https://doi.org/10.1016/0370-2693(81)90685-7; Citations: 391.

Reprinted from Physics Letter B, vol. 103, no. 2, pp. 124–128 (1981), Frits A. Berends, R. Kleiss, P. De Causmaecker, R. Gastmans and Tai Tsun Wu, Single Bremsstrahlung Processes in Gauge Theories, with permission from Elsevier.
Reprinted from The Lancet, 103, Frits A. Berends, R. Kleiss, P. De Causmaecker, R. Gastmans and Tai Tsun Wu, Single Bremsstrahlung Processes in Gauge Theories, 124–128, (1981), with permission from Elsevier.

[Pages 439–443]

11. Reprinted article 397, listed in Chapter 11:

Concept of the basic standard model and a relation between the three gauge coupling constants, *Int. J. Mod. Phys. A* **38** (2023) 28, 2350128, Tai Tsun Wu (Wisconsin U., Madison), Sau Lan Wu (Wisconsin U., Madison), DOI: https://doi.org/10.1142/S0217751X23501282; Downloaded 13,132 times.

Reprinted with permission from Tai Tsun Wu and Sau Lan Wu, International Journal of Modern Physics A, vol. 38, no. 28, 2350128 (36 pp.) (2023). World Scientific Publishing Company. https://www.worldscientific.com/doi/10.1142/S0217751X23501282.

[Pages 444–479]

1. Reprinted article 21, listed in Chapter 6:

IEEE TRANSACTIONS ON ANTENNAS AND PROPAGATION

The Cylindrical Antenna with Nonreflecting Resistive Loading

T. T. WU AND R. W. P. KING, FELLOW, IEEE

Abstract—The distribution of current along a center-driven cylindrical antenna is obtained when the material forming the antenna is resistive. The particular case is considered when the impedance per unit length of the antenna is a function of the distance from the end. A solution is obtained specifically when the current is represented by an outward traveling wave with no reflected wave. The admittance of the antenna and the far-field pattern is determined. Field patterns are evaluated for a wide range of lengths. These are characterized by a single major lobe with a very small minor lobe structure.

INTRODUCTION

FOR SOME PURPOSES, the directional and broadband properties of traveling-wave antennas are desirable. An example is the traveling-wave V-antenna. The first work on traveling-wave dipoles was reported by Altschuler[1] who inserted lumped resistors at a quarter wavelength from the ends of the antenna. Although this location of the resistors is not critical, the traveling-wave nature of the current diminishes as the frequency is changed so that the lumped resistors are no longer at the maximum of the current.

In a recent report[2] the distribution of current and the driving-point admittance were determined for a cylindrical antenna with a continuously distributed constant internal impedance per unit length. It is now proposed to investigate the cylindrical antenna with a variable internal impedance per unit length. In particular, it is desired to determine an axial distribution of the internal impedance for which a pure outward traveling wave exists on an antenna of finite length.

THE DIFFERENTIAL EQUATION AND ITS SOLUTION

The axial component $A_z(z)$ of the vector potential on the surface of a cylindrical antenna, that has the internal impedance per unit length $z^i(z)$, carries a total axial current $I_z(z)$, and is driven at $z=0$ by a delta-function generator with emf V_0^e, satisfies the one-dimensional wave equation in the form

$$\left(\frac{\partial^2}{\partial x^2}+k_0^2\right)A_z(z)=\frac{jk_0^2}{\omega}\left[z^i(z)I_z(z)-V_0^e\delta(z)\right] \quad (1)$$

if a time-dependence $e^{j\omega t}$ is assumed. The internal impedance per unit length $z^i(z)$ is expressed as a function of the axial coordinate z. It is given by

$$z^i(z)=\frac{1}{2\pi a d(z)\sigma(z)} \quad (2)$$

for a circular tube with constant radius a. In order to vary the impedance per unit length, it is assumed that the conductivity σ and the wall thickness d may be functions of location along the antenna. The vector potential on the surface of the antenna is

$$A_z(z)=\frac{\mu_0}{4\pi}\int_{-h}^{h}I_z(z')K(z,z')dz' \quad (3)$$

where

$$K(z,z')=\frac{e^{-jk_0r}}{r} \quad (4)$$

with

$$r=\sqrt{(z-z')^2+a^2} \quad (5)$$

Since the ratio of vector potential to current along an antenna is approximately constant, it is possible to set

$$\int_{-h}^{h}I_z(z')K(z,z')dz' \doteq I_z(z)\Psi \quad (6)$$

where Ψ is the value where the current $I_z(z)$ has a maximum.

With (3)–(6) it follows that

$$\left(\frac{\partial^2}{\partial z^2}+k_0^2\right)4\pi\mu_0^{-1}A_z(z) \doteq \frac{j4\pi k_0}{\zeta_0}\left[z^i(z)I_z(z)-V_0^e\delta(z)\right] \quad (7)$$

may be approximated by

$$\left(\frac{\partial^2}{\partial z^2}+k_0^2\right)I_z(z)=\frac{j4\pi k_0}{\zeta_0\Psi}\left[z^i(z)I_z(z)-V_0^e\delta(z)\right] \quad (8)$$

With the notation

$$f(z)=\frac{4\pi}{\zeta_0\Psi}z^i(z) \quad (9)$$

Manuscript received August 10, 1964. The work reported in this paper was prepared under Grant NsG 579 at the Gordon McKay Lab., Harvard University, Cambridge, Mass., for the National Aeronautics and Space Administration.

The authors are with the Gordon McKay Lab., Harvard University, Cambridge, Mass.

[1] Altshuler, E. E., The traveling-wave linear antenna, *IRE Trans. on Antennas and Propagation*, vol AP-9, Jul 1961, pp 324–329.

[2] King, R. W. P., and W. T. Tsun, The imperfectly conducting cylindrical transmitting antenna, Tech Rept 440, Cruft Lab., Harvard University, Cambridge, Mass., Mar 1964.

where $\zeta_0 = \sqrt{\mu_0/\epsilon_0} = 120\pi$ ohms, this equation becomes

$$\left[\frac{\partial^2}{\partial z^2} + k_0^2 - jk_0 f(z)\right] I(z) = -\frac{j4\pi k_0}{\zeta_0 \Psi} V_0^e \delta(z) \quad (10)$$

Except at the driving point $z=0$, the current must satisfy the differential equation

$$\left[\frac{\partial^2}{\partial z^2} + k_0^2 - jk_0 f(z)\right] I_z(z) = 0 \quad (11)$$

It is readily verified by direct substitution in (11) that when

$$f(z) = \frac{2}{h - |z|} \quad (12)$$

so that (11) becomes

$$\left(\frac{\partial^2}{\partial z^2} + k_0^2 - \frac{j2k_0}{h - |z|}\right) I_z(z) = 0 \quad (13)$$

a solution is

$$I_z(z) = C(h - |z|)e^{-jk_0|z|} \quad (14)$$

Note, in particular, that a solution of the form $e^{jk|z|}$ does not satisfy the equation.

The expression (14) represents a wave of current traveling in the direction of increasing $|z|$, that is, from the generator toward both ends. There is no reflected wave traveling in the opposite direction.

The Impedance and Expansion Parameter

If the current has the form (14), it follows that the vector potential is given by

$$4\pi\mu_0^{-1} A_z(z) = \Psi I_z(z) = \Psi C(h - |z|)e^{-jk_0|z|} \quad (15)$$

The scalar potential satisfies the Lorentz condition

$$\phi(z) = j\frac{\omega}{k_0^2}\frac{\partial A_z(z)}{\partial z} \quad (16)$$

Also, by symmetry, $\phi(-z) = -\phi(z)$. For $z \geq 0$

$$\phi(z) = \frac{j\omega\mu_0}{4\pi k_0^2}\Psi C e^{-jkz}[-1 - jk_0(h - z)] \quad (17)$$

$$\phi(+0) = \frac{j\omega\mu_0}{4\pi k_0^2}\Psi C(1 + jk_0 h) = \frac{-j}{4\pi\omega\epsilon_0}\Psi C(1 + jk_0 h) \quad (18)$$

If the driving voltage is defined by

$$V_0^e = \phi(+0) - \phi(-0) = 2\phi(+0) \quad (19)$$

it follows that

$$C = \frac{j2\pi\omega\epsilon_0 V_0^e}{\Psi(1 + jk_0 h)} \quad (20)$$

Hence,

$$I(z) = \frac{2\pi V_0^e}{\zeta_0\Psi(1 - j/k_0 h)}\left(1 - \frac{|z|}{h}\right)e^{-jk_0|z|} \quad (21)$$

The driving-point admittance is

$$Y_0 = \frac{2\pi}{\zeta_0\Psi}\frac{1}{1 - j/k_0 h} \quad (22)$$

The impedance is a resistance in series with a capacitance

$$Z_0 = R_0 - j/\omega C_0 \quad (23)$$

where

$$R_0 = \frac{\Psi\zeta_0}{2\pi} = 60\Psi \text{ ohms} \quad (24)$$

and

$$C_0 = \epsilon_0 h \quad (25)$$

Note that when $k_0 h \gg 1$, $R_0 \gg 1/\omega C_0$.

The parameter Ψ is defined in terms of the function

$$\Psi(z) = \frac{\int_0^h (h - z')e^{-jkz'}\left[\frac{e^{-jkr_1}}{r_1} + \frac{e^{-jkr_2}}{r_2}\right]dz'}{(h - z)e^{-jkz}} \quad (26)$$

where

$$r_1 = \sqrt{(z' - z)^2 + a^2} \qquad r_2 = \sqrt{(z' + z)^2 + a^2}$$

Since $I(z)$ and $A_z(z)$ both have maximum amplitudes at $z=0$, it is desirable to define $\Psi = \Psi(0)$. That is,

$$\Psi = 2\int_0^h \left(1 - \frac{z'}{h}\right)e^{-jkz'}\frac{e^{-jkr_0}}{r_0}dz' \quad (27)$$

where

$$r_0 = \sqrt{z'^2 + a^2}$$

Since $k_0 a \ll 1$, and $a \ll h$, no serious error is made by setting $kz' \doteq kr_0$ in the exponent in the first integral and $r_0 \doteq z'$ in the second integral.

$$\Psi \doteq 2\int_0^h \frac{e^{-j2kr_0}}{r_0}dz' - \frac{2}{h}\int_0^h e^{-j2kz'}dz' \quad (28)$$

With $A = ka$, it follows that

$$\Psi \doteq 2\left[\sinh^{-1}\frac{h}{a} - C(2A, 2kh) - jS(2A, 2kh)\right] + \frac{1}{kh}(1 - e^{-j2kh}) \quad (29)$$

Specifically, when $kh=\pi/2$, $h/a=75$, $\Omega=10$, $ka=1.57/75=0.021$, $2ka=0.042$,

$$\Psi \doteq 2[5.70 - 1.66 - j1.85] + j\frac{2}{\pi}(1+1)$$

$$= 8.08 - j2.43$$

Similarly, for a thin antenna with $h/a=11{,}013$, or $\Omega=20$, $ka=1.57/11,013=1.41\times10^{-4}$

$$\Psi \doteq 2[10.69 - 1.65 - j1.85] + j\frac{4}{\pi} = 18.08 - j2.43$$

In (29), $C(a, x)$ and $S(a, x)$ are the generalized sine and cosine integrals:

$$C(a, x) = \int_0^x \frac{1-\cos W}{W}du \qquad S(a, x) = \int_0^x \frac{\sin W}{W}du$$

where

$$W = (u^2 + a^2)^{1/2}$$

The Distributed Resistive Loading

The continuously varying resistive loading of the antenna is defined by (9) with (12). Thus

$$z^i(z) = \frac{\zeta_0\Psi}{8\pi}\frac{1}{h-|z|} = \frac{15\Psi}{h-|z|} \tag{30}$$

where the coefficient 15 is in ohms. With $\Omega=10$ to 20, the coefficient 15Ψ ohms ranges from $121-j36$ ohms to $272-j36$ ohms. At $\lambda=288$ m with $\Omega=20$, $h=72m$ so that

$$z^i(0) = 15\Psi/h = \frac{272-j36}{72} = 3.9 - j.5 \text{ ohms}/m$$

is the impedance per unit length at the driving point. In this case $a=6.54$ mm. With $z^i=\frac{1}{2}\pi a d\sigma$, $z_i=0.66$, 6.6, and 66 ohms/m for aluminum of thickness $d=10^{-6}$, 10^{-7}, and 10^{-8} m and $z^i=8.55$, 85.5, and 855 ohms/m for carbon of thickness $d=10^{-4}$, 10^{-5}, and 10^{-6} m. Thus, with thin layers of aluminum or carbon, a range from below 3.9 ohms/m at $z=0$ to very large values as $z\to h$ may be constructed.

Efficiency

The efficiency of the resistively loaded antenna is easily evaluated since

$$P_t = \frac{1}{2}|I(0)|^2R_0 = \frac{\zeta_0\Psi}{4\pi}|I(0)|^2$$

whereas the power dissipated in heat is

$$P_h = 2\int_0^h \frac{1}{2}|I(z)|^2 z^i(z)dz = \frac{\zeta_0\Psi}{8\pi}|I(0)|^2.$$

It follows that the antenna is 50 per cent efficient.

The Electromagnetic Field

The far-zone electric field is given by

$$E_\Theta^r = j\omega \sin\Theta\, A_z^r \tag{31}$$

where

$$A_z^r = \frac{e^{-jk_0r_0}}{4\pi\nu_0 r_0}\int_{-h}^{h} I_z(z')e^{jk_0z'\cos\Theta}dz' \tag{32}$$

With

$$I(z') = C(h - |z|)e^{-jk_0|z|} \tag{33}$$

this expression becomes

$$A_z^r = \frac{Ce^{-jk_0r_0}}{2\pi\nu_0 r_0}\int_0^h (h-z')e^{-jk_0z'}\cos(k_0z'\cos\Theta)dz' \tag{34}$$

The integral is readily evaluated with the following formulas:

$$\int e^{ax}\cos bx\,dx = \frac{e^{ax}}{a^2+b^2}(a\cos bx + b\sin bx) \tag{35a}$$

$$\int xe^{ax}\cos bx\,dx = \frac{e^{ax}}{a^2+b^2}\left\{\left[ax - \frac{a^2-b^2}{a^2+b^2}\right]\cos bx + \left[bx - \frac{2ab}{a^2+b^2}\right]\sin bx\right\} \tag{35b}$$

Thus,

$$\int_0^h (h-z')e^{az'}\cos bz'dz'$$
$$= \frac{e^{az'}}{a^2+b^2}\left\{\left[a(h-z') + \frac{a^2-b^2}{a^2+b^2}\right]\cos bz' + \left[b(h-z') + \frac{2ab}{a^2+b^2}\right]\sin bz'\right\}$$
$$= \frac{1}{(a^2+b^2)^2}\{[(a^2-b^2)\cos bh + 2ab\sin bh]e^{ah} - [ah(a^2+b^2) + a^2 - b^2]\} \tag{36}$$

With $a=-jk_0$, $b=k_0\cos\Theta$, $a^2+b^2=-k_0^2\sin^2\Theta$, $a^2-b^2=-k^2(1+\cos^2\Theta)$, it follows that

$$\int_0^h (h-z')e^{-jk_0z'}\cos(k_0z'\cos\Theta)dz'$$
$$= \frac{[-(1+\cos^2\Theta)\cos(k_0h\cos\Theta) - j2\cos\Theta\sin(k_0h\cos\Theta)]e^{-jk_0h} - jk_0h\sin^2\Theta + (1+\cos^2\Theta)}{k_0^2\sin^4\Theta} \tag{37}$$

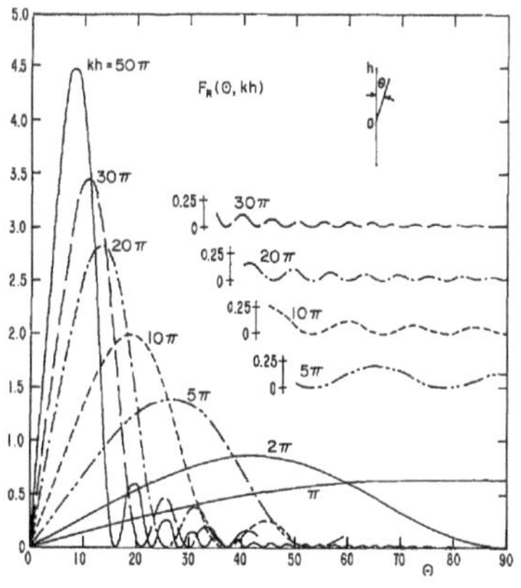

Fig. 1. Real part of far-field pattern of nonreflecting antenna.

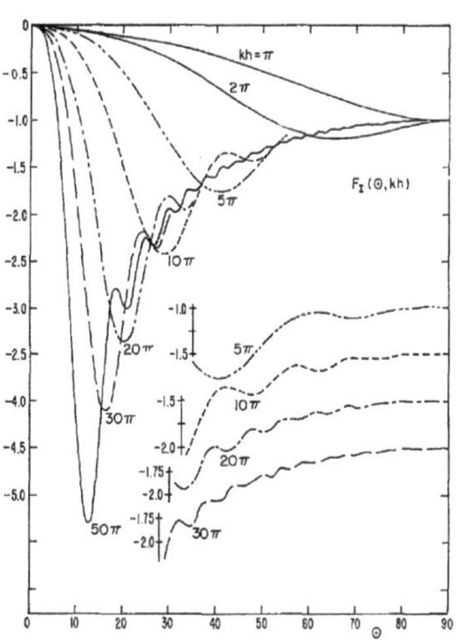

Fig. 2. Imaginary part of far-field pattern of nonreflecting antenna.

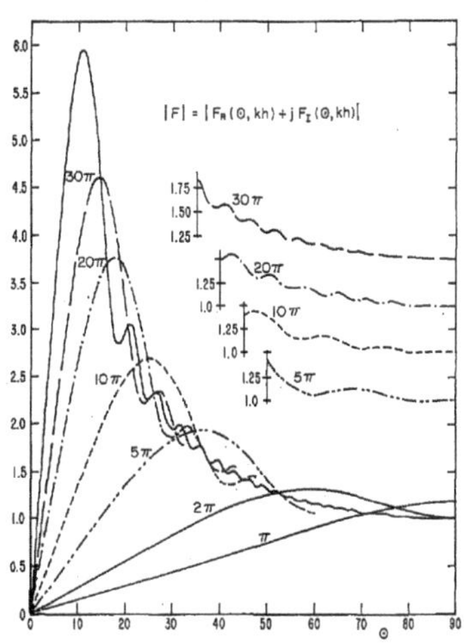

Fig. 3. Magnitude of far-field pattern of nonreflecting antenna.

The electric field is

$$E_\Theta{}^r = \frac{j\zeta_0 hCe^{-jk_0r}}{2\pi r}F(k_0h,\Theta) \tag{38a}$$

where the vertical field factor is

$$F(k_0h,\Theta) = \frac{-jk_0h\sin^2\Theta + (1+\cos^2\Theta) - [j2\cos\Theta\sin(k_0h\cos\Theta) + (1+\cos^2\Theta)\cos(k_0h\cos\Theta)]e^{-jk_0h}}{k_0h\sin^3\Theta} \tag{38b}$$

This function vanishes along the axis $\Theta=0$ and has the value

$$F\left(k_0h,\frac{\pi}{2}\right) = \frac{-j(k_0h-\sin k_0h)+(1-\cos k_0h)}{k_0h} \tag{39}$$

in the equatorial plane, $\Theta=\pi/2$. When $k_0{}^2h^2 \ll 1$,

$$F(k_0h,\Theta) = k_0h\sin\Theta \tag{40}$$

which is the same as for any electrically short antenna. The real and imaginary parts are

$$F_R(k_0h,\Theta) = \frac{(1+\cos^2\Theta)[1-\cos k_0h\cos(k_0h\cos\Theta)] - 2\cos\Theta\sin k_0h\sin(k_0h\cos\Theta)}{k_0h\sin^3\Theta} \tag{41a}$$

$$F_I(k_0h,\Theta) = \frac{-k_0h\sin^2\Theta - 2\cos\Theta\cos k_0h\sin(k_0h\cos\Theta) + (1+\cos^2\Theta)\sin k_0h\cos(k_0h\cos\Theta)}{k_0h\sin^3\Theta} \tag{41b}$$

When

$$k_0h = \frac{\pi}{2}$$

$$F_R\left(\frac{\pi}{2},\Theta\right) = \frac{1+\cos^2\Theta - 2\cos\Theta\sin\left(\frac{\pi}{2}\cos\Theta\right)}{\frac{\pi}{2}\sin^3\Theta} \tag{42a}$$

$$F_I\left(\frac{\pi}{2},\Theta\right) = \frac{-\frac{\pi}{2}\sin^2\Theta + (1+\cos^2\Theta)\cos\left(\frac{\pi}{2}\cos\Theta\right)}{\frac{\pi}{2}\sin^3\Theta} \tag{42b}$$

At

$$\Theta = \frac{\pi}{2},\qquad F_R\left(\frac{\pi}{2},\frac{\pi}{2}\right) = \frac{2}{\pi},\qquad F_I\left(\frac{\pi}{2},\frac{\pi}{2}\right) = \frac{2}{\pi}-1.$$

When $k_0h \gg 1$ $F(k_0h,\Theta) \to F_I(k_0h,\Theta) \to -\csc\Theta$, $\Theta \neq 0$.

Graphs of $F_R(k_0h,\Theta)$, $F_I(k_0h,\Theta)$, and $|F(k_0h,\Theta)|$ are given in Figs. 1–3 for a range of values of k_0h extending from $\pi/2$ to 50π. It is seen that $|F(k_0h,\Theta)|$ has one large maximum that is located at $\Theta=90°$ for $k_0h \leq \pi$ and moves toward $\Theta=0$ as k_0h is increased. With $k_0h=50\pi$ the maximum is near $\Theta=11°$. Of particular interest is the fact that minor lobes are little more than a small ripple on the broad tail of the major maximum for all values of $k_0h>\pi$.

Conclusion

Properties of a center-driven cylindrical antenna, characterized by a pure traveling wave of current, was investigated. Combination of two such antennas into traveling-wave V-antennas will be treated in another paper.

2. Reprinted article 70, listed in Chapter 7:

Nuclear Physics B107 (1976) 365- 380

DIRAC MONOPOLE WITHOUT STRINGS: MONOPOLE HARMONICS

Tai Tsun WU *
Gordon McKay Laboratory, Harvard University, Cambridge, Massachusetts 02138

and

Chen Ning YANG **
Institute for Theoretical Physics, State University of New York, Stony Brook, New York 11794

Received 17 February 1976

Using the ideas developed in a previous paper which are borrowed from the mathematics of fibre bundles, it is shown that the wave function ψ of a particle of charge Ze around a Dirac monopole of strength g should be regarded as a *section*. The section is without discontinuities. Thus the monopole *does not* possess strings of singularities in the field around it. The eigensections of the angular momentum operators are monopole harmonics which are explicitly exhibited.

1. Introduction

In this paper, and a later on classical Lagrangian dynamics, we study the formulation of Dirac's magnetic monopoles without strings. The two papers are, however, logically and technically independent, and may be read separately.

Very soon after Dirac's original paper [1], Tamm [2] studied the wave function of an electrically charged particle around a magnetic monopole. He introduced "generalized spherical harmonics" for such wave functions. These harmonics possess discontinuities or cusps. Later Fierz [3] discussed these harmonics from a different point of view†.

Since the space around a monopole is spherically symmetrical and without singularities, the wave function of a positon or electron around the monopole *should* have no singularities. An examination of this question using the concepts developed

* Work supported in part by the US ERDA under contract no. E(11-1)-3227.
** Work supported in part by the National Science Foundation under grant MPS74-13208 A01.
† There are many papers after the publication of refs. [1–3] dealing with magnetic monopoles and generalized spherical harmonics which are more or less similar to the essence of these papers. We shall not refer to these later papers here.

in a recent paper [4] shows that this is indeed the case. By a conceptual change, we shall look at the generalized spherical harmonics from a new view point and shall call them *monopole harmonics*. The monopole harmonics are *everywhere analytic* and possess no discontinuities or cusps at all. They form a complete orthonormal set and can be used as the basis of expansion of any wave function around the monopole.

In this new view point, the wave function of an electrically charged particle of charge Ze around a monopole of strength g *should not be thought of as an ordinary function.* It should instead be considered as a "*section*" characterized by a number q defined by

$$q = \tfrac{1}{2}DZ\,, \tag{1}$$

where $D = 2eg$ = monopole strength in Dirac's unit which is $(2e)^{-1}$. We have put

$$c = \hbar = 1\,.$$

D is an integer which may be positive, negative or zero. So is $2q$.

The concept of a section is familiar in the mathematics of fibre bundles. For the case in question, the wave function is mathematically [5] a section on a C^1 vector bundle, or a line bundle.

2. Wave function as a section

The basic new point is understandable as follows. The cusps and discontinuities arise because any choice of the vector potential $\boldsymbol{A}$ around the monopole must [4] have singularities. The situation is similar to that encountered in the choice of a coordinate system on the surface of a sphere, such as the longitude and latitude system. No choice is possible which does not have some singularities. Yet the geometry of the sphere is clearly without intrinsic singularities. To avoid introducing singularities in the coordinate system one divides the sphere into more than one overlapping region and defines a singularity-free coordinate system in each region. In the overlap one has singularity-free coordinate transformations between the different coordinate systems.

Imitating this method, we divide [4] the space outside of a magnetic monopole into two regions, R_a and R_b, and define a vector potential $(A_\mu)_\mathrm{a}$ in R_a and a vector potential $(A_\mu)_\mathrm{b}$ in R_b. Using spherical coordinates r, θ, ϕ with the monopole at the origin we choose

$$R_\mathrm{a}: \quad 0 \leqslant \theta < \tfrac{1}{2}\pi + \delta\,, \qquad 0 < r\,, \qquad 0 \leqslant \phi < 2\pi\,, \tag{2}$$

$$R_\mathrm{b}: \quad \tfrac{1}{2}\pi - \delta < \theta \leqslant \pi\,, \qquad 0 < r\,, \qquad 0 \leqslant \phi < 2\pi\,, \tag{3}$$

$$R_\mathrm{ab}: \quad \tfrac{1}{2}\pi - \delta < \theta < \tfrac{1}{2}\pi + \delta\,, \qquad 0 < r\,, \qquad 0 \leqslant \phi < 2\pi \quad \text{(overlap)}, \tag{4}$$

where we choose δ such that $0 < \delta \leqslant \frac{1}{2}\pi$.

The vector potentials are chosen to be

$$(A_r)_a = (A_\theta)_a = 0 , \qquad (A_\phi)_a = \frac{g}{r \sin\theta}(1 - \cos\theta) ,$$

$$(A_r)_b = (A_\theta)_b = 0 , \qquad (A_\phi)_b = \frac{-g}{r \sin\theta}(1 + \cos\theta) , \tag{5}$$

where A_r, A_θ, A_ϕ are the projections of A in the three local orthogonal directions. One has

$$(A_\mu)_a = (A_\mu)_b + \frac{i}{Ze} S_{ab} \frac{\partial S_{ab}^{-1}}{\partial x^\mu} , \tag{6}$$

where

$$S = S_{ab} = e^{2iq\phi} = \text{transition function} . \tag{7}$$

S is the gauge transformation phase factor for changing from $(A_\mu)_b$ to $(A_\mu)_a$ in the overlap R_{ab},

$$\psi_a = S_{ab} \psi_b , \tag{8}$$

where ψ_a and ψ_b are the wave function of a particle of charge Ze in R_a and R_b, respectively. A function ξ which assumes values ξ_a and ξ_b in R_a and R_b and satisfies

$$\xi_a = S_{ab} \xi_b = e^{2iq\phi} \xi_b \tag{9}$$

in the overlap R_{ab} is called a *section. ψ is thus a section.*

Let the charged particle interact with the monopole and with a potential $V(r)$ which is spherically symmetrical. We assume $V(r)$ to be without singularities for $r > 0$. Then

$$\frac{1}{2m} (\boldsymbol{p} - Ze\boldsymbol{A})^2 \psi + V\psi = E\psi , \tag{10}$$

meaning

$$\frac{1}{2m} (\boldsymbol{p} - Ze\boldsymbol{A}_a)^2 \psi_a + V\psi_a = E\psi_a , \qquad \text{in } R_a , \tag{10a}$$

$$\frac{1}{2m} (\boldsymbol{p} - Ze\boldsymbol{A}_b)^2 \psi_b + V\psi_b = E\psi_b , \qquad \text{in } R_b . \tag{10b}$$

It is obvious that these equations are compatible with (8) because of (6).

3. Hilbert space of sections

It is clear that if ξ is a section, $x\xi$ is also a section. Also $(p - ZeA)_x\xi$ is a section. Thus $\boldsymbol{r}$ and $\boldsymbol{p} - Ze\boldsymbol{A}$ are operators on the Hilbert space of sections. We define the scalar product of two sections ξ, η as

$$(\eta, \xi) \equiv \int \eta^* \xi \, \mathrm{d}^3 r \,. \tag{11}$$

(The question of convergence at $r = \infty$ and $r = 0$ is here ignored.) This integral is well defined because in the overlap R_{ab}

$$\eta_{\mathrm{b}}^* \xi_{\mathrm{b}} = \eta_{\mathrm{a}}^* \xi_{\mathrm{a}} \,.$$

It is clear that $\boldsymbol{r}$ and $\boldsymbol{p} - Ze\boldsymbol{A}$ are Hermitian operators. Following Fierz [3] we shall now try to construct angular momentum operators.

Define

$$\boldsymbol{L} = \boldsymbol{r} \times (\boldsymbol{p} - Ze\boldsymbol{A}) - \frac{q\boldsymbol{r}}{r} \,. \tag{12}$$

It is clear that L_x, L_y, L_z are Hermitian operators on the Hilbert space of sections. The following commutation rules can be easily verified:

$$[L_x, x] = 0 \,, \qquad [L_x, y] = iz \,, \qquad [L_x, z] = -iy \,,$$

$$[L_x, p_x - ZeA_x] = 0 \,, \qquad [L_x, p_y - ZeA_y] = i(p_z - ZeA_z) \,,$$

$$[L_x, p_z - ZeA_z] = -i(p_y - ZeA_y) \,. \tag{13}$$

It follows from these that

$$[L_x, L_y] = iL_z \,, \qquad \text{etc.} \tag{14}$$

Eq. (13), together with its consequence (14), show that L_x, L_y, L_z are *the angular momentum operators* [3]. We emphasize that neither the Hilbert space, nor these operators, possess any "singularities". (The singularities of A_{a} and A_{b} are not real singularities because they occur outside of R_{a} and R_{b}, respectively.)

4. Monopole harmonics $Y_{q,l,m}$

Since $[r^2, \boldsymbol{L}] = 0$, we can diagonalize r^2 and study operators $\boldsymbol{L}$ for fixed r^2. I.e. we shall study sections of the form

$$\delta(r^2 - r_0^2)\xi \,,$$

where ξ is a section dependent only on angular coordinates θ and ϕ.

$\boldsymbol{L}$ operates then on "angular sections". In the rest of this paper except sect. 11, we shall be dealing with angular sections only.

Eq. (14) shows that $[\boldsymbol{L}^2, L_z] = 0$. Simultaneous diagonalization produces the familiar multiplets with eigenvalues $l(l+1)$ and m,

$$\boldsymbol{L}^2 Y_{q,l,m} = l(l+1)\, Y_{q,l,m}\,; \; L_z Y_{q,l,m} = m Y_{q,l,m}\,, \tag{15}$$

where $l = 0, \frac{1}{2}, 1, \ldots$ and, for each value of l, m ranges from $-l$ to $+l$ in integral steps of increment. The $Y_{q,l,m}$ are the eigensections which we shall call *monopole harmonics.* We shall show later that the allowed values of l and m are

$$l = |q|, |q|+1, |q|+2, \ldots, \qquad m = -l, -l+1, \ldots, l\,, \tag{16}$$

and that each of these l, m combinations occur exactly once. We shall choose each Y normalized so that

$$\int_0^{\pi} \sin\theta\, \mathrm{d}\theta \int_0^{2\pi} |Y_{q,l,m}|^2\, \mathrm{d}\phi = 1\,. \tag{17}$$

(Notice that in R_{ab}, $|(Y_{q,l,m})_{\mathrm{a}}|^2 = |(Y_{q,l,m})_{\mathrm{b}}|^2$.) Different $Y_{q,l,m}$ (for a fixed q) are orthogonal, a fact one easily proves in the usual way from (15). We shall choose the phases of $Y_{q,l,m}$ such that the matrix elements of L_z, L_y, L_z between the Y's conform to the convention adopted in ch. 2 of Edmonds' book [6]. In particular

$$(L_x + iL_y)\, Y_{q,l,m} = (l-m)^{1/2}(l+m+1)^{1/2}\, Y_{q,l,m+1}\,. \tag{18}$$

These monopole harmonics will be explicitly exhibited. Each is analytic. That is, $(Y_{q,l,m})_{\mathrm{a}}$ is analytic in R_{a} and $(Y_{q,l,m})_{\mathrm{b}}$ is analytic in R_{b}. The set of all monopole harmonics for a fixed q forms a complete set of sections, as we shall see.

5. Explicit expressions for $Y_{q,l,m}$

Stating from (12) one easily verifies

$$\boldsymbol{L}^2 = [\boldsymbol{r} \times (\boldsymbol{p} - Ze\boldsymbol{A})]^2 + q^2\,, \tag{19}$$

$$m Y_{q,l,m} = L_z Y_{q,l,m} = (-i\partial_\phi - q)\, Y_{q,l,m}\,, \qquad \text{in } R_{\mathrm{a}}\,,$$

$$m Y_{q,l,m} = L_z Y_{q,l,m} = (-i\partial_\phi + q)\, Y_{q,l,m}\,, \qquad \text{in } R_{\mathrm{b}}\,. \tag{20}$$

Eq. (20) shows that

$$Y_{q,l,m} = \Theta_{q,l,m}(\theta)\, \mathrm{e}^{i(m+q)\phi} \text{ in } R_{\mathrm{a}}\,,$$

$$Y_{q,l,m} = \Theta_{q,l,m}(\theta)\, \mathrm{e}^{i(m-q)\phi} \text{ in } R_{\mathrm{b}}\,. \tag{21}$$

The condition for a section, (9), shows that $[\Theta_{q,l,m}(\theta)]_a = [\Theta_{q,l,m}(\theta)]_b$ in the overlap. They are, in fact, the same function. Apply (19) to $Y_{q,l,m}$. An explicit evaluation of the operator $[\boldsymbol{r} \times (\boldsymbol{p} - Ze\boldsymbol{A})]^2$ acting on $Y_{q,l,m}$ gives

$$[l(l+1) - q^2]\,\Theta_{q,l,m} = \left[-\frac{1}{\sin\theta}\frac{\partial}{\partial\theta}\sin\theta\frac{\partial}{\partial\theta} + \frac{1}{\sin^2\theta}(m + q\cos\theta)^2\right]\Theta_{q,l,m}\,. \tag{22}$$

Writing $\cos\theta = x$, this gives

$$[l(l+1) - q^2]\,\Theta = -(1-x^2)\,\Theta'' + 2x\Theta' + \frac{1}{1-x^2}(m+qx)^2\,\Theta\;,$$

$$-1 \leqslant x \leqslant 1\;, \tag{23}$$

where prime means differentiation with respect to x. This equation can be treated in the usual way, through analyzing the indical equations at $x = \pm 1$. We shall, however, pursue a different method which yields the normalization constant and phase factor automatically.

Before proceeding we note that since Y is single valued in each region, (21) shows that

$$m - q = \text{integer}\;.$$

Thus

$$l - q = \text{integer}\;. \tag{24}$$

Now (19) shows that

$$l(l+1) \geqslant q^2\;. \tag{25}$$

Eqs. (24) and (25) show that the allowed values of l are among those given in (16).

We shall now show that each value of l in (16) is allowed, by constructing, for each of them, the explicit function $\Theta_{q,l,m}$:

$$\Theta_{q,l,-l} = N_{q,l}\sqrt{1-x}^{\,l-q}\sqrt{1+x}^{\,l+q}\,, \qquad l - |q| = \text{integer} \geqslant 0\,, \tag{26}$$

where

$$N_{q,l} = \left[\frac{(2l+1)!}{4\pi(2^{2l})(l-q)!\,(l+q)!}\right]^{1/2} > 0\,. \tag{27}$$

To show this one substitutes (26) into (23) and verifies that the latter is satisfied. The factor $N_{q,l}$ is inserted so that $Y_{q,l,-l}$ is normalized in the sense of (17).

Repeated application of (18) onto the monopole harmonics $Y_{q,l,-l}$ (given by (21) and (26)) leads to, (for l, m satisfying (26)), the explicit expression for $Y_{q,l,m}$

given below. (As stated above, this method leads to automatically normalized $Y_{q,l,m}$ starting from normalized $Y_{q,l,-l}$.)

$$(Y_{q,l,m})_a = M_{q,l,m}(1-x)^{\alpha/2}(1+x)^{\beta/2} P_n^{\alpha,\beta}(x)\, e^{i(m+q)\phi} ,$$

$$(Y_{q,l,m})_b = (Y_{q,l,m})_a\, e^{-2iq\phi} , \tag{28}$$

where

$$\alpha = -q-m, \qquad \beta = q-m, \qquad n = l+m, \qquad x = \cos\theta , \tag{29}$$

$$M_{q,l,m} = 2^m \left[\frac{2l+1}{4\pi} \frac{(l-m)!\,(l+m)!}{(l-q)!\,(l+q)!}\right]^{1/2} , \tag{30}$$

and $P_n^{\alpha,\beta}(x)$ are [7] the Jacobi polynomials,

$$P_n^{\alpha,\beta}(x) = \frac{(-1)^n}{2^n n!}(1-x)^{-\alpha}(1+x)^{-\beta} \frac{d^n}{dx^n}[(1-x)^{\alpha+n}(1+x)^{\beta+n}] , \tag{31}$$

which are defined if

$$n,\ n+\alpha,\ n+\beta \text{ and } n+\alpha+\beta \text{ are all integers} \geqslant 0 . \tag{32}$$

Eq. (28) will be proved in appendix A, and some properties of the Jacobi polynomials will be discussed in appendix B.

6. Completeness of monopole harmonics

For a given q (q may be negative) the set of $Y_{q,l,m}$ with l, m satisfying (16) form a complete set of orthonormal sections. I.e. every continuous section (i.e. a section satisfying (9), with ξ_a and ξ_b being continuous in R_a and R_b) can be expanded as a series

$$\sum_{l,m} a_{l,m} Y_{q,l,m} .$$

Proof: According to appendix C, $Y_{q,l,m}$ can be expressed in terms of $P_\nu^{|\alpha|,|\beta|}(x)$. Now for fixed q = integer or half-integer, and $q+m$ = integer, there are four possible cases:

$$\alpha \geqslant 0,\ \beta \geqslant 0,\ \text{so that } -m \geqslant |q| \quad \text{and} \quad \nu = l+m , \tag{33}$$

$$\alpha \geqslant 0,\ \beta \leqslant 0,\ \text{so that } |m| \leqslant -q, \quad q \leqslant 0 \quad \text{and } \nu = l+q , \tag{34}$$

$$\alpha \leqslant 0,\ \beta \geqslant 0,\ \text{so that } |m| \leqslant q,\ q \geqslant 0 \text{ and } \nu = l-q , \tag{35}$$

$$\alpha \leqslant 0,\ \beta \leqslant 0,\ \text{so that } m \geqslant |q| \quad \text{and } \nu = l-m . \tag{36}$$

In case (33), the allowed values of l, according to (16), are $l = |m|, |m| + 1, \ldots$ which are precisely

$$\nu = 0, 1, 2, \ldots . \tag{37}$$

In case (34), the allowed values of l according to (16) are $l = -q, -q + 1, \ldots$ which are also precisely (37). Continuing this way we conclude that given q = integer or half-integer, $q + m$ = integer, the allowed values of l according to (16) are always precisely those given by (37).

Now for fixed $|\alpha|, |\beta|$, the Jacobi polynomials $P_\nu^{|\alpha|,|\beta|}$, ($\nu = 0, 1, 2, \ldots$) form [7] a complete set. The exponential functions $e^{i\phi(m+q)}$, ($m + q$ = all integers) also form a complete set. It can be proved from these results that $Y_{q,l,m}$ forms a complete set of sections for fixed q.

7. Examples and analyticity of $Y_{q,l,m}$

For the case $q = 0$, $\alpha = \beta$, and (31) shows that

$$P_{l+m}^{-m,-m} = \frac{(-1)^m}{2^m} \frac{l!}{(l+m)!} (1 - x^2)^{m/2} P_l^m , \tag{38}$$

Table 1
Examples of $\sqrt{4\pi} Y_{q,l,m}$ in region a

q	l	m	$(\sqrt{4\pi} Y_{q,l,m})_a$
$\frac{1}{2}$	$\frac{1}{2}$	$\frac{1}{2}$	$-e^{i\phi}\sqrt{1-x}$
	$-\frac{1}{2}$	$-\frac{1}{2}$	$e^0\sqrt{1+x}$
	$\frac{3}{2}$	$\frac{3}{2}$	$\sqrt{3/2}\, e^{2i\phi}\sqrt{1+x}(1-x)$
	$\frac{3}{2}$	$\frac{1}{2}$	$-\sqrt{1/2}\, e^{i\phi}\sqrt{1-x}(1+3x)$
	$\frac{3}{2}$	$-\frac{1}{2}$	$-\sqrt{1/2}\, e^0\sqrt{1+x}(1-3x)$
	$\frac{3}{2}$	$-\frac{3}{2}$	$\sqrt{3/2}\, e^{-i\phi}\sqrt{1-x}(1+x)$
1	1	1	$\sqrt{3/4}\, e^{2i\phi}(1-x)$
	1	0	$-\sqrt{3/2}\, e^{i\phi}\sqrt{1-x^2}$
	1	-1	$\sqrt{3/4}\, e^0 (1+x)$

$x = \cos\theta$. To obtain $Y_{q,l,m}$ in R_b apply (9).

where P^l_m is the associated Legendre function. Substitution of (38) into (28) shows that

$$Y_{0,l,m} = \text{usual spherical harmonics } Y_{l,m} \,.$$

We tabulate in table 1 a few of the monopole harmonics for $q = \frac{1}{2}$, 1, These examples illustrate the fact that $Y_{q,l,m}$ *is analytic everywhere.* I.e., $(Y_{q,l,m})_a$ is analytic in R_a and $(Y_{q,l,m})_b$ is analytic in R_b. For example, $(Y_{\frac{1}{2}\frac{1}{2}\frac{1}{2}})_a$ is clearly analytic in R_a, which includes the point $\theta = 0$, and

$$(Y_{\frac{1}{2}\frac{1}{2}\frac{1}{2}})_{b\,1/2)_b} = \sqrt{1-\cos\theta}/\sqrt{4\pi} \tag{39}$$

is clearly analytic in R_b which includes the point $\theta = \pi$.

8. Zeros of $Y_{q,l,m}$

Table 1 shows that each of the $Y_{q,l,m}$ exhibited has at least one zero. This is in fact a special case of a general topological theorem that for $q \neq 0$, any continuous section must have at least one zero. This theorem can be proved as follows. If ξ is a continuous section and has no zeros, trace the value of $\xi_a|\xi_a|^{-1}$ in *the complex plane* as one goes along the parallel $r = 1$, $\theta = \theta_0$, from $\phi = 0 \to 2\pi$. $\xi_a|\xi_a|^{-1}$ (= the phase of ψ_a) describes a loop which is confined to the unit circle. As θ_0 changes, the loop is *continuously* distorted, remaining always on the unit circle. As $\theta_0 \to 0$, the loop shrinks to a point. Thus for any θ_0 satisfying $0 < \theta_0 < \frac{1}{2}\pi + \delta$ (see (2)), the loop is always shrinkable, along the circle, to a point. The same is true for the loop described by $\xi_b|\xi_b|^{-1}$ for $\frac{1}{2}\pi - \delta < \theta_0 < \pi$. Now take $\theta_0 = \frac{1}{2}\pi$. These two last statements together contradict (9) if $q \neq 0$.

9. Global gauge transformation on $Y_{q,l,m}$

The monopole harmonics exhibited above are for a special gauge [4] in which the regions R_a, R_b and the vector potential A_μ were chosen to be that given in (2) → (5). One can make global gauge transformations [4] which change the regions, the vector potential A_μ, and the value of $Y_{q,l,m}$ in a coordinated manner.

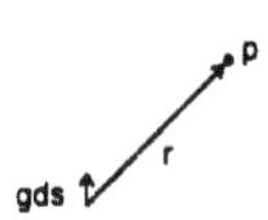

Fig. 1. Pseudomagnetic field produced by current segment ds. It is equal to $g\boldsymbol{ds} \times \boldsymbol{r}r^{-3}$ where g = current.

A lateral wave is the electromagnetic wave near the interface between two different planar media such as air and sea water. Although most antennas are located near an interface, the effect of an interface is often ignored or described as "ground clutter." In many situations, an understanding of the effect of the interface is essential. After the publication of the book Professor Wu extended the work in a number of directions such as non-planar interfaces.

6) The Two-Dimensional Ising Model, Harvard University Press, 1973. Second edition Dover, 2014.

Barry M. McCoy, Stony Brook University
Tai Tsun Wu, Harvard University

1. INTRODUCTION

Statistical mechanics is an old and venerable branch of physics which has received the attention of many physicists from the time of Gibbs. Since then it has developed in several directions so that at present it is possible to distinguish at least three different theoretical approaches to the subject: (1) the foundational approach, which is concerned with establishing the general properties of, and proving existence theorems for, statistical mechanical systems by rigorous mathematical means, (2) the phenomenological approach, which is concerned with correlating and quantitatively explaining the results of experiments by any available method, and (3) the model-building approach, which attempts to gain insight into practical situations by studying simple models in which at least some physically interesting quantities may be exactly computed. Each of these approaches has made such valuable contributions to our understanding of statistical mechanics that it is neither feasible nor desirable to separate them completely. However, because each approach has developed such a large body of literature, it is likewise not possible to give an adequate treatment of all of them in a single book. Therefore, while we will attempt to make this book self-contained by giving a brief discussion of the foundation of statistical mechanics, and while we will attempt to place the book in a somewhat broader context by making contact with the existing experimental situation, we will concentrate our efforts on the study of certain solvable models.

The number of exactly solvable problems in a field depends on the complexity of the subject. For example, there are innumerable solvable problems in classical mechanics, whereas, at the other extreme, very few problems in relativistic quantum field theory have ever been exactly netic field of the monopole in R_a and thus we may use $(A')_a$ as the vector potential in R_a.

Using $(A'_\mu)_a$ and $(A_\mu)_b$ as the vector potentials requires a global gauge transformation from $(A_\mu)_a$ and $(A_\mu)_b$. In other words, we can find a transformation phase factor $T_{a'a}$ such that

$$(A'_\mu)_a = (A_\mu)_a + \frac{i}{Ze} T_{a'a} \frac{\partial T_{a'a}^{-1}}{\partial x^\mu} \text{ in } R_a . \tag{44}$$

For any section ξ,

$$\xi_{a'} = T_{a'a} \xi_a . \tag{45}$$

Comparison of (44) and (42) shows that

$$T_{a'a} = \exp(-i Zeg\,\Omega) = \exp(-i q\Omega) , \tag{46}$$

where Ω is the solid angle subtended by circuit $C + (-L)$ in fig. 2c at the point where $T_{a'a}$ is evaluated. Actually Ω is defined and is continuous not just in R_a but in all space outside of a surface bordered by the closed circuit $C + (-L)$. Take the surface to be the shaded area in fig. 2c. On the surface, but outside of the border $C + (-L)$, Ω increases discontinuously by 4π in going from above the diagram to underneath the diagram. Since D is an integer, $T_{a'a}$ is single valued.

The transition function in the overlap region R_{ab} is now

$$S_{a'b} = T_{a'a} S_{ab} = \exp(2iq\phi - iq\Omega) . \tag{47}$$

We have thus defined completely the new gauge: regions R_a, R_b and transition function $S_{a'b}$. We have also defined the gauge field in this new gauge: $(A'_\mu)_a$, $(A_\mu)_b$. We have, in addition, exhibited the global gauge transformation $T_{a'a}$ between the old gauge and the new.

Notice that the continuation of $(A'_\mu)_a$ into R_b yields singularities not on L, but on C.

If we had taken another half curve C' outside of R_a, we would have gotten another new vector potential $(A''_\mu)_a$. Any linear combination

$$(A'''_\mu)_a = \alpha(A_\mu)_a + \alpha'(A'_\mu)_a + \alpha''(A''_\mu)_a , \tag{48}$$

with

$$\alpha + \alpha' + \alpha'' = 1 ,$$

where α, α' and α'' are positive or negative real numbers, is also a possible vector potential in R_a. Notice that the continuation of $(A'''_\mu)_a$ into R_b has, in general, singularities on L, C and C'. Thus the position of the singularities of $(A_\mu)_a$, when continued into space not covered by R_a, is in general quite arbitrary.

10. Rotation of coordinate axes

A rotation of coordinate axes generates [6] a linear combination of the usual spherical harmonics,

$$Y_{0,l,m}(\theta',\phi') = \sum_{m'=-l}^{l} Y_{0,l,m'}(\theta,\phi)\mathcal{D}^{(l)}_{m'm}\,, \tag{49}$$

where $\mathcal{D}$ depends on the rotation. Does this hold also for the case $q \neq$ o? Define

$$Z_{q,l,m}(\theta',\phi') = \sum_{m'=-l}^{l} Y_{q,l,m'}(\theta,\phi)\mathcal{D}^{(l)}_{m'm}\,, \tag{50}$$

for $q \neq 0$. $Z_{q,l,m}$ is $Y_{q,l,m}(\theta'\phi')$ but in *a different gauge*, because the vector potential A_μ has not yet been changed to that which conforms with convention (5) for the new coordinate system. If one performs a global gauge transformation on $Z_{q,l,m}$ by first changing $(A_\mu)_\mathrm{a}$ so that its singularities after continuation become the new negative z-axis, and then changing $(A_\mu)_\mathrm{b}$ so that its singularities after continuation become the new positive z-axis then $Z_{q,l,m} \to Y_{q,l,m}(\theta',\phi') \times$ (phase factor which is independent of m).

11. Schrödinger equation

It is simple to show by explicit evaluation, and with the aid of (19) that

$$(\boldsymbol{p} - Ze\boldsymbol{A})^2 = -\frac{1}{r^2}\frac{\partial}{\partial r}\left(r^2\frac{\partial}{\partial r}\right) + \frac{1}{r^2}\left[\boldsymbol{r}\times(\boldsymbol{p} - Ze\boldsymbol{A})\right]^2$$

$$= -\frac{1}{r^2}\frac{\partial}{\partial r}\left(r^2\frac{\partial}{\partial r}\right) + \frac{1}{r^2}\left[\boldsymbol{L}^2 - q^2\right]. \tag{51}$$

The Hamiltonian in (10) thus commutes with $\boldsymbol{L}^2$ and L_z. Hence in solving (10) we can choose specific eigenvalues for $\boldsymbol{L}^2$ and L_z. I.e. we take

$$\psi = R(r)Y_{q,l,m}\,, \tag{52}$$

obtaining

$$\left[-\frac{1}{2mr^2}\frac{\partial}{\partial r}\left(r^2\frac{\partial}{\partial r}\right) + \frac{l(l+1) - q^2}{2mr^2} + V - E\right]R = 0\,. \tag{53}$$

For the case that $V = 0$ this equation was solved by Tamm [2] who found that R is a Bessel function, if $E > 0$,

$$R = \frac{1}{\sqrt{kr}}\,J_\mu(kr)\,, \tag{54}$$

where

$$\mu = \sqrt{l(l+1) - q^2 + \tfrac{1}{4}} = \sqrt{(l+\tfrac{1}{2})^2 - q^2} > 0 \,, \qquad (55)$$

$$k = \sqrt{2mE} \,.$$

If $E \leqslant 0$, (53) has no meaningful solution.

It is a pleasure to thank Professor Shiing-shen Chern for enlightening us on the mathematical concepts of fibre bundles and sections.

Appendix A

Proof of (28)

A straight forward computation shows that in R_a

$$L_x + iL_y = e^{i\phi}\left[-\sqrt{1-x^2}\,\frac{\partial}{\partial x} + i\frac{x}{\sqrt{1-x^2}}\frac{\partial}{\partial \phi} - q\sqrt{\frac{1-x}{1+x}}\right], \qquad (A.1)$$

where

$$x = \cos\theta \,.$$

Substitute this into (18) and use (21). One obtains, in R_a,

$$\Theta_{q,l,m+1} = \frac{1}{\sqrt{(l-m)(l+m+1)}}\left[-\sqrt{1-x^2}\,\frac{d}{dx} - \frac{mx}{\sqrt{1-x^2}} - q\frac{1}{\sqrt{1-x^2}}\right]\Theta_{q,l,m}$$

$$= \frac{-\sqrt{1-x^2}^{\,m+1}}{\sqrt{(l-m)(l+m+1)}}\left[\frac{d}{dx} + q\,\frac{1}{1-x^2}\right]\sqrt{1-x^2}^{\,-m}\,\Theta_{q,l,m} \,. \qquad (A.2)$$

Repeated application of (A.2) gives

$$\Theta_{q,l,m} = (\mathrm{const})\,(1-x^2)^{m/2}\left[\frac{d}{dx} + q\,\frac{1}{1-x^2}\right]^{l+m}(1-x^2)^{l/2}\,\Theta_{q,l,-l} \,. \qquad (A.3)$$

Now use

$$\frac{d}{dx} + q\,\frac{1}{1-x^2} = \sqrt{\frac{1-x}{1+x}}^{\,q}\,\frac{d}{dx}\,\sqrt{\frac{1+x}{1-x}}^{\,q} \,. \qquad (A.4)$$

One obtains eq. (28) from (A.3). (The constant in (A.3) can be evaluated explicitly in the repeated application of (A.2).)

Appendix B

Some properties of $P_n^{\alpha,\beta}$

$P_n^{\alpha,\beta}(x)$ as defined by (31) and (32) is a polynomial of degree n. It satisfies

$$P_n^{\alpha,\beta}(-x) = (-1)^n P_n^{\beta,\alpha}(x)\,, \tag{B.1}$$

$$P_n^{\alpha,\beta}(x) = 2^{-n} \sum_{\lambda=0}^{n} \frac{(n+\alpha)!}{\lambda!(n+\alpha-\lambda)!} \frac{(n+\beta)!}{(n-\lambda)!(\beta+\lambda)!} (x-1)^{n-\lambda}(x+1)^{\lambda}, \tag{B.2}$$

in which $m!$ is defined to be ∞ when $m < 0$ and $0! = 1$. To prove (B.2) we arrange the square bracket in (31) into a product of $2n + \alpha + \beta$ factors, each being $(1-x)$ or $(1+x)$, and choose λ factors $(1-x)$ and $n - \lambda$ factors $(1+x)$ for differentiation. (B.2) then follows

We shall now show

$$P_{n+\alpha}^{-\alpha,\beta} = 2^{-\alpha}(x-1)^{\alpha} \frac{n!\,(n+\alpha+\beta)!}{(n+\beta)!\,(n+\alpha)!} P_n^{\alpha,\beta}\,, \tag{B.3}$$

$$P_{n+\beta}^{\alpha,-\beta} = 2^{-\beta}(x+1)^{\beta} \frac{n!\,(n+\alpha+\beta)!}{(n+\beta)!\,(n+\alpha)!} P_n^{\alpha,\beta}\,, \tag{B.4}$$

$$P_{n+\alpha+\beta}^{-\alpha,-\beta} = 2^{-\alpha-\beta}(x-1)^{\alpha}(x+1)^{\beta} P_n^{\alpha,\beta}\,. \tag{B.5}$$

To show (B.3) we use (B.2),

$$2^{n+\alpha} P_{n+\alpha}^{-\alpha,\beta} = \sum_{\lambda=0}^{n+\alpha} \frac{n!}{\lambda!(n-\lambda)!} \frac{(n+\alpha+\beta)!}{(n+\alpha-\lambda)!(\beta+\lambda)!} (x-1)^{n+\alpha-\lambda}(x+1)^{\lambda}$$

$$= \sum_{\lambda=0}^{n} \text{same} = (x-1)^{\alpha} \frac{n!(n+\alpha+\beta)!}{(n+\alpha)!\,(n+\beta)!} 2^n P_n^{\alpha,\beta}\,,$$

which leads to (B.3). Eq. (B.4) can be proved similarly. Eq. (B.5) can be proved by using (B.3) and (B.4) in succession.

Define

$$R_n^{\alpha,\beta} \equiv (1-x)^{\alpha/2}(x+1)^{\beta/2}\, 2^{-(\alpha+\beta)/2} P_n^{\alpha,\beta}\,. \tag{B.6}$$

Then (B.3), (B.4) and (B.5) together show that

$$R_n^{\alpha,\beta} = (-1)^{(\alpha-|\alpha|)/2} R_\nu^{|\alpha|,|\beta|}\,, \tag{B.7}$$

where

$$\nu = n + \tfrac{1}{2}(\alpha+\beta-|\alpha|-|\beta|)\,. \tag{B.8}$$

Using (28) as the definition of $Y_{q,l,m}$, we obtain by utilizing (B.5)

$$Y^*_{q,l,m} = (-1)^{q+m} Y_{-q,l,-m} , \tag{B.9}$$

which is a useful formula. It is correct in both regions R_a and R_b. If we take $q = 0$, (B.6) reduces to the usual formula for the complex conjugate of $Y_{l,m}$.

Appendix C

Alternative expression for $Y_{q,l,m}$

Using (28) and (B.6) we obtain

$$(Y_{q,l,m})_a = M_{q,l,m} 2^{(\alpha+\beta)/2} R_n^{\alpha,\beta} e^{i(m+q)\phi} . \tag{C.1}$$

Now use (B.7) to obtain

$$(Y_{q,l,m})_a = (\text{const}) (1-x)^{|\alpha|/2} (1+x)^{|\beta|/2} P_\nu^{|\alpha|,|\beta|} e^{i(m+q)\phi} , \tag{C.2}$$

where ν is given by (B.8), and n, α and β are given by (29).

Appendix D

Clebsch-Gordan coefficients

We shall define the usual Clebsch-Gordan coefficients

$$\langle lml'm' | ll'jm_j \rangle \tag{D.1}$$

as in ref. [6]. Some usage of these coefficients for combining sections will be discussed below:

(a) Consider the product of two sections $Y_{q,l,m}(\theta, \phi) Y_{q',l',m'}(\theta, \phi)$ *of the same argument* θ, ϕ. The result is clearly a section with $q'' = q + q'$. The usual vector addition theorem applies and we have

$$\sum_{mm'} Y_{q,l,m} Y_{q',l',m'} \langle lml'm' | ll'jm_j \rangle = K Y_{q+q',j,m_j} , \tag{D.2}$$

where K depends on q, l, q', l', and j but not on m. Notice that sometimes K is zero. For example, for $Y_{q,l,m} Y_{q,l,m'}$, it is well known that the CG coefficients are symmetrical (with respect to $m \leftrightarrow m'$) for $j = 2l -$ even integer and antisymmetrical for $j = 2l -$ odd integer. For the latter case clearly $K = 0$. Notice also that if $j < |q + q'|$, then the right-hand side of (D.2) must vanish, since Y_{q+q',j,m_j} does not then exist.

For example, for the case $Y_{1,1,m}Y_{0,1,m'}$ the final j value is, *a priori*, 2, 1 or 0. But the case $j = 0$ vanishes since $Y_{1,0,m_j}$ does not exist. This can indeed be checked with the aid of tables 1, and the appropriate values of $Y_{0,1,m'}$ and the Clebsch-Gordan coefficients

$$Y_{1,1,1}(\sqrt{1-x^2}\,e^{-i\phi}) - Y_{1,1,0}(\sqrt{2}x) + Y_{1,1,-1}(-\sqrt{1-x^2}\,e^{i\phi}) = 0\,. \qquad \text{(D.3)}$$

Similarly, $Y_{1,1,m}Y_{1,1,m'}$ can be linearly combined, *a priori*, to give $Y_{2,2,m_j}$ $Y_{2,1,m}$ and $Y_{2,0,m}$. But the latter two do not exist, giving rise to the following identity which can be checked with table 1

$$Y_{1,1,1}Y_{1,1,-1} - Y_{1,1,0}Y_{1,1,0} + Y_{1,1,-1}Y_{1,1,1} = 0\,. \qquad \text{(D.4)}$$

(b) For a problem with two particles of different charges Ze, $Z'e$ moving in the field of a magnetic monopole, the wave function is a "double" section with respect to both $\boldsymbol{r}$ and $\boldsymbol{r}'$. Then

$$\sum_{mm'} Y_{q,l,m}(\theta,\phi)\,Y_{q',l',m'}(\theta',\phi')\langle ll'jm_j|lml'm'\rangle = F_{q,q',j,m_j}\,, \qquad \text{(D.5)}$$

is a double section that transforms under a simultaneous rotation of θ and ϕ like Y_{j,m_j} does. One has to, however, remember that after the rotation one is using a different gauge, as discussed before in sect. 10.

(c) For a particle with spin $\boldsymbol{S}$, the total angular momentum is

$$\boldsymbol{J} = \boldsymbol{L} + \boldsymbol{S} = \boldsymbol{r} \times (\boldsymbol{p} - Ze\boldsymbol{A}) - \frac{q\boldsymbol{r}}{r} + \boldsymbol{S}\,. \qquad \text{(D.6)}$$

The addition of $\boldsymbol{L}$ and $\boldsymbol{S}$ is achieved with the Clebsch-Gordan coefficients in the usual way with no difficulty.

References

[1] P.A.M. Dirac, Proc. Roy. Soc. A133 (1931) 60.
[2] Ig. Tamm, Z. Phys. 71 (1931) 141.
[3] M. Fierz, Helv. Phys. Acta 17 (1944) 27.
[4] Tai Tsun Wu and Chen Ning Yang, Phys. Rev. D12 (1975) 3845.
[5] S.S. Chern, Complex manifolds without potential theory (Van Nostrand, 1967).
[6] A.R. Edmonds, Angular momentum in quantum mechanics (Princeton University Press, 1960).
[7] Bateman Manuscript Project, Higher transcendental functions, ed. A. Erdelyi (McGraw-Hill, 1953) vol. 2, p. 168; ref. [6], p. 57–58.

3. Reprinted article 87, listed in Chapter 7:

REMARKS ON YANG–MILLS THEORY

TAI TSUN WU*
Gordon McKay Laboratory, Harvard University, Cambridge, MA 02138, USA
and
Theory Division, CERN, CH-1211 Geneva 23, Switzerland

The Yang–Mills theory is the most important development in physics in the second half of this century. Some of the highlights are briefly summarized.

During my first year as a graduate student at Harvard, Professor Yang came to give a Monday Physics Colloquium. It was on the newly conceived Yang–Mills theory. Since I had not yet learned quantum field theory, I was not able to follow his discussions. It is my understanding that in 1954 Professor Yang gave just two seminars on this new theory, the first one at the Institute for Advanced Study, and this was the second one.

If I remember correctly, Professor Yang gave this Colloquium in May of 1954. If this is correct, then it was almost exactly forty-five years ago. It was the first time I met Professor Yang.

Shortly after carrying out this work, Yang and Mills gave a summary at the Washington meeting of the American Physical Society[1] and also submitted their famous paper for publication in the *Physical Review.*[2] The abstract of their talk is reproduced as Fig. 1; it took place on April 30, 1954, with Professor Freeman Dyson presiding.

The title and abstract of Ref. 2 is shown in Fig. 2, and some of the basic equations from this paper are shown in Fig. 3. There are a number of remarkable features in this paper besides the fundamental concept of non-Abelian gauge fields; let me mention just the following two.

First, after the Introduction section, Gauge Transformation is discussed before the Field Equations, as seen from Fig. 3. This is the authors' way of emphasizing the importance of gauge transformation.

Secondly, near the end of this paper, just after their Fig. 1 which shows the self-interaction of the Yang–Mills b field, there is a paragraph discussing its mass. This paragraph is reproduced as Fig. 4. We shall return to this prophetic paragraph later.

*Presented at *Symmetries and Reflections*, a Symposium in honor of Professor C. N. Yang, May 22, 1999. Work supported in part by the United States Department of Energy under Grant No. DE-FG02-84ER40158.

M7. Isotopic Spin Conservation and a Generalized Gauge Invariance.* C. N. YANG† AND R. MILLS, *Brookhaven National Laboratory*— **The conservation of isotopic spin points to the existence of a fundamental invariance law similar to the conservation of electric charge. In the latter case, the electric charge serves as a source of electromagnetic field; an important concept in this case is that a gauge invariance which is closely connected with (1) the equation of motion of the electromagnetic field, (2) the existence of a current density, and (3) the possible interactions between a charged field and the electromagnetic field. We have tried to generalize this concept of gauge invariance to apply to isotopic spin conservation. It turns out that a very natural generalization is possible. The field that plays the role of the electromagnetic field is here a vector field that satisfies a nonlinear equation even in the absence of other fields. (This is because unlike the elecromagnetic field this field has an isotopic spin and consequently acts as a source of itself.) The existence of a current density is automatic, and the interaction of this field with any fields of arbitrary isotopic spin is of a definite form (except for possible terms similar to the anomalous magnetic moment interaction terms in electrodynamics).**

*Work performed under the auspices of the U. S. Atomic Energy Commission.

† On leave of absence from the Institute for Advanced Study, Princeton, New Jersey.

F. J. Dyson Presiding
April 30, 1954

Fig. 1. Reference 1 in its entirety.

Generalization to other Lie groups is straightforward. Therefore, the Yang–Mills theory gives a principle for writing down the interactions of elementary particles.

It is natural to ask the question: Which Lie groups are appropriate for physics? When Yang and Mills wrote their papers, there was no way for them to prefer one Lie group over another. This important question was answered in the 1960's by Glashow, Weinberg and Salam[3] for electroweak interactions and in 1973 by Fritzsch, Gell-Mann and Leutwyler[4] from strong interactions. The answers are SU(2) $\times$ U(1) for electroweak and SU(3) for strong interactions.

Aside from this important answer about the choice of Lie groups, some of the developments, both theoretical and experimental, of the Yang–Mills theory are shown in Fig. 5. For example, twenty years after 1954, the global formulation of Yang–Mills theory was accomplished.[5] This makes conection not only with non-integrable phase factors but also with fiber bundles. The resulting translation between the gauge-field and the fiber-bundle terminologies is reproduced in Table 1. The group SU(2) is readily replaced by other non-Abelian Lie groups.

Conservation of Isotopic Spin and Isotopic Gauge Invariance*

C. N. Yang† and R. L. Mills
Brookhaven National Laboratory, Upton, New York
(Received June 28, 1954)

It is pointed out that the usual principle of invariance under isotopic spin rotation is not consistent with the concept of localized fields. The possibility is explored of having invariance under local isotopic spin rotations. This leads to formulating a principle of isotopic gauge invariance and the existence of a b field which has the same relation to the isotopic spin that the electromagnetic field has to the electric charge. The b field satisfies nonlinear differential equations. The quanta of the b field are particles with spin unity, isotopic spin unity, and electric charge $\pm e$ or zero.

Fig. 2. The title and abstract of Ref. 2.

Table 1. Translation of terminology.

Gauge Field Terminology	Bundle Terminology
gauge (or global gauge)	principal coordinate bundle
gauge type	principal fiber bundle
gauge potential b_μ^k	connection on a principal fiber bundle
S_{ba}	transition function
phase factor Φ_{QP}	parallel displacement
field strength $F_{\mu\nu}^k$	curvature
source J_μ^k	?
electromagnetism	connection on a U(1) bundle
isotopic spin gauge field	connection on a SU(2) bundle
Dirac's monopole quantization	classification of U(1) bundle according to first Chern class
electromagnetism without monopole	connection on a trivial U(1) bundle
electromagnetism with monopole	connection on a nontrivial U(1) bundle

On the experimental side, twenty-five years after 1954, the Yang–Mills particle was observed for the first time. It was the one for strong interactions, and this feat was accomplished by the TASSO Collaboration at DESY.[6] Nearly thirty years after 1954, the Yang–Mills particle for electroweak interactions was observed for the first time by the UA(1) Collaboration at CERN.[7] These pioneering experimental observations completed the list of all the vector bosons of the standard model.[3,4]

ISOTOPIC GAUGE TRANSFORMATION

$$\psi = S\psi', \tag{1}$$

$$S(\partial_\mu - i\epsilon B'_\mu)\psi' = (\partial_\mu - i\epsilon B_\mu)\psi. \tag{2}$$

Combining (1) and (2), we obtain the isotopic gauge transformation on B_μ:

$$B'_\mu = S^{-1}B_\mu S + \frac{i}{\epsilon}S^{-1}\frac{\partial S}{\partial x_\mu}. \tag{3}$$

Define now

$$F_{\mu\nu} = \frac{\partial B_\mu}{\partial x_\nu} - \frac{\partial B_\nu}{\partial x_\mu} + i\epsilon(B_\mu B_\nu - B_\nu B_\mu). \tag{4}$$

One easily shows from (3) that

$$F'_{\mu\nu} = S^{-1}F_{\mu\nu}S. \tag{5}$$

FIELD EQUATIONS

$$\mathcal{L} = -\frac{1}{4}\mathbf{f}_{\mu\nu}\cdot\mathbf{f}_{\mu\nu} - \bar{\psi}\gamma_\mu(\partial_\mu - i\epsilon\tau\cdot\mathbf{b}_\mu)\psi - m\bar{\psi}\psi. \tag{11}$$

One obtains from this the following equations of motion:

$$\begin{aligned} \partial\mathbf{f}_{\mu\nu}/\partial x_\nu + 2\epsilon(\mathbf{b}_\nu \times \mathbf{f}_{\mu\nu}) + \mathbf{J}_\mu &= 0, \\ \gamma_\mu(\partial_\mu - i\epsilon\tau\cdot\mathbf{b}_\mu)\psi + m\psi &= 0, \end{aligned} \tag{12}$$

where

$$\mathbf{J}_\mu = i\epsilon\bar{\psi}\gamma_\mu\tau\psi. \tag{13}$$

Fig. 3. A few equations taken from Ref. 2.

We next come to the question of the mass of the b quantum, to which we do not have a satisfactory answer. One may argue that without a nucleon field the Lagrangian would contain no quantity of the dimension of a mass, and that therefore the mass of the b quantum in such a case is zero. This argument is however subject to the criticism that, like all field theories, the b field is beset with divergences, and dimensional arguments are not satisfactory.

Fig. 4. The discussion on the mass of the Yang–Mills b particle in Ref. 2.

1954 YANG-MILLS THEORY

Twenty years later

Integral formalism

non-integrable phase factor

fiber bundles

EXPERIMENT

Twenty-five years later

First experimental observation of a

Yang-Mills particle

Yang-Mills particles for strong inter.

– TASSO Collaboration at DESY

Thirty years later

Experimental observation of

Yang-Mills particles for electroweak

– UA(1) Collaboration at CERN

Fig. 5. Some of the developments of the Yang–Mills theory since 1954.

Let me return to the interesting paragraph of the Yang–Mills paper shown in Fig. 4. When Professor Yang gave the seminar on this new theory at the Institute for Advanced Study in February of 1954, Pauli repeatedly asked the question: What is the mass of this field B_μ? No definitive answer was possible at that time, and by now it is clear why. The Yang–Mills particle for strong interactions, now often called the gluon, is massless, while the Yang–Mills particles Z and W for electroweak interactions are massive, about 80–91 GeV/c^2. Therefore, both possibilities, massless and massive, can happen for Yang–Mills particles and actually do. This is the reason why the question raised by Pauli could not have been answered at that time, and also why the paragraph of Fig. 4 can only be called prophetic.

In summary, since 1954 particle physicists have been learning about and developing the Yang–Mills theory, which permeates throughout particle physics. The Yang–Mills theory is the most important and deepest development in physics for the second half of the twentieth century.

Acknowledgment

I would like to thank the Theory Division of CERN for its kind hospitality.

References

1. C. N. Yang and R. L. Mills, *Phys. Rev.* **95**, 631 (1954).
2. C. N. Yang and R. L. Mills, *Phys. Rev.* **96**, 191 (1954).
3. S. L. Glashow, *Nucl. Phys.* **22**, 579 (1961); S. Weinberg, *Phys. Rev. Lett.* **19**, 1264 (1967); A. Salam, *Proc. Eighth Nobel Symp.*, May 1968, ed. N. Svartholm (Wiley, 1968) p. 367.
4. H. Fritzsch, M. Gell-Mann and H. Leutwyler, *Phys. Lett.* **B47**, 365 (1973).
5. T. T. Wu and C. N. Yang, *Phys. Rev.* **D12**, 3845 (1975).
6. B. H. Wiik, *Proc. Int. Conf. Neutrinos, Weak Interactions and Cosmology*, 18–22 June 1979, p. 113; P. Söding, *Proc. European Society Int. Conf. High Energy Physics*, Geneva, Switzerland, 27 June–4 July, 1979, p. 271; TASSO Collaboration, R. Brandelik *et al.*, *Phys. Lett.* **B86**, 243 (1979).
7. UA(1) Collaboration, G. Arnison *et al.*, *Phys. Lett.* **B122**, 103, and **B126**, 398 (1983).

4. Reprinted article 110, listed in Chapter 8:

THE

PHYSICAL REVIEW

A journal of experimental and theoretical physics established by E. L. Nichols in 1893

SECOND SERIES, VOL. 184, NO. 5 25 AUGUST 1969

Anharmonic Oscillator

CARL M. BENDER*†
Department of Physics, Harvard University, Cambridge, Massachusetts 02138

AND

TAI TSUN WU
Division of Engineering and Applied Physics, Harvard University, Cambridge, Massachusetts 02138

(Received 4 February 1969)

We consider the anharmonic oscillator defined by the differential equation $(-d^2/dx^2+\frac{1}{4}x^2+\frac{1}{4}\lambda x^4)\Phi(x)=E(\lambda)\Phi(x)$ and the boundary condition $\lim_{x\to\pm\infty}\Phi(x)=0$. This model is interesting because the perturbation series for the ground-state energy diverges. To investigate the reason for this divergence, we analytically continue the energy levels of the Hamiltonian H into the complex λ plane. Using WKB techniques, we find that the energy levels as a function of λ, or more generally of λ^α, have an infinite number of branch points with a limit point at $\lambda=0$. Thus, the origin is not an isolated singularity. Level crossing occurs at each branch point. If we choose $\alpha=\frac{1}{3}$, the resolvent $(z-H)^{-1}$ has no branch cut. However, for all z it has an infinite sequence of poles which have a limit point at the origin. The anharmonic oscillator is of particular interest to field theoreticians because it is a model of $\lambda\varphi^4$ field theory in one-dimensional space-time. The unusual and unexpected properties exhibited by this model may give some indication of the analytic structure of a more realistic field theory.

I. INTRODUCTION

IN this paper, we discuss the familiar anharmonic oscillator.[1] We are concerned with the eigenvalue problem defined by the differential equation

$$(-d^2/dx^2+\tfrac{1}{4}x^2+\tfrac{1}{4}\lambda x^4)\Phi(x)=E(\lambda)\Phi(x) \quad (1.1)$$

and the associated boundary condition

$$\lim_{x\to\pm\infty}\Phi(x)=0. \quad (1.2)$$

We propose to answer the following questions within the context of the anharmonic oscillator:

(i) Is the perturbation series for the ground-state energy, which is a power series in the coupling constant λ, convergent for any $\lambda\neq0$?

(ii) If not, does the ground-state energy, considered as a function of complex λ or more generally as a function of some fractional power α of λ, have an isolated singularity at $\lambda=0$?

(iii) Is the resolvent $(z-H)^{-1}$, considered as a function of λ^α for fixed z, analytic at $\lambda=0$? If not, is the point $\lambda=0$ an isolated singularity?

The answer to all these questions is *no*. More precisely, the energy levels, which are originally defined only for positive values of λ, can be analytically continued into the complex λ plane. This analytic continuation has an infinite number of branch points, which have a limit point at the origin $\lambda=0$. Moreover, level crossing occurs at each branch point. If α is chosen to be $\frac{1}{3}$, the resolvent has no branch cut. However, for all z, $(z-H)^{-1}$ has an infinite number of poles which have a limit point at the origin. These qualitative analytic properties of the energy levels and of the resolvent were totally unexpected and are, we feel, most unusual and exciting.

This paper will adhere to the following outline: In Sec. II (and Appendices A–E), we inspect in great detail the properties of the perturbation series for the ground-state energy in our model. However, this direct inspection gives no clue as to the nature of the singularities in the λ plane which cause this divergence. To

* National Science Foundation Predoctoral Fellow. Present address: Institute for Advanced Study, Princeton, N. J. 08540.
† Supported in part by the Office of Naval Research under Contract No. Nonr-1866(55).
[1] The techniques used in this paper to analyze the anharmonic oscillator were first introduced by C. M. Bender and T. T. Wu, Phys. Rev. Letters 21, 406 (1968).

understand these singularities better, in Sec. III, we analytically continue the energy levels into the λ plane and derive some exact properties of this continuation. In particular, we derive an exact condition which gives the locations of the singularities in the λ plane.

To apply this condition we must know the wave function in coordinate space. To calculate the wave function approximately, in Sec. IV, we present a technique using zeroth-order WKB methods in the complex plane. (In Appendix F, we make the calculation of the wave function more precise by using first-order WKB techniques.)

In Sec. V, we use the results of Sec. IV to determine the singularities of the resolvent. In Sec. VI (and Appendix G), we combine the results of Secs. III and IV to find the approximate locations of the singularities in the analytic continuation of the energy levels. Section VI contains an exhaustive qualitative description of the analytic continuation of the energy levels $E(\lambda)$. The major result of this paper is that the energy levels of the anharmonic oscillator for a given positive real λ are the positive real values of $E(\lambda)$ on each of the infinite number of branches of a Riemann surface. Each energy level corresponds to a sheet of this Riemann surface.

Although it is not necessary to understand any field theory to read this paper (aside from Appendices A–C), it is important to point out that the anharmonic oscillator is a simple model field theory in one-dimensional space-time. This field theory is defined by the Hamiltonian

$$H=\tfrac{1}{2}\dot{\varphi}^2+\tfrac{1}{2}m^2\varphi^2+\lambda\varphi^4 \tag{1.3}$$

and the commutation relations

$$[\varphi,\dot{\varphi}]=i. \tag{1.4}$$

[In Appendix A, we prove rigorously that Eqs. (1.1) and (1.2) are consequences of Eqs. (1.3) and (1.4). In fact, Eqs. (1.1) and (1.2) are the coordinate representation of H.]

Since this field-theory model involves no space dimensions, there are no asymptotic states and therefore no particle scattering. Also, there is only one degree of freedom. In fact, all this theory describes is a universe which sits at one point and oscillates. Nevertheless, this model has value to the field theoretician. It enables one to conjecture with slight supporting evidence[2] that many of the fascinating properties we have observed are also present in a more realistic field theory. This model may give us some idea of the vast underlying complexity of such a theory.

The model was originally chosen for study because it had a well-defined but divergent perturbation series. We believe that the divergence of perturbation series in field theory was first discussed by Dyson.[3] Since then, many papers have been written on this topic. A particularly interesting investigation was carried out by Jaffe.[4] Jaffe proved that in two-dimensional space-time the perturbation series for Green's functions in self-interacting boson field theories diverge even though these series exist term by term. However, as far as we know, no techniques have been discovered to elucidate the singularities in field theories with divergent perturbation series. In this paper, we develop for the first time methods to describe the actual singularity structure of a particular field theory.

Having achieved the results of this paper, we hope in the future to apply the new analytical tools of Secs. III–VI to other models of increasing complexity in order to better understand how singularities depend upon the form of interaction. Additional zero-dimensional bose field theories which ought to be considered are self-interactions such as φ^6, φ^{2N}, and $\cos\varphi$ and couplings such as $\varphi_1^2\varphi_2^2$. Hopefully, these WKB techniques will be advanced to the point where they may be used to analyze higher-dimensional theories.

II. GROUND-STATE-ENERGY PERTURBATION SERIES

The perturbation series for the ground-state energy $E_0(\lambda)$ in this model is a power series in the coupling constant λ. This power series takes the form

$$E_0(\lambda)=\tfrac{1}{2}m+\sum_{n=1}^{\infty} mA_n(\lambda/m^3)^n. \tag{2.1}$$

Two methods for calculating the A_n in this equation are discussed below. Each method reveals some important properties of Eq. (2.1).

Method 1. In general, in a field theory the perturbation series for the ground-state energy is the sum of all connected Feynman diagrams having no external legs. Hence, Feynman diagrams can be used to calculate the A_n. (As an illustration of Feynman-diagram techniques, we calculate A_1, A_2, and A_3 in Appendix B.)

Moreover, an integral representation for these Feynman diagrams provides upper and lower bounds on the terms A_n in the perturbation series. We outline below the procedure for estimating these upper and lower bounds. (Detailed proofs for the following statements are given in Appendix C.)

(i) From the form of the Feynman integral representation, it follows that in the perturbation series all diagrams with the same number of vertices add in phase.

(ii) The integral representation places *uniform* bounds on the contribution of each diagram having n vertices.

[2] A. M. Jaffe, in his thesis [Princeton University, 1965 (unpublished)], was able to analytically continue the resolvent for $\lambda\varphi^4$ field theories of any space dimension into the cut λ plane. Jaffe did not predict the properties of the resolvent when $|\arg\lambda|\geq 180°$ (off the first sheet) but, as far as he did predict, his work agrees with ours.

[3] F. J. Dyson, Phys. Rev. 85, 631 (1952).

[4] A. M. Jaffe, Commun. Math. Phys. 1, 127 (1965). This paper contains a very complete list of references.

(iii) The maximum and minimum number of diagrams with n vertices is found to be $(2n-1)!!8^n$ and $(n-1)!3^{-n}$.

Combining the above three results implies that $A_n(-1)^{n+1}$ are positive numbers bounded below and above by

$$A\Gamma(n)B^n < A_n(-1)^{n+1} < C\Gamma(\tfrac{3}{2}n)D^n, \tag{2.2}$$

where A, B, C, and D are positive constants.[5] Equation (2.2) implies that, although the perturbation series [Eq. (2.1)] is finite in every order, the ground-state energy is not an analytic function of λ^α about $\lambda=0$ for any α.

Method 2. The A_n can also be calculated from a difference equation. To derive this difference equation we use the differential equation [Eq. (1.1)] which is

$$(-d^2/dx^2+\tfrac{1}{4}x^2+\tfrac{1}{4}\lambda x^4)\Phi(x)=E(\lambda)\Phi(x), \tag{2.3}$$

and the boundary conditions [Eq. (1.2)]

$$\lim_{x\to\pm\infty}\Phi(x)=0. \tag{2.4}$$

In Eq. (2.3), we substitute

$$\Phi(x)=\sum_{n=0}^{\infty}\lambda^n e^{-x^2/4}B_n(x), \tag{2.5}$$

where $B_0=1$ and $B_n(x)$ are the polynomials in x to be determined for $n=1, 2, 3, \cdots$.

We then combine Eqs. (2.1), (2.3), and (2.5) to get

$$\sum_{n=0}^{\infty}\lambda^n[xB_n'(x)-B_n''(x)+\tfrac{1}{4}x^4B_{n-1}(x)]$$
$$=\Big(\sum_{n=1}^{\infty}\lambda^nA_n\Big)\Big(\sum_{m=0}^{\infty}\lambda^mB_m(x)\Big). \tag{2.6}$$

Finally, we let $x=\sqrt{2}y$ and

$$B_i(y)=\sum_{j=1}^{2i}y^{2j}B_{i,j}(-1)^i. \tag{2.7}$$

The result is

$$2jB_{i,j}=(j+1)(2j+1)B_{i,j+1}+B_{i-1,j-2}$$
$$-\sum_{p=1}^{i-1}B_{i-p,1}B_{p,j}. \tag{2.8}$$

Equation (2.8) is the desired difference equation because the numbers $B_{i,j}$ are related to the A_n by the simple equation

$$A_n=-B_{n,1}. \tag{2.9}$$

[5] If the Hamiltonian were Wick-ordered, there would be fewer diagrams of order n because a line could not have both ends connected to the same vertex. Thus, the terms in the Wick-ordered perturbation series would be slightly smaller than those of our non-Wick-ordered series. However, the estimate [Eq. (2.2)] would hold for both perturbation series.

Equation (2.8) was used on a computer to calculate the first 75 terms in the ground-state-energy perturbation series [Eq. (2.1)].[6] From these results we have determined that their detailed asymptotic growth is

$$A_n\sim(-1)^{n+1}(6/\pi^3)^{1/2}\Gamma(n+\tfrac{1}{2})3^n. \tag{2.10}$$

[In Appendix D we discuss the details of this computer calculation and list the first 75 terms in the perturbation series to 12-place accuracy. In Appendix E, we present the calculation of Eq. (2.10).]

Numerical investigation of the difference equation [Eq. (2.8)] brings to light the following exact additional properties of the A_n which are not immediately evident from the Feynman-diagram treatment of the perturbation series:

(i) All A_n are rational fractions.
(ii) If we define

$$I_n\equiv(-1)^{n+1}4^nA_n, \tag{2.11}$$

then the I_n are all positive *integers*.
(iii) Every integer I_n is divisible by 3.

We list below the first nine terms in the perturbation series in exact fraction form to illustrate the above characteristics of the A_n.

$$A_1=\tfrac{3}{4}, \tag{2.12a}$$
$$A_2=-21/8, \tag{2.12b}$$
$$A_3=333/16, \tag{2.12c}$$
$$A_4=-30\,885/128, \tag{2.12d}$$
$$A_5=916\,731/256, \tag{2.12e}$$
$$A_6=-65\,518\,401/1024, \tag{2.12f}$$
$$A_7=2\,723\,294\,673/2048, \tag{2.12g}$$
$$A_8=-1\,030\,495\,099\,053/32\,768, \tag{2.12h}$$
$$A_9=54\,626\,982\,511\,455/65\,536. \tag{2.12i}$$

III. ANALYTIC CONTINUATION OF ENERGY LEVELS

In Sec. II and associated appendices, we studied the properties of the ground-state-energy perturbation series [Eq. (2.1)]. However, this study did not reveal the reason for the divergence of the series. In other words, the singularity structure of $E_0(\lambda)$ in the complex λ plane is as yet unknown. In this section, we define the analytic continuation of the energy levels, which are the eigenvalues of the differential equation [Eq. (1.1)], into the λ plane. Then we present some heuristic derivations of some properties of this continuation.

[6] One author (CMB) wishes to acknowledge partial financial support for this calculation from Dr. B. M. McCoy.

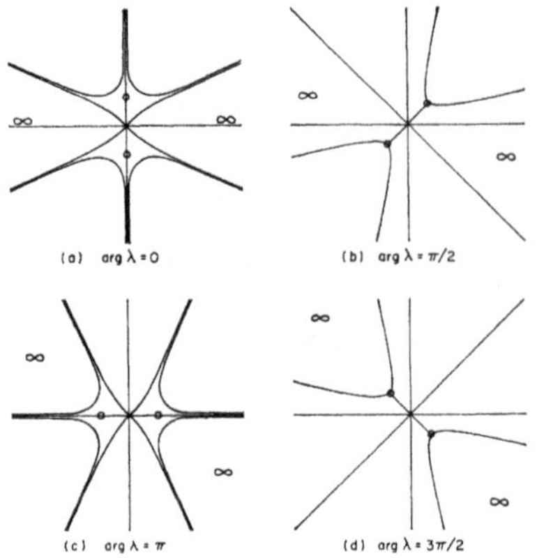

FIG. 1. Curves in the complex x plane where $\mathrm{Re}\,\lambda^{-1}[(1+\lambda x^2)^{1/2}-1]=0$ for various values of $\arg\lambda$. The circles denote the turning points at $x=\pm i\lambda^{-1/2}$. The symbol ∞ indicates the sectors in which the boundary condition [Eq. (3.4)] applies.

We shall need only Eqs. (1.1) and (1.2) which are

$$(-d^2/dx^2+\tfrac{1}{4}x^2+\tfrac{1}{4}\lambda x^4)\Phi(x)=E(\lambda)\Phi(x) \qquad (3.1)$$

and

$$\lim_{x\to\pm\infty}\Phi(x)=0. \qquad (3.2)$$

Equations (3.1) and (3.2) yield the asymptotic behavior of $\Phi(x)$ for large $|x|$ along the real axis:

$$\Phi(x)\sim\exp(-\tfrac{1}{6}\lambda^{1/2}|x|^3). \qquad (3.3)$$

Equation (3.3) implies that Eq. (3.2), the boundary condition, holds not only on the real axis but also for complex x in the sector $|\arg x|<\frac{1}{6}\pi$.

Until now λ has been positive. When λ is complex, $\Phi(x)$ must satisfy the general boundary condition

$$\lim_{|x|\to\infty}\Phi(x)=0 \quad\text{when } |\arg(\pm x)+(\tfrac{1}{6}\arg\lambda)|<\tfrac{1}{6}\pi, \qquad (3.4)$$

which follows from Eqs. (3.2) and (3.3).

We can now continue $E(\lambda)$ into the λ plane. We *define* the analytic continuation $E(\lambda)$ by the differential equation [Eq. (3.1)] in which $\Phi(x)$ satisfies the boundary condition in Eq. (3.4).

We shall use Eqs. (3.1) and (3.4) to find some exact properties of the analytic continuation of $E(\lambda)$:

Property (i). We may analytically continue $E(\lambda)$ to all points λ (except $\lambda=0$) where $\int_C\Phi^2(x,\lambda)dx\neq0$. The integral is taken along an appropriate path C in the complex x plane.

Proof. Normalize $\Phi(x)$ by choosing

$$\Phi(0,\lambda)=1 \text{ for an even-parity wave function} \qquad (3.5a)$$

and

$$(d/dx)\Phi(0,\lambda)=1 \text{ for an odd-parity wave function}. \qquad (3.5b)$$

Vary λ by

$$\lambda\to\lambda+\delta\lambda. \qquad (3.6)$$

Then,

$$E(\lambda)\to E(\lambda)+\delta E(\lambda) \qquad (3.7)$$

and

$$\Phi(x,\lambda)\to\Phi(x,\lambda)+\delta\Phi(x,\lambda). \qquad (3.8)$$

$\delta\Phi$ has the property that

$$\delta\Phi(0,\lambda)=0 \quad\text{or}\quad \delta\Phi'(0,\lambda)=0$$

for even or odd parity, respectively. $\delta\Phi(x,\lambda)$ obeys the same boundary conditions at ∞ as $\Phi(x,\lambda)$ does, namely, Eq. (3.4).

Substituting Eqs. (3.6), (3.7), and (3.8) into Eq. (3.1) gives

$$(d^2/dx^2-\tfrac{1}{4}x^2-\tfrac{1}{4}\lambda x^4)\delta\Phi(x,\lambda)-\tfrac{1}{4}\delta\lambda x^4\Phi(x,\lambda)$$
$$=-E(\lambda)\delta\Phi(x,\lambda)+\delta E(\lambda)\Phi(x,\lambda). \qquad (3.9)$$

Multiply Eq. (3.9) by $\Phi(x,\lambda)$. Then integrate by parts along a path C in the complex x plane such that each end of the path goes to ∞ in a sector in which the boundary condition [Eq. (3.4)] applies. [See Fig. 1 for the location of the sectors in which Eq. (3.4) applies for various choices of $\arg\lambda$.] The result of this manipulation is

$$0=\tfrac{1}{4}\delta\lambda\int_C x^4\Phi^2(x,\lambda)dx-\delta E(\lambda)\int_C\Phi^2(x,\lambda)dx. \qquad (3.10)$$

Hence, $E(\lambda)$ is analytic (its derivative with respect to complex λ exists) at all points except those for which

$$\int_C\Phi^2(x,\lambda)dx=0. \qquad (3.11)$$

Equation (3.11) will be used in Sec. VI as a condition for determining the location of the singularities of $E(\lambda)$.

Property (ii). At points where $E(\lambda)$ is analytic, there is no degeneracy in the energy.

Proof. Suppose $\Phi_1(x,\lambda)$ and $\Phi_2(x,\lambda)$ are two linearly-independent wave-function solutions to Eq. (3.1) each associated with energy E. We find that the derivative of the Wronskian $W'[\Phi_1(x),\Phi_2(x)]$ vanishes. Hence the Wronskian W is a constant. Because Φ_1 and Φ_2 obey the boundary condition [Eq. (3.4)], the constant must be 0. A vanishing Wronskian implies that Φ_1 and Φ_2 are not linearly-independent. Hence there is no degeneracy.

Property (iii). If there is a point λ_0 in the λ plane where $E_1(\lambda)\to E_2(\lambda)$ as $\lambda\to\lambda_0$, then $E_1(\lambda)$ is singular at $\lambda=\lambda_0$.

Proof. If $\Phi_i(x,\lambda)$ are the wave functions associated with $E_i(\lambda)$ for $i=1$ and 2, then because of property (ii)

$\Phi_1(x,\lambda) \to \Phi_2(x,\lambda)$ as $\lambda \to \lambda_0$. Consider the equation

$$\Phi_1(x,\lambda)(-d^2/dx^2+\tfrac{1}{4}x^2+\tfrac{1}{4}\lambda x^4)\Phi_2(x,\lambda) = E_2\Phi_1(x,\lambda)\Phi_2(x,\lambda). \quad (3.12)$$

Integrating Eq. (3.12) over the contour C in the complex x plane and integrating by parts gives

$$(E_1-E_2)\int_C \Phi_1(x,\lambda)\Phi_2(x,\lambda)dx=0. \quad (3.13)$$

Thus, when $\lambda\neq\lambda_0$,

$$\int_C \Phi_1(x,\lambda)\Phi_2(x,\lambda)dx=0. \quad (3.14)$$

Letting $\lambda\to\lambda_0$ in Eq. (3.14) gives

$$\int_C \Phi_2{}^2(x,\lambda)dx=0, \quad (3.15)$$

which is just Eq. (3.11). As a result of property (i), $E_1(\lambda)$ is singular at $\lambda=\lambda_0$. Thus, level crossing only occurs at singularity points of $E(\lambda)$.

Property (iv). The singularities of $E(\lambda)$ are of the square-root type for $\lambda\neq 0$.

Proof. Suppose $E(\lambda)$ has a singularity at $\lambda=\lambda_0\neq 0$. Let

$$\epsilon=(\lambda-\lambda_0)^{1/2}. \quad (3.16)$$

Then,

$$(-d^2/dx^2+\tfrac{1}{4}x^2+\tfrac{1}{4}\lambda_0 x^4+\tfrac{1}{4}\epsilon^2x^4)\Phi(x,\epsilon)=E(\epsilon)\Phi(x,\epsilon). \quad (3.17)$$

Expand

$$\Phi(x,\epsilon)=\Phi_0(x)+\epsilon\Phi_1(x)+\epsilon^2\Phi_2(x)+\cdots \quad (3.18)$$

and

$$E(\epsilon)=\alpha+\epsilon\beta+\epsilon^2\gamma+\cdots. \quad (3.19)$$

If we let $\Theta(x)=-d^2/dx^2+\frac{1}{4}x^2+\frac{1}{4}\lambda_0x^4$, then

$$\Theta(x)\Phi_0(x)=\alpha\Phi_0(x), \quad (3.20a)$$

$$\Theta(x)\Phi_1(x)=\alpha\Phi_1(x)+\beta\Phi_0(x), \quad (3.20b)$$

and

$$\Theta(x)\Phi_2(x)+\tfrac{1}{4}x^4\Phi_0(x)=\alpha\Phi_2(x)+\beta\Phi_1(x)+\gamma\Phi_0(x). \quad (3.20c)$$

Equation (3.20b) gives no useful information, but multiplying Eq. (3.20c) by $\Phi_0(x)$ and integrating gives

$$\beta=\int_C \tfrac{1}{4}x^4\Phi_0{}^2(x)dx \Big/ \int_C \Phi_1(x)\Phi_0(x)dx. \quad (3.21)$$

β exists if the denominator in Eq. (3.21) does not vanish[7]; that is,

$$\int_C \Phi_1(x)\Phi_0(x)dx\neq 0. \quad (3.22)$$

An argument similar to the one given in the proof of property (i) implies that λ_0 is a square-root-type singularity point of $E(\lambda)$. Since a square-root singularity has associated with it a double Riemann sheet, energy-level crossing must take place at all points $\lambda_0\neq 0$ satisfying Eq. (3.11).

IV. WKB SOLUTION

In this section, we shall approximate the wave function $\Phi(x)$ using zeroth-order WKB techniques.[8] We only consider the case where $|\lambda|$ is small. We will need Eqs. (1.1) and (1.2) which are

$$(-d^2/dx^2+\tfrac{1}{4}x^2+\tfrac{1}{4}\lambda x^4)\Phi(x)=E(\lambda)\Phi(x) \quad (4.1)$$

and

$$\lim_{x\to\pm\infty}\Phi(x)=0. \quad (4.2)$$

The zeroth-order WKB solution to Eq. (4.1) is[8]

$$\Phi=\text{const}\times\mu^{-1/2}\exp\left(\pm\int^x \mu(x)dx\right), \quad (4.3)$$

where

$$\mu(x)=[-E(\lambda)+\tfrac{1}{4}x^2+\tfrac{1}{4}\lambda x^4]^{1/2}. \quad (4.4)$$

For large $|x|$, Eq. (4.3), the zeroth-order WKB solution, becomes

$$\Phi(x)\sim\text{const}\times(x^2+\lambda x^4)^{-1/4}\times\exp\{(-6\lambda)^{-1}[(1+\lambda x^2)^{3/2}-1]\}. \quad (4.5)$$

Given $\arg\lambda$, we use Eq. (4.5) to locate in the complex x plane the turning points and the sectors in which the general boundary condition [Eq. (3.4)] applies. The turning points are at $\pm i\lambda^{-1/2}$. The boundaries of the above sectors are curves given by

$$\text{Re}\{\lambda^{-1}[(1+\lambda x^2)^{3/2}-1]\}=0.$$

The sectors and turning points are plotted in Fig. 1 for $\arg\lambda=0, \frac{1}{2}\pi, \pi$, and $\frac{3}{2}\pi$.

We now begin the WKB analysis. Near the origin $x=0$, the differential equation [Eq. (4.1)] becomes

$$(d^2/dx^2+E-\tfrac{1}{4}x^2)\Phi(x)=0, \quad (4.6)$$

which is the defining equation for the parabolic cylinder function $D_{E-\frac{1}{2}}(x)$.[9] The correct physical solution to Eq. (4.6) is

$$\Phi(x)=C[D_{E-\frac{1}{2}}(x)\pm D_{E-\frac{1}{2}}(-x)] \quad (4.7)$$

---

[7] To zeroth order in the WKB expansion discussed in Secs. IV and VI, it can be shown that for those points which satisfy Eq. (3.11), Eq. (3.22) is indeed true. If Eq. (3.22) were not true, we would have to reexpand $E(\lambda)$ and $\Phi(x,\lambda)$ in terms of $(\lambda-\lambda_0)^{1/3}=\epsilon$. Physically, this would mean that three energy levels cross at λ_0 instead of two as is the case. One would expect intuitively that the chance of three levels crossing at the same point λ_0 is small.

[8] For a review of zeroth-order WKB techniques, see A. Messiah, *Quantum Mechanics* (John Wiley & Sons, Inc., New York, 1961), Vol. 1, pp. 232–233. For a discussion of WKB techniques in the complex plane see T. T. Wu, Phys. Rev. **143**, 1110 (1966).

[9] A. Erdélyi, W. Magnus, F. Oberhettinger, and F. G. Tricomi, *Bateman Manuscript Project* (McGraw-Hill Book Co., New York, 1953), Vol. 2, pp. 116–123. In the future we will refer to this reference as BMP.

for even- or odd-parity wave functions. C is a multiplicative constant.

When the phase of λ is less than 270°, we can asymptotically connect the parabolic cylinder function in Eq. (4.7) to the WKB solution in Eq. (4.5). The connection restricts the possible values of E to

$$E=2n+\tfrac{1}{2} \quad \text{for even parity} \qquad (4.8a)$$

and

$$E=2n+\tfrac{3}{2} \quad \text{for odd parity}, \qquad (4.8b)$$

where $n=0, 1, 2, \cdots$.

However, Eq. (4.7) cannot be directly connected to Eq. (4.5) when the phase of λ is near 270° because a turning point lies in the path of the connection [see Fig. 1(d)]. Instead, we use the following procedure:

We define new variables $r\equiv xe^{i\pi/4}$, $\rho\equiv\lambda e^{-3\pi i/2}$, and $\epsilon\equiv 4iE$. In terms of these variables, Eq. (4.1) becomes

$$[d^2/dr^2+\tfrac{1}{4}(-\epsilon+r^2-\rho r^4)]\Phi(r)=0. \qquad (4.9)$$

We assume that $|\lambda|$ is so small that $|\rho\epsilon|\ll 1$.

We define

$$r_0\equiv\{[1-(1-4\rho\epsilon)^{1/2}](2\rho)^{-1}\}^{1/2}\sim\epsilon^{1/2} \qquad (4.10a)$$

and

$$r_1\equiv\{[1+(1-4\rho\epsilon)^{1/2}](2\rho)^{-1}\}^{1/2}\sim\rho^{-1/2}. \qquad (4.10b)$$

r_1 is approximately the position of the turning point. $|\rho\epsilon|\ll 1$ implies that $|r_0|\ll|r_1|$.

Next, we identify four regions in the complex x plane: region A in which r is near the origin, $0<|r|\ll|r_0|\ll|r_1|$; region B in which $|r|\ll|r_1|$ but $|r|\gg|r_0|$; region C in which r is near the turning point, $|r|\lesssim|r_1|$ but $|r|\gg|r_0|$; and region D in which r is at the turning point, $|r|\sim|r_1|$.

In order to carry out the WKB procedure when $\arg\lambda$ is near 270°, approximate solutions to Eq. (4.9) must be found in regions A, B, C, and D. Then these solutions must be connected asymptotically at the boundaries of the regions.

Region A. In region A, Eq. (4.9) becomes approximately (that is, to zeroth order in λ)

$$(d^2/dr^2-\tfrac{1}{4}\epsilon-\tfrac{1}{4}r^2)\Phi_A(r)=0. \qquad (4.11)$$

Equation (4.11) is the transformed defining equation for a parabolic cylinder function.[9] The even-parity solution to this equation is

$$\Phi_A(r)=C[D_{-\frac{1}{2}-i\epsilon/4}(re^{-i\pi/4})+D_{-\frac{1}{2}-i\epsilon/4}(-re^{-i\pi/4})]. \qquad (4.12)$$

(In this section, we will treat even-parity wave functions in detail; only the final results will be given for odd-parity solutions.)

To compute the asymptotic expansion of Φ_A in region A, we need the leading term in the asymptotic expansion of $D_\nu(z)$,[9] which is

$$D_\nu(z)\sim z^\nu e^{-\frac{1}{4}z^2} \quad \text{for } -\tfrac{3}{4}\pi<\arg z<\tfrac{3}{4}\pi \qquad (4.13a)$$

and

$$D_\nu(z)\sim z^\nu e^{-\frac{1}{4}z^2}-[(2\pi)^{1/2}/\Gamma(-\nu)]e^{\nu\pi i}z^{-\nu-1}e^{\frac{1}{4}z^2}$$
$$\text{for } \tfrac{1}{4}\pi<\arg z<(5/4)\pi. \qquad (4.13b)$$

Equation (4.13) yields the asymptotic behavior of $\Phi_A(r)$ in region A for large $|r|$; that is, for r near region B. After much algebraic manipulation we have the desired asymptotic expansion:

$$\Phi_A(r)\sim\frac{C_1}{\sqrt{r}}\left\{\frac{1}{\Gamma(\frac{1}{4}-\frac{1}{8}i\epsilon)}\exp[\tfrac{1}{4}i(r^2-\epsilon\ln r+\tfrac{1}{2}\epsilon\ln 2-\tfrac{1}{2}\pi)]\right.$$
$$\left.+\frac{1}{\Gamma(\frac{1}{4}+\frac{1}{8}i\epsilon)}\exp[-\tfrac{1}{4}i(r^2-\epsilon\ln r+\tfrac{1}{2}\epsilon\ln 2-\tfrac{1}{2}\pi)]\right\}, \qquad (4.14)$$

where

$$C_1=\frac{C2\pi e^{\pi\epsilon/16}2^{-i\epsilon/8}}{\Gamma(\frac{3}{4}+\frac{1}{8}i\epsilon)}. \qquad (4.15)$$

Region B. In regions B and C, we need the zeroth-order WKB solution[8] to Eq. (4.9) which is

$$\Phi_{WKB}(x)=(-\epsilon+r^2-\rho r^4)^{-1/4}$$
$$\times\left\{C_2\exp\left[\tfrac{1}{2}i\int_{r_0}^{r}(-\epsilon+r^2-\rho r^4)^{1/2}dr\right]\right.$$
$$\left.+C_3\exp\left[-\tfrac{1}{2}i\int_{r_0}^{r}(-\epsilon+r^2-\rho r^4)^{1/2}dr\right]\right\}. \qquad (4.16)$$

In Eq. (4.16), we choose to integrate from the reference point r_0. We are free to choose the reference point, but this choice must be maintained for the WKB solution in region C as well as in region B.

For r in region B, the choice of r_0 and r_1 in Eq. (4.10) allows us to simplify Eq. (4.16). To zeroth order in ρ,

$$(-\epsilon+r^2-\rho r^4)\sim r^2-\epsilon. \qquad (4.17)$$

Using Eq. (4.17), we easily evaluate the integrals and expand the expression in front of the curly brackets in Eq. (4.16). The result is

$$\Phi_B(x)\sim(1/\sqrt{r})\{C_2\exp[\tfrac{1}{4}i(r^2-\epsilon\ln r-\tfrac{1}{2}\epsilon+\epsilon\ln(\tfrac{1}{2}\sqrt{\epsilon}))]$$
$$+C_3\exp[-\tfrac{1}{4}i(r^2-\epsilon\ln r-\tfrac{1}{2}\epsilon+\epsilon\ln(\tfrac{1}{2}\sqrt{\epsilon}))]\}. \qquad (4.18)$$

Region C. In region C, we need the zeroth-order WKB solution [Eq. (4.16)] to Eq. (4.9). The WKB solution in region C has the same constants C_2 and C_3 as the solution in region B. However, the integrals for region C are more difficult to perform than those for region B because of the approximations associated with region C which are $|r_0|\ll|r|\lesssim|r_1|$. We use the following decomposition to evaluate the integrals:

$$\tfrac{1}{2}i\int_{r_0}^{r} dr(-\epsilon+r^2-\rho r^4)^{1/2}$$
$$=-\tfrac{1}{2}i\rho^{1/2}\int_{r}^{r_1}[(r^2-r_0^2)(r_1^2-r^2)]^{1/2}dr$$
$$+\tfrac{1}{2}i\rho^{1/2}\int_{r_0}^{r_1}[(r^2-r_0^2)(r_1^2-r^2)]^{1/2}dr. \quad (4.19)$$

To do the first integral, we introduce a change of variables

$$R=r_1-r. \quad (4.20)$$

Then, the integral becomes

$$-\tfrac{1}{2}i\rho^{1/2}(2r_1)^{1/2}(r_1^2-r_0^2)^{1/2}\int_0^R R^{1/2}dR$$
$$\sim -i3^{-1}2^{1/2}\rho^{-1/4}R^{3/2}. \quad (4.21)$$

In the second integral (the definite integral), we let $K^2=1-r_0^2r_1^{-2}$. Then,

$$\tfrac{1}{2}i\rho^{1/2}\int_{r_0}^{r_1}[(r^2-r_0^2)(r_1^2-r^2)]^{1/2}dr$$
$$=\tfrac{1}{6}ir_1^3[(1+r_0^2r_1^{-2})E(k)-2r_0^2r_1^{-2}K(k)]. \quad (4.22)$$

$E(k)$ and $K(k)$ are the standard elliptic integrals.[10] Expanding $E(k)$ and $K(k)$ for k near 1 gives[11]

$$E(k)\sim 1+\frac{r_0^2}{4r_1^2}\left(4\ln 2-1-2\ln\frac{r_0}{r_1}\right) \quad (4.23a)$$

and

$$K(k)\sim \ln(4r_1/r_0). \quad (4.23b)$$

Thus,

$$\tfrac{1}{2}i\rho^{1/2}\int_{r_0}^{r_1} dr[(r^2-r_0^2)(r_1^2-r^2)]^{1/2}$$
$$\sim\frac{i}{6\rho}-\tfrac{1}{8}i\epsilon\left[\ln\frac{1}{\rho\epsilon}+4\ln 2+1\right]. \quad (4.24)$$

We also note that

$$(-\epsilon+r^2-\rho r^4)^{-1/4}\sim\rho^{-1/4}(r_1^2-r_0^2)^{-1/4}(2r_1)^{-1/4}R^{-1/4}$$
$$\sim 2^{-1/4}\rho^{1/8}R^{-1/4}. \quad (4.25)$$

Equations (4.19)–(4.25) imply that near the turning point the WKB solution [Eq. (4.16)] in region C is asymptotically

$$\Phi_C(r)\sim 2^{-1/4}\rho^{1/8}R^{-1/4}\Bigg\{C_2\exp\Bigg[-\tfrac{1}{3}i\sqrt{2}\rho^{-1/4}R^{3/2}$$
$$+\frac{i}{6\rho}-\tfrac{1}{8}i\epsilon\left(\ln\frac{1}{\rho\epsilon}+4\ln 2+1\right)\Bigg]$$
$$+C_3\exp\Bigg[+\tfrac{1}{3}i\sqrt{2}\rho^{-1/4}R^{3/2}$$
$$-\frac{i}{6\rho}+\tfrac{1}{8}i\epsilon\left(\ln\frac{1}{\rho\epsilon}+4\ln 2+1\right)\Bigg]\Bigg\}. \quad (4.26)$$

Region D. In region D, we use Eq. (4.20) to change variables from r to R in the differential equation [Eq. (4.9)] and find that to zeroth order in ρ

$$(d^2/dR^2+\tfrac{1}{2}R\rho^{-1/2})\Phi_D(R)=0. \quad (4.27)$$

Equation (4.27) is an Airy differential equation.[12] The solution to Eq. (4.27) is

$$\Phi_D(R)=D(\tfrac{1}{3}y)^{1/2}K_{1/3}(2y^{3/2}), \quad (4.28)$$

where

$$y=-(18\sqrt{\rho})^{-1/3}R. \quad (4.29)$$

There is only one undetermined constant D in Eq. (4.28) because only the $K_{1/3}$ function obeys the general boundary conditions at ∞ [Eq. (3.4)].

To determine the behavior of $K_{1/3}(2y^{3/2})$ near the turning point, we note that[13]

$$K_{1/3}(2e^{3\pi i/2}(-y)^{3/2})=-\tfrac{1}{2}i\pi\{e^{i\pi/6}H_{1/3}^{(1)}(2(-y)^{3/2})$$
$$+e^{-i\pi/6}H_{1/3}^{(2)}(2y^{3/2})\}. \quad (4.30)$$

Then we use the asymptotic expansions[14] for $H^{(1)}$ and $H^{(2)}$ and find that in region D near the turning point

$$\Phi_D\sim C_4R^{-1/4}[\exp(\tfrac{1}{3}i\sqrt{2}\rho^{-1/4}R^{3/2}-\tfrac{1}{4}i\pi)$$
$$+\exp(-\tfrac{1}{3}i\sqrt{2}\rho^{-1/4}R^{3/2}+\tfrac{1}{4}i\pi)], \quad (4.31)$$

where

$$C_4=D\pi^{1/2}3^{-1/3}2^{-11/12}\rho^{1/24}. \quad (4.32)$$

We have thus completed the determination of the asymptotic behavior of Φ in regions A, B, C, and D. Next we must connect Φ across regions A and B and across C and D. (Φ in region B is already connected to Φ in region C because the same WKB solution is used in both regions.) Connecting Eq. (4.14) in region A to Eq. (4.18) in region B and Eq. (4.26) in region C to Eq. (4.31) in region D gives two independent determinations of the ratio C_2/C_3. These are, respectively,

$$\frac{C_2}{C_3}=\frac{\Gamma(\frac{1}{4}+\frac{1}{8}i\epsilon)}{\Gamma(\frac{1}{4}-\frac{1}{8}i\epsilon)}\exp\left[\frac{i}{2}\left(\frac{\epsilon}{2}+\frac{\epsilon}{2}\ln\frac{8}{\epsilon}-\frac{\pi}{2}\right)\right] \quad (4.33a)$$

and

$$\frac{C_2}{C_3}=\exp\left[i\left(\frac{\pi}{2}-\frac{1}{3\rho}+\frac{\epsilon}{4}\ln\frac{16}{\rho\epsilon}+\frac{\epsilon}{4}\right)\right]. \quad (4.33b)$$

---

[10] BMP, Vol. 2, pp. 317–318.

[11] To expand $E(k)$ and $K(k)$ near $k=1$, we use

$$E(k)=\tfrac{1}{2}\pi\,{}_2F_1(-\tfrac{1}{2},\tfrac{1}{2};1;k^2)\quad\text{and}\quad K(k)=\tfrac{1}{2}\pi\,{}_2F_1(\tfrac{1}{2},\tfrac{1}{2};1;k^2).$$

[See BMP, Vol. 2, p. 318, Eqs. (5) and (6).] Then we consult BMP, Vol. 1, p. 110, Eq. (12), where we find the appropriate expansion for the hypergeometric functions. These expansions may be simplified by using BMP, Vol. 1, pp. 1–20.

[12] BMP, Vol. 2, p. 22.

[13] BMP, Vol. 2, p. 5, Eq. (15) and p. 80, Eq. (43).

[14] BMP, Vol. 2, p. 85, Eqs. (1) and (2).

Combining Eqs. (4.33a) and (4.33b) gives a single transcendental equation which contains ρ and ϵ only.

$$\frac{\Gamma(\frac{1}{4}-\frac{1}{8}i\epsilon)}{\Gamma(\frac{1}{4}+\frac{1}{8}i\epsilon)}=\exp\left[\frac{5i\pi}{4}+\frac{i}{3\rho}+\frac{i\epsilon}{4}\ln\frac{\rho}{2}\right]. \quad (4.34)$$

Because Φ_D goes exponentially to 0 at ∞ as required by Eq. (4.5), we have connected the origin to ∞ via regions A, B, C, and D. We have found that Eq. (4.34) is the one requirement which must be obeyed for this connection to be valid. One should note that since Eq. (4.9) is a second-order differential equation, there are two undetermined constants in each region A, B, C, and D. An equation such as Eq. (4.34) is posible only because parity considerations restrict Φ_A to one constant C_1 and the boundary conditions [Eq. (3.4)] at ∞ limit Φ_D to one constant C_4.

If we rewrite Eq. (4.34) in terms of E instead of ϵ, we get

$$\frac{\Gamma(\frac{1}{4}+\frac{1}{2}E)}{\Gamma(\frac{1}{4}-\frac{1}{2}E)}=\exp\left(\frac{i}{3\rho}+\frac{5\pi i}{4}-E\ln\frac{\rho}{2}\right). \quad (4.35a)$$

For odd-parity wave functions the equivalent result is

$$\frac{\Gamma(\frac{3}{4}+\frac{1}{2}E)}{\Gamma(\frac{3}{4}-\frac{1}{2}E)}=\exp\left(\frac{i}{3\rho}-\frac{5\pi i}{4}-E\ln\frac{\rho}{2}\right). \quad (4.35b)$$

We may regard these results as approximate equations to zeroth order in λ relating the low-lying energy levels to complex λ near $\arg\lambda=270°$.

Appendix F contains a WKB calculation similar to the calculation in this section but to first order in λ. The result of Appendix F is more precise than Eq. (4.35), but it reduces to Eq. (4.35) when terms to first order in λ are neglected.

V. RESOLVENT

As a simple application of Eq. (4.35), we calculate in this section the locations of the singularities of the resolvent of the Hamiltonian H [Eq. (1.3)]. The resolventof the Hamiltonian is conventionally defined as $(z-H)^{-1}$. We will prove that $(z-H)^{-1}$ has poles when $\arg\lambda$ is near 270°.

It is clear that $(z-H)^{-1}$ has a pole in the λ plane whenever the following equations are satisfied:

$$\frac{\Gamma(\frac{1}{4}+\frac{1}{2}z)}{\Gamma(\frac{1}{4}-\frac{1}{2}z)}=e^{5\pi i/4}e^{i/3\rho}e^{-z\ln(\rho/2)} \quad (5.1a)$$

for even parity;

$$\frac{\Gamma(\frac{3}{4}+\frac{1}{2}z)}{\Gamma(\frac{3}{4}-\frac{1}{2}z)}=e^{-5\pi i/4}e^{i/3\rho}e^{-z\ln(\rho/2)} \quad (5.1b)$$

for odd parity.

We solve each of the above equations separately for ρ by taking the logarithm of both sides. Given z, it is always possible to choose a sufficiently small $|\rho|$ so that the $i/3\rho$ term dominates the $z\ln(2/\rho)$ term. Thus, the resolvent has poles at

$$\lambda=e^{3\pi i/2}\left[-3i\ln\frac{\Gamma(\frac{1}{4}+\frac{1}{2}z)}{\Gamma(\frac{1}{4}-\frac{1}{2}z)}+6\pi N+\frac{1}{4}\pi\right]^{-1}, \quad (5.2a)$$

provided that z is not an even integer $+\frac{1}{2}$, and at

$$\lambda=e^{3\pi i/2}\left[-3i\ln\frac{\Gamma(\frac{3}{4}+\frac{1}{2}z)}{\Gamma(\frac{3}{4}-\frac{1}{2}z)}+6\pi N-\frac{1}{4}\pi\right]^{-1}, \quad (5.2b)$$

provided that z is not an odd integer $+\frac{1}{2}$. In Eq. (5.2), N is any sufficiently large positive integer.

Equation (5.2) implies that for both parities $(z-H)^{-1}$ has an infinite sequence of poles approaching the origin. These correspond to an infinite sequence of choices for large positive N.

That the resolvent has poles was a surprise to us and, as far as we know, this had not even been conjectured. Jaffe[2] has proved that for negative z the resolvent is analytic in the cut λ plane, the cut extending from the origin to $-\infty$ along the negative real axis. This is entirely consistent with our results which maintain that no singularities appear in the resolvent until the phase of the coupling constant reaches nearly $\pm 270°$. Since the analytic continuation of the resolvent disappears through Jaffe's cut at $\arg\lambda=180°$, Jaffe observed no singularities.

VI. QUALITATIVE DESCRIPTION OF $E(\lambda)$

In this section, we use the results of Secs. III and IV to discover the properties of the analytic continuation of the energy levels $E(\lambda)$ as a function of complex λ. For purposes of clarity, this section is divided into three parts. Part A gives a calculation of the locations in the λ plane of the branch points of $E(\lambda)$. Part B describes in detail the paths traced out by the energy levels in the complex E plane as λ moves about the complex λ plane. Part C contains a discussion of level crossing.

A. Locations of Branch Points

We find the coordinates in the λ plane of the branch points of $E(\lambda)$ by solving a pair of simultaneous equations. The first of these is Eq. (4.35) which resulted from matching the zeroth-order WKB solution across the boundaries of regions A, B, C, and D, as was done in Sec. IV. We repeat Eq. (4.35) here:

$$\frac{\Gamma(\frac{1}{4}+\frac{1}{2}E)}{\Gamma(\frac{1}{4}-\frac{1}{2}E)}=\exp\left(\frac{i}{3\rho}+\frac{5\pi i}{4}-E\ln\frac{\rho}{2}\right) \quad \text{for even parity}, \quad (6.1a)$$

$$\frac{\Gamma(\frac{3}{4}+\frac{1}{2}E)}{\Gamma(\frac{3}{4}-\frac{1}{2}E)}=\exp\left(\frac{i}{3\rho}-\frac{5\pi i}{4}-E\ln\frac{\rho}{2}\right) \quad \text{for odd parity}. \quad (6.1b)$$

The second of the simultaneous equations is obtained by explicitly performing the integral in Eq. (3.11) which is

$$0=\int_C dx\,\Phi^2(x,\lambda)\,. \tag{6.2}$$

Equation (6.2) is an exact condition which is satisfied by those λ and only those λ which are branch points of $E(\lambda)$. Now that the approximate WKB wave function is known from Sec. IV, it is possible to evaluate this integral and in turn to use the result to find the branch points. The evaluation is very lengthy and is done in Appendix G. The result of this calculation is

$$0=-\psi(E+\tfrac{1}{2})+\frac{\pi}{2}\cot\left(\frac{1}{4}-\frac{E}{2}\right)\pi+\ln\frac{4}{\rho}\quad\text{for even parity} \tag{6.3a}$$

and

$$0=-\psi(E+\tfrac{1}{2})+\frac{\pi}{2}\cot\left(\frac{3}{4}-\frac{E}{2}\right)\pi+\ln\frac{4}{\rho}\quad\text{for odd parity}\,, \tag{6.3b}$$

where ψ is the logarithmic derivative of the Γ function. Equation (6.3) now embodies the condition for λ to be a singularity of $E(\lambda)$. Because Eqs. (6.3) and (6.1) are both correct to zeroth order in λ and since these are the pair of simultaneous equations that will be solved to find the singularities of $E(\lambda)$, the positions of the singularities will be known to zeroth order in λ.

Equations (6.1) and (6.3) must be put into a form favorable for simultaneous solution. To do so we use the following procedure:

(i) In this calculation, we consider $|\rho|$ to be small and $\arg\rho$ to be near 0° (because $\arg\lambda$ is near 270°). We may express this by

$$\mathrm{Re}(1/\rho)=O(\Lambda) \tag{6.4a}$$

and

$$\mathrm{Im}(1/\rho)=O(\ln\Lambda)\,, \tag{6.4b}$$

where Λ is very large.

(ii) We also consider only low-lying energy levels (that is, levels near the ground state) because the derivation of Eq. (6.1) in Sec. IV is valid only for such levels. We may express this by

$$|E|=O(1)\,. \tag{6.5}$$

(iii) Define η by

$$E=2n+\tfrac{1}{2}+\eta\quad\text{for even parity} \tag{6.6a}$$

and

$$E=2n+\tfrac{3}{2}+\eta\quad\text{for odd parity}\,, \tag{6.6b}$$

where $n=0,1,2,\cdots$.

(iv) Take the logarithm of both sides of Eq. (6.1), substitute Eq. (6.6), and apply the Legendre duplication formula[15] to get

$$\ln\Gamma(-n-\tfrac{1}{2}\eta)-\ln\Gamma(n+\tfrac{1}{2}\eta)+\ln\Gamma(2n+\eta)+\ln(2\sqrt{\pi})-(2n+\eta)\ln 2=\frac{i}{3\rho}+\frac{5\pi i}{4}-E\ln\frac{\rho}{2}-2\pi Ni \tag{6.7a}$$

and

$$\ln\Gamma(-n-\tfrac{1}{2}\eta)-\ln\Gamma(n+\tfrac{1}{2}\eta)+\ln\Gamma(2n+\eta)+\ln(2\sqrt{\pi})-(2n+\eta)\ln 2+\ln(\tfrac{1}{2}+n+\tfrac{1}{2}\eta)=\frac{i}{3\rho}-\frac{5\pi i}{4}-E\ln\frac{\rho}{2}-2\pi Ni\,. \tag{6.7b}$$

N in this equation is an integer referring to the multiple-valued logarithmic function.

(v) Expand the $\ln\Gamma$ functions using[16]

$$\ln\Gamma(1+z)=-\gamma z+\sum_{m=2}^{\infty}(-1)^m\xi(m)(z^m/m)\,. \tag{6.8}$$

The result is

$$\ln(\tfrac{1}{2}\eta)+\ln(2n)!+\tfrac{1}{2}\ln(2\pi)+\eta\left(-\gamma+\sum_{k=1}^{2n}\frac{1}{k}\right)+\sum_{m=2}^{\infty}\frac{(-1)^m\eta^m}{m}\left\{\xi(m)-\xi(m)\left[\frac{1}{2^m}+\frac{1}{(-2)^m}\right]-\sum_{k=1}^{2n}\frac{1}{k^m}\right\}=\frac{i}{3\rho}+\frac{\pi i}{4}-(n\pi i)-2\pi Ni-(2n+\eta+\tfrac{1}{2})\ln\frac{\rho}{4} \tag{6.9a}$$

and

$$\ln(\tfrac{1}{2}\eta)+\ln(2n+1)!+\tfrac{1}{2}\ln(2\pi)+\eta\left(-\gamma+\sum_{k=1}^{2n+1}\frac{1}{k}\right)+\sum_{m=2}^{\infty}\frac{(-1)^m\eta^m}{m}\left\{\xi(m)-\xi(m)\left[\frac{1}{(2)^m}+\frac{1}{(-2)^m}\right]-\sum_{k=1}^{2n+1}\frac{1}{k^m}\right\}=\frac{i}{3\rho}+\frac{\pi i}{4}-(n\pi i)-2\pi(N+1)i-(2n+\eta+\tfrac{3}{2})\ln\frac{\rho}{4}\,. \tag{6.9b}$$

(vi) Substitute Eq. (6.6) into Eq. (6.3) and expand the function using[17]

$$\psi(1+z)=-\gamma+\sum_{m=2}^{\infty}(-1)^m\xi(m)z^{m-1}\,. \tag{6.10}$$

The result is

$$\frac{1}{\eta}+\left(-\gamma+\sum_{k=1}^{2n}\frac{1}{k}\right)+\sum_{m=2}^{\infty}(-1)^m\eta^{m-1}\left[-\sum_{k=1}^{2n}\frac{1}{k^m}+\xi(m)-\left(\frac{1}{2^m}+\frac{1}{(-2)^m}\right)\xi(m)\right]=-\ln(\tfrac{1}{4}\rho) \tag{6.11a}$$

---

[15] BMP, Vol. 1, p. 5, Eq. (15).

[16] BMP, Vol. 1, p. 45, Eq. (2). $\xi(m)$ is the Riemann zeta function.

[17] BMP, Vol. 1, p. 45, Eq. (5).

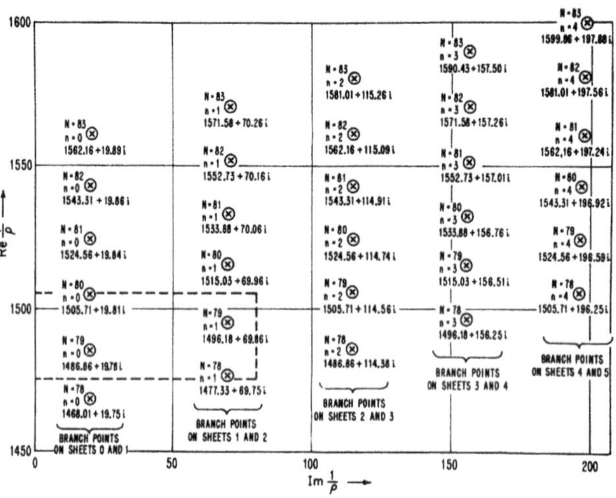

FIG. 2. A portion of the $1/\rho$ plane on which are plotted the even-parity branch points. The branch points are labeled by the integers n ranging from 0 through 4 and N ranging from 78 to 83. Next to each branch point are its coordinates correct to the nearest hundredth. The region of the graph surrounded by the dashed line is enlarged in Fig. 4 and will be discussed in connection with level crossing.

and

$$\frac{1}{\eta}+\left(-\gamma+\sum_{k=1}^{2n+1}\frac{1}{k}\right)+\sum_{m=2}^{\infty}(-1)^m\eta^{m-1}\left[-\sum_{k=1}^{2n+1}\frac{1}{k^m}+\xi(m)-\left(\frac{1}{2^m}+\frac{1}{(-2)^m}\right)\xi(m)\right]=-\ln(\tfrac{1}{4}\rho). \qquad (6.11\text{b})$$

Equations (6.9) and (6.11) are now in a form amenable to approximate simultaneous solution. The expansions in Eqs. (6.9) and (6.11) are power series in η and are most susceptible to techniques of approximation if $|\eta|\ll 1$. η is indeed small as will be seen. N is a large integer.

The simultaneous solution consists of using Eqs. (6.9) and (6.11) alternatively to provide successively more accurate approximations to $1/\rho$ and $1/\eta$. First, we use Eq. (6.9). To zeroth order in η, the locations of the singularities in the analytic continuation $E(\lambda)$ are approximately

$$1/\rho=6\pi N \qquad (6.12\text{a})$$

and

$$1/\rho=6\pi(n+1). \qquad (6.12\text{b})$$

Next we use Eqs. (6.12) and (6.11) to find $1/\eta$ to zeroth order in η:

$$\frac{1}{\eta}=\ln(24\pi N)+\gamma-\sum_{k=1}^{2n}\frac{1}{k} \qquad (6.13\text{a})$$

and

$$\frac{1}{\eta}=\ln[24\pi(N+1)]+\gamma-\sum_{k=1}^{2n+1}\frac{1}{k}. \qquad (6.13\text{b})$$

Remembering that N is large, we see that Eq. (6.13) verifies the earlier assumption that $|\eta|$ is small. We use Eqs. (6.13) and (6.9) and iterate once more to find the locations of the singularities of $E(\lambda)$ to first order in η:

$$1/\rho=6\pi N+3\pi n-3\pi/4-3i[\ln(2n)!+\tfrac{1}{2}\ln(\pi/2)-\ln\ln N-(2n+\tfrac{1}{2})\ln(24\pi N)-1] \qquad (6.14\text{a})$$

and

$$1/\rho=6\pi(N+1)+3\pi n+3\pi/4-3i\{\ln(2n+1)!+\tfrac{1}{2}\ln(\pi/2)-\ln\ln N-(2n+\tfrac{3}{2})\times\ln[24\pi(N+1)]-1\}. \qquad (6.14\text{b})$$

Equation (6.14) is the equation we have sought. It lists the coordinates of the branch points of $E(\lambda)$ parametrized by two integers N (large) and n (small).

Before discussing this equation, it is appropriate to make some observations here on the structure of Eqs. (6.9) and (6.11), the simultaneous equations that were used to derive Eq. (6.14). We note that Eq. (6.11) is the derivative of Eq. (6.9) with respect to η. This is true because Eq. (6.3) is the logarithmic derivative of Eq. (6.1) with respect to E. However, the significance of this relationship is not obvious.

An explanation for this relationship comes from a broader interpretation of Eqs. (4.35) than we have chosen to use. We might have assumed that Eqs. (4.35) were more than just equations relating E and λ for small λ. We might have treated them as approximate (to zeroth order in λ) analytic continuations of E as a function of λ. Based on this assumption, the condition for locating those points in the λ plane where the energy

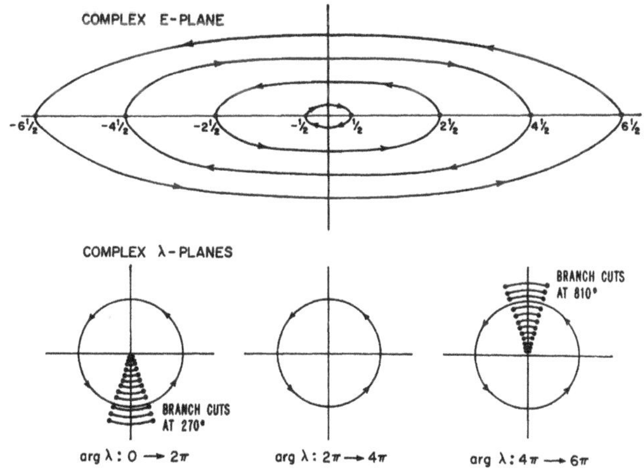

FIG. 3. A rough picture of a tour of the λ plane with the associated image of $E(\lambda)$ in the E plane. $|\lambda|$ is held fixed while $\arg\lambda$ ranges from 0 to 6π. The λ plane is drawn as three planes to show the full rotation of $\arg\lambda$. The branch points and arc-shaped branch lines are indicated schematically. The graph gives the paths of the even-parity energy levels only. The odd-parity levels behave similarly.

is twofold degenerate would be that the derivative of Eq. (6.9) with respect to E (or η) must be a valid equation. Because it gives the correct answer, this argument would have constituted a "derivation" of Eq. (6.11) and would have been a short cut past Appendix G.

We did not use the above argument on grounds of rigor. It is not clear what is meant by an "approximate zeroth-order analytic continuation." It is difficult to understand how one could approximate an exact concept like analyticity by zeroth-order techniques. The laborious method of getting Eq. (6.11) by deriving and evaluating $\int_C \Phi^2(x)dx$ in Sec. III and Appendix G was presented in this paper to avoid these mathematical imprecisions.

Having made this point we return to analyze Eq. (6.14). We note that as the integer N takes on a sequence of values approaching ∞, the associated sequence of branch points in the λ plane given by Eq. (6.14) begins at a finite distance from the origin and has a limit point at the origin. (Remember that $\rho=\lambda e^{-3\pi i/2}$.) That this sequence of branch points has a limit point at the origin explains the divergence of the perturbation series discussed earlier in Sec. II. Note that the origin in the λ plane is not an isolated singularity of $E(\lambda)$.

It is useful to plot the results of Eq. (6.14) on a graph. Figure 2 shows a portion of the $1/\rho$ plane on which are plotted the even-parity branch points labeled by the integer n ranging from 0 through 4 and N ranging from 78 to 83. Next to each branch point are its coordinates correct to two decimal places. The region of the graph surrounded by the dashed line is blown up in Fig. 4 and will be discussed in detail in Part C of this section in connection with level crossing.

It is clear from Fig. 2 that the branch points lie in a sequence of sequences. Each sequence of branch points is labeled by n and each sequence approaches the origin. A glance at Fig. 4 shows that the first ($n=0$) sequence of points lies at an angle of about $\frac{3}{4}°$ less than 270° in the λ plane. The next ($n=1$) sequence lies approximately $2\frac{1}{2}°$ less than 270°. Each successive sequence is rotated by a small angle toward the real axis in the λ plane.

However, we suspect that as n increases, these sequences bunch up before they reach 180° and that no branch points have a complex argument of less than 180°. If there were such branch points then, by the argument of Sec. V, the resolvent of the Hamiltonian would have poles in the cut λ plane. Jaffe[2] proved that this is not so. Unfortunately, Eq. (6.14) only holds for small n so this bunching phenomenon cannot be illustrated graphically.[18]

B. Paths in E Plane

Having found the branch points of $E(\lambda)$, we can give a qualitative picture of the complex function $E(\lambda)$. We will do this by tracing the paths of the energy levels in the complex E plane as λ moves about the complex λ plane.

We begin our analysis by observing that both differential equations [Eqs. (4.1) and (4.9)] are real in

[18] In Sec. VI C, it will be shown that the WKB solution [Eq. (6.1)] is valid in two semidisk-shaped regions of the λ plane bounded by $\arg\lambda=270°\pm90°$ and $\arg\lambda=810°\pm90°$. In the rest of the λ plane, the WKB solution is superseded by Eq. (4.8). Since the existence of branch points is a phenomenon associated only with Eq. (6.1), branch points must lie throughout the region in which Eq. (6.1) holds. Therefore, we expect that the bunching phenomenon occurs at the boundaries of the region; that is, at $\arg\lambda=180°$, 360°, 720°, and 900°.

their respective variables. From this arise some symmetry properties of $E(\lambda)$ which are

$$E(\lambda)=E^*(\lambda^*)=-E^*(e^{3\pi i}\lambda^*)=-E(e^{3\pi i}\lambda). \quad (6.15)$$

We also note that since the Hamiltonian [Eq. (1.3)] is Hermitian and positive definite, the energy levels lie along the positive real axis in the energy plane when λ is positive real. Similarly, when $\arg\lambda=270°$, Eq. (4.9) indicates that the energy levels are pure imaginary. At $\arg\lambda=270°$, the energy levels lie on both the positive and negative imaginary axes.

Equation (4.8) indicates that for fixed $|\lambda|$ the energy levels lie near the real axis in the energy plane as $\arg\lambda$ increases from 0 until nearly 270°. However, for $\arg\lambda$ in the neighborhood of 270° the full energy spectrum moves rapidly toward the imaginary axis.

Combining the symmetry properties [Eq. (6.1)] with the results of the previous two paragraphs enables us to predict the paths of the energy levels in the complex E plane when $|\lambda|$ is held fixed and $\arg\lambda$ is varied. (These paths are given in Fig. 3.) We predict that:

(i) When $\arg\lambda=0$, the energy spectrum lies along the positive real axis.

(ii) As $\arg\lambda$ increases toward $\frac{1}{4}(6\pi)$, each energy level traces out a path in the complex E plane which approaches the imaginary axis. The energy levels lie along the imaginary axis at $\arg\lambda=\frac{1}{4}(6\pi)$.

(iii) At $\arg\lambda=\frac{1}{2}(6\pi)$, the energy levels lie on the negative real axis at points equal to the negative of their original values for $\arg\lambda=0$.

(iv) As $\arg\lambda$ increases from $\frac{1}{2}(6\pi)$ to 6π, each energy level traces a path reflectionally symmetric about the origin to the path described above. At $\arg\lambda=6\pi$ the energy levels return to the values on the real axis that they had for $\arg\lambda=0$.

Thus, as λ goes in a circle three times about the origin in the λ plane, starting at the real axis, each energy level in the E plane traces a closed path starting and ending on the positive real axis. Moreover, it will be shown in Part C of this section that the paths in the complex E plane are concentric and that the directions of successive concentric paths alternate between clockwise and counterclockwise.

Because the energy levels return to their original values when $\arg\lambda$ is increased by 6π, in addition to the branch-point singularities discussed in Part A of this section there is also a $\lambda^{1/3}$ singularity at the origin. However, the divergence of perturbation theory discussed in Sec. II is in no way related to this singularity. If it were, we could then expand $E(\lambda)$ in powers of $\lambda^{1/3}$ and get a convergent power series. But there is no convergent power-series representation of $E(\lambda)$ about the origin in any fractional power of λ. Perturbation theory diverges because of the sequences of branch points discussed in Part A of this section.

How do these branch points affect the curves in the E plane in Fig. 3? Surprisingly, the answer to this question is "*not at all.*" To understand why this is we argue as follows:

(i) As proved in Sec. III, each of these singularities is a square-root type. (In the neighborhood of each branch point, there is a double-sheeted Riemann surface.) Level crossing (twofold degeneracy) must occur at each branch point. A branch line is connected to each singularity.

(ii) The symmetry properties listed in Eq. (6.15) imply that there is another set of branch points symmetrically reflected about 270° in the λ plane. Moreover, all these singularities (those in Fig. 2 and those reflected through 270°) also must occur at −270° (which is the same as 810° because of the cube-root singularity at the origin).

(iii) The branch lines join the branch points on either side of 270° and on either side of 810° pairwise; that is, each point is joined to its reflected image. These branch cuts are shaped like circular arcs and are very short because the branch points lie a short distance from either side of 270° or 810°. The cuts form two sequences of concentric arcs with each sequence of arcs approaching the origin. These cuts are schematically displayed on Fig. 3.

(iv) In the circular tour of the λ plane shown on Fig. 3 it is clear that none of the branch cuts was crossed because $|\lambda|$ was held fixed.

This reasoning explains why the branch points could not affect the curves in the E plane in Fig. 3. In addition, it shows that the presence of the branch points can never be detected as long as $|\lambda|$ is held fixed.

Therefore, to exhibit the delicate intricacy of this theory caused by the array of cuts in Fig. 3, we must make a new tour of the λ plane in which we cross the branch cuts. We still increase $\arg\lambda$ from 0 to 6π. However, when $\arg\lambda$ is near 270° (or 810°) we vary $|\lambda|$ so that the path in the λ plane crosses a branch line. If we follow an energy level as it traces a path in the complex E plane, we find that it returns to the positive real axis when $\arg\lambda=6\pi$ but as a different energy level from the original. This is what is meant by level crossing.

A complete description of level crossing is given in Part C of this section. Nevertheless, it should now be clear that the energy levels are completely interrelated by the structure of the branch cuts. That is, one can get to any given energy level from any other energy level by continuing around the appropriate branch points.

As a corollary of this, once one knows the ground-state energy of the anharmonic oscillator, one knows all the energy levels by analytic continuation. Or, in other words, *the physical energy levels of the anharmonic oscillator for a given positive real λ are the positive real values of $E(\lambda)$ on each of the infinite number of branches of a Riemann surface. Each energy level corresponds to a sheet of this Riemann surface.* This is the major result of our paper.

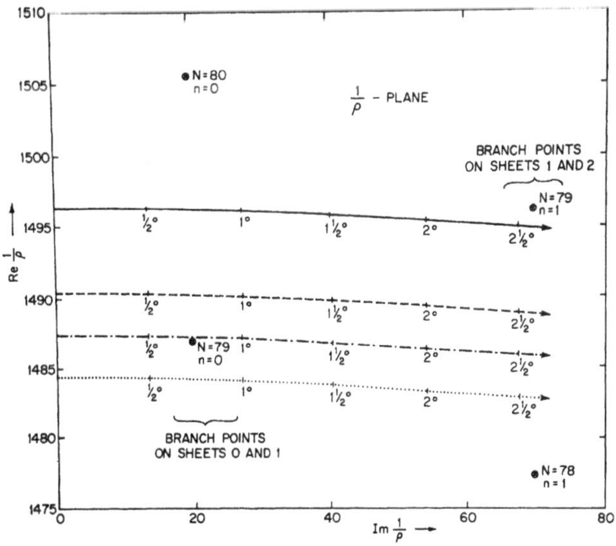

FIG. 4. An enlargement of a small portion of the $1/\rho$ plane of Fig. 2. The ($N=78$, $n=1$), ($N=79$, $n=0$), ($N=79$, $n=1$), and ($N=78$, $n=1$) branch points are precisely located. Four short circular arcs of unequal radii are drawn and the angular rotation along each arc is given in degrees. The arcs are indicated by a solid line, dashed line, dash-dotted line, and dotted line. The radii of these arcs to the nearest hundredth are 1496.50, 1490.50, 1487.50, and 1484.50, respectively. The images of these arcs in the E plane are plotted on Fig. 5 in connection with level crossing.

C. Level Crossing

We have shown that level crossing takes place at the branch points in the λ plane and we know the positions of these branch points. But the details of level crossing are not yet fully explained. In particular, in this part of Sec. VI we will answer these questions: (a) Which levels cross at which branch points? (b) What does level crossing look like graphically?

To give a precise answer to these questions, we carry out an intensive program of numerical calculation which enables us to construct Figs. 4 and 5. Figures 4 and 5 will give a beautiful pictorial answer to the questions we have posed. The numerical calculation involves the following procedure:

(i) We take a sequence of circular paths of varying radii in the $1/\rho$ plane. (This is equivalent to and easier to handle numerically than a sequence of circles in the λ plane.) The image of each of these circles in the E plane is a set of energy-level curves like those crudely drawn in Fig. 3. As in Fig. 3, we will consider the even-parity case only. (The odd-parity case is qualitatively similar.)

(ii) By symmetry [see Eq. (6.15) and the accompanying discussion] we need only to vary $\arg\lambda$ from 0° to 270°. Moreover, until $\arg\lambda$ is very nearly 270°, the energy levels stay extremely close to the values given in Eq. (4.8a). Therefore, we only need to consider the portions of the energy-level curves for λ near 270° (that is, where $1/\rho$ is approximately real).

(iii) We choose the sequence of radii in the $1/\rho$ plane so that a branch point is crossed. We then use Eq. (6.1a) to plot points accurately in the E plane and expect to observe the process of level crossing. We choose four values for the radii of the curves in the $1/\rho$ plane. To the nearest hundredth, these values of $|1/\rho|$ are 1496.50, 1490.50, 1487.50, and 1484.50. These curves are drawn on Fig. 4. (Figure 4 is an enlarged version of a part of Fig. 2. That portion which is enlarged is indicated on Fig. 2 by a dashed line.) The branch point that is crossed is the one labeled $N=79$, $n=0$.

(iv) We then choose points along the curves in Fig. 4 and solve the transcendental equation [Eq. (6.1)] numerically for complex E to the nearest hundredth. To solve Eq. (6.1a) numerically we take the logarithm of both sides. This gives

$$\ln\Gamma(\tfrac{1}{4}+\tfrac{1}{2}E)-\ln\Gamma(\tfrac{1}{4}-\tfrac{1}{2}E) = \frac{i}{3\rho}+\frac{5\pi i}{4}+E\ln\frac{2}{\rho}+2\pi i M. \quad (6.16)$$

M is an integer referring to the many-valued logarithmic function. M occurs for the same reason that N did in Eq. (6.7).

(v) We choose the branch line for the logarithmic function in Eq. (6.16) to lie along the negative real axis. We then break up Eq. (6.16) into its real and imaginary parts and, given $1/\rho$, solve the resulting

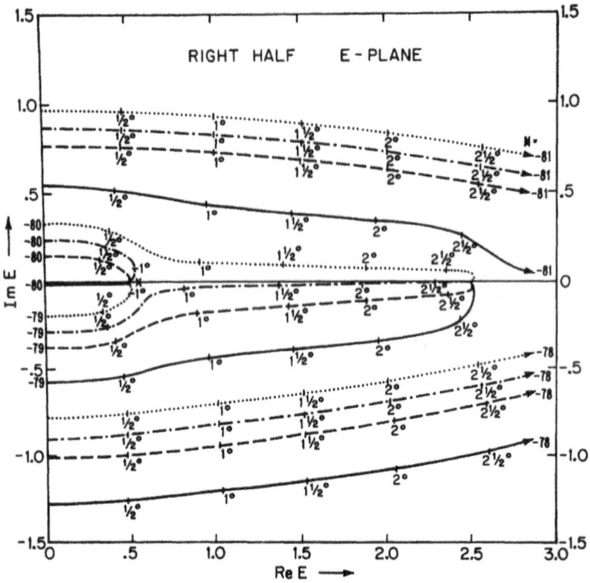

FIG. 5. The complex E plane, showing to the nearest hundredth the images of the circular arcs in the $1/\rho$ plane of Fig. 4. These curves exhibit level crossing. The precise point where this level crossing occurs is marked with an **X**. **X** is the image of the branch point ($N=79$, $n=0$). The energy-level curves shown correspond with the four lowest energy levels. The curves are labeled by their M values [see Eq. (6.16)] and are represented by dotted, dot-dashed, dashed, and solid lines to correspond with the values of $|1/\rho|$ on Fig. 4 (1484.50, 1487.50, 1490.50, and 1496.50). The numbers in degrees along each curve indicate the rotation in the $1/\rho$ plane of the preimage points along the curves on Fig. 4.

pair of simultaneous transcendental equations for E. The solution is a painstaking trial-and-error process involving tables of $\ln\Gamma(z)$ functions. Fortunately, this mammoth task is tractable because $\arg(1/\rho)$ is small, ranging from 0 to approximately 3° (see Fig. 4). Several hundred values for E have been plotted on Fig. 5 and the energy-level curves have been drawn and labeled by their M values. We have plotted the energy-level curves connected to the four lowest energy levels.

This completes the description of the construction of Figs. 4 and 5. We list below our observations and conclusions with regard to these graphs.

As $|1/\rho|$ gets smaller, we observe that the points where the energy-level curves meet the imaginary axis in the E plane move upward. {This fact can be proved from the differential equation [Eq. (4.9)] independently of the numerical calculations involved in the graphing of Fig. 5.} Therefore, as $|1/\rho|$ decreases, either we will find that (a) level crossing occurs (that is, the curves associated with a given M value become associated with different energy values along the real E axis) or else that (b) the curves all remain attached to the same energy values on the real E axis and stretch higher and higher up the imaginary axis. The enormous distortion of the curves in the latter possibility (b) makes the former (a) a good probability.

Level crossing is indeed found on inspection of Fig. 5. (For each value of M the curves on Fig. 5 are represented in order of decreasing $|1/\rho|$ by solid lines, dashed lines, dash-dotted lines, and dotted lines.) We observe the following example of level crossing.

The solid $M=-79$ line is connected to $E=2\frac{1}{2}$ along the real axis. There is just a hint of inflection as this curve passes below $E=\frac{1}{2}$. Its companion, the solid $M=-80$ line, goes directly to $E=\frac{1}{2}$ and the dash-dotted line inflects strongly. Meanwhile, the companion $M=-80$ lines belly increasingly outward above $E=\frac{1}{2}$ to meet the inflecting $M=-79$ lines. As the curves in the $1/\rho$ plane pass through the $N=79$, $n=0$ branch point, the energy levels flip and the dotted $M=-79$ line is suddenly and discontinuously connected to $E=\frac{1}{2}$ and the companion $M=-80$ line to $E=2\frac{1}{2}$. *This is a precise pictorial description of level crossing.*

It is interesting to determine where the curves connecting $F=\frac{1}{2}$ and $E=2\frac{1}{2}$ cross. To determine this crossing point we have calculated the image of the branch point $N=79$, $n=0$ in the E plane. The crossing point is labeled on Fig. 5 by an **X**.

In addition to the above level crossing we also observe an incipient level crossing at $E=2\frac{1}{2}$. As we look at the lines in the opposite order (dotted, dot-dashed, dashed, and solid), we see on Fig. 4 that a level crossing is about to take place at the $N=79$, $n=1$

branch point. The solid lines $M=-81$ and $M=-79$ are ready to cross near $E=2\frac{1}{2}$.

As $|\lambda|$ increases and the circle in the λ plane passes through the appropriate branch points, each pair of adjacent levels flips repeatedly. *Only adjacent levels can cross.* For example, the zeroth and first energy levels cross at $n=0$ branch points, the $n=1$ and $n=2$ energy levels cross at $n=1$ branch points, and so on. In general, *the nth sequence of branch points is common to and lies only on the nth and $(n+1)$st sheets of the Riemann surface of $E(\lambda)$.* (Remember that the nth sheet is the sheet which contains the nth energy level for real positive λ.) This is equivalent to saying that *the nth and $(n+1)$st energy levels can only cross at the nth sequence of branch points.*

Figure 5 evokes an additional, unexpected, and most welcome observation. Recall that near $\lambda=270°$ the WKB equation [Eq. (6.1a)] describes the paths of the energy levels in the E plane. Near $\arg\lambda=0$, Eq. (4.8a) describes these paths. Therefore, at some value of $\arg\lambda$ between 270° and 0°, Eq. (6.1a) is no longer accurate and is superseded by Eq. (4.8a). The angle at which this changeover occurs defines the angular opening about 270° (and 810°) in which the WKB solution is valid. The numerical investigations used to plot Fig. 5 give answers to two important questions: (a) At what angle does the changeover occur? (b) How accurate is the WKB result?

To answer these questions, we notice that the curves on Fig. 5 all approach the values predicted by Eq. (4.8a) which are $E=\frac{1}{2}, 2\frac{1}{2}, 4\frac{1}{2}, \cdots$. However, at some angle in the $1/\rho$ plane the WKB result [Eq. (6.1a)] no longer directs the energy-level curves toward these values. At that angle the curves begin to veer away. Therefore, the angle at which the curves are closest to the points $E=\frac{1}{2}, 2\frac{1}{2}, 4\frac{1}{2}, \cdots$ is the answer to question (a). Furthermore, the *distance* of closest approach of the WKB curves to these points is a useful measure of how accurate the WKB solution is and will answer question (b). The WKB result is accurate if the distance of closest approach is small.

It is easy to calculate the changeover angle and the distance of closest approach. We expand $\ln\Gamma(\frac{1}{4}-\frac{1}{2}E)$ in a Taylor series[16] about $E=\frac{1}{2}, 2\frac{1}{2}, 4\frac{1}{2}, \cdots$, and insert the series into Eq. (6.16). We determine from the resulting equation that $\arg(1/\rho)=90°$ is the condition for closest approach. That is, the WKB result applies in an angular opening of 90° on either side of $\arg\lambda=270°$ (and 810°). This answers question (a). This angular opening is much wider than originally anticipated and shows that the WKB solution is valid in a very large region of the λ plane.

Furthermore, the changeover angle at $\arg\lambda=180°$ ties in beautifully with three earlier results of this paper, First, Fig. 1(c) in Sec. IV shows that it is exactly at $\arg\lambda=180°$ that the WKB turning point enters the sector in which the general boundary condition [Eq. (3.4)] applies. This is just the point at which we would expect Eq. (6.1a), the WKB equation, to become valid. Second, a changeover angle of 180° indicates that the bunching phenomenon discussed in Part A of Sec. VI takes place at 180°.[18] Third, when combined with the arguments of Sec. V, a changeover angle of 180° implies that the poles of the resolvent occur when $\arg\lambda$ is larger than 180°. This is consistent with Jaffe's findings.[2]

A short numerical calculation reveals that the distance of closest approach to $E=\frac{1}{2}$ is the amazingly small number $\frac{1}{3}\times 10^{-153}$. This distance is much smaller than originally anticipated. It is gratifying to observe that, in answer to question (b), the WKB solution is extremely accurate.

ACKNOWLEDGMENTS

We wish to thank Professor A. M. Jaffe for many interesting discussions. We are most grateful to Mrs. Carl M. Bender for her careful editorial reading of this paper.

APPENDIX A

In this appendix, we use field theory to derive the coordinate representation given by Eqs. (1.1) and (1.2) which are

$$(-d^2/dx^2+\tfrac{1}{4}x^2+\tfrac{1}{4}\lambda x^4)\Phi(x)=E(\lambda)\Phi(x) \tag{A1}$$

and

$$\lim_{x\to\pm\infty}\Phi(x)=0. \tag{A2}$$

A quick derivation follows from substituting

$$\varphi=2^{-1/2}x \quad\text{and}\quad \dot{\varphi}=-i2^{1/2}d/dx \tag{A3}$$

into the Hamiltonian H [Eq. (1.3)] which is

$$H=\tfrac{1}{2}\dot{\varphi}^2+\tfrac{1}{2}m^2\varphi^2+\lambda\varphi^4. \tag{A4}$$

We choose $m=1$ without loss of generality. Then the wave function $\Phi(x)$ satisfies the differential equation [Eq. (A1)]. Equation (A2) results from the requirement that $\Phi(x)$ be normalizable.

A more rigorous derivation follows from the assumption that in Fock space the energy eigenstates of H are normalizable.

In the interaction picture, the representation of the field φ in terms of creation and annihilation operators on Fock space is

$$\varphi=[1/(2m)^{1/2}](ae^{-imt}+a^\dagger e^{imt}), \tag{A5a}$$

$$\dot{\varphi}=(\tfrac{1}{2}m)^{1/2}(-iae^{-imt}+ia^\dagger e^{imt}), \tag{A5b}$$

$$[a,a^\dagger]=1. \tag{A5c}$$

In terms of the creation and annihilation operators, the Hamiltonian is

$$H=m(a^\dagger a+\tfrac{1}{2})+(g/4m^2)(a+a^\dagger)^4. \tag{A6}$$

Let $|E\rangle$ be an eigenstate of H of energy E. Then

$$H|E\rangle = E|E\rangle. \tag{A7}$$

The Fock representation consists of the states

$$|n\rangle = [1/\sqrt{(n!)}](a^\dagger)^n|0\rangle, \quad n=0,1,2,\cdots \tag{A8a}$$

where

$$\langle n|m\rangle = \delta_{nm}. \tag{A8b}$$

Assuming that $\{|n\rangle\}$ are a complete set of states,

$$|E\rangle = \sum_{n=0}^{\infty} a_n|n\rangle\sqrt{(n!)}. \tag{A9}$$

Equations (A6)–(A9) can be combined to give a difference equation that the $\{a_n\}$ satisfy:

$$\begin{aligned} Ea_n = m(n+\tfrac{1}{2})a_n + (g/4m^2)[&a_{n-4} + 4(n-2)a_{n-2} \\ &+ 6n(n-1)a_n + 4n(n+1)(n+2)a_{n+2} \\ &+ (n+1)(n+2)(n+3)(n+4)a_{n+4}]. \end{aligned} \tag{A10}$$

Furthermore, since $|E\rangle$ is assumed normalizable,

$$\sum_{n=0}^{\infty} |a_n|^2 n! < \infty. \tag{A11}$$

From Eq. (A11) it follows that as $n\to\infty$, $|a_n|\to 0$ faster than $1/\sqrt{(n!)}$. Thus, if we define the generating function $f(z)$ by

$$f(z) = \sum_{n=0}^{\infty} a_n z^n, \tag{A12}$$

then $f(z)$ is entire.

The difference equation [Eq. (A10)] becomes a differential equation satisfied by $f(z)$:

$$Ef(z) = \frac{g}{4m^2}\left(z + \frac{d}{dz}\right)^4 f(z) + mz\frac{d}{dz}f(z) + \tfrac{1}{2}mf(z). \tag{A13}$$

The differential equation [Eq. (A13)] and the side condition that $f(z)$ is entire embody the assumption that $|E\rangle$ is an eigenstate of H of finite norm and that $\{|n\rangle\}$ are a complete set of states.

Now let

$$f(z) = e^{-z^2/2}h(z). \tag{A14}$$

Then,

$$Eh(z) = (g/4m^2)h''''(z) + m[zh'(z) - z^2h(z)] + \tfrac{1}{2}mh(z). \tag{A15}$$

Next we show that $h(z)$ must go to 0 rapidly along the imaginary axis. To do this, we apply the Cauchy formula to Eq. (A12):

$$a_n = \frac{1}{2\pi i}\oint \frac{f(z)}{z^{n+1}}dz = \frac{1}{2\pi i}\oint \frac{e^{-z^2/2}g(z)}{z^{n+1}}. \tag{A16}$$

Suppose that $g(z)$ did not go to 0 along the imaginary axis. In fact, suppose for simplicity that $g(z)$ were constant. Then, since the major contribution to this integral would come from the region where the path of integration crosses the imaginary axis, we calculate (using the method of steepest descent) that $a_n \sim A/\sqrt{n}$ for large n. This contradicts Eq. (A11) and the entirety of $f(z)$. Hence, $h(z)$ must go to 0 along the imaginary axis.

To see how fast $h(z)\to 0$, we consider the differential equation for $h(z)$ [Eq. (A15)]. By systematically investigating pairs of terms, we find that the dominant behavior at large z is given by

$$(g/4m^2)h''''(z) - mz^2h(z) = 0. \tag{A17}$$

Thus,

$$h(z) \sim \exp[C\tfrac{2}{3}(4m^2/g)^{1/4}z^{3/2}], \tag{A18}$$

where $C=1, -1, i, -i$. Since there are four solutions to the fourth-order differential equation [Eq. (A17)], we have them all. Combining Eq. (A18) with our knowledge that $h(z)$ goes to 0 along the imaginary axis, we learn that $h(z)$ goes to 0 along the imaginary axis extremely rapidly. Hence, $h(z)$ may be Fourier-transformed along the imaginary axis. We thus define (for real x)

$$h(z) = \int_{-\infty}^{\infty} H(x)e^{zx}dx. \tag{A19}$$

Fourier-transforming Eq. (A15) according to Eq. (A19) gives

$$mH''(x) + mxH'(x) + (E + \tfrac{1}{2}m)H(x) - (g/4m^2)x^4H(x) = 0. \tag{A20}$$

Finally, we let

$$H(x) = e^{-x^2/4}\Phi(x). \tag{A21}$$

This substitution [Eq. (A21)] simplifies (A20) to

$$m\Phi''(x) + E\Phi(x) - \tfrac{1}{4}mx^2\Phi(x) - (g/4m^2)x^4\Phi(x) = 0. \tag{A22}$$

Equation (A22) implies that the dominant behavior of Φ for large x is $e^{|x|^3}$ or $e^{-|x|^3}$. The former is not possible because the inverse transform would not exist and $h(z)$ would be undefined. Hence, $\Phi(x)\to 0$ as $|x|\to\infty$.

We set $m=1$ as we did earlier in this appendix and Eq. (A22) becomes

$$-\Phi''(x) + \tfrac{1}{4}x^2\Phi(x) + \tfrac{1}{4}\lambda x^4\Phi(x) = E\Phi(x), \tag{A23}$$

and

$$\lim_{|x|\to\infty}\Phi(x) = 0. \tag{A24}$$

Equations (A23) and (A24) are precisely Eqs. (A1) and (A2). We have thus rigorously derived the coordinate representation and the boundary condition satisfied by the wave function Φ.

This connection between field theory and the wave function in coordinate space will be used in Sec. II and Appendices B and C.

APPENDIX B

In this appendix, we use Feynman rules to calculate A_1, A_2, and A_3. These are the first three terms beyond the constant zero-point energy term in the perturbation series for the ground-state energy [Eq. (2.1)]. The Feynman rules for the anharmonic oscillator considered as a field theory are

$$(E^2-m^2+i\epsilon)^{-1} \quad \text{for a propagator}, \tag{B1a}$$

$$24\lambda \quad \text{for a vertex}, \tag{B1b}$$

$$i(2\pi)^{-1}\int_{-\infty}^{\infty} dE \quad \text{for a loop integration}. \tag{B1c}$$

The symmetry numbers are the same as in the (3+1)-dimensional theory.[19]

The nth term in the perturbation series for the ground-state energy is the sum of all connected n-vertex Feynman diagrams with no external legs. (Each diagram is of course multiplied by its appropriate symmetry number.) Figure 6 shows all such diagrams of order $n=1$, 2, and 3 with their symmetry numbers.[20]

We find the contribution of each of the diagrams by applying the Feynman rules [Eq. (B1)] and by evaluating the resulting simple contour integrals. This calculation yields

$$\begin{aligned}
(1a)&=(24)\tfrac{1}{4}(\lambda/m^3)m,\\
(2a)&=(24)^2(-1/16)(\lambda/m^3)^2m,\\
(2b)&=(24)^2(-1/32)(\lambda/m^3)^2m,\\
(3a)&=(24)^3(3/128)(\lambda/m^3)^3m,\\
(3b)&=(24)^3(5/512)(\lambda/m^3)^3m,\\
(3c)&=(24)^3(3/512)(\lambda/m^3)^3m,\\
(3d)&=(24)^3(1/64)(\lambda/m^3)^3m.
\end{aligned} \tag{B2}$$

Next, we combine the symmetry numbers listed on Fig. 6 with Eq. (B2) and find that[21]

$$A_1=(24/4)\times\tfrac{1}{8}=\tfrac{3}{4}, \tag{B3a}$$

$$A_2=-\frac{(24)^2}{16\times16}-\frac{(24)^2}{32\times48}=-\frac{21}{8}, \tag{B3b}$$

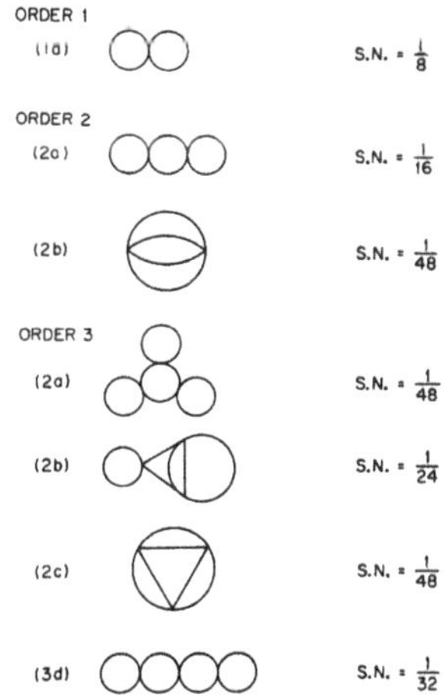

FIG. 6. All connected Feynman diagrams of order $n=1$, 2, and 3 with no external legs. To the right of each diagram is its symmetry number (S.N.).

$$A_3==\frac{(24)^3\times3}{128\times48}+\frac{(24)^3\times5}{512\times24}+\frac{(24)^3\times3}{512\times48}+\frac{(24)^3}{64\times32}=\frac{333}{16}. \tag{B3c}$$

The results in Eqs. (B3a)–(B3c) agree with Eqs. (2.12a)–(2.12c). This agreement illustrates the equivalence of the Feynman-diagram and difference-equation techniques for computing the numbers A_n.

APPENDIX C

Equation (2.2) gave without proof upper and lower bounds on $(-1)^{n+1}A_n$,[22] the terms in the perturbation series for the ground-state energy [Eq. (2.1)]. In this appendix, we rigorously derive these bounds using a Feynman-integral representation.

To develop the notation of the Feynman-integral representation, we restate here the ground-state-energy perturbation series [Eq. (2.1)]:

$$E_0(\lambda)=\tfrac{1}{2}m+\sum_{n=1}^{\infty} m(\lambda/m^3)^n A_n. \tag{C1}$$

A_n is the sum of all connected diagrams of order n (diagrams with n vertices) with no external legs.

[19] T. T. Wu, Phys. Rev. **125**, 1436 (1962).

[20] The interaction we are discussing is not Wick-ordered. If it were Wick-ordered, only diagrams (2b) and (3c) would contribute to the ground-state-energy perturbation series. All other diagrams would be excluded because they have at least one line emerging from and returning to the same vertex.

[21] In the definition of the ground-state-energy perturbation series [Eq. (2.1)], which is

$$E_0(\lambda)=\tfrac{1}{2}m+\sum_{n=1}^{\infty}(\lambda/m^3)^n mA_n,$$

the factors of m and $(\lambda/m^3)^n$ have been isolated from the A_n.

[22] We find upper and lower bounds on $(-1)^{n+1}A_n$ because $(-1)^{n+1}A_n$ are positive quantities. This is evident from Eq. (C3).

Rewriting the perturbation series [Eq. (C1)] in terms of individual diagrams gives

$$E_0(\lambda)=\tfrac{1}{2}m+\sum_{n=1}^{\infty} m(\lambda/m^3)^n \sum I, \tag{C2}$$

where the summation $\sum I$ is over all connected diagrams of order n with no external legs.

I is the numerical value of each diagram. In terms of a Feynman-integral representation for non-Wick-ordered connected diagrams with no external legs, I is given by

$$I=(\text{symmetry number})(\tfrac{1}{2})^w(24)^n(-1)^{n+1}$$
$$\times(1/4\pi)^{(n+1-w)/2}\Gamma(\tfrac{3}{2}n-\tfrac{1}{2}w-\tfrac{1}{2})$$
$$\times\int_0^{\infty}\prod_{i=1}^{N}\frac{dx_i}{\sqrt{x_i}}\{\delta(1-\sum_{i=1}^{N}x_i)/[D(1/x_i)]^{1/2}\}. \tag{C3}$$

In Eq. (C3), (i) n = order of diagram = number of vertices. In Eq. (C3), $n\geq 2$. For $n=1$, A_1 is calculated in Appendix B. (ii) w = number of lines returning to the same vertex from which they emerged (w would be zero if the interaction were Wick ordered). (iii) N = number of lines $=2n$. (iv) x_i = the Feynman parameters. (v) A skeleton of a diagram is a simply connected subdiagram. $D(1/x_i)$ = the sum over all skeletons of a diagram of

$$\left(\prod_{i=1}^{N}\frac{1}{x_i}\right)^{-1}.$$

The product is taken over all lines in the skeleton.[23]

We use Eq. (C3) to establish bounds on $(-1)^{n+1}A_n$ by proving a sequence of theorems.

Theorem 1. Let

$$L\equiv\int_0^{\infty}\prod_{i=1}^{N}\frac{dx_i}{\sqrt{x_i}}\{\delta(1-\sum_{i=1}^{N}x_i)/[D(1/x_i)]^{1/2}\}. \tag{C4}$$

Then, (i) L is uniformly bounded above and (ii) L is uniformly bounded below. (A uniform bound on diagrams means that the bound depends only on the number of vertices and not on the individual diagrams.)

Proof of part (i). We substitute $y=\sqrt{x_i}$ into the definition of L [Eq. (C4)] and get

$$L=\int_0^{\infty}\prod_{i=1}^{N}2dy_i\frac{\delta(1-\sum y_i^2)}{[D(1/y_i^2)]^{1/2}}. \tag{C5}$$

Let K_N be the surface area of a unit sphere in N dimensions. Then,

$$L\leq K_N\sup\frac{1}{[D(1/y_i^2)]^{1/2}}\bigg|_{\Sigma y_i^2=1}. \tag{C6}$$

We evaluate the factors in Eq. (C6) separately.

By a straightforward evaluation of the surface-area integral, we get for the first factor

$$K_N=2\pi^{\frac{1}{2}N}/\Gamma(\tfrac{1}{2}N). \tag{C7}$$

In order to find the maximum of $[D(1/y_i^2)]^{-1/2}|_{\Sigma y_i^2=1}$, we note that $0\leq y_i^2\leq 1$ for each $1\leq i\leq N$. We can overestimate a maximum by letting every $y_i=1$. Then

$$\frac{1}{[D(1/y_i^2)]^{1/2}}\bigg|_{\Sigma y_i^2=1}\leq\frac{1}{\sqrt{D(1)}}$$
$$\leq\frac{1}{\sqrt{(\text{number of skeletons})}}. \tag{C8}$$

Clearly, the smallest number of skeletons a graph could have is 1. Hence,

$$\sup\frac{1}{[D(1/y_i^2)]^{1/2}}\bigg|_{\Sigma y_i^2=1}=1. \tag{C9}$$

Combining Eqs. (C6), (C7), and (C9) and using $N=2n$ yields

$$L\leq 2\pi^n/\Gamma(n). \tag{C10}$$

This is the desired estimate for Theorem 1, part (i).

Proof of part (ii).[24] We make the substitution $\alpha_i=1/x_i$ in Eq. (C4) and integrate over the δ function. The result is that

$$L=\int_0^{\infty}\cdots\int_0^{\infty}\prod_{i=2}^{N}\frac{d\alpha_i}{\alpha_i^{3/2}}\alpha_1^{1/2}\frac{1}{[D(\alpha_i)]^{1/2}}\bigg|_{\Sigma(1/\alpha_i)=1}. \tag{C11}$$

Hence,

$$L\geq\int_{2N}^{3N}\cdots\int_{2N}^{3N}\prod_{i=2}^{N}\frac{d\alpha_i}{\alpha_i^{3/2}}\alpha_1^{1/2}\frac{1}{[D(\alpha_i)]^{1/2}}\bigg|_{\Sigma(1/\alpha_i)=1}. \tag{C12}$$

So

$$L\geq\left\{\int_{2N}^{3N}\prod_{i=2}^{N}\frac{d\alpha_i}{\alpha_i^{3/2}}\right\}\{\inf(\sqrt{\alpha_1})|_{2N\leq\alpha_i\leq 3N}\}$$
$$\times\left\{\inf\frac{1}{[D(\alpha_i)]^{1/2}}\bigg|_{\Sigma(1/\alpha_i)=1,\ 2N\leq\alpha_i\leq 3N}\right\}. \tag{C13}$$

We treat each of the factors in (C13) separately.

We evaluate the first factor exactly:

$$\int_{2N}^{3N}\prod_{i=2}^{N}\frac{d\alpha_i}{\alpha_i^{3/2}}=\left(\frac{2\sqrt{3}-2\sqrt{2}}{(6N)^{1/2}}\right)^{N-1}\geq N^{(1-N)/2}(\tfrac{1}{10})^{N-1}. \tag{C14}$$

The smallest possible value for the second factor is obtained when $\alpha_i=3N$. Then,

$$\inf(\sqrt{\alpha_1})=\left(\frac{3}{2+1/N}\right)^{1/2}\geq 1. \tag{C15}$$

[23] Precise definitions of "skeleton" and "$D(1/x_i)$" are given in Jaffe's paper. See Ref. 4.

[24] The proof of Theorem 2, part (ii) closely follows the proof given by Jaffe. See Ref. 4. However, Jaffe's other arguments do not apply here because of counting and duplication problems associated with diagrams having no external legs.

The minimum value for the third term in Eq. (C13) is found by computing the maximum value for $D(\alpha_i)|_{\Sigma\alpha_i=1,\ 2N\leq\alpha_i\leq 3N}$. But

$$\sup D(\alpha_i)=(\text{max number of skeletons})\times(\text{max value for each skeleton}). \quad \text{(C16)}$$

An overestimation of the maximum number of skeletons a graph of n vertices has is 4^{n-1}. Also, the maximum value for each skeleton is achieved when all the $\alpha_i=3N$. Putting this into Eq. (C16) yields

$$\sup D(\alpha_i)=4^{n-1}(3N)^{n-1}. \quad \text{(C17)}$$

Hence,

$$\inf\left\{\frac{1}{[D(\alpha_i)]^{1/2}}\bigg|_{\Sigma(1/\alpha_i)=1,\ 2N\leq\alpha_i\leq 3N}\right\}=(12N)^{(1-n)/2}. \quad \text{(C18)}$$

We combine Eqs. (C14), (C15), and (C18) with Eq. (C13) to get

$$L\geq(N^{(1-N)/2})(\tfrac{1}{10})^{N-1}(12N)^{(1-n)/2}. \quad \text{(C19)}$$

Substituting $N=2n$,

$$L\geq(2n)^{(1-2n)/2}(10)^{1-2n}(24n)^{(1-n)/2}. \quad \text{(C20)}$$

This is the desired estimate for Theorem 1, part (ii). Note that the bounds given in Eq. (C10) and in Eq. (C20) are uniform. That is, they depend on n but not on the individual diagrams.

Theorem 2. The symmetry number for diagrams is uniformly bounded (i) above and (ii) below.

Proof of part (i). Since all symmetry numbers are less than 1, an upper bound on symmetry numbers for all diagrams is 1.

Proof of part (ii). A small symmetry number corresponds to a diagram having much symmetry. The class of connected diagrams having the greatest degree of symmetry per number of vertices is shown in Fig. 7. If n is the number of vertices, the symmetry numbers for diagrams belonging to this infinite class is exactly

$$\text{S.N.}=1/(8\times 6^{n/2}). \quad \text{(C21)}$$

Equation (C21) serves as a lower bound on all symmetry numbers. This completes Theorem 2.

Combining the results of Theorems 1 and 2 places upper and lower bounds on $(-1)^{n+1}I$. Those bounds are

$$(-1)^{n+1}I\leq(\tfrac{1}{2})^w(24)^n\left(\frac{1}{(4\pi)^{1/2}}\right)^{n+1-w}\times\Gamma(\tfrac{3}{2}n-\tfrac{1}{2}w-\tfrac{1}{2})\frac{2\pi^n}{\Gamma(n)} \quad \text{(C22)}$$

and

$$(-1)^{n+1}I\geq\frac{1}{8}\frac{1}{6^{n/2}}(\tfrac{1}{2})^w(24)^n\left(\frac{1}{(4\pi)^{1/2}}\right)^{n+1-w}\times\Gamma(\tfrac{3}{2}n-\tfrac{1}{2}w-\tfrac{1}{2})(2n)^{(1-2n)/2}\times 10^{1-2n}(24n)^{(1-n)/2}. \quad \text{(C23)}$$

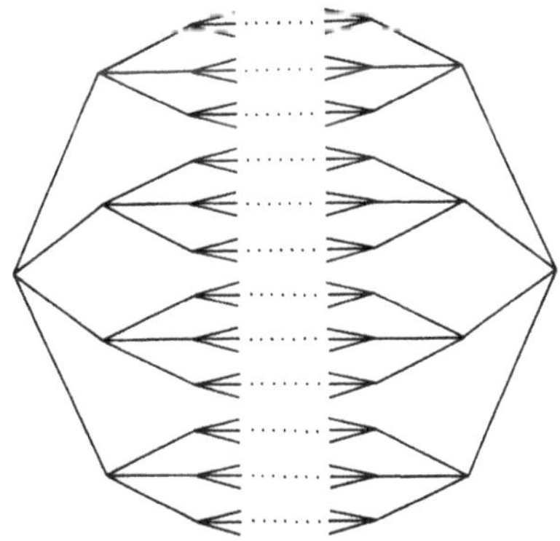

FIG. 7. The infinite class of diagrams having maximum symmetry per number of vertices. These diagrams have the smallest symmetry numbers for diagrams with a comparable number of vertices. All diagrams in this class start from a single vertex, each of whose four lines branches repeatedly and symmetrically into three new lines; after an arbitrary number of branchings, the lines symmetrically recombine into a single vertex.

To find bounds on $(-1)^{n+1}A_n$, we must make the estimates for $(-1)^{n+1}I$ independent of w. Clearly, for a connected diagram of n vertices, w can range from 0 to n. Therefore, in Eq. (C22) we take $w=0$ and in Eq. (C23) we take $w=n$. After simplification, the estimates [Eqs. (C22) and (C23)] become

$$(-1)^{n+1}I\leq\frac{1}{\sqrt{\pi}}(12\sqrt{\pi})^n\frac{\Gamma(\tfrac{3}{2}n-\tfrac{1}{2})}{\Gamma(n)} \quad \text{(C24)}$$

and

$$(-1)^{n+1}I\geq\frac{5}{2}\left(\frac{3}{\pi}\right)^{1/2}\left(\frac{1}{200}\right)^n\Gamma(n-\tfrac{1}{2})n^{(2-3n)/2}. \quad \text{(C25)}$$

We now return to the integral representation [Eq. (C3)] and observe that for a given order n the contributions from all diagrams add in phase. Therefore, upper and lower bounds on $(-1)^{n+1}A_n$, the terms in the perturbation series, may be obtained once we place bounds on the number of diagrams in each order.

Theorem 3. The number of diagrams of order n is (i) bounded above by $(4n-1)!!$ and (ii) bounded below by $n!/3^n$.

Proof of part (i). We establish an upper bound on the number of diagrams by constructing them as follows.

We begin with n vertices. None of the vertices are connected, so each has four free lines emerging from it. There are $4n$ free lines.

We pick any line and connect it to one of the other $4n-1$ lines. There are now $4n-2$ lines. We pick and connect another line in one of $4n-3$ ways. We continue until there are no free lines. Therefore, there are at most $(4n-1)!!$ diagrams.

Proof of part (ii). In order to establish a lower bound on the number of diagrams, we establish a lower bound on the number of diagrams belonging to a particular class. We consider only those diagrams in which all the vertices lie on a continuous closed line (a circular path). We construct such a class of diagrams as follows.

Draw a circle and place all the n vertices on it. The circle uses up two of the four lines at each vertex. This leaves two free (unconnected) lines extending from every vertex. Now start with any vertex and pick one of the two free lines emerging from it. Join this line to a line coming from any of the n vertices (including itself). Connect the last free line coming from this new vertex to one of the remaining $n-1$ vertices. Continue this process, going sequentially from vertex to vertex until all pairs of free lines are used up. Since we use both lines at each vertex one right after the other, the number of available vertices keeps decreasing by one. Of course, at any time the process may close on itself if the free line is connected to the last line available at the original vertex. If this happens, we simply start over at any other free vertex, which will have two available lines, and continue.

Clearly, there are $n!$ ways to carry out the above process, but there will be many duplications. To avoid these duplications, we divide by three at each vertex.

We conclude that there are at least $n!/3^n$ diagrams. This establishes Theorem 3.

Combining the results of Theorem 3 with Eqs. (C24) and (C25) places upper and lower bounds on $(-1)^{n+1}A_n$:

$$(-1)^{n+1}A_n \leq \frac{(4n-1)!!\Gamma(\frac{3}{2}n-\frac{1}{2})}{\Gamma(n)} \frac{(12\sqrt{\pi})^n}{\sqrt{\pi}} \qquad \text{(C26)}$$

and

$$(-1)^{n+1}A_n \geq \frac{n!\Gamma(n-\frac{1}{2})}{n^{(3n-2)/2}}(600)^{-n}\frac{5}{2}\left(\frac{3}{\pi}\right)^{1/2}. \qquad \text{(C27)}$$

Simplifying Eqs. (C26) and (C27) yields

$$AB^n\Gamma(\tfrac{1}{2}n) \leq (-1)^{n+1}A_n \leq CD^n\Gamma(\tfrac{3}{2}n). \qquad \text{(C28)}$$

Equation (C28) is the final result. Note that it agrees with the bounds in Eq. (2.2).

APPENDIX D

In this appendix, we discuss the details of the computer calculation of the first 75 terms in the ground-state-energy perturbation series [Eq. (2.1)] which is

$$E_0(\lambda) = \tfrac{1}{2}m + \sum_{n=1}^{\infty} mA_n(\lambda/m^3)^n. \qquad \text{(D1)}$$

In the calculation, the difference equation [Eq. (2.8)] which is

$$2jB_{i,j} = (j+1)(2j+1)B_{i,j+1} + B_{i-1,j-2} - \sum_{p=1}^{i-1}(B_{i-p,1})(B_{p,j}) \qquad \text{(D2)}$$

and Eq. (2.9) which is

$$A_i = -B_{i,1} \qquad \text{(D3)}$$

were used on an IBM 7094 in double-precision mode. (The results of this computer calculation are listed in Table I. The absolute value of A_n is listed to the right of the order n. The number following the comma is the power of 10 multiplying the decimal.)

The computer program had three interesting features.

(i) To calculate A_n it was necessary to know all $B_{i,j}$ for $j=1, 2, \cdots, 2n$ and $i=1, \cdots, n$. Because $B_{i,j}$ has two indices, the limiting factor in this calculation was not time but memory capacity. Seventy-five was the maximum number of terms in the perturbation series

TABLE I. The absolute values of the first 75 terms A_n in the perturbation series for the ground-state energy [Eq. (2.1)]. The number following the comma is the power of 10 multiplying the decimal. The signs of A_n alternate beginning with plus, i.e., $A_n=(-)^{n-1}|A_n|$.

n	$\vert A_n\vert$		n	$\vert A_n\vert$	
1	0.750000000000,	0	39	0.559961162001,	64
2	0.262500000000,	1	40	0.664186377517,	66
3	0.208125000000,	2	41	0.807715625319,	68
4	0.241289062500,	3	42	0.100646863922,	71
5	0.358098046875,	4	43	0.128429408835,	73
6	0.639828134766,	5	44	0.167730490075,	75
7	0.132973372705,	7	45	0.224085871222,	77
8	0.314482146928,	8	46	0.306093131234,	79
9	0.833541603263,	9	47	0.427287814796,	81
10	0.244789407028,	11	48	0.609277658120,	83
11	0.789333316003,	12	49	0.887045954492,	85
12	0.277387769635,	14	50	0.131804184041,	88
13	0.105564665831,	16	51	0.199796510697,	90
14	0.432681068354,	17	52	0.308853507485,	92
15	0.190081719760,	19	53	0.486698406263,	94
16	0.891210175364,	20	54	0.781543229859,	96
17	0.444255088999,	22	55	0.127844051221,	99
18	0.234646430681,	24	56	0.212959346691,	101
19	0.130915026105,	26	57	0.361127834038,	103
20	0.769399985382,	27	58	0.623214611884,	105
21	0.475124077343,	29	59	0.109419743747,	108
22	0.307579295113,	31	60	0.195392812615,	110
23	0.208301009434,	33	61	0.354775753462,	112
24	0.147290492138,	35	62	0.654807077714,	114
25	0.108552296005,	37	63	0.122820914441,	117
26	0.832483627531,	38	64	0.234056056576,	119
27	0.663329371112,	40	65	0.453052532884,	121
28	0.548392431330,	42	66	0.890541554378,	123
29	0.469782420784,	44	67	0.177719768349,	126
30	0.416502699806,	46	68	0.359993943833,	128
31	0.381734895502,	48	69	0.740009711275,	130
32	0.361299554016,	50	70	0.154336989562,	133
33	0.352778001982,	52	71	0.326515091790,	135
34	0.355023393768,	54	72	0.700567351807,	137
35	0.367917476478,	56	73	0.152414090414,	140
36	0.392301600257,	58	74	0.336160314738,	142
37	0.430055097492,	60	75	0.751507982642,	144
38	0.484327278330,	62			

which could be computed without using additional memory space on disks or tapes.

(ii) The numbers $B_{i,j}$ grow exponentially large or small with i for j near 1 or near $2i$. As a result, the capacity of the 7094 to manipulate these large and small exponents in floating point would be exceeded before $i=20$. Therefore, it was necessary to scale the $B_{i,j}$. One particularly useful scaling is

$$C_{i,j}=(1/100^{i-j})B_{i,j}. \tag{D4}$$

Equation (D4) is adequate for $n\lesssim 100$.

(iii) The structure of the difference equation [Eq. (D2)] indicated that a huge number of manipulations would have to be performed in which numbers with greatly differing exponents would be added together. This would cause round-off errors which on a computer are cumulative. Therefore, although the calculation was done to 16 decimal places (double-precision mode), one must take care to determine the number of decimal places which are really significant.

If we assume that the round-off errors are purely random (not systematic), then the cumulative error grows as the square root of the number of operations. Because there are approximately 10^8 operations performed in the calculation of A_{75}, we expect that the error is about 10^4 in 10^{16}. That is, the last four decimal places (of the sixteen) are random.

We can check that there are indeed 12 significant figures in the table of A_n as follows: Divide out the dominant asymptotic behavior of the A_n [see Eq. (2.10)] and make a table of the resulting numbers Q_n:

$$Q_n\equiv\frac{A_n}{(\sqrt{6})\pi^{-3/2}\Gamma(n+\frac{1}{2})3^n}. \tag{D5}$$

Then compute a table of the successive differences:

$$Q_n^{(1)}\equiv Q_n-Q_{n-1}, \tag{D6a}$$

$$Q_n^{(2)}\equiv Q_n^{(1)}-Q_{n-1}^{(1)}, \tag{D6b}$$

and so on.

The $Q_n^{(i)}$ are regularly monotonic as i increases until they become random. By inspection we find that $Q^{(9)}$ is random and, as expected, it is nonzero in the last four decimal places (of the sixteen). This verifies that the Q_n and hence the A_n are correct to 12 places.

APPENDIX E

There is no known mathematically rigorous proof of the detailed asymptotic growth of the A_n as given in Eq. (2.10). However, in this appendix, we justify this equation using numerical methods. The raw data used in this appendix are contained in Table I.

We approach the problem by finding the rough growth of A_n. To do this, we construct a table of the ratios

$$R_n=A_{n+1}/A_n \tag{E1}$$

from the values of A_n listed in Table I. As $n\to 75$, we observe that to a good approximation $R_n\to 3(n+\frac{1}{2})$. This indicates that A_n grows as $3^n\Gamma(n+\frac{1}{2})$.

Next we try to fit A_n very accurately by

$$A_n\sim 3^n\Gamma(n+\tfrac{1}{2})\left(a_0+\frac{a_1}{n}+\frac{a_2}{n^2}+\frac{a_3}{n^3}+\cdots\right). \tag{E2}$$

[The $\sim$ notation in Eq. (E2) is justified because the term in square brackets is an asymptotic (nonconvergent) series.] Note that it would not be possible to fit A_n by a series such as that in Eq. (E2) if A_n did not grow as $3^n\Gamma(n+\frac{1}{2})$; that is, if the rough growth of A_n were, say, $(2.9)^n\Gamma(n+\frac{1}{2})$ or $3^n\Gamma(n+0.6)$.

To evaluate the a_i in Eq. (E2), we use Table I to calculate

$$P_n=A_n/3^n\Gamma(n+\tfrac{1}{2}). \tag{E3}$$

Then we terminate the series in Eq. (E2) at a_k and solve for a_i, $i=0, 1, \cdots, k$, in the resulting set of simultaneous equations:

$$P_n=a_0+\frac{a_1}{n}+\frac{a_2}{n^2}+\cdots+\frac{a_k}{n^k}, \tag{E4}$$

where $n=75-k, 75-k+1, \cdots, 75$.

We solve for a_0 in Eq. (E4) by multiplying each equation by n^k, taking successive differences, and using the beautiful identity:

$$\sum_{i=0}^{k}(n-i)^k\binom{k}{i}(-1)^i=k!. \tag{E5}$$

a_0 may thus be given exactly in closed form by

$$a_0=\sum_{i=0}^{k}(-1)^iP_{75-i}\binom{k}{i}(75-i)^k\Big/k!. \tag{E6}$$

Once a_0 is known, we can reduce Eq. (E4) to a new set of equations in which P_n is replaced by $(P_n-a_0)n$ and k by $k-1$. We solve as before for the new a_0 which is the old a_1. Repeating this process k times gives all a_i.

If we terminate the series at a_6, we find that to four parts in 10^9

$$a_0=(\sqrt{6})/\pi^{3/2}. \tag{E7}$$

We also find that

$$a_1/a_0=1.3194528, \tag{E8a}$$

$$a_2/a_0=1.9366649, \tag{E8b}$$

$$a_3/a_0=7.1004516, \tag{E8c}$$

$$a_4/a_0=27.445970, \tag{E8d}$$

$$a_5/a_0=665.57303. \tag{E8e}$$

Apparently, as i increases, the growth of the numbers a_i is so rapid that when $n\sim 75$, the term a_6/n^6 is larger than a_5/n^5. This suggests that the asymptotic series in Eq. (E4) is an optimal approximation of P_n when it is terminated at a_5/n^5. (More terms could be included, giving a more accurate approximation to P_n, only if n were larger.)

Since we keep six terms in the asymptotic series, we expect that P_n should be represented accurately and uniformly to a few parts in 10^9. This turns out to be the case for n near 75 down to well below 50. Because a_0 is $(\sqrt{6})\pi^{-3/2}$ to four parts in a billion and since the magnitude of this error is exactly what we would expect, we assert that a_0 *is* $(\sqrt{6})\pi^{-3/2}$.

Combining this value for a_0 with Eq. (E2) gives Eq. (2.10) which is the result we had set out to derive in this appendix, namely,

$$A_n \sim 3^n \Gamma(n+\tfrac{1}{2})(\sqrt{6})\pi^{-3/2}. \tag{E9}$$

One might note that although a_0 is known to four parts in 10^9, a_i/a_0 in Eq. (E8) is known less and less accurately as i increases. a_1/a_0 is accurate to seven or eight places; a_2/a_0 is accurate to five or six places; and a_5/a_0 is accurate to one place at most.

To increase the accuracy of a_i, A_n would have to be calculated for n much larger than 75. As is always the case with asymptotic series, nothing is gained by calculating A_n more precisely for n near 75.

Aside from simplicity, there is no compelling reason why P_n should be expanded into an asymptotic series of the form in Eq. (E2). One might prefer to use the more general series

$$P_n \sim \alpha_0 + \frac{\alpha_1}{(n+\epsilon)} + \frac{\alpha_2}{(n+\epsilon)^2} + \frac{\alpha_3}{(n+\epsilon)^3} + \cdots, \tag{E10}$$

where ϵ is any number which is small compared to n. Then a simple formula relates the a_i to the α_i:

$$\alpha_0 = a_0, \tag{E11a}$$

$$\alpha_1 = a_1, \tag{E11b}$$

$$\alpha_2 = a_2 + \epsilon a_1, \tag{E11c}$$

$$\alpha_3 = a_3 + 2\epsilon a_2 + \epsilon^2 a_1, \tag{E11d}$$

$$\alpha_4 = a_4 + 3\epsilon a_3 + 3\epsilon^2 a_2 + \epsilon^3 a_1, \tag{E11e}$$

and, in general,

$$\alpha_n = \sum_{i=1}^{n} a_i \epsilon^{n-i} \binom{n-1}{i-1}. \tag{E11f}$$

This completes our remarks on these numerical calculations. However, before concluding this appendix, we will discuss an example of how investigations of this sort might lead to a better understanding of the singularity structure of the ground-state energy $E_0(\lambda)$ in the complex plane.

Example. As a crude illustration of how one might recover the analytic properties of $E_0(\lambda)$ from Eq. (E2), we neglect a_i for $i \geq 1$. Then Eqs. (E2) and (2.1) give

$$E_0(\lambda) \sim \sum_{n=0}^{\infty} \lambda^n (-1)^n (\sqrt{6})\pi^{-3/2} \Gamma(n+\tfrac{1}{2}) 3^n. \tag{E12}$$

Then, ignoring the possibility that two unequal functions can have the same asymptotic expansion, we substitute the expression

$$\Gamma(n+\tfrac{1}{2}) = \int_0^\infty e^{-t} t^{n-\frac{1}{2}} dt \tag{E13}$$

into Eq. (E12). Interchanging summation and integration, summing the series, and letting $s=\sqrt{t}$ gives a convergent integral representation for E_0:

$$E_0(\lambda) \cong \frac{2\sqrt{6}}{\pi^{3/2}} \int_0^\infty \frac{e^{-s^2}}{1+3\lambda s^2} ds. \tag{E14}$$

For λ real and positive, this integral can be evaluated explicitly with the result that

$$E_0(\lambda) \cong (2/\pi\lambda)^{1/2} \exp(1/3\lambda)\, \mathrm{Erfc}[(3\lambda)^{-1/2}], \tag{E15}$$

where

$$\mathrm{Erfc}(x) = \frac{2}{\sqrt{\pi}} \int_0^\infty e^{-t^2} dt. \tag{E16}$$

More importantly, the dispersion-type integral in Eq. (E14) defines a nontrivial singularity structure in the λ plane—a cut along the negative real λ axis. The jump across the cut is easily calculated. Unfortunately, this result does not come close to giving the true singularity structure of $E_0(\lambda)$ as derived in Secs. III, IV, and VI of this paper. Nevertheless, it is hoped that future calculations of this sort will lead to positive results.

APPENDIX F

In this appendix, we present the WKB calculation of Eq. (4.35) to first order in λ. A successful calculation would be a significant result because:

(i) Our continued ability to match the asymptotic expansions across the boundaries in regions A, B, C, and D would give added assurance that the application of WKB techniques to the problem of locating singularities in $E(\lambda)$ is valid.

(ii) It is most important to make sure that the first-order result we obtain here differs from the zeroth-order result only by terms of order λ. This would imply that the error in Eq. (4.35) is indeed small, as is tacitly assumed in Secs. IV, V, and VI.

(iii) As a byproduct, this calculation would define the boundaries of regions A, B, C, and D more precisely than the zeroth-order calculation does. This added precision would result from the greater number of terms that would be matched across the boundaries.

In this appendix, we will follow the same procedure as in the zeroth-order calculation in Sec. IV. However, we keep terms of order λ as well as of order 1. We neglect terms of order λ^2 or smaller. We again use the differential equation [Eq. (4.9)] which is

$$[d^2/dr^2+\tfrac{1}{4}(-\epsilon+r^2-\rho r^4)]\Phi(r)=0, \qquad \text{(F1)}$$

and recall the definitions $\rho=\lambda e^{-3\pi i/2}$, $\epsilon=4iE$, and $r=xe^{i\pi/4}$.

We define the four regions A, B, C, and D as in Sec. IV. That is, in region A, $0<|r|\ll|r_0|$; in region B, $|r_0|\ll|r|\ll|r_1|$; in region C, $|r_0|\ll|r|\lesssim|r_1|$; and in region D, $|r|\sim|r_1|$.

Also, for the first-order calculation r_0 and r_1 must be given more exactly than for the zeroth-order calculation. From Eq. (4.10), we have

$$r_0=\{[1-(1-4\rho\epsilon)^{1/2}](2\rho)^{-1}\}^{1/2}\sim(\sqrt{\epsilon})(1+\tfrac{1}{2}\rho\epsilon) \qquad \text{(F2a)}$$

and

$$r_1=\{[1+(1-4\rho\epsilon)^{1/2}](2\rho)^{-1}\}^{1/2}\sim(1/\sqrt{\rho})(1-\tfrac{1}{2}\rho\epsilon). \qquad \text{(F2b)}$$

As in Sec. IV, we consider separately the regions A, B, C, and D.

Region A. We solve Eq. (F1) to first order in λ. To do this for region A, we find a transformation which to first order in λ turns Eq. (F1) into a parabolic cylinder function differential equation.[25] The transformation which works is

$$w=x+\lambda(\tfrac{1}{8}x^3+\tfrac{3}{4}Ex). \qquad \text{(F3)}$$

Then, the even-parity solution to (D1) is

$$\Phi_A=C(1-\tfrac{3}{16}\lambda x^2)[D_\nu(w)+D_\nu(-w)], \qquad \text{(F4)}$$

where

$$\nu=E-\tfrac{1}{2}-\tfrac{3}{2}\lambda(E^2+\tfrac{1}{4}). \qquad \text{(F5)}$$

[In this appendix, we do the WKB calculation for even-parity wave functions only. The differences for odd parity involve the algebraic manipulations in region A.]

We compute the asymptotic expansion of Φ_A in region A for r near region B. The first two terms in the asymptotic expansion[26] for $D_\nu(w)$ are

$$D_\nu(w)\sim w^\nu e^{-\frac{1}{4}w^2}\left[1+\frac{\nu(1-\nu)}{2w^2}\right] \quad \text{for } -\tfrac{3}{4}\pi<\arg w<\tfrac{3}{4}\pi \qquad \text{(F6a)}$$

and

$$D_\nu(w)\sim w^\nu e^{-\frac{1}{4}w^2}\left[1+\frac{\nu(1-\nu)}{2w^2}\right]-\frac{(2\pi)^{1/2}}{\Gamma(-\nu)}\times e^{\nu\pi i}w^{-\nu-1}e^{\frac{1}{4}w^2}\left[1+\frac{(1+\nu)(2+\nu)}{2w^2}\right] \quad \text{for } \tfrac{1}{4}\pi<\arg w<(5/4)\pi. \qquad \text{(F6b)}$$

Equations (F2)–(F6) are used to expand Φ_A into its asymptotic form. We expand to first order in powers of ρ and r_0^2/r^2. We keep terms of the form ρ and r_0^2/r^2 and neglect terms of the form ρ^2, $\rho r_0^2/r^2$, and r_0^4/r^4. These approximations are justified because they give the correct asymptotic connection with Φ_B. As we had expected, these approximations define the extent of region A more clearly than the zeroth-order calculation did.

After considerable simplification,[27] the asymptotic expansion for Φ_A is

$$\Phi_A\sim C2\pi 2^{\frac{1}{2}\nu+\frac{1}{4}}e^{\pi i/8}e^{\pi i\nu/4}\left(1+\frac{3\rho r^2}{16}\right)\left(1+\frac{\epsilon}{4r^2}+\frac{\rho r^2}{16}+\frac{17\rho\epsilon}{64}\right)\left[1-\frac{\rho\epsilon}{32}\left(\frac{37}{8}+\frac{\epsilon^2}{32}\right)\right]$$
$$\times\left\{\frac{r^\nu}{\Gamma(\frac{1}{2}+\frac{1}{2}\nu)}\exp\left[\frac{ir^2}{4}\left(1-\frac{3\rho\epsilon}{8}-\frac{\rho r^2}{4}\right)-\frac{i\pi}{8}\right]\exp\left[\frac{i}{4}\left(\frac{3\rho\epsilon^2}{16}+\frac{\rho\epsilon r^2}{8}+\frac{\epsilon^2}{8r^2}-\frac{2}{2r^2}\right)\right]2^{-\frac{1}{2}\nu-\frac{1}{4}}\right.$$
$$\left.+\frac{r^{-\nu-1}}{\Gamma(-\frac{1}{2}\nu)}\exp\left[\frac{-ir^2}{4}\left(1-\frac{3\rho\epsilon}{8}-\frac{\rho r^2}{4}\right)+\frac{i\pi}{8}\right]\exp\left[-\frac{i}{4}\left(\frac{3\rho\epsilon^2}{16}+\frac{\rho\epsilon r^2}{8}+\frac{\epsilon^2}{8r^2}-\frac{3}{2r^2}\right)\right]2^{\frac{1}{2}\nu+\frac{1}{4}}\right\}. \qquad \text{(F7)}$$

This is the desired result for region A.

Region B. The first-order WKB solution[28] to the differential equation [Eq. (F1)] is

$$\Phi_{\text{WKB}}=\frac{\sqrt{\mu}}{(1-\frac{1}{8}\mu'^2+\frac{1}{4}\mu''\mu)^{1/2}}\left(C_2\exp\left\{i\left[\int^r\left(\frac{1}{\mu}-\frac{\mu'^2}{8\mu}\right)dr+\frac{\mu'}{4}\right]\right\}C_3\exp\left\{-i\left[\int^r\left(\frac{1}{\mu}-\frac{\mu'^2}{8\mu}\right)dr+\frac{\mu'}{4}\right]\right\}\right), \qquad \text{(F8)}$$

[25] BMP, Vol. 2, p. 116, Eq. (1).

[25] BMP, Vol. 2, pp. 116–123. There is a serious misprint not on the errata sheet which must be corrected before one can do the first-order WKB calculation. Equation (2), p. 123, should read

$$D_\nu(z)=z^\nu e^{-z^2/4}\left[\sum_{n=0}^{N}\frac{(-\frac{1}{2}\nu)_n(\frac{1}{2}-\frac{1}{2}\nu)_n}{n!(-z^2/2)^n}+O|z^2|^{-N-1}\right]-\frac{(2\pi)^{1/2}}{\Gamma(-z)}e^{\nu\pi i}z^{-\nu-1}e^{z^2/4}\left[\sum_{n=0}^{N}\frac{(1+\frac{1}{2}\nu)_n(\frac{1}{2}+\frac{1}{2}\nu)_n}{n!(z^2/2)^n}+O|z^2|^{-N-1}\right] \text{ for } \pi/4<\arg z<5\pi/4.$$

Equation (3), p. 123, has a similar misprint but it is not needed in this paper. The equation is derived correctly in E. T. Whittaker and G. N. Watson, *A Course of Modern Analysis* (Cambridge University Press, Cambridge, 1952).

[27] BMP, Vol. 1, pp. 1–5.

[28] See Ref. 8. For first-order WKB techniques, see A. Messiah, pp. 233–234.

where

$$\mu=2(-\epsilon+r^2-\rho r^4)^{-1/2}. \tag{F9}$$

We would encounter a serious difficulty with this first-order calculation if we were to proceed as we did in Sec. IV. If we fixed the constants C_2 and C_3 in Eq. (F8) by defining the lower endpoint (the reference point) in the integrals to be r_0, we would find that the integrals become divergent regardless of the upper endpoint. The same would be true if we picked r_1 to be the reference point. In order to avoid this difficulty, we carry out the following procedure.

We treat $-\frac{1}{8}\int dr\,\mu'^2/\mu$ as an indefinite integral and integrate by parts repeatedly. This isolates the divergences at $r=r_0$ and r_1. After some tedious algebra, the result is

$$-\frac{1}{8}\int^r dr\frac{\mu'^2}{\mu}=-\frac{1}{6}\left(\rho r^3-\frac{r}{2}\right)(-\epsilon+r^2-\rho r^4)^{-3/2}$$
$$-\frac{1}{12\epsilon(1-4\rho\epsilon)}(-\epsilon+r^2-\rho r^4)^{-1/2}$$
$$\times[\rho r^3(1-12\rho\epsilon)-r(1-8\rho\epsilon)]$$
$$+\frac{\rho}{12\epsilon(1-4\rho\epsilon)}\int_{r_0}^{r}\frac{4\epsilon+(1-12\rho\epsilon)r^2}{(-\epsilon+r^2-\rho r^4)^{1/2}}dr+\text{const}. \tag{F10}$$

In Eq. (F10),

$$\epsilon=r_1^2r_0^2/(r_1^2+r_0^2), \tag{F11}$$

$$\rho=(r_1^2+r_0^2)^{-1}. \tag{F12}$$

Equation (F10) is an exact equation. Equations (F11) and (F12) are also exact and follow from the definitions of r_0 and r_1 [Eq. (4.10)].

We determine the constant in Eq. (F10) to first order by evaluating the integral $-\frac{1}{8}\int^r dr\,\mu'^2/\mu$ using a second method. Using Eq. (F9), we have

$$\int^r -\frac{1}{8}\frac{\mu'^2}{\mu}dr$$
$$=-\int^r dr(-\epsilon+r^2-\rho r^4)^{-5/2}(\rho r^3-\tfrac{1}{2}r)^2. \tag{F13}$$

Expanding the right-hand side of Eq. (F13) in region B to first order, we find that

$$\int^r -\frac{1}{8}\frac{\mu'^2}{\mu}dr\sim-\frac{1}{4}\int^r dr\left(r^{-3}+\frac{5}{2}\epsilon r^{-5}-\frac{3}{2}\rho r^{-1}-\frac{5}{4}r^{-3}\rho\epsilon\right)$$
$$\sim\frac{1}{8r^2}+\frac{3\rho}{8}\ln\frac{2r}{r_0}. \tag{F14}$$

As in region A, the first-order asymptotic expansion in region B means that we expand in powers of ρ and and r_0^2/r^2. We keep terms of the form ρ and r_0^2/r^2 and neglect terms of the form r_0^4/r^4, ρ/r^2, and ρ^2.

Finally, we return to Eq. (F10) and approximate that equation to first order in region B with the result that

$$\int^r -\frac{1}{8}\frac{\mu'^2}{\mu}dr\sim\text{const}+\frac{1}{8r^2}+\frac{3\rho}{8}\ln\frac{2r}{r_0}+\frac{1}{12\epsilon}-\frac{3\rho}{8}. \tag{F15}$$

Comparing Eqs. (F14) and (F15) determines that the constant in Eq. (F13) gives

$$\text{const}=3\rho/8-1/12\epsilon. \tag{F16}$$

We have thus avoided the divergences which come from the endpoints in the evaluation of $-\frac{1}{8}\int dr\,\mu'^2/\mu$. As a result, Eq. (F10) with the constant of integration given in Eq. (F16) is a well-defined and consistent integral evaluation valid in both regions B and C. Hence, the constants C_2 and C_3 are fixed.

We now calculate to first order the asymptotic expansion of $\Phi_{\rm WKB}$ [Eq. (F8)] in region B.

First, we expand $\frac{1}{4}\mu'$.

$$\tfrac{1}{4}\mu'=(-\epsilon+r^2-\rho r^4)^{-3/2}(\rho r^3-\tfrac{1}{2}r) \tag{F17}$$

$$\sim-\frac{1}{2r^2}+\frac{\rho}{4}. \tag{F18}$$

Second, we expand $(1-\frac{1}{8}\mu'^2+\frac{1}{4}\mu\mu'')^{-1/2}$ to first order and get simply

$$(1-\tfrac{1}{8}\mu'^2+\tfrac{1}{4}\mu\mu'')^{-1/2}\sim1. \tag{F19}$$

Third, we expand $\sqrt{\mu}$. To first order,

$$\sqrt{\mu}\sim\left(\frac{2}{r}\right)^{1/2}\left(1+\frac{3}{16}\rho r^2\right)\left(1+\frac{\epsilon}{4r^2}+\frac{\rho r^2}{16}+\frac{17\rho\epsilon}{64}\right). \tag{F20}$$

Fourth, we must evaluate $\int_{r_0}^{r}(1/\mu)dr$. Note that there is no difficulty with endpoint divergences in this integral. Therefore, we treat it just as we did in Sec. IV except that we retain first-order terms. From Eq. (F9), we have

$$\int_{r_0}^{r}\frac{1}{\mu}dr=\frac{1}{2}\int_{r_0}^{r}(-\epsilon+r^2-\rho r^4)^{1/2}dr. \tag{F21}$$

We expand the integral in Eq. (F21) and get

$$\int_{r_0}^{r}\frac{1}{\mu}dr\sim\frac{\sqrt{\rho}}{2}\int_{r_0}^{r}dr(r^2-r_0^2)^{1/2}r_1-\left(\frac{r^2}{2r_1}\right). \tag{F22}$$

Substituting $r=r_0\cosh\Theta$ into Eq. (F22) and integrating gives

$$\int_{r_0}^{r}\frac{1}{\mu}dr\sim\frac{1}{4}\left\{-\epsilon\left(1+\frac{3}{8}\rho\epsilon\right)\ln\frac{2r}{r_0}+r^2\left(1-\frac{1}{4}\rho\epsilon\right)\right.$$
$$\left.-\frac{r^4\rho}{4}-\frac{1}{2}\epsilon\left(1+\frac{9}{16}\rho\epsilon\right)+\frac{\epsilon^2}{8r^2}\right\}. \tag{F23}$$

We combine Eqs. (F14), (F18)–(F20), and (F23). These five results are needed to give us the asymptotic expansion of Φ_{WKB} (F8) in region B:

$$\Phi_B \sim \left(\frac{2}{r}\right)^{1/2}\left(1+\frac{3}{16}\rho r^2\right)\left(1+\frac{\epsilon}{4r^2}+\frac{\rho r^2}{16}+\frac{17\rho\epsilon}{64}\right)$$

$$\times\left(C_2\exp\left\{i\left[\frac{1}{4}\left(-\epsilon(1+\tfrac{3}{8}\rho\epsilon)\ln\frac{2r}{r_0}+r^2(1-\tfrac{1}{4}\rho\epsilon)-\tfrac{1}{4}r^4\rho\right.\right.\right.\right.$$

$$\left.\left.\left.-\tfrac{1}{2}\epsilon\left(1+\frac{9}{16}\rho\epsilon\right)+\frac{\epsilon^2}{8r^2}\right)-\frac{1}{2r^2}+\frac{\rho}{4}+\frac{1}{8r^2}+\frac{3\rho}{8}+\frac{2r}{r_0}\right]\right\}$$

$$\left.+C_3\exp\{-i[\cdots]\}\right). \quad \text{(F24)}$$

Region C. We calculate to first order the asymptotic expansion of Φ_{WKB} [Eq. (F8)] in region C. For this purpose it is convenient to use a new variable R:

$$R \equiv r_1 - r_0. \quad \text{(F25)}$$

Expansions in region C are done in powers of ρ and R/r_1. For first order we keep terms of the form ρ and R/r_1 and neglect terms of the form R^2/r_1^2, $R\rho/r_1$, and ρ^2.

The components of the asymptotic expansion of Φ_{WKB} [Eq. (F8)] are computed in turn:

First,

$$\mu^{1/2} \sim \frac{2^{1/4}\rho^{1/8}}{R^{1/4}}\left[1+\frac{\rho}{4}\left(\frac{5}{2}\epsilon+\frac{5}{2}\frac{R}{\rho^{1/2}}\right)\right]. \quad \text{(F26)}$$

Second, as in region B,

$$(1-\tfrac{1}{8}\mu'^2+\tfrac{1}{4}\mu''\mu)^{-1/2} \sim 1. \quad \text{(F27)}$$

Third,

$$\tfrac{1}{4}\mu' \sim \frac{R^{-3/2}\rho^{1/4}}{24\sqrt{2}}\left(6+\frac{15}{2}\rho\epsilon-\frac{15}{2}R\sqrt{\rho}\right). \quad \text{(F28)}$$

Fourth, we decompose the integral $\int_{r_0}^{r}(1/\mu)dr$ as follows:

$$\int_{r_0}^{r}\frac{1}{\mu}dr = \int_{r_0}^{r_1}\frac{1}{\mu}dr - \int_{r}^{r_1}\frac{1}{\mu}dr. \quad \text{(F29)}$$

The first integral in Eq. (F29) will be expressed later in terms of elliptic integrals [see Eq. (F47)]. For now we label it P:

$$P = \int_{r_0}^{r_1}\frac{1}{\mu}dr. \quad \text{(F30)}$$

The second integral in Eq. (F29) may be approximated easily, with the result that

$$-\int_{r}^{r_1}\frac{1}{\mu}dr \sim -\tfrac{1}{3}\sqrt{3}\rho^{-1/4}R^{3/2}\left(1-\frac{5}{4}\rho\epsilon-\frac{3}{4}(\sqrt{\rho})R\right). \quad \text{(F31)}$$

Fifth, we use Eq. (F10) to approximate $\int -\frac{1}{8}(\mu'^2/\mu)dr$. We decompose the integral in Eq. (F10) using the same method as in Eq. (F29) and define the definite integral as Q:

$$Q = \int_{r_0}^{r_1}\frac{\rho dr}{12\epsilon(1-4\rho\epsilon)}\frac{4\epsilon+(1-12\rho\epsilon)r^2}{(-\epsilon+r^2-\rho r^4)^{1/2}}. \quad \text{(F32)}$$

Then, after some heavy algebra,

$$\int^{r}-\frac{1}{8}\frac{\mu'^2}{\mu}dr = -\frac{3\rho}{8}-\frac{1}{12\epsilon}+\frac{10R^{-3/2}\rho^{1/4}R\sqrt{\rho}}{24\sqrt{2}}$$

$$-\frac{\rho^{1/4}R^{-3/2}}{24\sqrt{2}}\left(1+\frac{5}{4}\rho\epsilon-\frac{5}{4}(\sqrt{\rho})R\right)+Q. \quad \text{(F33)}$$

We have now determined Φ_C, the asymptotic expansion of Φ_{WKB} in region C:

$$\Phi_C \sim \frac{2^{1/4}\rho^{1/8}}{R^{1/4}}\left(1+\frac{5}{8}\rho\epsilon\right)\left[1+\frac{5}{8}(\sqrt{\rho})R\right]$$

$$\times\left(C_2\exp\left\{i\left[\frac{R^{-3/2}\rho^{1/4}}{24\sqrt{2}}\left(5+\frac{25}{4}\rho\epsilon+\frac{15}{4}R\sqrt{\rho}\right)+P\right.\right.\right.$$

$$\left.\left.+Q+\tfrac{3}{8}\rho-\frac{1}{12\epsilon}-\tfrac{1}{3}\sqrt{2}\rho^{-1/4}R^{3/2}[1-\tfrac{3}{2}(\tfrac{5}{6}\rho\epsilon+\tfrac{1}{2}\rho^{1/2}R)]\right]\right\}$$

$$\left.+C_3\exp\{-i[\cdots]\}\right), \quad \text{(F34)}$$

where P and Q are defined in Eqs. (F30) and (F32), respectively.

Region D. In region D, we substitute Eq. (F25) into the differential equation (F1) and transform the equation to

$$0 = \left\{\frac{d^2}{dR^2}+\tfrac{1}{2}\rho^{-1/2}R\left[1-\rho\left(\frac{5}{2}\epsilon+\frac{5}{2}\frac{R}{\rho^{1/2}}\right)\right]\right\}\Phi(R). \quad \text{(F35)}$$

Equation (F35) must be solved to first order and Φ_D is its solution. We solve Eq. (F35) by finding a transformation of variable which to first order changes Eq. (F35) into an Airy equation.[29] The correct change of variables is

$$R = \rho^{1/2}[z+\rho(\tfrac{3}{2}+\tfrac{5}{6}\epsilon z+\tfrac{1}{2}z^2)]. \quad \text{(F36)}$$

Then,

$$\Phi_D = \{\exp\tfrac{1}{2}\rho[z-\rho(\tfrac{5}{6}\epsilon+\tfrac{1}{2}z^2)]\}\chi. \quad \text{(F37)}$$

χ in Eq. (F37) satisfies this Airy equation to first order:

$$\{d^2/dz^2+\tfrac{1}{2}\rho z\}\chi = 0. \quad \text{(F38)}$$

The solution to the Airy equation [Eq. (F38)] is

$$\chi(z) = [-\tfrac{1}{3}(18\rho^{-1})^{-1/3}z]^{1/2}$$

$$\times K_{1/3}(2(-z)^{3/2}(18\rho^{-1})^{-1/2}). \quad \text{(F39)}$$

[29] BMP, Vol. 2, Sec. 7.3.7, p. 22.

As in Sec. IV, to calculate the asymptotic behavior of Φ_D we need to use Eq. (4.31) which is[30]

$$K_{1/3}(e^{3\pi i/2}x)=-\tfrac{1}{2}i\pi[e^{i\pi/6}H_{1/3}^{(1)}(x)+e^{-i\pi/6}H_{1/3}^{(2)}(x)]. \quad \text{(F40)}$$

Also, we need the asymptotic expansions of $H_{1/3}^{(1)}$ and $H_{1/3}^{(2)}$ to first order[31]:

$$H_{1/3}^{(1)}(x)\sim\frac{\sqrt{2}}{\sqrt{(\pi x)}}e^{i(x-5\pi/12)}\left(1+\frac{5}{i72x}\right) \quad \text{(F41a)}$$

and

$$H_{1/3}^{(2)}(x)\sim\frac{\sqrt{2}}{\sqrt{(\pi x)}}e^{-i(x-5\pi/12)}\left(1-\frac{5}{72ix}\right). \quad \text{(F41b)}$$

We combine Eqs. (F36), (F37), (F39), (F40), and (F41), and after some heavy algebra, we have the asymptotic expansion of Φ_D:

$$\Phi_D(R)\sim\frac{2^{-11/12}\pi^{1/2}3^{-1/3}\rho^{1/24}D}{R^{1/4}}\left(1+\frac{5\rho\epsilon}{24}+\frac{5R\sqrt{\rho}}{8}\right)$$
$$\times(\exp\{i[\tfrac{1}{3}\sqrt{2}\rho^{-1/4}R^{3/2}[1-\tfrac{3}{4}(\tfrac{5}{8}\rho\epsilon+\tfrac{1}{2}R\sqrt{\rho})]$$
$$-\tfrac{1}{4}\pi-(5/24\sqrt{2})\rho^{1/4}R^{-3/2}[1+\tfrac{3}{4}(\tfrac{5}{6}\rho\epsilon+\tfrac{1}{2}R\sqrt{\rho})]]\}$$
$$+\exp\{-i[\cdots]\}). \quad \text{(F42)}$$

This completes the calculation of the asymptotic expansions for Φ_A, Φ_B, Φ_C, and Φ_D. It is necessary to asymptotically connect Φ_A with Φ_B and Φ_C with Φ_D as was done in the zeroth-order calculation in Sec. IV.

First, we match Φ_A in Eq. (F7) with Φ_B in Eq. (F24) and determine the ratio of C_2 to C_3:

$$\frac{C_2}{C_3}=(\tfrac{1}{8}r_0^2)^{\nu+1/2}e^{-i\pi/4}\frac{\Gamma(-\frac{1}{2}\nu)}{\Gamma(\frac{1}{2}+\frac{1}{2}\nu)}e^{i\epsilon/4}\times\exp\left[i\rho\left(1+\frac{15}{64}\epsilon^2\right)\right], \quad \text{(F43)}$$

where ν is given in Eq. (F5).

Second, we match Φ_C in Eq. (F34) with Φ_D in Eq. (F42) and, once again, determine the ratio of C_2 to C_3:

$$\frac{C_2}{C_3}=e^{i\pi/2}\exp\left[-2i\left(P+Q+\frac{3\rho}{8}-\frac{1}{12\epsilon}\right)\right], \quad \text{(F44)}$$

where P and Q are given in Eqs. (F30) and (F32), respectively.

It only remains to calculate the integrals P and Q. Both integrals may be treated by substituting

$$z^2=(r_1^2-r^2)/(r_1^2-r_0^2). \quad \text{(F45)}$$

[30] BMP, Vol. 2, p. 5, Eq. (15) and p. 80, Eq. (43).

[31] BMP, Vol. 2, p. 85, Eqs. (1) and (2).

After some algebra both P and Q are found to be linear combinations of $K(k)$ and $E(k)$. These are the complete elliptic integrals of the first and second kind.[32] $K(k)$ and $E(k)$ are defined by

$$K(k)\equiv\int_0^1\frac{dz}{(1-z^2)^{1/2}(1-k^2z^2)^{1/2}} \quad \text{(F46a)}$$

and

$$E(k)\equiv\int_0^1\frac{dz(1-k^2z^2)^{1/2}}{(1-z^2)^{1/2}}, \quad \text{(F46b)}$$

where in our case

$$k^2\equiv1-r_0^2/r_1^2. \quad \text{(F46c)}$$

Using Eq. (F46), P and Q may be written as

$$P=\frac{(\sqrt{\rho})r_1^3}{2}\left[\frac{2-k^2}{3}E(k)+2\left(\frac{k^2-1}{3}\right)K(k)\right] \quad \text{(F47)}$$

and

$$Q=\frac{\sqrt{\rho}}{12r_1\epsilon(1-4\rho\epsilon)}[4\epsilon K(k)+(1-12\rho\epsilon)r_1^2E(k)]. \quad \text{(F48)}$$

We expand $K(k)$ and $E(k)$ to first order for k near 1 by means of the following formulas[33]:

$$K(k)=\tfrac{1}{2}\pi\ {}_2F_1(\tfrac{1}{2},\tfrac{1}{2};1;k^2) \quad \text{(F49a)}$$

and

$$E(k)=\tfrac{1}{2}\pi\ {}_2F_1(-\tfrac{1}{2},\tfrac{1}{2};1;k^2). \quad \text{(F49b)}$$

We then expand the hypergeometric functions in Eq. (F49) to first order for k near 1[34]:

$$K(k)\sim\ln\frac{4r_1}{r_0}+\frac{r_0^2}{4r_1^2}\left(\ln\frac{4r_1}{r_0}-1\right) \quad \text{(F50a)}$$

and

$$E(k)\sim1+\frac{r_0^2}{4r_1^2}\left(2\ln\frac{4r_1}{r_0}-1\right)+\frac{3}{32}\frac{r_0^4}{r_1^4}\left(2\ln\frac{4r_1}{r_0}-\frac{13}{6}\right). \quad \text{(F50b)}$$

Equation (F2) allows us to write Eq. (F50) in terms of ρ and ϵ:

$$K(k)\sim\left(\ln\frac{4}{\sqrt{(\rho\epsilon)}}\right)\left(1+\frac{\rho\epsilon}{4}\right)-\frac{5\rho\epsilon}{4} \quad \text{(F51a)}$$

and

$$E(k)\sim1-\frac{\rho\epsilon}{4}-\frac{77}{64}\rho^2\epsilon^2+\rho\epsilon\left(\frac{1}{2}+\frac{19}{16}\rho\epsilon\right)\ln\left(\frac{4}{\sqrt{(\rho\epsilon)}}\right). \quad \text{(F51b)}$$

Finally, we combine Eqs. (F47), (F48), and (F51) to

[32] BMP, Vol. 2, p. 317, Eqs. (1) and (2).

[33] BMP, Vol. 2, p. 318, Eqs. (5) and (6).

[34] For the expansion of $K(k)$ and $E(k)$ see Ref. 11.

get the first-order evaluation of P and Q:

$$P\sim-\frac{\epsilon}{4}\left(\ln\frac{4}{\sqrt{(\rho\epsilon)}}\right)\left(1+\frac{3\rho\epsilon}{8}\right)+\frac{1}{6\rho}-\frac{\epsilon}{8}+\frac{17}{128}\rho\epsilon^2 \quad \text{(F52)}$$

and

$$Q\sim\frac{1}{12\epsilon}-\frac{35\rho}{48}+\frac{3\rho}{8}\left(\ln\frac{4}{\sqrt{(\rho\epsilon)}}\right). \quad \text{(F53)}$$

Combining Eqs. (F52) and (F53) with (F44), we have

$$\frac{C_2}{C_3}=e^{i\pi/2}\exp\left\{i\left[\frac{\epsilon}{4}-\frac{1}{3\rho}+\rho\left(\frac{17}{24}-\frac{17}{64}\epsilon^2\right)\right.\right.$$
$$\left.\left.+\left(\ln\frac{16}{\rho\epsilon}\right)\left(\frac{\epsilon}{4}-\frac{3\rho}{8}+\frac{3\rho\epsilon^2}{32}\right)\right]\right\}. \quad \text{(F54)}$$

We have thus calculated two independent expressions for C_2/C_3. Setting these expressions in Eqs. (F43) and (F54) equal gives an even-parity version of Eq. (4.35) correct to first order instead of zeroth order as previously computed in Sec. IV. Thus,

$$\frac{\Gamma(\frac{1}{2}+\frac{1}{2}\nu)}{\Gamma(-\frac{1}{2}\nu)}=\exp\left[(\nu+\tfrac{1}{2})\ln\frac{2}{\rho}+i\left(\frac{5\pi}{4}+\frac{1}{3\rho}\right)\right.$$
$$\left.+i\rho\left(\frac{7}{24}-\frac{61E^2}{4}\right)\right], \quad \text{(F55)}$$

where

$$\nu=E-\tfrac{1}{2}+\tfrac{3}{2}i\rho(E^2+\tfrac{1}{4}).$$

This is the result we have sought. It is reassuring to observe that Eq. (F55) reduces to the zeroth-order calculation [Eq. (4.35)] if we neglect $\frac{3}{2}i\rho(E^2+\frac{1}{4})$ compared to $(E-\frac{1}{2})$ in ν and $\rho(7/24+E^2 61/4)$ compared to $5\pi/4$.

Considering the zeroth- and first-order results, it is tempting to conjecture that to all orders the WKB result corresponding to Eq. (F55) takes the form

$$\frac{\Gamma(\frac{1}{2}+\frac{1}{2}\nu(\rho))}{\Gamma(-\frac{1}{2}\nu(\rho))}=\exp\left\{[\nu(\rho)+\tfrac{1}{2}]\ln\frac{2}{\rho}\right.$$
$$\left.+i\left(\frac{5\pi}{4}+\frac{1}{3\rho}\right)+i\alpha(\rho)\right\}, \quad \text{(F56)}$$

where $\nu(\rho)$ and $\alpha(\rho)$ are power series in ρ:

$$\nu(\rho)=\sum_{n=0}^{\infty}\nu_n\rho^n, \quad \text{(F57)}$$

with $\nu_0=E-\frac{1}{2}$ and $\nu_1=\frac{3}{2}i(E^2+\frac{1}{4})$;

$$\alpha(\rho)=\sum_{n=1}^{\infty}\alpha_n\rho^n, \quad \text{(F58)}$$

with $\alpha_1=7/24-61E^2/4$.

APPENDIX G

In Sec. VI, the result of performing the integral $\int_C\Phi_{\rm WKB}^2(z)dz$ was stated in Eq. (6.3). In this appendix, we demonstrate explicitly the evaluation of this integral which we call J. We will calculate J only for even-parity wave functions Φ. The odd-parity case is similar and presents no new difficulties.

We chose the contour C to be the straight line $z=re^{-i\pi/4}$, where r runs from $-\infty$ to ∞. Because the WKB solutions have definite parity about $r=0$, we simplify J to

$$J=2e^{-i\pi/4}\int_{r=0}^{\infty}dr\,\Phi_{\rm WKB}^2(re^{-i\pi/4}). \quad \text{(G1)}$$

In Eq. (G1) we break the region of integration into four pieces:

(I) $0\le r\le\bar{r}_0$,

(II) $\bar{r}_0\le r\le\bar{r}_1$,

(III) $\bar{r}_1\le r\le r_2$,

(IV) $r_2\le r$,

where

$$r_0<\bar{r}_0\ll\bar{r}_1<r_1<r_2. \quad \text{(G2)}$$

The $<$ sign in Eq. (G2) means that we disregard r_0 compared to $\bar{r}_0$ to zeroth order.

$\Phi_{\rm WKB}$ is approximated by Φ_A [Eq. (4.12)] in region I, by the oscillating WKB solution [Eq. (4.16)] in region II, by Φ_D [Eq. (4.28)] in region III, and by the exponentially decreasing WKB solution [Eq. (4.5)] in region IV. The integrals in each of these regions must be treated separately using

$$J=J_{\rm I}+J_{\rm II}+J_{\rm III}+J_{\rm IV}. \quad \text{(G3)}$$

Region I. In region I, we evaluate

$$J_{\rm I}=2e^{-i\pi/4}\int_{r=0}^{\bar{r}_0}dr\,\Phi_A^2(re^{-i\pi/4}). \quad \text{(G4)}$$

Inserting Eq. (4.12), the definition of Φ_A, into Eq. (G4) we get

$$J_{\rm I}=2e^{-i\pi/4}C^2\int_{r=0}^{\bar{r}_0}[D_\nu(re^{-i\pi/4})+D_\nu(-re^{-i\pi/4})]^2dr, \quad \text{(G5)}$$

where $\nu=-\frac{1}{2}-\frac{1}{4}i\epsilon$ and $\epsilon=4iE$.

We expand the integrand in Eq. (G5) using identities of the parabolic cylinder functions[35] so that

$$J_{\rm I}=2e^{-i\pi/4}C^2\left[(1-e^{2\nu\pi i})\int_{r=0}^{\bar{r}_0}D_\nu^2(re^{-i\pi/4})dr\right.$$
$$+\left(\frac{\sqrt{(2\pi)}}{\Gamma(-\nu)}\right)^2e^{\pi i(\nu+1)}\int_{r=0}^{\bar{r}_0}D_{-\nu-1}^2(re^{i\pi/4})dr$$
$$\left.+2(1+e^{\pi i\nu})\int_{r=0}^{\bar{r}_0}D_\nu(re^{-i\pi/4})D_\nu(-re^{-i\pi/4})dr\right]. \quad \text{(G6)}$$

[35] BMP, Vol. 2, p. 117, Eq. (7).

We evaluate the integrals in Eq. (G6) by introducing a new function F into Eq. (G4) as follows:

$$J_{\rm I}=2e^{-i\pi/4}\left\{\int_{r=0}^{\bar{r}_0} dr[\Phi_A{}^2(re^{-i\pi/4})-F(r)] +\int_{r=0}^{\bar{r}_0}\right\} dr\, F(r). \quad \text{(G7)}$$

$F(r)$ is the zeroth-order asymptotic approximation to those terms in $\Phi_A{}^2(re^{-i\pi/4})$ which would have divergent tails if integrated to $r=\infty$. This definition of $F(r)$ implies that a good approximation to $J_{\rm I}$ is

$$J_{\rm I}\sim 2e^{-i\pi/4}\left\{\int_{r=0}^{\infty} dr[\Phi_A{}^2(re^{-i\pi/4})-F(r)] +\int_{r=0}^{\bar{r}_0} dr\, F(r)\right\}. \quad \text{(G8)}$$

All the integrations in Eq. (G8) can be done easily.

Explicitly these integrals are

$$J_{\rm I}\sim 2e^{-i\pi/4}C^2\left\{(1-e^{2\nu\pi i})\int_{r=0}^{\infty} D_\nu{}^2(re^{-i\pi/4})dr +\left(\frac{\sqrt{(2\pi)}}{\Gamma(-\nu)}\right)^2 e^{\pi i(\nu+1)}\int_{r=0}^{\infty} D_{-\nu-1}{}^2(e^{i\pi/4}r)dr+2(1+e^{\pi i\nu}) \times\int_{r=0}^{\infty} dr[D_\nu(re^{-i\pi/4})D_\nu(-re^{-i\pi/4})-F(r)] +\int_{r=0}^{\bar{r}_0} F(r)dr\right\}. \quad \text{(G9)}$$

Asymptotically, $D_\nu(re^{-i\pi/4})D_\nu(-re^{-i\pi/4})$ behaves as $[\sqrt{(2\pi)}/\Gamma(-\nu)]r^{-1}e^{i\pi/4}$ for large r.[36] However, if we were to choose the counterterm $F(r)$ to have this form, we would introduce a spurious pole at $r=0$. Therefore, we pick a form for $F(r)$ with the same asymptotic behavior at $r=\infty$ but with no pole in the region of integration. We thus define

$$F(r)\equiv\frac{1}{r+1}\frac{\sqrt{(2\pi)}}{\Gamma(-\nu)}e^{i\pi/4}. \quad \text{(G10)}$$

After rotating the contours in the complex plane by $\pi/4$ and $-\pi/4$, the first two integrals in Eq. (G9) become known forms.[37] In order to perform the third integral in Eq. (G9) which is

$$\int_{r=0}^{\infty}\left[D_\nu(re^{-i\pi/4})D_\nu(-re^{-i\pi/4}) -\frac{\sqrt{(2\pi)}}{\Gamma(-\nu)}(r+1)^{-1}e^{i\pi/4}\right]dr, \quad \text{(G11)}$$

we introduce a small positive ϵ into Eq. (G11) as follows:

$$\lim_{\epsilon\to 0}\int_{r=0}^{\infty}\left[D_{\nu-\epsilon}(re^{-i\pi/4})D_\nu(-re^{-i\pi/4}) -\frac{\sqrt{(2\pi)}}{\Gamma(-\nu)}\frac{1}{(r+1)^{\epsilon+1}}e^{i\pi/4}\right]dr. \quad \text{(G12)}$$

After rotating the contour in Eq. (G12) by $\pi/4$, the integrals become known forms.[38] We evaluate the integrals in Eq. (G12) and take the limit as $\epsilon\to 0$. After considerable simplification[39] the final result of this evaluation is

$$J_{\rm I}\sim\frac{4C^2\sqrt{(2\pi)}\cos(\frac{1}{2}\nu\pi)e^{\nu\pi i/2}}{\Gamma(-\nu)} \times[2\ln\bar{r}_0-\psi(E+\tfrac{1}{2})+\tfrac{1}{2}\pi\cot(\tfrac{1}{4}-\tfrac{1}{2}E)\pi], \quad \text{(G13)}$$

where $\psi(x)\equiv d\ln\Gamma(x)/dx$.

Region II. In region II, we evaluate

$$J_{\rm II}\sim 2e^{-i\pi/4}\int_{r=\bar{r}_0}^{\bar{r}_1} dr(-\epsilon+r^2-\rho r^4)^{-1/2}2C_2C_3. \quad \text{(G14)}$$

To get this equation, we have substituted Eq. (4.16) into Eq. (G1) and squared the expression. In squaring, we have disregarded two small oscillatory terms because their contribution to the integral could only become significant if r were near r_0 or r_1. The choice of $\bar{r}_0$ and $\bar{r}_1$ as endpoints of the integration region prevents r from getting near r_0 or r_1.

To do the integral in Eq. (G14) we decompose the region of integration into three pieces:

$$J_{\rm II}\sim 4C_2C_3e^{-i\pi/4}\left(\int_{r=r_0}^{r_1}-\int_{r=r_0}^{\bar{r}_0}-\int_{r=\bar{r}_1}^{r_1}\right) \times dr(-\epsilon+r^2-\rho r^4)^{-1/2}. \quad \text{(G15)}$$

By substituting $x^2=(r_1{}^2-r^2)/(r_1{}^2-r_0{}^2)$ and $k^2=1-r_0{}^2/r_1{}^2$, the first integral in Eq. (G13) is shown to be $K(k)$, the elliptic integral of the first kind.[40] The last two integrals are approximated by factoring $(-\epsilon+r^2-\rho r^4)=\rho(r^2-r_0{}^2)(r_1{}^2-r^2)$. Combining these results gives

$$J_{\rm II}\sim 4C_2C_3e^{-i\pi/4}\left[-\ln\frac{2\bar{r}_0}{r_0}-\left(2-\frac{2\bar{r}_1}{r_1}\right)^{1/2} +\frac{1}{r_1\sqrt{\rho}}K(k)\right]. \quad \text{(G16)}$$

We now expand $K(k)$ for k near 1 [see Eq. (4.23b) for

[36] BMP, Vol. 2, pp. 122–123, Eqs. (1) and (2).
[37] BMP, Vol. 2, p. 122, Eq. (22).
[38] BMP, Vol. 2, p. 122, Eq. (21).
[39] BMP, Vol. 1, pp. 1–5 and 16.
[40] BMP, Vol 2, p. 317, Eq. (1).

this expansion] and find that the contribution from region II is

$$J_{\rm II}\sim 4C_2C_3e^{-i\pi/4}[\ln(2/\bar{r}_0\sqrt{\rho})-\rho^{1/4}\sqrt{2}(r_1-\bar{r}_1)^{1/2}]. \quad \text{(G17)}$$

Regions III and IV. Using Eqs. (4.20), (4.28), and (4.29), we can write down the integral contribution from region III:

$$J_{\rm III}=2e^{-i\pi/4}\int_{r=\bar{r}_1}^{r_2}dr\frac{D^2}{3}$$
$$\times K_{1/3}^2[2(18\sqrt{\rho})^{-1/2}(r-r_1)^{3/2}](18\sqrt{\rho})^{-1/3}(r-r_1). \quad \text{(G18)}$$

We notice that the contribution from region IV can be included in Eq. (G18) by increasing the upper limit of integration from r_2 to ∞. This is true because the $K_{1/3}$ Bessel function for region III and the WKB solution for region IV [Eq. (4.5)] have similar exponentially decreasing tails. Hence,

$$J_{\rm III}+J_{\rm IV}=2e^{-i\pi/4}\int_{r=\bar{r}_1}^{\infty}dr\frac{D^2}{3}$$
$$\times K_{1/3}^2(2(18\sqrt{\rho})^{-1/2}(r-r_1)^{3/2})(18\sqrt{\rho})^{-1/3}(r-r_1). \quad \text{(G19)}$$

We simplify Eq. (G19) by substituting $s=r-r_1$:

$$J_{\rm III}+J_{\rm IV}=\frac{2e^{-i\pi/4}D^2(18\sqrt{\rho})^{-1/3}}{3}$$
$$\times\int_{s=\bar{r}_1-r_1}^{\infty}sds\,K_{1/3}^2(2(18\sqrt{\rho})^{-1/2}s^{3/2}). \quad \text{(G20)}$$

Next, as we did in region I, we introduce a counterterm $G(s)$ in Eq. (G21) and extend the region of integration in Eq. (G22). This procedure facilitates the evaluation of the integrals.

$$J_{\rm III}+J_{\rm IV}=\tfrac{2}{3}e^{-i\pi/4}D^2(18\sqrt{\rho})^{-1/3}$$
$$\times\left\{\int_{s=0}^{\infty}sds\,K_{1/3}^2(2(18\sqrt{\rho})^{-1/2}s^{3/2})\right.$$
$$+\int_{s=\bar{r}_1-r_1}^{0}ds[sK_{1/3}^2(2(18\sqrt{\rho})^{-1/2}s^{3/2})-G(s)]$$
$$\left.+\int_{s=\bar{r}_1-r_1}^{0}G(s)ds\right\}. \quad \text{(G21)}$$

This may be approximated by

$$J_{\rm III}+J_{\rm IV}\sim\tfrac{2}{3}e^{-i\pi/4}D^2(18\sqrt{\rho})^{-1/3}$$
$$\times\left\{\int_{s=0}^{\infty}sds\,K_{1/3}^2(2(18\sqrt{\rho})^{-1/2}s^{3/2})\right.$$
$$+\int_{s=-\infty}^{0}ds[sK_{1/3}^2(2(18\sqrt{\rho})^{-1/2}s^{3/2})-G(s)]$$
$$\left.+\int_{s=\bar{r}_1-r_1}^{0}G(s)ds\right\}. \quad \text{(G22)}$$

Equation (G22) is a good approximation to Eq. (G21) because $G(s)$ is chosen to behave asymptotically as $K_{1/3}^2$ for s near $-\infty$.

We reduce the first integral Eq. (G22) to a known form by substituting

$$x=2(18\sqrt{\rho})^{-1/2}s^{3/2}, \quad \text{(G23)}$$

and by using[41]

$$\int_0^{\infty}K_{1/3}^2(x)x^{1/3}dx=2^{-5/3}3\Gamma^2(\tfrac{2}{3}). \quad \text{(G24)}$$

We simplify the second integral in Eq. (G22) by substituting Eq. (G23) again and by observing that[42]

$$K_{1/3}(xe^{\frac{1}{2}\pi i})=(-\pi i/\sqrt{3})[J_{1/3}(x)+J_{-1/3}(x)]. \quad \text{(G25)}$$

Three Bessel function terms arise from plugging the square of Eq. (G25) into the second term in Eq. (G22). The techniques for integrating only one of these three terms will be presented here because the other two terms can be integrated similarly. Investigation of Eq. (G25) shows that in terms of the variable x the counterterm G is proportional to $x^{-2/3}$. As in region I, a limiting procedure must be used to do the counterterm integration as follows.

We insert a positive ϵ and the first of the three terms becomes

$$\lim_{\epsilon\to 0}\int_0^{\infty}\left[J_{1/3}(x)J_{1/3}(x(1+\epsilon))x^{1/3}\right.$$
$$\left.-\frac{(\cos x\epsilon)x^{-2/3}}{(1+\epsilon)^{1/2}\pi}\right]dx. \quad \text{(G26)}$$

The first integral in Eq. (G26) is a Weber-Schafheitlin integral.[43] The second may be evaluated by changing the contour. This gives a Γ function. The results are

$$\lim_{\epsilon\to 0}\left(2^{1/3}(1+\epsilon)^{-5/3}\frac{F(1,\frac{2}{3};\frac{4}{3};(1+\epsilon)^{-2})}{\Gamma(\frac{4}{3})\Gamma(\frac{1}{3})}\right.$$
$$\left.-\frac{\Gamma(\frac{1}{3})\epsilon^{-1/3}\cos\frac{1}{6}\pi}{\pi}\right). \quad \text{(G27)}$$

We transform[44] the hypergeometric function in Eq. (G27) as follows:

$$F(1,\tfrac{2}{3};\tfrac{4}{3};(1+\epsilon)^{-2})=\frac{\Gamma(\frac{4}{3})\Gamma(-\frac{1}{3})}{\Gamma(\frac{1}{3})\Gamma(\frac{2}{3})}$$
$$\times(1+\epsilon)^2F(1,\tfrac{1}{3};\tfrac{4}{3};-2\epsilon-\epsilon^2)+\frac{\Gamma(\frac{4}{3})\Gamma(\frac{1}{3})}{\Gamma(\frac{2}{3})}(1+\epsilon)^{2/3}$$
$$\times\left[1-\frac{1}{(1+\epsilon)^2}\right]^{-1/3}F(\tfrac{1}{3},0;\tfrac{2}{3};-2\epsilon-\epsilon^2). \quad \text{(G28)}$$

---

[41] BMP, Vol. 2, p. 93, Eq. (36).

[42] BMP, Vol. 2, p. 5, Eq. (15), p. 80, Eq. (43), and p. 4, Eqs. (5) and (6).

[43] BMP, Vol. 2, Sec. 7.7.4, pp. 51–52.

[44] BMP, Vol. 1, p. 109, Eq. (4).

We simplify[39] Eq. (G28) and insert it into Eq. (G27). We can then take the limit as $\epsilon \to 0$ because the terms containing $\epsilon^{-1/3}$ cancel. The result is that

$$\lim_{\epsilon\to 0}\int_0^\infty dx\left(J_{1/3}(x)J_{1/3}((\epsilon+1)x)x^{1/3}-\frac{(\cos x\epsilon)x^{-2/3}}{\pi x(1+\epsilon)^{1/2}}\right) = -\frac{3\times 2^{1/3}}{\Gamma^2(\frac{1}{3})}. \quad \text{(G29)}$$

The second term $(J_{1/3}J_{-1/3})$ and the third term $(J_{-1/3}J_{-1/3})$ each contribute 0 to the second term in Eq. (G22).

Thus, the first two integrals in Eq. (G22) cancel. After substituting for G, we are left with the third term in Eq. (G22) which is

$$J_{\rm III}+J_{\rm IV}\sim -ie^{-i\pi/4}\frac{2D^2}{3}\left(\frac{\sqrt{\rho}}{3}\right)^{1/3}\pi\int_{s=\bar{r}_1-r_1}^{0} x^{-2/3}dx\,, \quad \text{(G30)}$$

where $x=2(18\sqrt{\rho})^{-1/2}s^{3/2}$. We evaluate Eq. (G30) and find that

$$J_{\rm III}+J_{\rm IV}\sim \pi D^2 e^{-i\pi/4}2^{7/6}\rho^{1/12}3^{-2/3}(r_1-\bar{r}_1)^{1/2}. \quad \text{(G31)}$$

We have now evaluated all of the integrals in Eq. (G1). Finally, we must relate the constants C, C_1, C_2, C_3, C_4, and D in Eqs. (G13), (G17), and (G31). To do so we use Eqs. (4.14)–(41.6), (4.32), (4.33), and (4.35) and show algebraically that

$$C^2=\frac{\Gamma^2(\frac{3}{4}-\frac{1}{2}E)e^{-\pi iE/2}2^{-E}C_1^2}{(2\pi)^2}, \quad \text{(G32a)}$$

$$C_2C_3=\frac{C_1^2}{\Gamma(\frac{1}{4}-\frac{1}{2}E)\Gamma(\frac{1}{4}+\frac{1}{2}E)}, \quad \text{(G32b)}$$

$$D^2=\frac{\rho^{1/6}2^{4/3}3^{2/3}C_1^2}{\pi\Gamma(\frac{1}{4}+\frac{1}{2}E)\Gamma(\frac{1}{4}-\frac{1}{2}E)}. \quad \text{(G32c)}$$

We may pick an arbitrary value for C_1^2 because we are free to choose the over-all normalization of the original WKB wave function. We let

$$C_1^2=\tfrac{1}{2}e^{i\pi/4}\Gamma(\tfrac{1}{4}+\tfrac{1}{2}E)\Gamma(\tfrac{1}{4}-\tfrac{1}{2}E). \quad \text{(G33)}$$

Substituting Eq. (G33) into Eq. (G32) and simplifying,[39] the results are

$$C^2=\frac{\Gamma(-\nu)e^{-\pi iE/2}(\frac{1}{2}\pi)^{1/2}e^{i\pi/4}}{4\pi\cos\frac{1}{2}\pi\nu}, \quad \text{(G34a)}$$

$$C_2C_3=\tfrac{1}{2}e^{i\pi/4}, \quad \text{(G34b)}$$

$$D^2=(18\sqrt{\rho})^{1/3}e^{i\pi/4}/\pi. \quad \text{(G34c)}$$

Combining Eq. (G34) with Eqs. (G13), (G17), and (G31) shows that the contributions to J are

$$J_{\rm I}\sim 2\ln\bar{r}_0-\psi(E+\tfrac{1}{2})+\tfrac{1}{2}\pi\cot(\tfrac{1}{4}-\tfrac{1}{2}E)\pi\,, \quad \text{(G35a)}$$

$$J_{\rm II}\sim -2\ln\bar{r}_0+\ln(4/\rho)-2\rho^{1/4}\sqrt{2}(r_1-\bar{r}_1)^{1/2}, \quad \text{(G35b)}$$

$$J_{\rm III}+J_{\rm IV}\sim 2\sqrt{2}\rho^{1/4}(r_1-\bar{r}_1)^{1/2}. \quad \text{(G35c)}$$

Referring to Eq. (G3), we add together the three expressions in Eq. (G35) to find the total contribution to J which is

$$J=-\psi(E+\tfrac{1}{2})+\tfrac{1}{2}\pi\cot(\tfrac{1}{4}-\tfrac{1}{2}E)\pi+\ln(4/\rho). \quad \text{(G36)}$$

Note that $\bar{r}_0$ and $\bar{r}_1$ have dropped out of this final expression. This completes the evaluation of J for even parity.

For odd parity the result is

$$J=-\psi(\tfrac{1}{2}+E)+\tfrac{1}{2}\pi\cot(\tfrac{3}{4}-\tfrac{1}{2}E)\pi+\ln(4/\rho). \quad \text{(G37)}$$

The only differences between the even- and odd-parity derivations occur in the algebraic manipulations in region I.

5. Reprinted article 125, listed in Chapter 8:

VOLUME 24, NUMBER 25 PHYSICAL REVIEW LETTERS 22 JUNE 1970

LIMIT OF CROSS SECTIONS AT INFINITE ENERGY*

Hung Cheng†
Department of Mathematics, Massachusetts Institute of Technology, Cambridge, Massachusetts 02139

and

Tai Tsun Wu
Gordon McKay Laboratory, Harvard University, Cambridge, Massachusetts 02138
(Received 15 May 1970)

At infinite energy, we predict: (1) σ_{tot} approaches infinity; (2) the ratio of the real part to the imaginary part of the forward elastic amplitude approaches zero; (3) $\sigma_{\text{el}}/\sigma_{\text{tot}}$ approaches $\frac{1}{2}$; (4) the width of diffraction peak approaches zero; its product with σ_{tot} is a constant. We give theoretical evidence based on massive quantum electrodynamics as well as experimental evidence in support of these predictions, and a physical picture for high-energy scattering.

With the construction of larger and larger accelerators, it is hoped that the mysteries of high-energy scattering will unfold in the near future. It is therefore particularly urgent that theoretical progress will match the experimental progress in stride. On the basis of our results in quantum electrodynamics,[1-4] we shall make in this Letter a number of predictions on some of the fundamental questions in high-energy physics.[5] While these predictions cannot be rigorously proven in a mathematical way, the reasoning that leads to them is based on physical insights learned from field theory, and are sufficiently convincing to warrant attention. For clarity of presentation, we shall first list these predictions. Theoretical evidence in support of them will follow next. Experimental evidence as well as a physical picture for high-energy scattering will be given lastly.

For any hadron-hadron scattering process at extremely high energies, we make the following predictions:

Forward direction.—Let us denote the forward elastic scattering amplitude by $\mathfrak{M}(s, 0)$, where s is the square of the c.m. energy of the system. We predict:

$$\frac{\text{Re}\mathfrak{M}(s,0)}{\text{Im}\mathfrak{M}(s,0)} = \frac{\pi}{\ln|\mathcal{S}|} + O((\ln|\mathcal{S}|)^{-2}), \tag{1}$$

and

$$\sigma_{\text{tot}} = 2\pi R^2 + O(\ln|\mathcal{S}|), \tag{2}$$

where

$$\mathcal{S} = \left\{ \frac{(-s)^a}{[\ln(-s)]^2} + \frac{(-u)^a}{[\ln(-u)]^2} \right\}^{1/a}, \tag{3}$$

with u the third Mandelstam variable and a a positive constant, and

$$R = R_0 \ln|\mathcal{S}|, \tag{4}$$

with R_0 a constant independent of the energy.

Nonforward directions.—The usual diffraction peak is expected for the elastic-scattering amplitude around the forward direction. Furthermore, let Γ be the value of $-t$ at which the first dip occurs, where t is the negative of the momentum-transfer squared. We predict:

$$\Gamma\sigma_{\text{tot}} = 2\pi^3\beta_1^2 + O((\ln|\mathcal{S}|)^{-1}), \tag{5}$$

independent of the processes,

$$\sigma_{\text{el}} = \pi R^2 + O(\ln|\mathcal{S}|), \tag{6}$$

and

$$\sigma_{\text{el}}/\sigma_{\text{tot}} = \tfrac{1}{2} + O((\ln|\mathcal{S}|)^{-1}). \tag{7}$$

In (5), $\beta_1 = 1.2197$ is the first zero of $J_1(\beta\pi)$. Thus

$$2\pi^3\beta_1^2 = 92.254 = 35.92 \text{ mb (BeV}/c)^2. \tag{8}$$

Predictions (2), (5), (6), and (7) are model independent and are believed to be firm, while (1) is a specific result of quantum electrodynamics, and may be modified in strong interactions.

The above predictions mark a drastic departure from current concepts of high-energy scattering. This is why we were slow in recognizing its full significance when the relevant formula was first found.[2] After more than a year of deliberation, we now realize that the above predictions constitute the only answer supported by all theoretical evidences from field theory. We shall next write down the basic formula and list all the evidence.

At high energies, the fermion-fermion elastic-scattering amplitude was found to be[1,2]

$$\tfrac{1}{2}im^{-2}s\delta_{12}\delta_{1'2'}\int d\vec{x}_\perp e^{i\vec{\Delta}\cdot\vec{x}_\perp}(1-e^{-A}). \tag{9}$$

In (9), m is the mass of the fermion, δ_{12} and $\delta_{1'2'}$ are Kronecker deltas of the spins, $\vec{\Delta}$ is

VOLUME 24, NUMBER 25 PHYSICAL REVIEW LETTERS 22 JUNE 1970

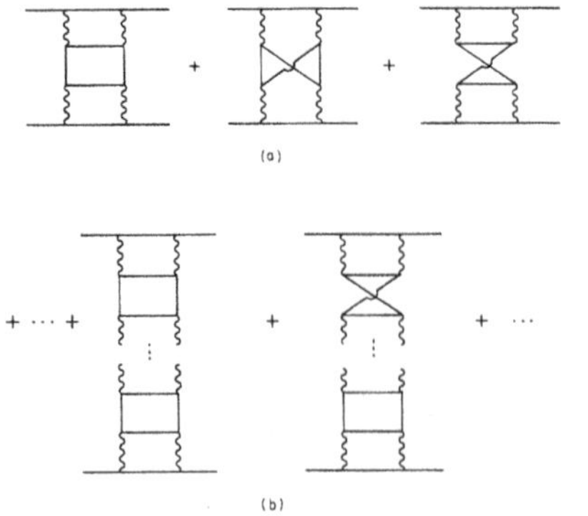

FIG. 1. The one-tower diagrams.

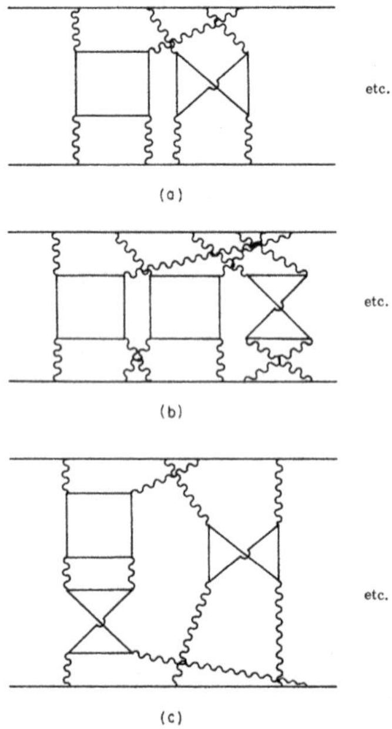

FIG. 2. Examples of the multitower diagrams.

the momentum transfer, and A is related to the Fourier transform of the asymptotic amplitude due to the sum of one-tower diagrams illustrated in Fig. 1. Specifically, if we expand $(1-e^{-A})$ in (9) into a Taylor series in A, the term in (9) proportional to A is the high-energy amplitude due to the sum of the one-tower diagrams. In fact, the A^N term in (9) is the high-energy amplitude for the sum of the N-tower diagrams. [Two of the two-tower diagrams are illustrated in Figs. 2(a) and 2(c), and a three-tower diagram is illustrated in Fig. 2(b).]

Equation (9) was obtained by summing up the leading terms in all multitower diagrams. Although this method has no mathematical justification, we believe that (9) is correct by the following steps of reasoning:

(1) The one-tower diagrams with n loops are the lowest-order diagrams which yield terms of the order of $s(\ln s)^n$.[6] By the optical theorem, the corresponding cross section for the production of n pairs is proportional to $(\ln s)^n$ for large s. Summing up these cross sections over n, we get a total cross section of the order of $s^a(\ln s)^{-2}$, where $a=11\alpha^2\pi/32$.[7-9] The fact that this result is larger than any power of $\ln s$ cannot be blamed on the process of summing only the leading terms. This is because the cross section for n-pair production is always positive. By choosing n sufficiently large, the corresponding n-pair production cross section is already larger than any given power of $\ln s$. Thus the sum must also be larger than any power of $\ln s$.

(2) To understand this large result $[s^a(\ln s)^{-2}]$, we must first recognize that the appearance of the logarithmic factors is due to the creation of low-energy pairs in the c.m. system.[3] It therefore bears a strong resemblance to the familiar infrared divergence in quantum electrodynamics. We recall that a charged particle in an external field can emit soft photons, and the number of logarithmic divergences is equal to the number of soft photons emitted. The scattering cross section is thus infinite, and experimental resolutions must be introduced to obtain finite answers. High-energy scattering processes closely parallel this situation in the following way: If the kinetic energy of the charged particle is T and the mass of the photon is λ, then the maximum number of photons that the charged particle can emit is kinematically limited to T/λ. Only in the physical case of massless photons can an infinite number of photons be emitted. In a high-energy scattering process, the maximum number of pairs that are kinematically allowed to be created in the c.m. system is ω/m, where ω is the

c.m. energy of one of the incident particles. In the limit of infinite energy, this number is also infinite. Thus the limit of $\omega \to \infty$ parallels that of $\lambda \to 0$. Since it is well known that the cross section for a charged particle scattered by an external field is more divergent than any power of $\ln\lambda$ due to the emission of an arbitrary number of soft photons, the cross section for high-energy scattering must be larger than any power of $\ln\omega$ due to the creation of slow particles in the c.m. system, i.e., due to pionization.[3] Thus we believe that the previously obtained answer of the order of $s^a/(\ln s)^2$ is correct. In strong interactions, the numerical value of a may be of the order of 1.

(3) This result $s^a/(\ln s)^2$ is not to be interpreted as a violation of the Froissart bound.[10] Rather, it ought to be regarded as a realization of the strongly absorptive "potential" with a coupling constant increasing with energy, as conceived in Froissart's original paper.[10] Thus, in a two-particle scattering process, if the interaction takes place at a sufficiently close transverse distance from the center of the target, the incident particle creates slow particles in the c.m. system and is lost to the beam. If the transverse distance involved is large, the incident particle does not necessarily create pionization products and may survive. Mathematically, this is achieved by including not only the one-tower diagrams but also all of the multitower diagrams as well. The resulting amplitude is (9). In the limit $s \to \infty$ and $|\vec{x}_\perp| = O(\ln s)$, we have[11]

$$A \sim bi^{-1}s^a e^{-\mu|\vec{x}_\perp|}, \tag{10}$$

where b and μ are real constants. In fact, $\mu < 2\lambda$, where λ is the mass of the vector meson.

(4) The inclusion of the multitower diagrams is not arbitrary, but has a physical basis. Since (9) is in the form of the impact-parameter representation, the quantity $1-e^{-A}$ is the opacity at the distance and energy at which A is evaluated. For a large fixed s, we may choose $|\vec{x}_\perp|$ sufficiently large so that A is very small. Then $1-e^{-A} \sim A$. Thus at sufficiently large transverse distances, A is proportional to the Fourier transform of the high-energy scattering amplitude. Since very little scattering occurs at this distance, the contribution of any diagram to A is always small. In fact, the diagrams with higher thresholds in the t channel are in general expected to contribute less to A for large $|\vec{x}_\perp|$. Thus the important diagrams are those with two-vector-meson cuts in the t channel. The one-tower diagrams are the leading diagrams of this kind, and the inclusion of other diagrams of this kind merely modifies the kernel $\mathcal{K}$ in Ref. 8. We therefore conclude that (10) is correct when $|\vec{x}_\perp|$ is large, although the constants a, b, and μ should be modified if the coupling is strong. If we now fix $|\vec{x}_\perp|$ and increase s, the right-hand side of (10) increases. This means that the interaction extends into larger and larger transverse distances as the energy increases. When the energy is sufficiently high or the transverse distance is not large enough, A becomes appreciable, and we may not approximate $1-e^{-A}$ by A. This means that, no matter what happens to an incident particle, it is, at most, taken from the beam and cannot yield a large cross section that increases as some power of the energy. In this way the Froissart bound is obeyed. As shown in Ref. 2, this exponentiation is achieved by the exchange of two, three, or more towers. It is interesting to note that, while the amplitude from the sum of one-tower diagrams is so large that unitarity is violated, the amplitude from the sum of two-tower diagrams is even larger. Fortunately, these two amplitudes are of opposite sign. In this way unitarity is restored.

(5) Finally, we emphasize that (9) has (i) all leading terms, (ii) unitarity, (iii) analyticity, and (iv) crossing symmetry.

The above considerations lead to a very dramatic physical picture. This picture is perfectly general and holds for all hadron-hadron scattering processes. At extremely high energies, a particle acts like a Lorentz-contracted pancake which can be roughly separated into the following two regions: (i) a black core (completely absorptive) with a radius R given by (4) which expands with energy [this core contributes the leading terms in Eqs. (1), (2), (5), (6), and (7)]; (ii) a gray fringe (partially absorptive) which extends further out and contributes to the next-order terms in these equations.

It is perhaps instructive to compare the present physical picture with those from the Regge-pole model[12-16] and the droplet model.[17-19] In each case, a particle is pictured either as a sphere or a disk, depending on the coordinate system used. In the Regge-pole model, the disk becomes larger and more transparent as energy increases; in the droplet model, the properties of the disk are independent of the energy at high energies; in the present physical picture, the disk becomes larger and more absorptive as energy increases. In particular, both in the Regge-

VOLUME 24, NUMBER 25 PHYSICAL REVIEW LETTERS 22 JUNE 1970

pole model and in the present physical picture, the diffraction peak shrinks in width as energy increases. However, the present picture differs from both the Regge-pole model and the droplet model in the high-energy behavior of σ_{tot}, the ratio of σ_{el} to σ_{tot}, and also the position of the dip in elastic scattering.

While much about the hard core is model independent, it is not so for the gray fringe. For the purpose of comparing with experiment, we must remember that, for proton-proton scattering at 30 BeV, the quantity $\ln[s/(\ln s)^2]$ is only about 1.3 if the energy scale is taken to be due to the rest mass of the proton. Thus the contribution of the gray fringe is comparable with that of the black core in this energy range; indeed the black core may not be black yet. Without a precise knowledge of the gray fringe, our predictions cannot be tested against the existing experimental data. For example, in view of our results from quantum electrodynamics, we can take A to be of the form (10). With this assumption, the contribution from the gray fringe can then be obtained precisely. Instead of dwelling on the possible assumptions, we shall briefly comment on some of the important aspects in the trends of high-energy experiments.

Total cross sections.—The total cross sections of $\pi^- $-$p$, π^--n, K^--p, and K^--n have all become energy independent in the region from 30 to 65 BeV.[20] No upward trend is noted yet, although the experimental data may be in agreement with slowly rising cross sections.[20]

The ratio $\mathrm{Re}\mathfrak{M}(s,0)/\mathrm{Im}\mathfrak{M}(s,0)$.—This ratio is still negative at 25 BeV for both $\pi^\pm$-p [21] and p-p [22] scattering. For both the π^+-p and π^--p cases, there is a definite trend for this ratio to rise above zero. The trend for the p-p case is not well established.

The product $\Gamma\sigma_{tot}$.—For proton-proton scattering at 19.2 BeV, this product is 39.5 mb $(\mathrm{BeV}/c)^2$,[23,24] which agrees quite well with (8). Since the diffraction peak continues to shrink[25] at 70 BeV, the agreement ought to improve at this energy.

The ratio σ_{el}/σ_{tot}.—Experimentally, in the region 8-20 BeV, this ratio is approximately 17% for both π^+-p and π^--p scattering, and 26% for p-p scattering.[26] By using the model of quantum electrodynamics, the terms of the orders of $\ln|s|$ and 1 in both (2) and (6) can be obtained explicitly. These terms are positive for σ_{tot} and negative for σ_{el}. As a result, at intermediate energies this ratio σ_{el}/σ_{tot} is much smaller than $\frac{1}{2}$, its value at infinite energy. At 20 BeV, this ratio is estimated to be very roughly in the neighborhood of 20%.

In conclusion, we emphasize the uniqueness of the present physical picture from our knowledge of high-energy behavior in field theory. Confrontation with the experimental data in the next few years will be most exciting.

---

*Work supported in part by the U. S. Atomic Energy Commission under Contract No. AT(30-1)-4101.

†Work supported in part by the National Science Foundation under Grant No. GP 13775.

[1]H. Cheng and T. T. Wu, Phys. Rev. Letters 22, 666 (1969).

[2]H. Cheng and T. T. Wu, Phys. Rev. 186, 1611 (1969). See in particular the last paragraph of this paper.

[3]H. Cheng and T. T. Wu, Phys. Rev. Letters 23, 1311 (1969).

[4]H. Cheng and T. T. Wu, Phys. Rev. Letters 24, 759 (1970).

[5]These predictions were first presented by H. Cheng, Bull. Am. Phys. Soc. 15, 564 (1970).

[6]H. Cheng and T. T. Wu, Phys. Rev. Letters 22, 1405 (1969).

[7]H. Cheng and T. T. Wu, Phys. Rev. D 1, 467 (1970).

[8]H. Cheng and T. T. Wu, Phys. Rev. (to be published).

[9]G. V. Frolov, V. N. Gribov, and L. N. Lipatov, Phys. Letters 31B, 34 (1970).

[10]M. Froissart, Phys. Rev. 123, 1053 (1961).

[11]The considerations involved here are quite complicated and will be submitted to the Physical Review. Very briefly, we first extend the results of Ref. 8 to nonforward directions, and find that the asymptotic behavior $s^{1+a}/(\ln s)^2$ for the amplitude holds in general, where a is independent of t. There is, however, a coefficient which depends on t but not on s. We then study the analyticity properties of this coefficient and find that there is a singularity for positive values of t, located precisely at the point where a Regge pole comes out from the moving cut (see Ref. 4).

[12]T. Regge, Nuovo Cimento 14, 951 (1959).

[13]G. F. Chew and S. C. Frautschi, Phys. Rev. Letters 7, 394 (1961).

[14]R. Blankenbecler and M. L. Goldberger, Phys. Rev. 126, 766 (1962).

[15]S. C. Frautschi, M. Gell-Mann, and F. Zachariasen, Phys. Rev. 126, 2204 (1962).

[16]For a more recent review, see H. M. Chan, in *Proceedings of the Fourteenth International Conference on High Energy Physics, Vienna, Austria, 1964*, edited by J. Prentki and J. Steinberger (CERN Scientific Information Service, Geneva, Switzerland, 1968).

[17]N. Byers and C. N. Yang, Phys. Rev. 142, 976 (1966).

[18]T. T. Chou and C. N. Yang, Phys. Rev. 170, 1591 (1966), and Phys. Rev. Letters 20, 1213 (1968), and Phys. Rev. 175, 1832 (1968).

[19]J. Benecke, T. T. Chou, C. N. Yang, and E. Yen, Phys. Rev. 188, 2159 (1969).

VOLUME 24, NUMBER 25 PHYSICAL REVIEW LETTERS 22 JUNE 1970

[20]J. V. Allaby et al., Phys. Letters 30B, 500 (1969).
[21]K. J. Foley et al., in *Proceedings of the Twelfth International Conference on High Energy Physics, Dubna, U. S. S. R., 1964* (Atomizdat., Moscow, U. S. S. R., 1966), and Phys. Rev. Letters 14, 74 (1965).
[22]A. E. Taylor et al., Phys. Letters 14, 54 (1965).
[23]S. J. Lindenbaum et al., Phys. Rev. Letters 7, 185 (1963).
[24]J. V. Allaby et al., Phys. Letters 28B, 67 (1968).
[25]G. B. Beznogikh et al., Phys. Letters 30B, 274 (1969).
[26]K. J. Foley et al., Phys. Rev. Letters 11, 425 (1963).

IMPLICATIONS OF LOCAL DUALITY IN MESON-DECUPLET SCATTERING*

M. J. King and Kameshwar C. Wali
Physics Department, Syracuse University, Syracuse, New York 13210
(Received 6 April 1970)

The consequences of the assumption that the imaginary parts of the direct-channel helicity amplitudes vanish in the forward (backward) scattering if the crossed t channel (u channel) is exotic are examined in the case of meson-decuplet scattering. Confining ourselves to a limited energy region above the threshold, we show that the solutions of the set of equations provided by different external helicities predict the existence of certain patterns of octet and decuplet baryons degenerate in mass and having spin and parity assignments $\frac{5}{2}^{\pm}$, $\frac{3}{2}^{\pm}$, $\frac{1}{2}^{\pm}$.

Recently several authors[1,2] have attempted to predict "exchange degeneracy" patterns for hadronic trajectories on the basis of crossing, SU(3) symmetry, and duality. The scattering amplitude is assumed to consist of two parts: (1) a primarily diffractive background, and (2) a resonant part, the imaginary part of which is expressible at high energies in terms of Regge trajectories in the crossed channels. Hence if a particular scattering amplitude is characterized by internal quantum numbers for which no resonances exist, the Regge trajectories in the crossed channels must exhibit exchange degeneracy so that the corresponding imaginary parts cancel.

In the present note, we investigate the consequences of duality in a local energy region in the case of of the reactions $P+D\to P+D$, where P and D represent, respectively, the octet of pseudoscalar mesons and the decuplet of $J^P=\frac{3}{2}^+$ baryons. Following Mandula, Weyers, and Zweig,[3] we assume that there is a region of s and small t on the one hand, and a region of s and small u on the other, within which the imaginary part can be calculated in two alternate ways, in terms of direct-channel resonances or in terms of Regge trajectories of the crossed channels. Therefore, if we select the s-channel reactions so that their t or u channels are characterized by exotic quantum numbers (viz., 27 in the t channel, 27 and 35 in the u channel), the imaginary parts due to s-channel resonances must add up to zero. We consider a set of direct-channel resonances degenerate in mass but with different spins and parities and examine the constraints on their coupling constants implied by the above requirement in the forward and the backward scattering. We show by including considerations of spin that the particles on leading trajectories must be accompanied by daughters with prescribed ratios of coupling constants between the parent and the daughter particles. We discuss the general pattern of the nontrivial solutions and compare the results with experiments in the case of nonstrange baryons.

The imaginary parts of the six independent s-channel helicity amplitudes free of kinematic singularities[4] can be written as

$$\mathrm{Im}\overline{G}^{R}_{0\mu;0\lambda}(s,\theta_s)=K(s)\sum_{JLL'R'}X^{RR'}(2L'+1)^{1/2}(2L+1)^{1/2}C(L'\tfrac{3}{2}J;0\mu)C(L\tfrac{3}{2}J;0\lambda)\chi_{R'}{}^{J}(L')\chi_{R'}{}^{J}(L)d_{\lambda\mu}{}^{J}(\theta_s), \quad (1)$$

where R characterizes the irreducible SU(3) representation of the crossed t or u channel and $X_{R'}{}^{R}$ are the elements of the corresponding crossing matrix. L,λ (L',μ) denote the orbital angular momentum and helicity of the initial (final) state, and $K(s)$ is a kinematic factor. The χ^J's are related to the partial widths through

$$\chi^J(L)=(\Gamma_{PD}{}^J\Gamma^J)^{1/2}\{[(W^*-W_{\mathrm{c.m.}})/\tfrac{1}{2}\Gamma]^2+1\}^{-1/2}, \quad (2)$$

where Γ^J is the total width and $\Gamma_{PD}{}^J$ the partial width of a resonance of angular momentum J and mass

6. Reprinted article 158, listed in Chapter 8:

PHYSICAL REVIEW D VOLUME 12, NUMBER 12 15 DECEMBER 1975

Concept of nonintegrable phase factors and global formulation of gauge fields

Tai Tsun Wu*
Gordon McKay Laboratory, Harvard University, Cambridge, Massachusetts 02138

Chen Ning Yang†
Institute for Theoretical Physics, State University of New York, Stony Brook, New York 11794
(Received 8 September 1975)

Through an examination of the Bohm-Aharonov experiment an intrinsic and complete description of electromagnetism in a space-time region is formulated in terms of a nonintegrable phase factor. This concept, in its global ramifications, is studied through an examination of Dirac's magnetic monopole field. Generalizations to non-Abelian groups are carried out, and result in identification with the mathematical concept of connections on principal fiber bundles.

I. MOTIVATION AND INTRODUCTION

The concept of the electromagnetic field was conceived by Faraday and Maxwell to describe electromagnetic effects in a space-time region. According to this concept, the field strenght $f_{\mu\nu}$ describes electromagnetism. It was later realized,[1] however, that $f_{\mu\nu}$ by itself does not, in quantum theory, completely describe all electromagnetic effects on the wave function of the electron. The famous Bohm-Aharonov experiment, first beautifully performed by Chambers,[2] showed that in a multiply connected region where $f_{\mu\nu}=0$ everywhere there are physical experiments for which the outcome depends on the loop integral

$$\frac{e}{\hbar c}\oint A_\mu dx^\mu \qquad (1)$$

around an unshrinkable loop. This raises the question of what constitutes *an intrinsic and complete description* of electromagnetism. In the present paper we wish to discuss this question and also its generalization to non-Abelian gauge fields.

An examination of the Bohm-Aharonov experiment indicates that in fact only *the phase factor*

$$\exp\left(\frac{ie}{\hbar c}\oint A_\mu dx^\mu\right), \qquad (2)$$

and *not the phase* (1), is physically meaningful. In other words, the phase (1) contains more information than the phase factor (2). But the additional information is not measurable. This simple point, probably implicitly recognized by many authors, is discussed in Sec. II. It leads to the concept of nonintegrable (i.e., path-dependent) phase factor as the basis of a description of electromagnetism.

This concept has been taken[3] as the basis of the definition of a gauge field. The discussions in Ref. 3, however, centered only on the local properties of gauge fields. To extend the concept to global problems we analyze in Sec. III the field produced by a magnetic monopole. We demonstrate how the quantization of the pole strength, a striking result due to Dirac,[4] is understood in this concept of electromagnetism. The demonstration is closely related to that in the original Dirac paper. Dirac discussed the phase factor of the wave function of an electron (which, among other things, depends on the electron energy). Our emphasis is on the nonintegrable electromagnetic phase factor (which does not depend on such quantities as the energy of the electron).

The monopole discussion leads to the recognition that in general the phase factor (and indeed the vector potential A_μ) can only be properly defined in each of many overlapping regions of space-time. In the overlap of any two regions there exists a gauge transformation relating the phase factors defined for the two regions. This discussion is made more precise in Sec. IV. It leads to the definition of global gauges and global gauge transformations.

In Sec. V generalizations to non-Abelian gauge groups are made. The special cases of SU_2 and SO_3 gauge fields are discussed in Secs. VI and VII. A surprising result is that the monopole types are quite different for SU_2 and SO_3 gauge fields and for electromagnetism.

The mathematics of these results is in fact well known to the mathematicians in *fiber bundle theory*. An identification table of terminologies is given in Sec. V. We should emphasize that our interest in this paper does not lie in the beautiful, deep, and general mathematical development in fiber bundle theory. Rather we are concerned with the necessary *concepts to describe the physics of gauge theories*. It is remarkable that these concepts have already been intensively studied as mathematical constructs.

Section VII discusses a "*gedanken*" generalized Bohm-Aharonov experiment for SU_2 gauge fields.

Unfortunately, the experiment is not feasible unless the mass of the gauge particle vanishes. In the last section we make several remarks.

II. DESCRIPTION OF ELECTROMAGNETISM

The Bohm-Aharonov experiment explores the electromagnetic effect on an electron beam (Fig. 1) in a doubly connected region where the electromagnetic field is zero. As predicted[1] by Aharonov and Bohm, the fringe shift is dependent on the phase factor (2), which is equal to

$$\exp\left(\frac{-ie}{\hbar c}\Omega\right),$$

where Ω is the magnetic flux in the cylinder. Thus two cases a and b for which

$$\Omega_a - \Omega_b = \text{integer} \times (hc/e) \tag{3}$$

give the same interference fringes in the experiment. This we shall state and prove as follows.

Theorem 1: If (3) is satisfied, no experiment outside of the cylinder can differentiate between cases a and b.

Consider first an electron outside of the cylinder. We look for a gauge transformation on the electron wave function ψ_a and the vector potential $(A_\mu)_a$ for case a, which changes them into the corresponding quantities for case b, i.e. we try to find $S = e^{-i\alpha}$ such that

$$S = S_{ab} = (S_{ba})^{-1},$$

$$\psi_b = S^{-1}\psi_a, \quad \text{or } \psi_b = e^{i\alpha}\psi_a, \tag{4}$$

$$(A_\mu)_b = (A_\mu)_a - \frac{i\hbar c}{e} S \frac{\partial S^{-1}}{\partial x^\mu}, \quad \text{or } (A_\mu)_b = (A_\mu)_a + \frac{\hbar c}{e}\frac{\partial \alpha}{\partial x^\mu}. \tag{5}$$

For this gauge transformation to be definable, S must be *single-valued,* but α itself need not be. Now $(A_\mu)_b - (A_\mu)_a$ is curlless; hence (5) can always be solved for α. But it is multiple-valued with an increment of

$$\Delta\alpha = \frac{e}{\hbar c}\oint [(A_\mu)_b - (A_\mu)_a]\, dx^\mu = \frac{e}{\hbar c}(\Omega_b - \Omega_a) \tag{6}$$

every time one goes around the cylinder. If (3) is satisfied, $\Delta\alpha = 2\pi \times$ integer and S is single-valued. Case a and case b outside of the cylinder are then gauge-transformable into each other, and no physically observable effects would differentiate them. The same argument obviously holds if one studies the wave function of an interacting system of particles provided the charges of the particles are all integral multiples of e. Thus we have shown the validity of Theorem 1.

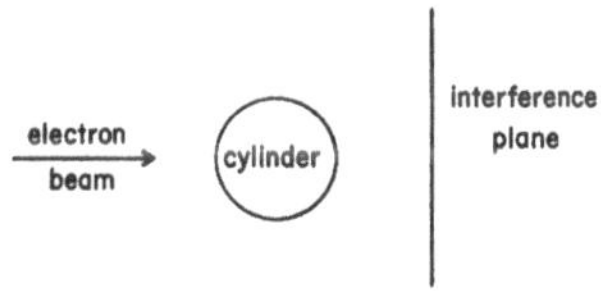

FIG. 1. Bohm-Aharonov experiment (Refs. 1, 2). A magnetic flux is in the cylinder. Outside of the cylinder the field strength $f_{\mu\nu} = 0$.

We conclude: (a) The field strength $f_{\mu\nu}$ underdescribes electromagnetism, i.e., different physical situations in a region may have the same $f_{\mu\nu}$. (b) The phase (1) overdescribes electromagnetism, i.e., different phases in a region may describe the same physical situation. What provides a complete description that is neither too much nor too little is the phase factor (2).

Expression (2) is less easy to use (especially when one makes generalizations to non-Abelian groups) as a fundamental concept than the concept of a phase factor for any path from P to Q

$$\Phi_{QP} = \exp\left(\frac{ie}{\hbar c}\int_P^Q A_\mu dx^\mu\right) \tag{7}$$

provided that an arbitrary gauge transformation

$$\exp\left(\frac{ie}{\hbar c}\int_P^Q A_\mu dx^\mu\right) \to \exp\left(\frac{ie}{\hbar c}a(Q)\right)\exp\left(\frac{ie}{\hbar c}\int_P^Q A_\mu dx^\mu\right)\exp\left(\frac{-ie}{\hbar c}a(P)\right) \tag{8}$$

does not change the prediction of the outcome of any physical measurements. Following Ref. 3, we shall call the phase factor (7) a nonintegrable (i.e., path-dependent) phase factor.

Electromagnetism is thus the gauge-invariant manifestation of a nonintegrable phase factor. We shall develop this theme further in the next section.

III. FIELD DUE TO A MAGNETIC MONOPOLE

The definition of a nonintegrable phase factor (7) in a general case may present problems. To illustrate the problem, let us study the magnetic monopole field of Dirac.[4] Consider a static magnetic monopole of strength $g \neq 0$ at the origin $\vec{r} = 0$ and take the region R of space-time under consideration to be all space-time minus the origin $\vec{r} = 0$. We shall now show the following:

Theorem 2: There does not exist a singularity-free A_μ over all R.

If a singularity-free A_μ does exist throughout R, consider the loop integral $\oint A_\mu dx^\mu$ for time $t=0$ around a circle at fixed spherical coordinates r and θ with azimuthal angle $\phi = 0 \to 2\pi$. This integral, denoted by $\Omega(r,\theta)$ for $r>0$, is equal to the magnetic flux through a cap bounded by the loop, or more explicitly $\Omega(r,\theta) = 2\pi g(1-\cos\theta)$. At $\theta = 0$, $\Omega(r,0)=0$. Increasing θ leads to a continuous increase in Ω till one approaches $\theta=\pi$, at which

$$\Omega(r,\pi) = 4\pi g\,. \tag{9}$$

But at $\theta = \pi$ the loop shrinks to a point. Therefore $\Omega(r,\pi)=0$ since A_μ has no singularity. We have thus reached a contradiction and Theorem 2 is proved.

With an A_μ which has singularities, the nonintegrable phase factor becomes undefined if the path goes through a singularity. This difficulty *must* be resolved in order to use a nonintegrable phase factor as a fundamental concept to describe electromagnetism. It can be resolved in the following way. Let us seek to divide R into two overlapping regions R_a and R_b and to define $(A_\mu)_a$ and $(A_\mu)_b$, each singularity-free in their respective regions, so that (i) their curls are equal to the magnetic field and (ii) in the overlapping region $(A_\mu)_a$ and $(A_\mu)_b$ are related by a gauge transformation. One possible choice is to take the regions to be

$$\begin{aligned} &R_a:\ 0\le\theta<\pi/2+\delta \quad 0<r\,,\quad 0\le\phi<2\pi\,,\ \text{all } t\\ &R_b:\ \pi/2-\delta<\theta\le\pi \quad 0<r\,,\quad 0\le\phi<2\pi\,,\ \text{all } t \end{aligned} \tag{10}$$

with an overlap extending throughout $\pi/2-\delta<\theta<\pi/2+\delta$. (We assume $0<\delta\le\pi/2$.) Take

$$\begin{aligned} &(A_t)_a=(A_r)_a=(A_\theta)_a=0\,,\quad (A_\phi)_a=\frac{g}{r\sin\theta}(1-\cos\theta),\\ &(A_t)_b=(A_r)_b=(A_\theta)_b=0\,,\quad (A_\phi)_b=\frac{-g}{r\sin\theta}(1+\cos\theta)\,. \end{aligned} \tag{11}$$

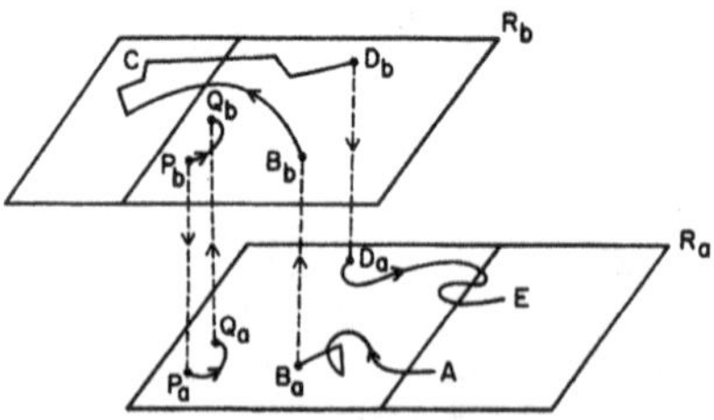

FIG. 2. Schematic diagram illustrating the relationship between R_a and R_b.

The gauge transformation in the overlap of the two regions is

$$S=S_{ab}=\exp(-i\alpha)=\exp\left(\frac{2ige}{\hbar c}\phi\right)\,. \tag{12}$$

This is an allowed gauge transformation if and only if S is single-valued, i.e.,

$$\frac{2ge}{\hbar c}=\text{integer}=D\,, \tag{13}$$

which is Dirac's quantization. With (13) we have

$$S_{ab}=\exp(iD\phi)\,. \tag{12'}$$

To define the phase factor for a path we refer to Fig. 2, where a point in the overlapping region, such as point P, is regarded as two points P_a and P_b. If a path is entirely within region a or b, we define Φ along the path by (7) with $(A_\mu)_a$ or $(A_\mu)_b$ in the integrand in the exponent. If the path $Q \leftarrow P$ is entirely within the overlapping region we have then two possible phase factors $\Phi_{Q_aP_a}$ and $\Phi_{Q_bP_b}$. It is easy to prove that

$$\Phi_{Q_bP_b}=S^{-1}(Q)\Phi_{Q_aP_a}S(P)\,, \tag{14}$$

i.e.,

$$\Phi_{Q_aP_a}S(P)=S(Q)\Phi_{Q_bP_b}\,, \tag{14'}$$

which merely states that $(A_\mu)_a$ and $(A_\mu)_b$ are related by a gauge transformation with the transformation factor (12).

For a path that crisscrosses in and out of the overlapping region, such as $A\to B\to C\to D\to E$ in Fig. 2, the definition of Φ is

$$\Phi_{EDCBA}=\Phi_{ED_a}S_{ab}(D)\Phi_{D_bCB_b}S_{ba}(B)\Phi_{B_aA}\,. \tag{15}$$

Notice that fixing the path but sliding the points B and D along it does not change Φ_{EDCBA} [because of formulas like (14')] so long as B and D remain in the overlapping region.

The phase factor so defined satisfies the group property, e.g.,

$$\begin{aligned} \Phi_{EDCBA}&=\Phi_{ED_a}\Phi_{D_aCBA}\\ &=\Phi_{ED_b}\Phi_{D_bCBA}\\ &=\Phi_{EDC}\Phi_{CBA}\,,\ \text{etc.} \end{aligned} \tag{16}$$

The relationship between the electromagnetic field and the phase factor around a loop is the same as usual. One only has to be careful that if the starting and terminating point A is in the overlapping region, the phase factor is taken to be $\Phi_{A_aBA_a}=\Phi_{A_bBA_b}$, and not $\Phi_{A_aBA_b}$ or $\Phi_{A_bBA_a}$. The phase factor around the loop is then equal to

$$\exp\left(\frac{ie}{\hbar c}\right)\Omega\,,$$

where Ω is the magnetic flux through a cap bordered by the loop. Notice that because of Dirac's quantization condition, the phase factor is the same whichever way one chooses the cap provided it does not pass through the point $\vec{r}=0$ (any t).

We have satisfactorily resolved the difficulty mentioned at the beginning of this section, provided Dirac's quantization condition (13) is satisfied. We shall now prove the following.

Theorem 3: If (13) is not satisfied (the above method of resolving the difficulty would not work since) there exists no division of R into overlapping regions $R_a, R_b, R_c, \ldots$ so that condition (i) and (ii) stated above, properly generalized to the case of more than two regions, would hold.

To prove this statement, observe that if such a division is possible, one could generalize (15) and arrive at a satisfactory definition of the phase factor. The phase factor around a loop is then a continuous function of the loop. Take the loop to be a parallel on the sphere r fixed, $t=0$, θ fixed, $\phi = 0 \to 2\pi$. The phase factor defined by the generalization of (15) is equal to

$$\exp\left[\frac{ie}{\hbar c}\Omega(r,\theta)\right] = \exp\left[\frac{ie}{\hbar c}2\pi g(1-\cos\theta)\right] . \qquad (17)$$

This is not equal to unity when $\theta = \pi$, since (13) is assumed to be invalid. Thus we have a contradiction.

Theorem 3 shows that if Dirac's quantization condition (13) is not satisfied, then the field of a magnetic monopole of strength g cannot be taken as a realizable physical situation in R. (Of course, if one excludes the half-line $x=y=0$, $z<0$, or any half-line starting from $\vec{r}=0$ leading to infinity, then it is possible to have any value for g.) This conclusion is the same as Dirac's, but viewed from a somewhat different point of emphasis.

IV. GENERAL DEFINITION OF GAUGE AND GLOBAL GAUGE TRANSFORMATION

Assuming that (13) holds, to round out our concept of a nonintegrable phase factor the question of the flexibility in the choice of the overlapping regions and the flexibility in the choice of A_μ in the regions must be faced. Both of these questions are related to gauge transformations.

Consider a gauge transformation ξ in R_b (ξ will be assumed to be many times differentiable, but not necessarily analytic), resulting in a new potential $(A_\mu)'_b$. We shall illustrate schematically the transformation by "elevating" the region b in Figure 3(a).

One could extend the region b. One could also contract it, provided the whole R remain covered.

One could create a new region by considering a subregion of b as an additional region R_c [Figure 3(b)], and define the gauge transformation connecting them as the identity transformation so that $(A_\mu)_c = (A_\mu)_b$. One can then "elevate" R_c and contract R_b, which results in Fig. 3(c).

Through operations of the kind mentioned in the last three paragraphs, which we shall call *distortions*, we arrive at a large number of possibilities, each with a particular choice of overlapping regions and with a particular choice of gauge transformation from the original $(A_\mu)_a$ or $(A_\mu)_b$ to the new A_μ in each region. Each of such possibilities will be called a *gauge* (or *global gauge*). This definition is a natural generalization of the usual concept, extended to deal with the intricacies of the field of a magnetic monopole.

For each choice of gauge there is a definition of a nonintegrable phase factor for every path. The group condition $\Phi_{C_cBA_a} = \Phi_{C_cB_b}\Phi_{B_bA_a}$ is always satisfied.

Notice that the original gauge we started with was characterized by (a) specifying [in (10)] the regions [R_a and R_b] and (b) specifying the gauge transformation factor (12′) in the overlap (between R_a and R_b). *It does not refer to any specific* A_μ. [A distortion may of course lead to no changes in characterizations (a) and (b). Thus two different gauges may share the same characterizations (a) and (b).] In the case of the monopole field, we had chosen the vector potential to be given by (11). But, in fact, we can attach to this gauge any $(A_\mu)_a$ and $(A_\mu)_b$ provided they are gauge-transformed into each other by (12′) in the region of overlap. (The resultant $f_{\mu\nu}$ is, of course, not a monopole field in general.) *Thus a gauge is a concept not tied to any specific vector potential.* We shall call the process of distortion leading from one gauge to another a *global gauge transformation*. It is also a concept not tied to any specific vector potential. It is a natural generalization of the usual gauge transformation.

The collection of gauges that can be globally gauge-transformed into each other will be said to

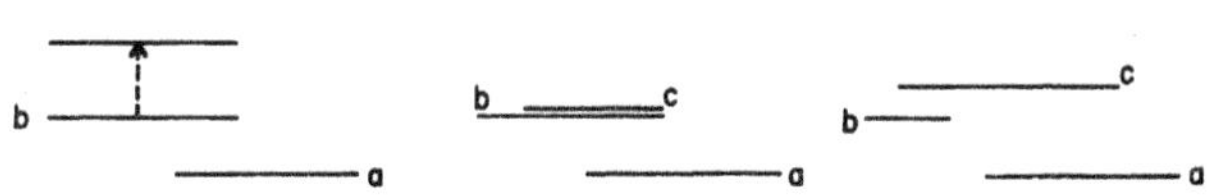

FIG. 3. Distortions allowed in gauge transformation.

belong to the same *gauge type*.

The phase factor around a loop starts and ends at the same point in the same region. Thus it does not change under any global gauge transformation, i.e. we have, for Abelian gauge fields, the following.

Theorem 4a: The phase factor around any loop is invariant under a global gauge transformation.

It follows trivially from this, by taking an infinitesimal loop, that

Theorem 5a: The field strength $f_{\mu\nu}$ is invariant under a global gauge transformation.

For a given value of D, the gauge defined by (10) and (12) will be denoted by $\mathcal{G}_D$. For $D \neq D'$, the relationship, or rather the lack of relationship, between $\mathcal{G}_D$ and $\mathcal{G}_{D'}$ is shown by Theorem 6.

Theorem 6: For $D \neq D'$, $\mathcal{G}_D$ and $\mathcal{G}_{D'}$ are not related by a global gauge transformation, i.e., they are not of the same gauge type.

To prove this theorem we use Theorem 7.

Theorem 7: Between two gauge fields defined on the same gauge there exists a continuous interpolating gauge field defined on the same gauge.

To prove Theorem 7, we simply make a linear interpolation between the two original gauge fields which we shall denote by $(A_\mu)^{(\alpha)}$ and $(A_\mu)^{(\beta)}$:

$$A^{(\gamma)} = t(A_\mu)^{(\alpha)} + (1-t)(A_\mu)^{(\beta)}, \quad 0 \le t \le 1. \tag{18}$$

In an overlap between regions a and b this interpolating vector potential assumes values $(A_\mu)_a^{(\gamma)}$ and $(A_\mu)_b^{(\gamma)}$ which are related by the proper gauge transformation belonging to this overlap. Thus we have proved Theorem 7.

Now go back to Theorem 6 and assume it to be invalid. Then we can gauge-transform the vector potential belonging to the monopole of strength $D'\hbar c/2e$ to the gauge $\mathcal{G}_D$. For this gauge we have then two monopole fields of different pole strengths. Using Theorem 7 we interpolate between them and obtain unquantized magnetic monopoles, which contradict Theorem 3.

Notice that although in this proof of Theorem 6 we have used two specific gauge fields, the theorem itself does not refer to any specific gauge fields at all.

By the same argument as used in the proof of Theorem 7, any gauge field defined on $\mathcal{G}_D$ must have a magnetic monopole of strength $D\hbar c/2e$ at the excluded point $\vec{r}=0$, in addition to possible fields produced by electric charges and currents. Thus the total magnetic flux around the origin $\vec{r}=0$ is equal to $(2\pi\hbar c/e)D$ for any gauge field defined on $\mathcal{G}_D$. We shall state this as a theorem and give another proof of it.

Theorem 8: Consider gauge $\mathcal{G}_D$ and define any gauge field on it. The total magnetic flux through a sphere around the origin $\vec{r}=0$ is *independent of the gauge field* and only depends on the gauge:

$$\iint f_{\mu\nu}\, dx^\mu\, dx^\nu = \frac{-i\hbar c}{e} \oint \frac{\partial}{\partial x^\mu}(\ln S_{ab})\, dx^\mu, \tag{19}$$

where S is the gauge transformation defined by (12) for the gauge $\mathcal{G}_D$ in question, and the integral is taken around any loop around the origin $\vec{r}=0$ in the overlap between R_a and R_b, such as the equator on a sphere $r=1$.

To prove this theorem we observe that the flux through the upper half of the sphere $r=1$ is equal to the following integral around the equator:

$$\oint (A_\mu)_a\, dx^\mu. \tag{20a}$$

The flux through the lower half is equal to a similar integral around the equator:

$$-\oint (A_\mu)_b\, dx^\mu. \tag{20b}$$

Hence

$$\text{total flux} = \oint [(A_\mu)_a - (A_\mu)_b]\, dx^\mu$$

$$= \frac{-i\hbar c}{e} \oint \frac{\partial}{\partial x^\mu}(\ln S_{ab})\, dx^\mu, \tag{21}$$

which completes the proof. Using (13) and (12), the right-hand side of (21) is equal to $4\pi g$, as expected.

If one starts with any gauge which is of the same gauge type as $\mathcal{G}_D$, and makes a global gauge transformation on it, the total flux is not changed by Theorem 5a. Thus (19), which depends only on the gauge, is in fact the same for all gauges of the same type. Notice that if there are more regions in a gauge than two, (19) should be replaced by a sum of line integrals along paths that are in the various overlaps between the regions. For a case of three regions there are three paths, which are illustrated in Fig. 4. Along each path the integral is of the form (19) with S denoting the gauge transformation factor, such as (12), between the two regions containing the path. To prove Theorem 8 in this case one need only add three loop integrals to-

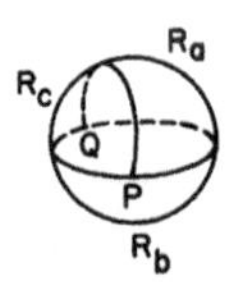

FIG. 4. Case of three regions for Theorem 8. The three paths from P to Q are in the three overlapping regions between (R_a, R_b), (R_b, R_c), and (R_c, R_a).

gether, each of the form of (20a) and (20b), and notice that along each path the integrand is always the difference of the vector potential A_μ between two regions, very much as in (21).

The first proof we gave above of Theorem 8 is easy and is "obvious" to a physicist. The second proof is more involved but is more intrinsic. The theorem is a special case of the Chern-Weil theorem which evolved from the famous Gauss-Bonnet-Allendoerfer-Weil-Chern theorem, a seminal development in contemporary mathematics.[5] We want to emphasize two consequences of the theorem. (i) The right-hand side of (19) is independent of the gauge field, and only depends on the gauge type. (ii) The right-hand side of (19) has as integrand the gradient of $\ln S$. Since S is single-valued, the integral must be equal to an *integral multiple* of a constant (in this case $2\pi i$). A remarkable fact is that these consequences remain valid in the general mathematical theorem, which is very deep.

V. GENERALIZATION TO NON-ABELIAN GAUGE FIELD

So far we have only considered electromagnetism and described it in terms of an Abelian gauge field that corresponds to the group U_1, or equivalently SO_2. On the basis of the discussions in the preceding section, the generalization to the non-Abelian case can be carried out without much difficulty. For a local region this has been done in Ref. 3. Extension to global considerations is our present focus of interest.

A gauge is defined by (a) a particular choice of overlapping regions and (b) a particular choice of *single-valued* gauge transformations S_{ab} in the overlapping regions. The choice of gauge transformations clearly must satisfy the following two conditions.

(1) In the overlapping region $R_a \cap R_b$, the gauge transformations S_{ba} from a to b and S_{ab} from b to a are related by

$$S_{ab}S_{ba}=1\,,$$

where 1 is the identity element of the gauge group.

(2) If three regions R_a, R_b, and R_c overlap, then there are gauge transformations $S_{ab}, S_{ba}, S_{ac}, S_{ca}, S_{bc}, S_{cb}$ so that

$$S_{ab}S_{bc}S_{ca}=1\,,\ \text{etc.}$$

in $R_a \cap R_b \cap R_c$.

As in the case of electromagnetism, both the concept of a gauge and the concept of a global gauge transformation are not tied to any specific gauge potentials, denoted in general by b_μ^k.

The *nonintegrable phase factor* for a given path is now an element of the gauge group. We shall still call it a phase factor. Since these phase factors do not in general commute with each other, Theorems 4a and 5a for the Abelian case need to be modified as follows.

Theorem 4: Under a global gauge transformation, the phase factor around any loop remains in the same class. The class does not depend on which point is taken as the starting point around the loop.

Theorem 5: The field strength $f_{\mu\nu}^k$ is covariant under a global gauge transformation.

Only theorem 4 is not immediately transparent. For a loop $ABCA$, under a gauge transformation[3]

$$\Phi_{ABCA} \to \Phi'_{ABCA} = \xi(A)\Phi_{ABCA}\xi^{-1}(A)\,.$$

Thus Φ'_{ABCA} and Φ_{ABCA} are in the same class. Also around the same loop if we change the starting point from A to C,

$$\Phi_{CABC}=\Phi_{CA}\Phi_{ABCA}\Phi_{AC}\,.$$

Hence changing the starting point does not change the class.

Theorem 4 defines *the class of a loop*. This concept is the generalization of the phase factor for electromagnetism around a loop with the magnetic flux as the exponent. It is a gauge-invariant concept.

These concepts have been extensively studied by the mathematicians in the framework of more general[6] mathematical constructs. A translation of terminology is given in Table I.

VI. CASE OF SU_2 GAUGE FIELD

For the SU_2 case we take the infinitesimal generators X_k to satisfy

$$X_1X_2 - X_2X_1 = X_3\,,\ \text{etc.} \tag{22}$$

and define the phase factor, as a generalization of (7), by[7]

$$\Phi_{QP} = \left[\exp\left(\int_P^Q \frac{-e}{\hbar c} b_\mu^k X_k dx^\mu\right)\right]_{\text{ordered}}, \tag{23}$$

i.e., we make the replacement

$$ieA_\mu \to -eb_\mu^k X_k\,, \tag{24}$$

or

$$A_\mu \to ib_\mu^k X_k\,. \tag{25}$$

[The subscript "ordered" means that, in the definition of the exponential in terms of a power series, the factors $b_\mu^k X_k$ are ordered along the path from P to Q with the factor $b_\mu^k(P)X_k$ at the right end of the product.] The algebraic operators X_k can be thought of as the collection of all irreducible representations of (22). The eigenvalue of iX_k with the

TABLE I. Translation of terminology.

Gauge field terminology	Bundle terminology
gauge (or global gauge)	principal coordinate bundle
gauge type	principal fiber bundle
gauge potential b^k_μ	connection on a principal fiber bundle
S_{ba} (see Sec. V)	transition function
phase factor Φ_{QP}	parallel displacement
field strength $f^k_{\mu\nu}$	curvature
source [a] J^k_μ	?
electromagnetism	connection on a $U_1(1)$ bundle
isotopic spin gauge field	connection on a SU_2 bundle
Dirac's monopole quantization	classification of $U_1(1)$ bundle according to first Chern class
electromagnetism without monopole	connection on a trivial $U_1(1)$ bundle
electromagnetism with monopole	connection on a nontrivial $U_1(1)$ bundle

[a] I.e., electric source. This is the generalization (see Ref. 3) of the concept of electric charges and currents.

minimum absolute value is $\pm\frac{1}{2}$. Therefore the minimum "charge" of all physical states can be read off from (24) by taking the 2×2 irreducible representation of X_k:

$$X_k = -\frac{i\sigma_k}{2}, \tag{26}$$

where σ_k are the Pauli matrices. Thus

$$\text{minimum "charge"} = \frac{e}{2}. \tag{27}$$

The particle of the gauge field belongs to the adjoint representation. Its "charges" are e, 0, and $-e$. Thus

$$\frac{\text{"charge" of gauge particle}}{\text{minimum "charge"}} = 2 \quad \text{for } SU_2. \tag{28}$$

We shall now try to define a Dirac monopole field as a special SU_2 field along only one isospin direction $k=3$, i.e., we define

$$b^1_\mu = b^2_\mu = 0, \quad b^3_\mu = A_\mu, \tag{29}$$

where A_μ is given in the two regions (10) by (11). In the overlapping region, transformation factor S of (12) and (14) now becomes

$$S_{ab} = \exp\left(-\frac{2ge}{\hbar c}\phi X_3\right) \tag{30}$$

by replacement (25). This is single-valued if and only if the quantization condition

$$\frac{eg}{\hbar c} = \text{integer} = D \tag{31}$$

is satisfied because for SU_2

$$\exp(4\pi X_3) = 1, \quad \exp(2\pi X_3) \neq 1,$$

which follows from the existence of half-integral representations such as (26).

The phase factor (30) describes a great circle, wound D times, on the manifold of SU_2 when ϕ varies from $0 \to 2\pi$. Such a circle can be continuously shrunk to the identity element, in contrast with the situation for electromagnetism. Thus, by a global gauge transformation S may be changed to $S'=1$, and the two regions a and b after the global gauge transformation can be *fused into one single region*. The gauge potential b^k_μ is then defined *everywhere* in R as a single region. Thus we have the following theorem.

Theorem 9: For the SU_2 gauge group, the gauges $\mathcal{G}_D$ for different D can be transformed into each other by global gauge transformations. The different monopole fields are therefore of the same type.

We shall only exhibit the global transformation for the case $\mathcal{G}_{-1}$ for which

$$S_{ba} = \exp(-2\phi X_3), \tag{32}$$

$$\frac{e}{\hbar c} = \frac{-1}{g}. \tag{33}$$

The gauge transformations we shall seek are illustrated in Fig. 5. We shall choose

$$\xi = \exp[\theta(X_1 \sin\phi - X_2\cos\phi)], \tag{34}$$

$$\eta = \exp[(\pi-\theta)(X_1\sin\phi - X_2\cos\phi)]\exp(\pi X_2). \tag{35}$$

It is easy to see that ξ is analytic in the coordinates x^μ at all points in R_a. (One only has to verify this statement at $\theta=0$, which is easily done.) Similarly η is analytic in R_b. ξ and η are therefore allowed gauge transformations in, respectively, R_a and R_b.

Now one can prove after some algebra that[8]

$$S'_{ba}=\eta S_{ba}\xi=1\,.$$

Thus after the gauge transformations ξ and η, which together form a global gauge transformation, regions R_a and R_b are related by the identity gauge transformation in their overlap, i.e., the two regions can be fused into one. To calculate the gauge potentials $b_\mu^{k'}$ after the global gauge transformation we use

$$\xi(Q)\left[1+\frac{1}{g}(b_\mu^k)'_a X_k dx^\mu\right]\xi^{-1}(P)=1+\frac{1}{g}(b_\mu^k)_a X_k dx^\mu$$
$$=1+\frac{1}{g}(A_\mu)_a X_3\,dx^\mu\,, \tag{36}$$

where A_μ is given by (11) and $Q=P+dx$. By choosing dx^μ to be along the t and r directions, one obtains $b_t^{k'}=b_r^{k'}=0$. By choosing dx^μ to be along the θ direction, one obtains

$$b_\theta^{1'}=\frac{-g}{r}\sin\phi\,,\quad b_\theta^{2'}=\frac{g}{r}\cos\phi\,,\quad b_\theta^{3'}=0\,. \tag{37}$$

Now take dx^μ to be along the ϕ direction. We obtain, to order $d\phi$,

$$1+\frac{1}{g}(b_\phi^k)'X_k\,r\sin\theta\,d\phi$$
$$=\xi^{-1}(Q)\xi(P)+\frac{1}{g}(A_\phi)_a r\sin\theta\,d\phi\;\xi^{-1}(P)X_3\xi(P)\,. \tag{38}$$

The first term on the right-hand side can be[8] computed in a straightforward manner:

$$\xi^{-1}(Q)\,\xi(P)=1-\sin\theta\,d\phi[(X_1\cos\phi+X_2\sin\phi)-X_3\tan\tfrac{1}{2}\theta]\,.$$

The second term also can be easily computed since

$$\xi^{-1}(P)\,X_3\xi(P)=\sin\theta(X_1\cos\phi+X_2\sin\phi)+X_3\cos\theta$$

and $(A_\phi)_a$ was given by (11). Finally one arrives at

$$b_\phi^{1'}=\frac{-g}{r}\cos\theta\cos\phi\,,$$
$$b_\phi^{2'}=\frac{-g}{r}\cos\theta\sin\phi\,, \tag{39}$$
$$b_\phi^{3'}=\frac{g}{r}\sin\theta\,.$$

Combining these results and remembering (33), we obtain

$$\frac{e}{\hbar c}b_\mu^{k'}X_k\,dx^\mu=\frac{-1}{r^2}\epsilon_{ikj}\,x^j dx^i X_k\,,$$

i.e.,

$$b_4^{k'}=0\,,\quad \frac{e}{\hbar c}b_i^{k'}=-\frac{1}{r^2}\epsilon_{ikj}\,x^j\,. \tag{40}$$

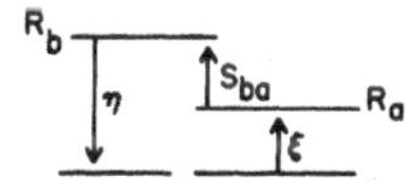

FIG. 5. A global transformation after which R_a and R_b can be fused.

Thus the new potential $b_\mu^{k'}$ is analytic in R_a. Because $\eta S_{ba}\xi=1$ the new potential (in the overlapping region) for R_b must be the same as (40). By analyticity (40) is seen to be valid throughout R. Notice that (40) is the same potential as one of the solutions [solution (12a)], for a sourceless gauge field, in Ref. 9.

The global gauge transformation that transforms $\mathcal{G}_D$ into $\mathcal{G}_0$ for $D\neq -1$ can be obtained by slightly modifying (34) and (35).

We shall discuss Theorem 9 further in the next section.

VII. CASE OF SO_3 GAUGE FIELD

We turn to SO_3, which is locally the same as SU_2, but for which

$$e^{2\pi X_k}=1\,. \tag{41}$$

Equations (22) to (25) remain unaltered. The minimum "charge" of all physical states is now

$$\text{minimum "charge"}=e\,, \tag{42}$$

giving

$$\frac{\text{"charge" of gauge particle}}{\text{minimum "charge"}}=1\quad\text{for } SO_3\,. \tag{43}$$

This last formula differentiates physically the SO_3 case from the SU_2 case.

We emphasize here a point already made in the literature[10] for electromagnetism: The local character of the gauge group is of course determined by the interactions (which determine the conservation laws). We want to ask what determines the global character. The global character (compact or noncompact in the case of electromagnetism, SU_2 or SO_3 in the isospin case) *is determined by* the representations for all states which *physically* exist. For example, in electromagnetism, if all charges are integral multiples of a single unit, the gauge group is compact,[10] because the group is *physically defined* as the simultaneous local phase factor change of *all* charge fields. There is then no physically definable meaning to the noncompact group. In the case of SU_2 or SO_3, if (43) is satisfied, then all representations of X_k physically realizable are integral representations.

Thus the simultaneous local changes of isospin phase factor of all physical systems *cannot differentiate* the group element $e^{2\pi X_k}$ from the identity. *Therefore*, the physical definition of $e^{2\pi X_k}$ is unity and the group must be SO_3.

Turning now to the monopole field for SO_3 we find that (30) is still correct. Equation (41) then leads to the quantization condition

$$\frac{2eg}{\hbar c} = \text{integer} = D \tag{44}$$

in order that S (as an element of SO_3) be a single-valued function of the coordinates x^μ in R.

As ϕ increases from 0 to 2π, the phase factor (30) describes a closed circuit in the group space of SO_3, starting from the identity element and returning to it. If one continuously traced the corresponding element of the group SU_2, one would have started from the identity and ended with the element that corresponds to

$$\begin{pmatrix} -1 & \\ & -1 \end{pmatrix}$$

in the 2×2 representation of SU_2 when D is odd. In such a case, no distortion of the closed circuit in SO_3 described by the phase factor (30) can shrink it to the identity element. This means that the gauge type for even D is not the same as that for odd D. By constructing explicit gauge transformations like (34) and (35) one can then complete the proof of the following theorem.

Theorem 10: For SO_3, all gauges $\mathcal{G}_D$ for D = even are of one type, and all gauges $\mathcal{G}_D$ for D = odd are of one type. These two types are different.

Summarizing the situation for U_1, SU_2, and SO_3 we find that in each case the "magnetic" monopole fields have quantized strengths. They belong to, respectively, infinitely many types for U_1 gauge group (electromagnetism), one type for the SU_2 gauge group, and two types for SO_3 gauge groups.

The physical meaning of these statements are as follows. In the SU_2 case, all magnetic monopole fields can be continuously changed into each other by the process of continuous changes[19] of "electric" sources. For example, starting with the "magnetic monopole" field for $\mathcal{G}_{-1}$ of Theorem 9 we can, by a gauge transformation, obtain the potentials b' [on $\mathcal{G}_0$] given in (40). We can then consider the potential (on $\mathcal{G}_0$): $b'' = \alpha b$, where $0 \le \alpha \le 1$. The gauge field for b'' is no longer electrically sourceless outside of the origin, but is magnetically sourceless except at the orgin, where it is not sourceless either magnetically or electrically. As α changes from 1 to 0 we thus have a continuous change of the original magnetic monopole field to empty space through a process during which there are continuous changes of electric charge-current distributions. Such a process is not possible for electromagnetism, by Theorem 6. (In the SO_3 case it is also not possible, although it is possible to change the magnetic monopole strength by two units by a similar process.) Thus the meaning of a magnetic monopole field in the non-Abelian case is quite different from that in electromagnetism.

It is not really surprising that in the case of electromagnetism one cannot change the magnetic monopole strength by changing electric sources: In the region R there are no magnetic monopoles. The continuity of magnetic lines of forces in R is guaranteed by the equation $\nabla \cdot \vec{H} = 0$. No continuous movement of magnetic lines of force could therefore increase or decrease the net total flux around the origin. That this state of affairs does not obtain for SU_2 and SO_3 is due to the fact that in general $\nabla \cdot \vec{H}^k \neq 0$ in the non-Abelian case, so that one cannot define the magnetic flux through a loop. However, we had seen before (Theorem 4) that in the case of a non-Abelian gauge field what takes the place of the magnetic flux is the *phase factor of a loop*. One may then ask what takes the place of the total magnetic flux outwards from a sphere around the origin $\vec{r} = 0$. To answer this question consider the loop

$$r = 1, \quad \theta = \text{fixed}, \quad \phi = 0 \to 2\pi. \tag{45}$$

As θ changes from 0 to π the phase factor of the loop changes and it describes a continuous circuit (in the space of the group) starting from and ending at the identity element. Clearly any other way of "looping" over the sphere only leads to a distortion of this circuit, without changing the starting and ending point. We shall call this circuit the *total circuit* for the gauge field around the origin $\vec{r} = 0$. It is a concept that replaces the total magnetic flux around $\vec{r} = 0$ in electromagnetism.

We can now prove the following generalization of Theorem 8.

Theorem 11: Consider region R and the group SU_2 or SO_3. Consider a gauge $\mathcal{G}$ and define any gauge field on it. The total circuit for the gauge field around the origin $\vec{r} = 0$ is independent of the gauge field and only depends on the gauge type of $\mathcal{G}$. For the case of $\mathcal{G}_D$,

$$\text{total circuit of the gauge field} \simeq [S_{ba}(\phi) \text{ for } \phi = 2\pi \to 0], \tag{46}$$

where $\simeq$ means "can be continuously distorted into."

This last formula is the generalization of (19).

To prove Theorem 11, consider first the loop (45). The phase factor in R_a and R_b will be denoted by $\Phi^a(\theta)$ and $\Phi^b(\theta)$. They are related in the overlap by

$$\Phi^b(\theta)=\Phi^a(\theta) \tag{47}$$

since $S(\phi=0)=I$. [One uses a generalization of (14).] Next consider the loop $L(\theta)$ which lies on the sphere $r=1$ with its projection onto the x-y plane given in Fig. 6. It consists of a first part $(BA)_1$ around the equator and a second part $(AB)_2$ not on the equator except for points A and B. It is clear that

$$[\text{loop}(45) \text{ for } \theta=0\to\pi/2]\simeq[L(\theta) \text{ for } \theta=0\to\pi/2]$$

because both sides "loop over" the upper hemisphere. Thus

$$[\Phi^a(\theta) \text{ for } \theta=0\to\pi/2]\simeq[\Phi^a_{L(\theta)} \text{ for } \theta=0\to\pi/2]$$
$$=[\Phi^a{}_{(AB)_2}\Phi^a_{(BA)_1} \text{ for } \theta=0\to\pi/2].$$

$\Phi^a_{(AB)_2}$ is continuous in θ, and

$$[\Phi^a_{(AB)_2} \text{ for } \theta=0\to\pi/2]\simeq \text{identity element}.$$

Thus

$$[\Phi^a(\theta) \text{ for } \theta=0\to\pi/2]\simeq[\Phi^a_{(BA)_1} \text{ for } \theta=0\to\pi/2]. \tag{48}$$

Similarly

$$[\Phi^b(\theta) \text{ for } \theta=\pi/2\to\pi]\simeq[\Phi^b_{(BA)_1} \text{ for } \theta=\pi/2\to\pi]. \tag{49}$$

At $\theta=\pi/2$, the left-hand sides of (48) and (49) match because of (47). Also the right-hand sides match. Thus we can take (48) and (49) in tandem, obtaining

$$\text{total circuit of gauge field}\simeq[\Phi^a_{(BA)_1} \text{ for } \theta=0\to\pi/2 \text{ followed by } S_B\Phi^a_{(BA)_1} \text{ for } \theta=\pi/2\to\pi], \tag{50}$$

where we have used

$$\Phi^b_{(BA)_1}=S_B\Phi^a_{(BA)_1}S^{-1}{}_A=S_B\Phi^a_{(BA)_1}. \tag{51}$$

Now

$$[\Phi^a_{(BA)_1} \text{ for } \theta=0\to\pi/2 \text{ followed by } \Phi^a_{(BA)_1} \text{ for } \theta=\pi/2\to\pi] \tag{52}$$

is a loop *that doubles back* on itself, i.e., (52) can be distorted to the identity element. Applying this fact to (50) one obtains

$$\text{total circuit of gauge field}\simeq[I \text{ for } \theta=0\to\pi/2 \text{ followed by } S_B \text{ for } \theta=\pi/2\to\pi]. \tag{53}$$

Now $S_B=S_{ba}(\phi=4\pi-4\theta)$. As $\theta=\pi/2\to\pi$, $S_B=S_{ba}(\phi)$ for $\phi=2\pi\to0$. Substitution into (53) leads to (46).

To complete the proof of Theorem 11 we need the generalization of (46) to gauges that contain more than two regions. This can be done without much difficulty, e.g., for the case that region b is further divided into regions c and d, as schematically illustrated in Fig. 7(a), (46) should be replaced by

$$\begin{aligned}\text{total circuit of gauge field}\simeq[&S_{dc}(B)S_{ca}(x) \text{ for } x=A\to A \text{ along direction of arrow},\\ &\text{followed by } S_{dc}(y)S_{ca}(A) \text{ for } y=B\to B \text{ along direction of arrow}].\end{aligned} \tag{54}$$

For the case that $\mathcal{G}$ has four regions a,b,c,d as illustrated in Fig. 7(b), (46) should be replaced by

$$\begin{aligned}\text{total circuit of gauge field}\simeq[&S_{da}(x) \text{ for } x=A\to B,\\ &\text{followed by } S_{db}(y)S_{ba}(x) \text{ for } x=B\to C,\ y=B\to D,\\ &\text{followed by } S_{dc}(D)S_{cb}(x)S_{ba}(C) \text{ for } x=D\to C,\\ &\text{followed by } S_{dc}(y)S_{ca}(x) \text{ for } x=C\to E,\ y=D\to E,\\ &\text{followed by } S_{da}(x) \text{ for } x=E\to A].\end{aligned} \tag{55}$$

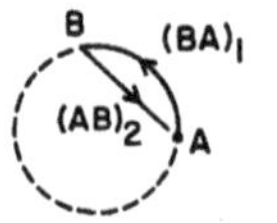

FIG. 6. Projection onto x-y plane of loop $L(\theta)$. The loop lies entirely on sphere $r=1$, and is in the upper (lower) hemisphere for $0 \le \theta \le \pi/2$ ($\pi/2 < \theta \le \pi$). The portion $(BA)_1$ lies on the equator. Coordinates for $A: r=1,\ \theta=\pi/2,\ \phi=0$. Coordinates for $B: r=1,\ \theta=\pi/2,\ \phi=h(\theta)$, where $h(\theta)=4\theta$ for $0 \le \theta \le \pi/2$ and $h(\theta)=4\pi-4\theta$ for $\pi/2 < \theta \le \pi$.

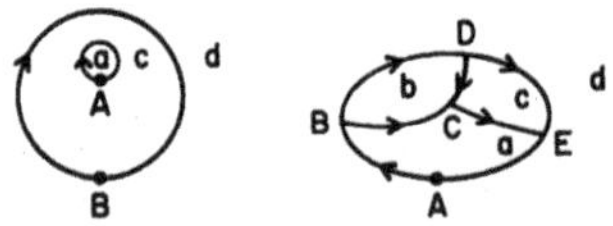

FIG. 7. Schematic diagrams for division lines in overlaps between three or more regions. The drawings are projections from the sphere $r=1$. The projection is from the south pole of the sphere onto the tangent plane at the north pole. The south pole is underneath the plane of the paper.

Notice that the right-hand sides of (54) and (55) are dependent only on the gauge type, and not on the specific gauge field.

VII. GENERALIZED BOHM-AHARONOV EXPERIMENT

The concept of an SU_2 gauge field was first discussed in 1954. In recent years many theorists, perhaps a majority, believe that SU_2 gauge fields do exist. However, so far there is *no experimental proof* of this theoretical idea, since conservation of isotopic spin only suggests, and does not require, the existence of an isotopic spin gauge field. What kind of experiment would be a definitive test of the existence of an isotopic spin gauge field? A generalized Bohm-Aharonov experiment would be.

If the gauge particle for isospin group SU_2 is massless, it is possible to design a *gedanken* generalized Bohm-Aharonov experiment as illustrated in Fig. 1. One constructs the cylinder of material for which the total I_z spin is not zero, e.g., a cylinder made of heavy elements with a neutron excess. One spins the cylinder around its axis, setting up a "magnetic" flux inside the cylinder, along the I_z "direction." If one scatters a proton beam around the cylinder, the fringe shift would be in the opposite direction from the corresponding shift observed with a neutron beam. To be more specific, imagine that one spins the cylinder clockwise. The magnetic flux would be emerging from the diagram towards the reader, since the cylinder has a net negative value for I_z. This means that for a proton (neutron) beam, the flux produces an increment (decrement) of path length counterclockwise around the cylinder. This increment (decrement) produces a net downward (upward) shift of the fringes, i.e., a shift toward the bottom (top) of the diagram.

If one scatters a coherent mixture of neutron and proton in a pure state, in the interference plane one would observe not only fluctuations of nucleon intensity, but also fluctuations of the neutron-proton mixing ratio. A variation of this phenomenon obtains if one imagines rotating a cylinder which has an average $\langle \vec{I} \rangle$ which is not zero, and is not in the I_z direction. A magnetic flux would then be set up which is in a "direction" other than I_z. Scattering a beam of protons would then produce some neutrons as well as protons in the interference plane. This implies, of course, that there is electric charge transfer between the beam and the cylinder together with the gauge field around it.

If the gauge particle has a finite mass $m>0$, then the experiment becomes difficult because the return flux would hug the outside surface of the cylinder, to a distance $\sim \hbar/mc$. Unless the fringe plane lies within this distance of the cylinder, the effect of the flux will be negligible.

IX. REMARKS

(a) From the viewpoint of the present paper, the electric charge and the magnetic charge play completely unsymmetrical roles. This matter deserves further comments. In the non-Abelian case it was in fact already pointed out[11] that the dual of an unquantized sourceless gauge field is not necessarily a gauge field. Thus the asymmetry between electric and magnetic phenomena is not due to the formalism, but is of an intrinsic nature in the non-Abelian case. In contrast, in the Abelian case the asymmetry is only formal since the electric and magnetic charges interact with the electromagnetic field in entirely symmetrical ways. In other words, one can use the phase factor Φ^M associated with the magnetic charges to describe the electromagnetic field, rather than the phase factor Φ discussed in the present paper. The mathematical relationship between these two kinds of phase factors (or between the associated vector potentials A_μ^M and A_μ) remains to be explored. So does the corresponding question in any second-quantized theory[12] of all the fields.

(b) In the proof of Theorem 9 we had shown explicitly how a magnetic monopole field for the SU_2 gauge group can be gauge-transformed into the solution (12a) of Ref. 9. Now a magnetic monopole field is not a gauge field at the origin $\vec{r}=0$ since it does not satisfy the Bianchi identity[3] at the origin. Thus, although solution (12a) of Ref. 9 is (electrically) sourceless at all points, including the origin, it is not a proper gauge field at the origin, a fact we did not realize before. All three solutions, (12a), (12d), and (12e), are, of course, of the same gauge type.

(c) In Sec. II it was emphasized that $f_{\mu\nu}$ underdescribes electromagnetism because of the Bohm-Aharonov experiment which involves a doubly connected space region. For non-Abelian cases, the field strength $f^k_{\mu\nu}$ underdescribes the gauge field even in a singly connected region. An example of this underdescription was given in Ref. 13.

(d) For the region of space-time outside of the cylinder of Fig. 1 there is only one gauge type. All electromagnetic fields in the region can be continuously distorted into each other by the movement of electric charges and currents inside and outside the cylinder.

(e) The phase factor for the group U_1 is the phase factor of the algebra of complex numbers. It is perhaps not accidental that such a phase factor provides the basis for the description of a physically realized gauge field—electromagnetism. Now the only possible more complicated division algebra is the *algebra of quaternions*. The phase factors of the quaternions form the group SO_3. It is tempting to speculate that such a phase factor provides the basis for the description of a physically realized gauge field—the SU_2 gauge field. Speculation about the possible relationship between quaternions and isospin has been made before.[14] Such speculations were, however, not made with reference to gauge fields. If one believes that gauge fields give the underlying basis for strong and/or weak interactions, then the fact that gauge fields are fundamentally *phase factors* adds weight to the speculation that quaternion algebra is the real basis of isospin invariance.

(f) It is a widely held view among mathematicians that the fiber bundle is a natural geometrical concept.[15] Since gauge fields, including in particular the electromagnetic field, are fiber bundles, *all gauge fields are thus based on geometry.*[16] To us it is remarkable that a geometrical concept formulated without reference to physics should turn out to be exactly the basis of one, and indeed maybe all, of the fundamental interactions of the physical world.

ACKNOWLEDGMENTS

It is a pleasure to thank Professor Shiing-shen Chern for correspondence and discussions. We are especially indebted to Professor J. Simons, whose lectures and patient explanations have revealed to us glimpses of the beauty of the mathematics of fiber bundles.

While we were making corrections on the draft of this paper, a report on the experimental discovery of a magnetic monopole[17] reached us.

Additional references to fiber bundles, monopoles and quaternions are given in footnote 18.

---

*Work supported in part by the U. S. ERDA under Contract No. AT(11-1)-3227.

†Work supported in part by the National Science Foundation under Grant No. MPS74-13208 A01.

[1]Y. Aharonov and D. Bohm, Phys. Rev. 115, 485 (1959). See also W. Ehrenberg and R. E. Siday, Proc. Phys. Soc. London B62, 8 (1949).

[2]R. G. Chambers, Phys. Rev. Lett. 5, 3 (1960).

[3]Chen Ning Yang, Phys. Rev. Lett. 33, 445 (1974). This paper introduced the formulation of gauge fields in terms of the concept of nonintegrable phase factors. The differential formulation of gauge fields for Abelian groups was first discussed by H. Weyl, Z. Phys. 56, 330 (1929); for non-Abelian groups it was first discussed by Chen Ning Yang and Robert L. Mills, Phys. Rev. 96, 191 (1954). See also S. Mandelstam, Ann. Phys. (N.Y.) 19, 1 (1962); 19, 25 (1962); I. Białynicki-Birula, Bull. Acad. Pol. Sci., Ser. Sci. Math. Astron. Phys. 11, 135 (1963); N. Cabibbo and E. Ferrari, Nuovo Cimento 23, 1146 (1962); R. J. Finkelstein, Rev. Mod. Phys. 36, 632 (1964); N. Christ, Phys. Rev. Lett. 34, 355 (1975); and A. Trautman, in *The Physicist's Conception of Nature*, edited by J. Mehra (Reidel, Boston, 1973), p. 179.

[4]P. A. M. Dirac, Proc. R. Soc. London A133, 60 (1931). Since this brilliant work of Dirac, there have been several hundred papers on the magnetic monopole. For a listing of papers until 1970, see the bibliography by D. M. Stevens, Virginia Polytechnic Institute Report No. VPI-EPP-70-6, 1970 (unpublished).

[5]See J. Milnor and J. Stasheff, *Characteristic Classes* (Princeton Univ. Press, Princeton, N.J., 1974); C. B. Allendoerfer and A. Weil, Trans. Am. Math. Soc. 53, 101 (1943); Shiing-shen Chern, Ann. Math. 45, 747 (1944). See also H. Weyl, Amer. J. Math. 61, 461 (1939), and Ref. 6 below.

[6]There are many books on fiber bundles. See e.g., N. Steenrod, *The Topology of Fibre Bundles* (Princeton Univ. Press, Princeton, N. J., 1951). For connection, see, e.g., S. Kobayashi and K. Nomizu, *Foundations of Differential Geometry* (Interscience, New York, Vol. I-1963, Vol. II-1969).

[7]The notation here is the same as that in Ref. 3, except for the normalization factor $e/\hbar c$ which was absorbed into b in Ref. 3. To avoid confusion with the azimuthal angle, we write Φ for the ϕ of Ref. 3. Notice that

$$\Phi_{(A+dx)A} = I - \frac{e}{\hbar c} b^k_\mu(x) X_k dx^\mu .$$

All formulas are the same as in Ref. 3, but the name for Φ_{QP} will now be "the phase factor from P to Q." (In Ref. 3 the same name applied to Φ_{PQ}.) The new name is in accordance with the usual convention of time ordering.

[8]Since the 2×2 representation is faithful, it is sufficient for computational purposes to use the representation (26) for X_k. This makes the algebra quite simple since one can apply the formula $e^{i\theta\sigma_k} = \cos\theta + i\,\sigma_k \sin\theta$.

[9]Tai Tsun Wu and Chen Ning Yang, in *Properties of Matter under Unusual Conditions*, edited by H. Mark and S. Fernbach (Wiley, New York, 1969), p. 349.

[10]Chen Ning Yang, Phys. Rev. D 1, 2360 (1970).

[11]Gu Chao-hao and Chen Ning Yang, Sci. Sin. 18, 483 (1975).

[12]P. A. M. Dirac, Phys. Rev. 74, 817 (1948); J. Schwinger, *Particles, Sources and Fields* (Addison-Wesley, Reading, Mass., Vol. 1-1970, Vol. 2-1973).

[13]Tai Tsun Wu and Chen Ning Yang, preceding paper, Phys. Rev. D 12, 3843 (1975).

[14]Cheng Ning Yang, comments after J. Tiomno's talk, session 9, *Proceedings of the Seventh Annual Rochester Conference on High-Energy Nuclear Physics, 1957* (Interscience, New York, 1957); Chen Ning Yang, in *The Physicist's Conception of Nature*, edited by J. Mehra (Reidel, Boston, 1973), p. 447.

[15]See, e.g., Shiing-shen Chern, Geometry of Characteristic Classes, Proceedings of the 13th Biennial Seminar, Canadian Mathematics Congress, 1972, p. 1.

[16]This is in sharp contrast with an interaction (if it exists), which is not related to gauge concepts.

[17]P. B. Price, E. K. Shirk, W. Z. Osborne, and L. S. Pinsky, Phys. Rev. Lett. 35, 487 (1975).

[18]There have been many papers on fiber bundles, monopoles, and quaternions in the physics literature. The following is only a partial list: Elihu Lubkin, Ann. Phys. (N.Y.) 23, 233 (1963); J. Math. Phys. 5, 1603 (1964); D. Finkelstein, J. M. Jauch, S. Schiminovich, and D. Speiser, J. Math. Phys. 4, 788 (1963); articles by J. A. Wheeler, B. S. DeWitt, A. Lichnerowicz, and C. W. Misner, in *Relativity, Groups and Topology*, edited by C. DeWitt and B. S. DeWitt (Gordon and Breach, New York, 1963); also C. Misner, K. Thorne, and J. A. Wheeler, *Gravitation* (Freeman, San Francisco, 1973); A. Trautman, Rep. Math. Phys. 1, 29 (1970); G. 't Hooft, Nucl. Phys. B79, 276 (1974); Hendricus G. Loos, Phys. Rev. D 10, 4032 (1974); J. Arafune, P. G. O. Freund, and C. J. Goebel, J. Math. Phys. 16, 433 (1975); B. Julia and A. Zee, Phys. Rev. D 11, 2227 (1975); M. K. Prasad and Charles M. Sommerfield, Phys. Rev. Lett. 35, 760 (1975).

[19]*Footnote added in proof.* Professor A. Lenard has raised an interesting question in this connection: In the continuous changes of "electric" sources, are sources quantized according to (44)? The answer to this question is "no," and requires explanations. The "electric" charges play two separate roles. They act as sources of gauge fields, and they also act as responders to gauge fields. In a physical situation the two roles are of course interrelated. In the discussion here, however, we separate the two roles. We therefore do not require quantization of electric charges as sources, but require quantization of electric charges as responders.

7. Reprinted article 233, listed in Chapter 9:

PHYSICAL REVIEW VOLUME 176, NUMBER 2 10 DECEMBER 1968

Theory of a Two-Dimensional Ising Model with Random Impurities. I. Thermodynamics

BARRY M. McCOY
Institute for Theoretical Physics, State University of New York at Stony Brook, Stony Brook, New York 11790
AND
TAI TSUN WU*
Gordon McKay Laboratory, Harvard University, Cambridge, Massachusetts 02138
(Received 9 July 1968)

Recent experiments demonstrate that at the Curie temperature the specific heat may be a smooth function of the temperature. We propose that this effect can be due to random impurities and substantiate our proposal by a study of an Ising model containing such impurities. We modify the usual rectangular lattice by allowing each row of vertical bonds to vary randomly from row to row with a prescribed probability function. In the case that this probability is a particular distribution with a narrow width, we find that the logarithmic singularity of Onsager's lattice is smoothed out into a function which at T_c is infinitely differentiable but not analytic. This function is expressible in terms of an integral involving Bessel functions and is computed numerically.

1. INTRODUCTION

THE free energy of a two-dimensional Ising model without a magnetic field was first obtained by Onsager[1] in 1944. One of the most striking features of Onsager's result is that the specific heat has a logarithmic singularity at the Curie temperature T_c. The spontaneous magnetization for this lattice was determined by Yang[2] in 1952 and shown to approach zero as $(T_c-T)^{1/8}$ as $T\rightarrow T_c^-$. In spite of the fact that these calculations are for a two-dimensional system, they form the basis of much of our understanding of ferromagnetic phase transitions. The precise form of the singularity in the specific heat of various magnetic systems is a matter of considerable debate, but the existence of the singularity is rarely questioned. Similarly, while measurements show that the $\frac{1}{8}$ power singularity of the spontaneous magnetization of the Ising model is too small to fit the experimental data, experiments have not yet revealed a necessity to assume anything other than a power-law singularity at T_c.

However, recent precise measurements of the specific heats of EuS [3] and Ni [4] indicate that in these systems, among others, if measurements are made close enough to T_c, the specific heat is seen not to diverge to infinity. In fact it does not even have a discontinuous or infinite first derivative. To a high degree of accuracy these specific heats are smooth functions of the temperature. This smooth behavior is markedly different from Onsager's result.

Perhaps the most interesting feature of this discrepancy between precise experiments and the intuition gained from Onsager's calculation is not that there is a discrepancy but that the discrepancy occurs when T is very close to T_c and has resisted detection for so long. Onsager's calculation was carried out for a perfect lattice, one in which *all* the vertical interactions and *all* the horizontal interactions are respectively the same. Virtually any difference between the real system and the idealized Ising model with nearest-neighbor interactions can be invoked to "explain" the experimental data. Some are more reasonable than others; one possible example is the finite size of the actual sample. It is the purpose here to study one aspect for the various differences, namely the presence of random impurities. By impurities we refer not only to the presence of foreign material but also to any physical property that makes the various lattice sites different from each other. An example is the presence of various isotopes in a sample, e.g., nickel contains roughly 68% of Ni^{58}, 26% of Ni^{60}, 1% of Ni^{61}, 4% of Ni^{62}, and 1% of Ni^{64}. The presence of impurities in this sense seems unavoidable in most actual magnetic systems. If these impurities distribute themselves through the system in a regular ordered fashion, then while the symmetry of the lattice would be reduced it would not be destroyed. With sufficient labor such an ordered sort of impurity can be studied in the Ising model. However, such an ordering of impurities does not always take place. Therefore, if we want to realistically study the effects of impurities in magnetic systems we may have to allow the impurities to be distributed at random throughout the lattice. The regularity of the system now has been not merely reduced but totally destroyed. A phase transition is a cooperative phenomena in which the entire system takes part. It is therefore not at all obvious that the highly regular Onsager lattice should possess a phase transition behavior that is in any way related to such an impure system.

In order to gain any insight into the possibility for

* Supported in part by National Science Foundation Grant No. GP-8019.
[1] L. Onsager, Phys. Rev. **65**, 117 (1944).
[2] C. N. Yang, Phys. Rev. **85**, 808 (1952).
[3] B. J. C. van der Hoeven, Jr., D. T. Teaney, and V. L. Moruzzi, Phys. Rev. Letters **20**, 719 (1968).
[4] P. Handler, D. Mapother, and M. Rayl, Phys. Rev. Letters **19**, 356 (1967).

the random impurities to change qualitatively the nature of the phase transition, it is necessary to study a simple model in detail. Such a model will be described shortly. At least in this specific model, and very probably under quite general circumstances, the logarithmic singularity in the specific heat is no longer present when there are random impurities. Instead, the specific heat is an infinitely differentiable function of the temperature. Thus, in the usual language, the phase transition is of infinite order. In particular, in the presence of random impurities, the critical exponents no longer have a precise meaning.

In this series of papers we consider the following modification of the two-dimensional Ising model. We retain the features of Onsager's rectangular lattice to the extent that all horizontal interactions are the same and that the vertical interaction between any site in the jth row and its nearest neighbor in the $j+1$th row is the same no matter what column these sites are in. However, $E_2(j)$ is allowed to randomly vary from row to row. More specifically we assume that, for $j\neq j'$, $E_2(j)$ and $E_2(j')$ are independent random variables with identical probability distributions $P(E_2)dE_2$.

Let us try to describe the model in greater detail. We are considering a collection of Ising lattices, each of which is specified by a particular set of interactions $\{E_2(j)\}$. We are interested in the thermodynamic limit where the size of these lattices becomes infinite. If, in the thermodynamic limit, the free energy of each lattice in our collection varied wildly from lattice to lattice our model would be useless. In that case the free energy of our random lattices would depend in detail on the arrangement of interactions. Fortunately, this is physically unreasonable and is indeed not the case. In the thermodynamic limit the free energy per site of each lattice does approach, with probability 1, the same value. Therefore, with probability 1 the Curie temperatures of any two lattices from this collection are the same. Furthermore, we expect the spontaneous magnetization of any two lattices to be the same with probability 1, because the spontaneous magnetization, like the free energy, is an average property of the entire lattice. However, not all quantities of interest have distributions which are so sharply peaked. For example, the spin-spin correlation function of neighboring spins *does* depend in detail on the local value of the interaction energies. For such quantities one needs more than an average value to characterize the result of a measurement made at an arbitrary position in the impure lattice.

The complete investigation of all aspects of this random Ising model is clearly beyond the scope of any one paper. In this paper we begin the investigation by considering the free energy in the absence of a magnetic field. In Sec. 2 we will formulate the mathematical problem to be solved and find a general formula for the critical temperature in terms of $P(E_2)$. In Sec. 3 we will derive several general properties of the integral equation found in Sec. 2. There is a great deal that can be said about the equations we derive because they depend on the arbitrary function $P(E_2)$. We wish in this paper to emphasize the physical consequences of random impurities. Therefore we specialize our treatment in Sec. 4 to a particular distribution $P(E_2)$ which has a narrow width (of order N^{-1}) that is particularly mathematically tractable. This distribution is in no way physically distinguished and our results are expected to be typical of a large class of narrow distributions. We show that while the critical temperature is shifted by a temperature on the order of N^{-1} the specific heat deviates appreciably (to order 1 as $N\rightarrow\infty$) from Onsager's specific heat only for $T-T_c\sim O(N^{-2})$. We furthermore find that at T_c the term of order 1 in the specific heat is *not* logarithmically divergent but is an *infinitely differentiable* function of T though it is *not* analytic. Finally we conclude in Sec. 5 with a discussion of the technical aspects of the calculation and outline several ways in which our calculations may be generalized.

2. FORMULATION OF THE PROBLEM

In this section we formulate the mathematical problem corresponding to the physical model described in the *Introduction*. We do this in two stages. The first stage is to relate the free energy to the solution of a pair of coupled recursion relations. This procedure has been discussed recently by us[5] for Onsager's lattice. That derivation of the recursion relation still holds for the class of lattices under consideration. We therefore only repeat as little of that derivation as is necessary to establish notation and refer the reader to the previous work for a detailed discussion. The final results are given in (2.9) and (2.11). The rest of this section is devoted to the study of the mathematical problem formed by (2.9) and (2.11) and is self-contained. We advise the interested, as opposed to the dedicated, reader[6] to omit the first stage except for the definitions.

We are interested in studying a particular class of rectangular two-dimensional Ising models with $\mathfrak{M}$ rows and $2\mathfrak{N}$ columns. We impose cyclic boundary conditions in the horizontal direction only. This class of systems is characterized by the Hamiltonians

$$\mathcal{E}=-E_1\sum_{j=1}^{\mathfrak{M}}\sum_{k=-\mathfrak{N}+1}^{\mathfrak{N}}\sigma_{j,k}\sigma_{j,k+1}-\sum_{j=1}^{\mathfrak{M}-1}E_2(j)\sum_{k=-\mathfrak{N}+1}^{\mathfrak{N}}\sigma_{j,k}\sigma_{j+1,k},\qquad(2.1)$$

where $\sigma_{j,k}$ is equal to $+1$ or -1, j and k label, respec-

[5] B. M. McCoy and T. T. Wu, Phys. Rev. **162**, 436 (1967). This paper is hereafter referred to as IV. The present notation differs from IV in a few trivial ways $\mathfrak{a}=-ia$, $\mathfrak{D}_n=iD_n$, and all other German letters are replaced by their Latin equivalents.

[6] The distinction seems to have been first clearly made by G. F. Chew, M. L. Goldberger, F. E. Low, and Y. Nambu, Phys. Rev. **106**, 1337 (1957).

tively, the row and column of each lattice site, and $k=\mathfrak{N}+1$ is identified with $k=-\mathfrak{N}+1$. As far as thermodynamics is concerned we may restrict E_1 and E_2 to be positive without loss of generality. We are interested in the limit $\mathfrak{N}\to\infty$ and $\mathfrak{M}\to\infty$; only in this limit will a phase transition occur. We complete the characterization by requiring that $E_2(j)$ be independent random variables with the probability density function $P(E_2)$.

Denote by $Z_{\{E_2\}}$ the partition function for the system described by (2.1) where the collection of bounds $\{E_2(j)\}$ is chosen at random according to the probability density function $P(E_2)$. We are interested in F_r, the free energy per site of the system in the thermodynamic limit. Under the assumption that with probability 1 this limit exists, F_r is defined as

$$F_r=-\beta^{-1}\lim_{\mathfrak{M}\to\infty,\mathfrak{N}\to\infty}(2\mathfrak{M}\mathfrak{N})^{-1}\ln Z_{\{E_2\}}. \quad (2.2)$$

Our class of lattices shares with Onsager's lattice the property of translational invariance in the horizontal direction. Therefore the calculation of Sec. 3 of IV may be taken over word for word to show that for any set of energies $E_2(j)$

$$Z_{\{E_2\}}{}^2=(2\cosh\beta E_1)^{4\mathfrak{M}\mathfrak{N}}\prod_{j=1}^{\mathfrak{M}-1}[\cosh\beta E_2(j)]^{4\mathfrak{N}} \times\prod_\theta|1+z_1e^{i\theta}|^{2\mathfrak{M}}\det C_{\{E_2\}}(\theta), \quad (2.3)$$

where $\prod_\theta$ is the product over the $2\mathfrak{N}$ roots of -1:

$$\theta=\prod(2n-1)/2\mathfrak{N}, \qquad n=1,2,\cdots 2\mathfrak{N}, \quad (2.4)$$

$$z_1=\tanh\beta E_1, \quad (2.5a)$$

$$z_2(j)=\tanh\beta E_2(j), \quad (2.5b)$$

and $\beta=(kT)^{-1}$. The $2\mathfrak{M}\times 2\mathfrak{M}$ matrix $C_{\{E_2\}}(\theta)$ is defined by

$$C_{j,j}=\begin{bmatrix} ia & b \\ -b & -ia\end{bmatrix}, \quad (2.6a)$$

$$C_{j,j+1}=-C_{j+1,j}{}^T=\begin{bmatrix} 0 & 0 \\ z_2(j) & 0\end{bmatrix} \quad (2.6b)$$

[compare with (3.9) of IV] and all other matrix elements are zero. Here we use

$$a(\theta)=-2z_1\sin\theta\,|1+z_1e^{i\theta}|^{-2}, \quad (2.7a)$$

$$b(\theta)=(1-z_1{}^2)\,|1+z_1e^{i\theta}|^{-2}. \quad (2.7b)$$

More explicitly $C_{\{E_2\}}(\theta)$ may be written as

$$\begin{bmatrix}
ia & b & & & & & & & \\
-b & -ia & z_2(1) & & & & & & \\
 & -z_2(1) & ia & b & & & & & \\
 & & -b & -ia & z_2(2) & & & & \\
 & & & -z_2(2) & \ddots & \ddots & & & \\
 & & & & \ddots & \ddots & \ddots & & \\
 & & & & & \ddots & \ddots & z_2(\mathfrak{M}-1) & \\
 & & & & & & -z_2(\mathfrak{M}-1) & ia & b \\
 & & & & & & & -b & -ia
\end{bmatrix}. \quad (2.8)$$

where $z_2(0)=0$ by definition. Therefore, the free energy F_r is given as

We may further follow the procedure of IV and define C_n to be the determinant of the $2n\times 2n$ random matrix of the form (2.8) and iD_n to be the corresponding $(2n-1)\times(2n-1)$ random determinant with the last row and column removed. Then $\det C_{\{E_2\}}(\theta)=C_{\mathfrak{M}}(\theta)$ and we obtain the recurrence relation for $n\geq 0$,

$$\begin{bmatrix} C_{n+1} & (\theta) \\ D_{n+1} & (\theta)\end{bmatrix}=\begin{bmatrix} a^2+b^2 & a \\ a & 1\end{bmatrix}\begin{bmatrix} 1 & 0 \\ 0 & z_2(n)^2\end{bmatrix}\begin{bmatrix} C_n & (\theta) \\ D_n & (\theta)\end{bmatrix} \quad (2.9)$$

together with the boundary condition

$$C_0(\theta)=1, \qquad D_0(\theta)=0, \quad (2.10)$$

$$F_r=-\beta^{-1}\Big\{\ln(2\cosh\beta E_1)+\int_0^\infty dE_2P(E_2)\ln\cosh\beta E_2 +(4\pi)^{-1}\int_{-\pi}^{\pi}d\theta\ln|1+z_1e^{i\theta}|^2 +(4\pi)^{-1}\int_{-\pi}^{\pi}d\theta\lim_{\mathfrak{M}\to\infty}\mathfrak{M}^{-1}\ln CA(\theta)\Big\}. \quad (2.11)$$

The object $C_n(\theta)$ is the first component of the vector we obtain by applying n random matrices of the form

$$\begin{bmatrix} a^2+b^2 & a\lambda \\ a & \lambda \end{bmatrix} \tag{2.12}$$

to the initial vector $\begin{bmatrix}1\\0\end{bmatrix}$. Here $\lambda \equiv z_2{}^2$ is a random variable with the normalized probability distribution function $\mu(\lambda)$ given by

$$\mu(\lambda)d\lambda = P(E_2)dE_2. \tag{2.13}$$

The general theory of such random matrix products, in particular the existence with probability 1 of the limit of the right-hand side of (2.11), has been presented recently in a rigorous manner by Furstenberg.[7]

We begin our evaluation of (2.11) by noting that

$$\lim_{\mathfrak{M}\to\infty}\mathfrak{M}^{-1}\ln C_{\mathfrak{M}}(\theta) = \lim_{\mathfrak{M}\to\infty}\mathfrak{M}^{-1}\sum_{n=0}^{\mathfrak{M}-1}\ln\frac{C_{n+1}}{C_n}. \tag{2.14}$$

Using the recursion relation (2.9) this may be rewritten as

$$\lim_{\mathfrak{M}\to\infty}\mathfrak{M}^{-1}\sum_{n=0}^{\mathfrak{M}-1}\ln\frac{C_{n+1}}{C_n}$$

$$= \lim_{\mathfrak{M}\to\infty}\mathfrak{M}^{-1}\sum_{n=0}^{\mathfrak{M}-1}\ln[a^2+b^2+a\lambda(n)D_n/C_n]. \tag{2.15}$$

This can be interpreted as an average over n. The only drawback is that while the $\lambda(n)$ are independent random variables the D_n/C_n are not at all independent. We now remark that because the matrix (2.12) is real and acts on a two-dimensional vector space,

$$x_n = C_n/D_n \tag{2.16}$$

may be thought of as the tangent of the angle which the vector $\begin{bmatrix}C_n\\D_n\end{bmatrix}$ makes with the D axis. From (2.9) we find

$$x_{n+1} = [(a^2+b^2)x_n + a\lambda(n)]/[ax_n+\lambda(n)]. \tag{2.17}$$

Because the space of angular dependence of these vectors is compact, as n becomes large the variable x_n will approach a limiting stationary distribution $\nu(x)$ that is independent of the initial vector.[7] This stationary distribution is characterized by the property that, if we apply a random matrix (2.12) to it and average the resulting distribution over $\mu(\lambda)$ of (2.13), $\nu(x)$ will transform into itself. Therefore, $\nu(x)$ satisfies the equation

$$\nu(x) = \int_{-\infty}^{\infty}dx'\int_0^1 d\lambda\,\delta\left(x - \frac{(a^2+b^2)x'+a\lambda}{ax'+\lambda}\right)\mu(\lambda)\nu(x'). \tag{2.18}$$

We may perform the λ integration to obtain

$$\nu(x) = \frac{b^2}{(x-a)^2}\int_{-\infty}^{\infty}dx'x'\mu\left[x'\frac{a^2+b^2-ax}{x-a}\right]\nu(x'). \tag{2.19}$$

[7] H. Furstenberg, Trans. Am. Math. Soc. **108**, 377 (1963).

The work of Furstenberg[7] guarantees that a solution to this equation exists and is unique.

Once we possess a stationary distribution $\nu(x)$ we may replace the average over n in (2.15) by an average over λ and over x. We thus arrive at the final result that with probability 1

$$\lim_{\mathfrak{M}\to\infty}\mathfrak{M}^{-1}\ln C_{\mathfrak{M}}(\theta)$$

$$= \int_{-\infty}^{\infty}dx\,\nu(x)\int_0^1 d\lambda\,\mu(\lambda)\ln(a^2+b^2+a\lambda x^{-1}). \tag{2.20}$$

An identical analysis may be carried out for the quantity $D_{\mathfrak{M}}$ and we find that with probability 1

$$\lim_{\mathfrak{M}\to\infty}\mathfrak{M}^{-1}\ln D_{\mathfrak{M}}(\theta)$$

$$= \int_{-\infty}^{\infty}dx\,\nu(x)\int_0^1 d\lambda\,\mu(\lambda)\ln(\lambda+ax). \tag{2.21}$$

Since $C_{\mathfrak{M}}$ and $D_{\mathfrak{M}}$ are the components of the same vector, their average rate of growth must each be separately equal to the average rate of growth of the vector itself. Therefore the right-hand sides of (2.20) and (2.21) must be equal so that

$$\int_{-\infty}^{\infty}dx\,\nu(x)\int_0^1 d\lambda\,\mu(\lambda)\ln(a^2+b^2+a\lambda x^{-1})$$

$$= \int_{-\infty}^{\infty}dx\,\nu(x)\int_0^1 d\lambda\,\mu(\lambda)\ln(\lambda+ax). \tag{2.22}$$

To prove this directly consider the difference d between the two sides of Eq. (2.22)

$$d = \int_{-\infty}^{\infty}dx\,\nu(x)\int_0^1 d\lambda\,\mu(\lambda)\ln\left(\frac{a^2+b^2+a\lambda x^{-1}}{\lambda+ax}\right). \tag{2.23}$$

This may be rewritten as

$$d = \int_{-\infty}^{\infty}dx\,\nu(x)\int_0^1 d\lambda\,\mu(\lambda)\ln\left(x\frac{a^2+b^2+a\lambda x^{-1}}{\lambda+ax}\right)$$

$$-\int_{-\infty}^{\infty}dx\,\nu(x)\ln x. \tag{2.24}$$

In the first integral we now replace the variable λ by

$$q = x(a^2+b^2+a\lambda x^{-1})/(\lambda+ax) \tag{2.25}$$

to obtain

$$d = \int_{-\infty}^{\infty}dq\int_{-\infty}^{\infty}dx\,\mu\left(x\frac{a^2+b^2-qa}{q-a}\right)\frac{xb^2}{(q-a)^2}\nu(x)\ln q$$

$$-\int_{-\infty}^{\infty}dx\,\nu(x)\ln x. \tag{2.26}$$

We now use the integral equation (2.19) on the first term of (2.26) to find that as expected $d=0$.

We may readily derive the expression for the critical temperature T_c in terms of $P(E_2)$ if we note that at $a=0$ the matrix (2.12) is diagonal. The only possible stationary distributions of vectors are those with the vectors parallel to the C or D axis. In the first case the vectors are multiplied by $b^2(0)$ under the action of (2.12) so we find

$$\lim_{\theta\to0}\lim_{\mathfrak{M}\to\infty}\mathfrak{M}^{-1}\ln C_{\mathfrak{M}}=\ln b(0)^2 \qquad (2.27)$$

and in the latter case the vectors are multiplied by λ so that

$$\lim_{\theta\to0}\lim_{\mathfrak{M}\to\infty}\mathfrak{M}^{-1}\ln C_{\mathfrak{M}}=\int_0^1 d\lambda\,\mu(\lambda)\ln\lambda. \qquad (2.28)$$

The free energy is related to the larger of these two expressions and the condition for $T=T_c$ is obtained by equating (2.27) and (2.28)

$$0=\int_0^1 d\lambda\,\mu(\lambda)\ln[\lambda b(0)^{-2}]$$

$$=\int_0^\infty dE_2P(E_2)\ln\{z_2^2[(1+z_1)/(1-z_1)]^2\} \qquad (2.29a)$$

or

$$2\beta E_1+\int_0^\infty dE_2P(E_2)\ln z_2=0. \qquad (2.29b)$$

If $P(E_2)=\delta(E_2-E_2^0)$ then (2.29) reads

$$z_2^0(1-z_1)/(1+z_1)=1, \qquad (2.30)$$

which is the known result for $T=T_c$ in the ferromagnetic Onsager lattice. The correctness of this criteria for locating T_c will be seen more explicitly in Sec. 4 where we demonstrate for a specific $\mu(\lambda)$ that at the T_c given by (2.29) the free energy F_r is not an analytic function of the temperature.

We note that (2.29) has the property that T_c is zero if $P(E_2)=p\delta(E_2)+\cdots$, where $0<p\leq1$. This is to be expected for our model. A δ function at $E_2=0$ means that in every random lattice there are, with probability 1, vertical bonds with zero strength a finite distance apart. But for our class of lattices the condition $E_2(j)=0$ cuts the lattice into two separate pieces. Therefore the term $p\delta(E_2)$ causes the lattice, with probability 1, to be cut up into an infinite number of strips of finite width and any two-dimensional Ising lattice that is not infinite in both dimensions has no critical temperature. However, we also note that if $\mu(\lambda)$ is bounded near $E_2=0$ and $E_1>0$ then T_c is greater than zero.

We finally remark that the argument leading to (2.29a) does not depend on the fact that E_2 rather than E_1 is random. We therefore see by a similar argument that if E_1 and E_2 are random with the joint probability density $P(E_1,E_2)$ there is a ferromagnetic phase transition at T_c determined from

$$\int_0^\infty dE_1\int_0^\infty dE_2P(E_1,E_2)\ln[z_2^{-1}(1-z_1)/(1+z_1)]=0. \qquad (2.31)$$

3. INTEGRAL EQUATION FOR $\nu(x)$

In this section we study several general properties of the integral equation (2.19). The limits of (2.19) have formally been written as $-\infty$ to $+\infty$, but over much of this range $\nu(x)$ vanishes identically. To see this, first consider the stationary distribution corresponding to the Onsager lattice where

$$\mu(\lambda)=\delta(\lambda-\bar\lambda). \qquad (3.1)$$

The stationary angular distribution of vectors will clearly be a δ function at that value of x which is unchanged by the application of the matrix (2.12) with $\lambda=\bar\lambda$. From (2.17) we see that the values of x which satisfy this eigenvector equation obey

$$x_0(\bar\lambda)=[(a^2+b^2)x_0(\bar\lambda)+a\bar\lambda]/[ax_0(\bar\lambda)+\bar\lambda], \qquad (3.2a)$$

which is also usefully expressed as

$$\bar\lambda=x_0(\bar\lambda)[a^2+b^2-ax_0(\bar\lambda)]/[x_0(\bar\lambda)-a]. \qquad (3.2b)$$

There are two solutions to (3.2) because the matrix (2.12) has two eigenvalues. To obtain the correct free energy we must choose that solution which has the larger eigenvalue. It may easily be seen from (2.12) or from (2.20) and (2.21) that the correct solution of (3.2) is

$$x_0(\bar\lambda)=(2a)^{-1}\{a^2+b^2-\bar\lambda+[(a^2+b^2-\bar\lambda)^2+4\bar\lambda a^2]^{1/2}\}, \qquad (3.3a)$$

with

$$x_0(\bar\lambda)^{-1}=(2a\bar\lambda)^{-1}$$
$$\times\{-(a^2+b^2-\bar\lambda)+[(a^2+b^2-\bar\lambda)^2+4\bar\lambda a^2]^{1/2}\}. \qquad (3.3b)$$

If we use (3.1), (3.3), and (2.21) in (2.11) we find that the free energy for Onsager's lattice is[1]

$$-\beta F_0=\ln(2\cosh\beta E_1\cosh\beta\bar E_2)$$
$$+(4\pi)^{-1}\int_{-\pi}^{\pi}d\theta\ln[\tfrac12|1+z_1e^{i\theta}|^2\{a^2+b^2+\bar z_2^2$$
$$+[(a^2+b^2-\bar z_2^2)^2+4a^2\bar z_2^2]^{1/2}\}]. \qquad (3.4)$$

From (3.3) we see that the eigenvector of the matrix (2.12) with the larger eigenvalue always lies in the range

$$ax_0(1)<ax<ax_0(0)=a^2+b^2. \qquad (3.5)$$

Consider any vector not in this range. Pick any matrix of the form (2.12) with $\bar\lambda$ not equal to 0 or 1 such that the eigenvector with the smaller eigenvalue does not lie in the direction of this vector. If we apply this matrix to the vector a sufficient number of times, the resultant vector will lie inside (3.5). But this is true not only if all the matrices correspond to $\bar\lambda$ but also if the matrices correspond to λ's lying in some neighborhood of $\bar\lambda$. Therefore, there is a nonzero probability that, after the application of (2.12) a finite number of times on a given vector, the resultant vector will lie in the range (3.5). Furthermore, it is clear from (2.17) and (3.2) that if x_n satisfied (3.5) then x_{n+1} does also. Therefore we conclude that no distribution of vectors that does not vanish outside the range $ax_0(1)<ax<(a^2+b^2)$ can be stationary.

In fact the same argument shows that if

$$\mu(\lambda)\equiv 0 \quad \text{unless} \quad \lambda_0'\leq\lambda\leq\lambda_0 \tag{3.6}$$

$$\nu(x)\equiv 0 \quad \text{unless} \quad ax_0(\lambda_0)<ax<ax_0(\lambda'_0). \tag{3.7}$$

To see what restriction (3.7) makes on the limits of integration in (2.19), it is convenient to plot in Fig. 1 for $a>0$, (1) the region in the $x-x'$ plane where the kernal of (2.19) is not zero; (2) the contours along which $\mu[x'(a^2+b^2-a)/(x-a)]$ is constant; and (3) the region where the kernal of (2.19) does not vanish and $\nu(x)$ and $\nu'(x')$ are different from zero. We then readily see that (2.19) may be more explicitly written (when $a>0$) as

$$\nu(x)=\frac{b^2}{(x-a)^2}\int_{\max[x_0(\lambda_0),\lambda_0'(x-a)/(a^2+b^2-ax)]}^{\min[\lambda_0(x-a)/(a^2+b^2-ax),x_0(\lambda_0')]} dx'x'\nu(x')\mu\left(x'\frac{a^2+b^2-ax}{x-a}\right) \tag{3.8}$$

for $x_0(\lambda_0)<x<x_0(\lambda_0')$ and $\nu(x)\equiv 0$ otherwise. If $a<0$ we clearly have

$$\nu(x,a)=\nu(-x,-a). \tag{3.9}$$

In the sequel we concentrate on the special case $\lambda_0'=0$ where (3.8) specializes to

$$\nu(x)=\frac{b^2}{(x-a)^2}\int_{x_0(\lambda_0)}^{\min\{\lambda_0[(x-a)/(a^2+b^2-ax)],(a^2+b^2)/a\}} dx'x'\nu(x')\mu\left(x'\frac{a^2+b^2-ax}{x-a}\right). \tag{3.8'}$$

An important difference between the Onsager lattice and all random lattices may be seen from Fig. 1. If $\mu(\lambda)=\delta(\lambda-\bar{\lambda})$ then, as we have seen earlier, $\nu(x)$ is given by a δ function at that value of x that lies within the interval (3.5) where the curve $x'=\bar{\lambda}(x-a)(a^2+b^2-ax)^{-1}$ intersects the line $x=x'$. This intersection is at the point $x_0(\bar{\lambda})$ given by (3.3). There are three distinct cases:

(a) if $b(0)^2>\bar{\lambda}(T>T_c)$ then as $\theta\rightarrow 0$

$$x_0(\bar{\lambda})\sim[b(0)^2-\bar{\lambda}]/a, \tag{3.10a}$$

(b) if $b(0)^2<\bar{\lambda}(T<T_c)$ then as $\theta\rightarrow 0$

$$x_0(\bar{\lambda})\sim\bar{\lambda}a/[\bar{\lambda}-b(0)^2], \tag{3.10b}$$

(c) and if $b(0)^2=\bar{\lambda}(T=T_c)$ then as $\theta\rightarrow 0$

$$x_0(\bar{\lambda})\rightarrow\pm\bar{\lambda}^{1/2}. \tag{3.10c}$$

We see from (3.10) that a δ-function distribution of λ leads to a δ-function stationary distribution. In the limit $\theta\rightarrow 0$, the δ-function stationary distribution remains at a finite value of x if $T=T_c$ but moves to zero (or infinity) if T is less than (or greater than) T_c.

Contrast the above case with the case of a very narrow $\mu(\lambda)$. As long as a is sufficiently far away from zero, a narrow $\mu(\lambda)$ must give use to a correspondingly narrow $\nu(x)$ because the projection on the x axis of the region of the $x=x'$ line where the kernel of (3.8) is appreciably different from zero is small. However, if θ gets close enough to zero and T is such that $\mu(\lambda)$ is different from zero in the region (however small) where $\lambda\sim b(0)^2$, then the above projection on the x axis becomes enormous. This dramatic broadening of $\nu(x)$ when $T\sim T_c$ and $\theta\sim 0$ will be exploited in the next section to obtain the dominant contribution to the specific heat for the particular narrow distribution

$$\mu(\lambda)=N\lambda_0^{-N}\lambda^{N-1} \tag{3.11}$$

for

$$0\leq\lambda\leq\lambda_0=\tanh^2\beta E_2^0$$

and

$$\mu(\lambda)=0 \qquad \text{otherwise.}$$

The above discussion, however, shows that there is nothing particular about this form of $\mu(\lambda)$, and we expect that the physical properties which (3.11) leads to will be qualitatively the same for a wide class of narrow distributions.

Before studying the power-law distribution (3.11) in detail, it is convenient to perform some manipulations on (3.8′) that are useful for any $\mu(\lambda)$ such that $\mu(\lambda)=0$ if $\lambda>\lambda_0$. We first note that it is possible to change variables so that the upper limit of (3.8′) is transformed from a curve and a straight line into two straight lines. To do this let

$$\eta=x_0(x-x_0)/(\lambda_0+x_0x), \tag{3.12}$$

where $x_0\equiv x_0(\lambda_0)$. We also introduce

$$B^2=\lambda_0(x_0-a)/x_0(\lambda_0+ax_0)$$
$$=\frac{a^2+b^2+\lambda_0-[(a^2+b^2-\lambda_0)^2+4a^2\lambda_0]^{1/2}}{a^2+b^2+\lambda_0+[(a^2+b^2-\lambda_0)^2+4a^2\lambda_0]^{1/2}} \tag{3.13}$$

so that

$$0\leq\eta\leq B^2\leq 1. \tag{3.14}$$

Define

$$\nu(x)=X(\eta)(d\eta/dx)$$
$$=X(\eta)[(1-\eta)^2x_0/(\lambda_0+x_0^2)], \tag{3.15}$$

so that

$$\int_0^{B^2} X(\eta)\,d\eta = 1. \tag{3.16}$$

Then (3.8′) may be rewritten as

$$X(\eta) = \frac{\lambda_0 B^2(\lambda_0 + x_0^2)}{(B^2 x_0^2 + \lambda_0\eta)^2} \int_0^{\min[B^2,\eta B^{-2}]} d\eta' X(\eta')$$

$$\times \frac{\eta'\lambda_0 + x_0^2}{1-\eta'}\,\mu\left[\frac{\eta'\lambda_0 + x_0^2}{1-\eta'}\,\frac{B^2-\eta}{\lambda_0^{-1}B^2x_0^2+\eta}\right]. \tag{3.17}$$

This equation may be cast into a somewhat simpler form when $\eta \leq B^4$ if we define

$$Y(\eta) = \int_0^{\eta} X(\eta')\,d\eta' \tag{3.18}$$

and integrate (3.17) over η to obtain

$$Y(\eta) = \lambda_0 B^2(\lambda_0 + x_0^2) \int_0^{\eta} d\eta_1 \int_0^{\eta_1 B^{-2}} d\eta' X(\eta')$$

$$\times \frac{\eta'\lambda_0 + x_0^2}{(1-\eta')(B^2x_0^2+\lambda_0\eta_1)^2}\,\mu\left[\frac{\eta'\lambda_0 + x_0^2}{1-\eta'}\,\frac{B^2-\eta_1}{\lambda_0^{-1}B^2x_0^2+\eta_1}\right]. \tag{3.19}$$

Interchange the order of integration and make the change of variable from η_1 to

$$\zeta = \frac{\eta'\lambda_0 + x_0^2}{1-\eta'}\,\frac{B^2-\eta_1}{\lambda_0^{-1}B^2x_0^2+\eta_1} \tag{3.20}$$

to obtain

$$Y(\eta) = \int_0^{B^{-2}\eta} d\eta' X(\eta')$$

$$\times \int_{[(\lambda_0\eta'+x_0^2)/(1-\eta')][(B^2-\eta)/(\lambda_0^{-1}B^2x_0^2+\eta)]}^{\lambda 0} d\zeta\,\mu(\zeta)$$

$$= Y(\eta B^{-2}) - \int_0^{\eta B^{-2}} d\eta' X(\eta')$$

$$\times \int_0^{[(\lambda_0\eta'+x_0^2)/(1-\eta')][(B^2-\eta)/(\lambda_0^{-1}B^2x_0^2+\eta)]} d\zeta\,\mu(\zeta). \tag{3.21}$$

Using the definition of Y, we may rewrite this as

$$\int_\eta^{\eta B^{-2}} d\eta' X(\eta') = \int_0^{\eta B^{-2}} d\eta' X(\eta')$$

$$\times \int_0^{[(\lambda_0\eta'+x_0^2)/(1-\eta')][(B^2-\eta)/(\lambda_0^{-1}B^2x_0^2+\eta)]} d\zeta\,\mu(\zeta). \tag{3.22}$$

It remains only to write out the special case (3.11).

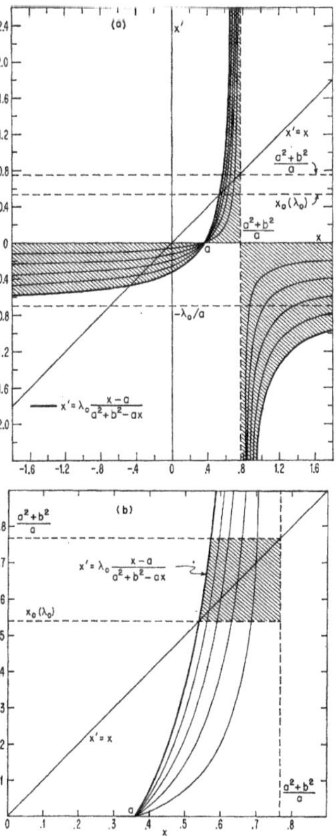

FIG. 1. (a) Contours along which the function

$$\mu(x'(a^2+b^2-ax)/(x-a))$$

is constant. The kernal of the integral equation for $\nu(x)$ is different from zero only in the shaded region. (b) Enlargement of 1(a). The shaded region is the only region in which the kernal of the integral equation does not vanish and $\nu(x)$ and $\nu(x')$ are different from zero.

We work the ζ integral in (3.22) and obtain

$$\int_\eta^{\eta B^{-2}} d\eta' X(\eta') = \int_0^{\eta B^{-2}} d\eta' X(\eta') N^{-1}$$

$$\times \left(\frac{\lambda_0\eta' + x_0^2}{1-\eta'}\,\frac{B^2-\eta}{\lambda_0^{-1}B^2x_0^2+\eta}\right)\mu\left(\frac{\lambda_0\eta' + x_0^2}{1-\eta'}\,\frac{B^2-\eta}{\lambda_0^{-1}B^2x_0^2+\eta}\right). \tag{3.23}$$

The right-hand side of this equation may be further simplified using the original integral equation (3.17). We obtain

$$\int_\eta^{\eta B^{-2}} d\eta' X(\eta') = \frac{(B^2-\eta)(B^2x_0^2+\eta\lambda_0)}{NB^2(\lambda_0+x_0^2)} X(\eta). \quad (3.24)$$

Finally, it is useful to make the exponential change of variable

$$\eta = e^{-\tau}, \quad (3.25)$$

where $-\ln B^2 < \tau < \infty$. Call

$$U(\tau) = \eta X(\eta), \quad (3.26)$$

so that

$$\int_{-\ln B^2}^{\infty} d\tau\, U(\tau) = 1. \quad (3.27)$$

Then we obtain

$$\int_\tau^{\tau+\ln B^2} d\tau' U(\tau') = -\frac{(B^2-e^{-\tau})(B^2x_0^2+\lambda_0 e^{-\tau})}{NB^2(\lambda_0+x_0^2)} e^\tau U(\tau). \quad (3.28)$$

4. POWER-LAW DISTRIBUTION

In this section we determine the dominant contribution to the specific heat of the lattice characterized by the power-law distribution (3.11). We consider only the case of large N and ignore all contributions to the specific heat which vanish as $N\to\infty$. Then the discussion of Sec. 3 shows that when a is away from zero $\nu(x)$ is given by a narrow distribution which will differ but little from the $\nu(x)$ of an Onsager problem. We are not interested in these small deviations and concentrate our attention on the opposite extreme when a is close to zero. From (2.29a) we find that $T=T_c$ if

$$\ln[b(0)^2\lambda_0^{-1}] = \ln B^2(0) = -N^{-1}. \quad (4.1)$$

When B^2 is close to this value and $a\sim 0$, the discussion of Sec. 3 shows that $\nu(x)$ is very broad. Therefore when x is of order 1, $\nu(x)$ [and hence $U(\tau)$] may be treated as slowly varying. Furthermore, (4.1) shows that the region of integration of (3.28) is of the order of N^{-1}. Hence to leading order in N^{-1} we may expand

$$U(\tau') \sim U(\tau) + (\tau'-\tau)U'(\tau) \quad (4.2)$$

and do the integral in (3.28) to obtain the approximate differential equation

$$\tfrac{1}{2}(\ln B^2)^2 (d/d\tau) U(\tau) + U(\tau)\ln B^2$$
$$= [-(B^2-e^{-\tau})(B^2x_0^2e^\tau+\lambda_0)/NB^2(\lambda_0+x_0^2)]U(\tau). \quad (4.3)$$

This is readily solved and we find

$$U(\tau) = C_N(B^2, x_0)\exp\{s\tau - te^{-\tau} - ue^\tau\} \quad (4.4)$$

where, by (3.3),

$$s = 2(\ln B^2)^{-2}[-\ln B^2 - N^{-1}(\lambda_0x_0^{-1}-x_0)/(\lambda_0x_0^{-1}+x_0)],$$
$$= 2(\ln B^2)^{-2}\{-\ln B^2 + N^{-1}(a^2+b^2-\lambda_0)$$
$$\times[(a^2+b^2-\lambda_0)^2+4\lambda_0a^2]^{-1/2}\}, \quad (4.5a)$$

$$t = 2(\ln B^2)^{-2}N^{-1}B^{-2}\lambda_0x_0^{-1}(\lambda_0x_0^{-1}+x_0)^{-1}$$
$$= (\ln B^2)^{-2}N^{-1}B^{-2}\{1-(a^2+b^2-\lambda_0)$$
$$\times[(a^2+b^2-\lambda_0)^2+4\lambda_0a^2]^{-1/2}\}, \quad (4.5b)$$

$$u = 2(\ln B^2)^{-2}N^{-1}B^2x_0(\lambda_0x_0^{-1}+x_0)^{-1}$$
$$= (\ln B^2)^{-2}N^{-1}B^2\{1+(a^2+b^2-\lambda_0)$$
$$\times[(a^2+b^2-\lambda_0)^2+4\lambda_0a^2]^{-1/2}\}, \quad (4.5c)$$

and $C_N(B^2, x_0)$ is an appropriate normalization constant. Therefore, when $T\sim T_c$ and $\theta\sim 0$, we obtain

$$X(\eta) \sim C_N(B^2, x_0^2)\eta^{-s-1}\exp(-t\eta - u/\eta). \quad (4.6)$$

This approximation to $X(\eta)$ forms the basis of all further considerations of this section.

To give insight into the structure of $X(\eta)$ we first remark that as $N\to\infty$ the power-law distribution (3.11) approaches $\delta(\lambda-\lambda_0)$. Therefore when $T\neq T_c$ and $\theta\neq 0$, the exact $\nu(x)$ must approach $\delta(x-x_0)$ as $N\to\infty$ [and hence $X(\eta)$ must approach $\delta(\eta)$]. Our approximation to $X(\eta)$ was derived under the assumption that $T-T_c$ and θ were small and hence it is not obvious that (4.6) will approach $\delta(\eta)$ as $N\to\infty$. To see that this is in fact the case we note that the maximum of $\eta X(\eta)$ occurs at

$$\eta_m = (2t)^{-1}[-s+(s^2+4tu)^{1/2}] \quad (4.7a)$$

and its width (in the sense of a steepest descent integral) is

$$w_\eta = \{(\partial^2/\partial_\eta^2)[s\ln\eta + t\eta + u\eta^{-1}]\}^{1/2} \mid \eta=\eta_m,$$
$$= \eta_m^{3/2}(2u - s\eta_m)^{-1/2}. \quad (4.7b)$$

If $T\neq T_c$ and $\theta\neq 0$ are fixed and $N\to\infty$, $s=-2(\ln B^2)^{-1}+O(N^{-1})$, $t=O(N^{-1})$, and $u=O(N^{-1})$ so that $\eta_m=O(N^{-1})$ and $w_\eta=O(N^{-1})$. Therefore $X(\eta)\to\delta(\eta-\eta_m)$ in the sense that in the evaluation of the normalization integral of $X(\eta)$ we may replace $X(\eta)$ by $\delta(\eta-\eta_m)$.

We are interested in the contributions of order 1 to the specific heat when N is large. This is found by using our approximation for $X(\eta)$ in the integral (2.20) which in terms of η is

$$\lim_{\mathfrak{M}\to\infty}\mathfrak{M}^{-1}\ln C_{\mathfrak{M}}(\theta) = \int_0^{B^2} d\eta\, X(\eta)\int_0^1 d\lambda\,\mu(\lambda)\ln[a^2+b^2+a\lambda(1-\eta)/(\eta\lambda_0x_0^{-1}+x_0)]. \quad (4.8)$$

Define $f(\eta)$ by

$$f(\eta) = \ln[a^2+b^2+a\lambda(1-\eta)/(\eta\lambda_0x_0^{-1}+x_0)]$$
$$= \ln\{a^2+b^2+2a^2\lambda[(a^2+b^2-\lambda_0)+(1+\eta)(1-\eta)^{-1}((a^2+b^2-\lambda_0)^2+4\lambda a^2)^{1/2}]^{-1}\}. \quad (4.9)$$

When $T \neq T_c$ and $\theta \neq 0$ are fixed and $N \to \infty$ we see that if $\eta = O(N^{-1})$

$$f(\eta) = \ln[a^2+b^2+a\lambda x_0^{-1}] + O(N^{-1}).$$

Therefore our approximation $X(\eta)$ may be replaced by $\delta(\eta-\eta_m)$ for the purpose of obtaining the leading order term of (4.8). This is the behavior expected of the exact $X(\eta)$ in this limit so we conclude that approximation (4.6) may be used for *all* T and θ for the purpose of obtaining the leading order term in the specific heat when $N \to \infty$.

In fact, for our limited purpose $X(\eta)$ is well approximated by $\delta(\eta-\eta_m)$ not only when $T \neq T_c$ and $\theta \neq 0$ are fixed and $N \to \infty$ but also when a is of the order of $N^{-\epsilon}$ and $b(0)^2-\lambda_0$ is of the order of $N^{-\gamma}$, where $0<\epsilon<1$ and $0<\gamma<1$. There are three cases to be considered. (a) $\gamma>\epsilon$, (b) $\gamma<\epsilon$ and $b(0)^2-\lambda_0<0$, and (c) $\gamma<\epsilon$ and $b(0)^2-\lambda_0>0$. We find

case (a),

$$\eta_m = O(N^{-1+\epsilon}), \qquad w_\eta = O(N^{-1+\epsilon/2}),$$
$$f(\eta) = \ln b^2 + O(N^{-\epsilon}),$$

case (b),

$$\eta_m = O(N^{-1-2\epsilon+3\gamma}), \qquad w_\eta = O(N^{-1-2\epsilon+5\gamma/2}),$$
$$f(\eta) = \ln b^2 + O(N^{-\gamma}), \text{ and}$$

case (c),

$$\eta_m = O(N^{-1+\gamma}), \qquad w_\eta = O(N^{-1+\gamma/2}),$$
$$f(\eta) = \ln b^2 + O(N^{-2\epsilon+\gamma}).$$

Since the variations in $f(\eta)$ caused by deviations in η on the order of w_η vanish as $N \to \infty$, we conclude that in these cases, for our purpose, $X(\eta)$ may be replaced by $\delta(\eta-\eta_m)$.

Because of the fact that even if $T \sim T_c$ the contribution of $\nu(x)-\delta(x-x_m)$ to the specific heat is very small for $|a| \gg N^{-1}$, it is useful to consider the contributions of $\delta(x-x_m)$ and $\nu(x)-\delta(x-x_m)$ separately. It is easily seen from (4.7a) and (3.12) that x_m obeys

$$x_m^2 a[1+b^{-2}G] - x_m[a^2+b^2-\lambda_0+G(1+2a^2b^{-2})] - \lambda_0 a[1-G(a^2+b^2)\lambda_0^{-1}b^{-2}] = 0, \quad (4.10\text{a})$$

where

$$G = -N^{-1}(\ln B^2)^{-1}[(a^2+b^2-\lambda_0)^2+4a^2\lambda_0]^{1/2}. \quad (4.10\text{b})$$

If we use $\delta(x-x_m)$ for $\nu(x)$ in (2.20) we find, correct to order N^{-1},

$$-\beta\bar{F}_r = \int_0^\infty dE_2 p(E_2) \ln(2\cosh\beta E_1 \cosh\beta E_2) + (4\pi)^{-1}\int_{-\pi}^{\pi} d\theta \int_{-\infty}^{\infty} dx\, \delta(x-x_m) \int_0^1 d\lambda\, \mu(\lambda)$$
$$\times \ln\{|1+z_1e^{i\theta}|^2[a^2+b^2+a\lambda x^{-1}]\} \sim \ln(2\cosh\beta E_1 \cosh\beta E_2^0) + (4\pi)^{-1}\int_{-\pi}^{\pi} d\theta$$
$$\times \ln\{|1+z_1e^{i\theta}|^2 \tfrac{1}{2}[1-G(a^2+b^2)\lambda_0^{-1}b^{-2}]^{-1}(a^2+b^2+\lambda_0-Gb^{-2}\lambda_0^{-1}[2(a^2+b^2)^2+\lambda_0(b^2+2a^2)]$$
$$+\{[a^2+b^2-\lambda_0+G(1+2a^2b^{-2})]^2+4a^2\lambda_0(1+b^{-2}G)[1-G(a^2+b^2)\lambda_0^{-1}b^{-2}]\}^{1/2})\}. \quad (4.11)$$

By comparing $\bar{F}_r$ with the Onsager free energy F_0 of (3.4), we conclude that to order N^{-1} $\bar{F}_r$ is the same as F_0 except that the critical temperature has been shifted from $b(0)^2-\lambda_0=0$ to

$$b(0)^2-\lambda_0 = -G(0) = -[b(0)^2-\lambda_0]N^{-1}[\ln b(0)^2\lambda_0^{-1}]^{-1}, \quad (4.12)$$

which is exactly the same equation for T_c as (4.1). Therefore over the range of T where $\nu(x)$ is well approximated by $\delta(x-x_m)$, F_r is dominated by a term which has an apparent logarithmic divergence at T_c.

To study the behavior very near T_c we write

$$F_r-\bar{F}_r \sim -(\beta 4\pi)^{-1}\int_{-\pi}^{\pi} \theta \int_{-\infty}^{\infty} dx\,[\nu(x)-\delta(x-x_m)]\ln[a^2+b^2+a\lambda_0 x^{-1}], \quad (4.13)$$

where the λ integral has been evaluated to $O(N^{-1})$. We consider only T close enough to T_c so that $b^2-\lambda_0<0$. The $\sim$ sign is to mean that both sides lead to the same specific heat to order 1 as $N \to \infty$. Furthermore we have seen that unless $a^2+b^2-\lambda_0=O(N^{-1})$ we cannot have a contribution of order 1 to the specific heat obtained from (4.13). Therefore, $ax_0^{-1}=O(N^{-1})$ and, since $|x_0|<|x|$, we may expand the logarithm of (4.13) to obtain

$$F_r-\bar{F}_r \sim -(\beta 4\pi)^{-1}\int_{-\pi}^{\pi} d\theta \int_{-\infty}^{\infty} dx[\nu(x)-\delta(x-x_m)]a\lambda_0(a^2+b^2)^{-1}x^{-1}$$
$$= (-\beta\pi)^{-1}\int_{-\pi}^{0} d\theta\, a^2\lambda_0(a^2+b^2)^{-1}\int_0^{B^2} d\eta\,[X(\eta)-\delta(\eta-\eta_m)]$$
$$\times\{a^2+b^2-\lambda_0+(1+\eta)(1-\eta)^{-1}[(a^2+b^2-\lambda_0)^2+4a^2\lambda_0]^{1/2}\}^{-1}. \quad (4.14)$$

To evaluate this approximately to order 1, we need to find the correct range of θ, η, and $T-T_c$. We have previously considered the case $a\gg N^{-1}$ and $|b^2-\lambda_0|\gg N^{-1}$ and seen via a steepest-descent integration that the contribution is negligible. In a similar fashion we may show that the cases (1) a is of order N^{-1}, (2) $a=o(N^{-1})$ and $s=O(N)$ give only contributions of an order N^{-1} to the specific heat. The only region that gives a contribution of order one to the specific heat as $N\to\infty$ is $a=O(N^{-2})$ and $T-T_c=O(N^{-2})$.

When $T-T_c=O(N^{-2})$ and $a=O(N^{-2})$, $X(\eta)$ and $\delta(\eta-\eta_m)$ do not cancel at all so that F_r and $\bar{F}_r$ must be evaluated separately. Because $\bar{C}_v{}^r=-T\bar{F}_r{}''(T)$ is to leading order in N the same as Onsager's specific heat, we may use Onsager's result[1] to find

$$\bar{C}_v{}^r\sim 4k\beta_c{}^2\pi^{-1}\{z_{1c}z_{2c}{}^0(1-z_{1c})^{-2}(1-z_{2c}{}^0)^{-2}[E_1(1-z_{1c})+E_2{}^0(1-z_{2c}{}^0)]^2$$
$$\times[\ln T_c/|T-T_c|-\ln\tfrac{1}{8}\beta_c(E_1(z_{1c}+z_{1c}{}^{-1})+E_2{}^0(z_{2c}{}^0+z_{2c}{}^{0-1}))]$$
$$-[E_1{}^2 4z_{1c}{}^2z_{2c}{}^{02}(1-z_{2c}{}^0)^{-4}\mathrm{gd}\,2\beta_cE_1+2E_1E_2{}^0+E_2{}^{02}4z_{1c}{}^2z_{2c}{}^{02}(1-z_{1c})^{-4}\mathrm{gd}\,2\beta_cE_2{}^0]\},\quad(4.15)$$

where gd stands for the Gudermannian [$\mathrm{gd}x=\tan^{-1}\sinh x$], the subscript c means $T=T_c$, and $z_{2c}{}^0=\tanh\beta_cE_2{}^0$.

We evaluate $C_v{}^r$ by splitting the θ integration into two regions, one for $0<|\theta|\le O(N^{-2})$ and the other $|\theta|\gg N^{-2}$. In the second region the integrand of (4.14) is small. Therefore, to leading order the parts of $C_v{}^r$ and $\bar{C}_v{}^r$ coming from angular integration over this second region are equal. It is easily seen that the large angles give a contribution to $C_v{}^r$ that is a constant. The point of separation between the two regions is not well defined but, as will shortly be verified, the precise choice of this cutoff does not affect the temperature-dependent part of $C_v{}^r$ but only the constant. Therefore, we may evaluate the leading contribution to $C_v{}^r$ as $N\to\infty$ when $T-T_c=O(N^{-2})$ up to a constant by integrating θ only up to $O(N^{-2})$. We then may determine the constant by the requirement that, when $|T-T_c|\gg N^{-2}$, $C_v{}^r\sim\bar{C}_v{}^r$ where $\bar{C}_v{}^r$ is given by (4.15).

To explicitly carry out this evaluation, we define

$$\phi=-8\lambda_0{}^{-1/2}z_1(1+z_1)^{-2}N^2\theta,\quad(4.16)$$

so that

$$a=\tfrac{1}{4}N^{-2}\lambda_0{}^{1/2}\phi+O(N^{-4}).\quad(4.17)$$

We further define δ by

$$\lambda_0{}^{-1}(1-z_1)^2(1+z_1)^{-2}-e^{-N^{-1}}=\tfrac{1}{2}N^{-2}\delta.\quad(4.18)$$

As $N\to\infty$ δ is to be of order 1. Explicitly,

$$\delta=(T-T_c)N^2 4k\beta_c{}^2(1+z_{2c}{}^0)z_{2c}{}^{0-1}$$
$$\times\{[E_1(1-z_{1c})+E_2{}^0(1-z_{2c}{}^0)]+O(N^{-1})\}.\quad(4.19)$$

We then have

$$C_v{}^r=-T(\partial^2F_r/\partial T^2)\sim-k\beta_c{}^3 16(1+z_{2c}{}^0)^2z_{2c}{}^{0-2}$$
$$\times\{E_1(1-z_{1c})+E_2{}^0(1-z_{2c}{}^0)\}^2N^4(\partial^2F_r/\partial\delta^2).\quad(4.20)$$

Furthermore,

$$a^2+b^2-\lambda_0=-\lambda_0N^{-1}[1-(2N)^{-1}(\delta+1)]+O(N^{-3})\quad(4.21)$$

and

$$\ln B^2=-N^{-1}[1-(2N)^{-1}\delta]+O(N^{-3}).\quad(4.22)$$

We then find

$$s=-\delta+O(N^{-1}),\quad(4.23a)$$
$$t=2N+O(1),\quad(4.23b)$$
$$u=(8N)^{-1}\phi^2+O(N^{-2}).\quad(4.23c)$$

Therefore,

$$C_v{}^r\sim-k\beta_c{}^2(8\pi)^{-1}(1+z_{1c})^2(1+z_{2c}{}^0)^2z_{2c}{}^{0-1}z_{1c}{}^{-1}\{E_1(1-z_{1c})+E_2{}^0(1-z_{2c}{}^0)\}^2N^{-1}$$
$$\times\frac{\partial^2}{\partial\delta^2}\int_0^{N^2}d\phi\,\phi^2\int_0^1 d\eta\,\eta^{\delta-1}\exp(-2N\eta-\phi^2/8N\eta)C_N(\phi,\delta)\left(1-\frac{1+\eta}{1-\eta}\right)^{-1}+\bar{K},\quad(4.24)$$

where the coefficient of N^2 is arbitrary and $\bar{K}$ is determined by the requirement that as $\delta\to\infty$, $C_v{}^r\to\bar{C}_v{}^r$.

The normalization constant $C_N(\phi,\delta)$ is determined from the requirement

$$1=C_N\int_0^1 d\eta\,\eta^{\delta-1}\exp(-2N\eta-\phi^2/8N\eta).\quad(4.25)$$

Letting $\xi=4N\phi^{-1}\eta$, we have

$$C_N{}^{-1}=(4N)^{-1}\phi\int_0^{4N\phi^{-1}}d\xi\,(\xi\phi/4N)^{\delta-1}$$
$$\times\exp[-\tfrac{1}{2}\phi(\xi+\xi^{-1})].\quad(4.26)$$

The upper limit may be extended to ∞ and we find

$$C_N{}^{-1}\sim 2(\phi/4N)^{\delta}K_\delta(\phi),\quad(4.27)$$

where $K_\delta(\phi)$ is the modified Bessel function of the third kind[8] of order δ. Similarly, we find

$$\int_0^1 d\eta\,\eta^{\delta-1}\exp(-2N\eta-\phi^2/8N\eta)[1-(1+\eta)(1-\eta)^{-1}]^{-1}$$
$$\sim-(\phi/4N)^{\delta-1}K_{\delta-1}(\phi).\quad(4.28)$$

If we also note that $(1-z_{1c}{}^2)(1-z_{2c}{}^{02})=4|z_{1c}||z_{2c}{}^0|+$

$O(N^{-1})$, we have

$$C_v^r \sim 4k\beta_c^2\pi^{-1}(1-z_{1c})^{-2}(1-z_{2c}^0)^{-2}z_{1c}z_{2c}^0 \times\{E_1(1-z_{1c})+E_2^0(1-z_{2c}^0)\}^2 \times\frac{\partial^2}{\partial\delta^2}\int_0^{N^2} d\phi\,\frac{\phi K_{\delta-1}(\phi)}{K_\delta(\phi)}+\bar{K}. \tag{4.29}$$

As $\phi\to\infty$,

$$K_{\delta-1}(\phi)/K_\delta(\phi)\to 1-\phi^{-1}(\delta-\tfrac{1}{2})+\tfrac{1}{2}\phi^{-2}(\delta^2-\tfrac{1}{4})+O(\phi^{-3}), \tag{4.30}$$

so there is a term in (4.29) that behaves as $\ln N^2$ (where we remember that the coefficient of N^2 is arbitrary). We explicitly extract this term to write

$$\frac{\partial^2}{\partial\delta^2}\int_0^{N^2} d\phi\,\frac{\phi K_{\delta-1}(\phi)}{K_\delta(\phi)}=\int_0^\infty d\phi\left\{\phi\,\frac{\partial^2}{\partial\delta^2}\,\frac{K_{\delta-1}(\phi)}{K_\delta(\phi)}-(\phi+1)^{-1}\right\}+\ln N^2+O(N^{-1}), \tag{4.31}$$

which, using the recurrence relation

$$K_{\delta-1}(\phi)=-\delta\phi^{-1}K_\delta(\phi)-(d/d\phi)K_\delta(\phi), \tag{4.32}$$

may be reexpressed as

$$\int_0^\infty d\phi\left[\frac{\partial^2}{\partial\delta^2}\ln K_\delta(\phi)-(\phi+1)^{-1}\right]+\ln N^2-1+O(N^{-1})=R(\delta)+\ln N^2-1+O(N^{-1}), \tag{4.33}$$

which defines the function $R(\delta)$.

For convenience we absorb the -1 term of (4.33) into $\bar{K}$ of (4.29) by defining

$$K=\bar{K}-4k\beta_c^2\pi^{-1}(1-Z_{1c})^{-2}(1-Z_{2c}^0)^{-2}Z_{1c}Z_{2c}^0\{E_1(1-Z_{1c})+E_2^0(1-Z_{2c}^0)\}^2. \tag{4.34}$$

To determine K we must expand (4.33) as $\delta\to\infty$. We use the expansion[9]

$$K_\delta\sim(\tfrac{1}{2}\pi)^{1/2}(\delta^2+\phi^2)^{-1/4}\exp[-(\delta^2+\phi^2)^{1/2}+\delta\sinh^{-1}(\delta/\phi)]\{1+[-\tfrac{1}{8}+(5/24)(1+\phi^2/\delta^2)^{-1}](\delta^2+\phi^2)^{-1/2}\}, \tag{4.35}$$

to find that as $\delta\to\infty$

$$\int_0^\infty d\phi\left[\frac{\partial^2}{\partial\delta^2}\ln K_\delta(\phi)-(\phi+1)^{-1}\right]+\ln N^2=\ln 2+\ln N^2\delta^{-1}-\tfrac{1}{8}\delta^{-2}+O(\delta^{-3}). \tag{4.36}$$

Therefore we find as $\delta\to\infty$

$$C_v^r(\delta)\to -4k\beta_c^2\pi^{-1}(1-z_{1c})^{-2}(1-z_{2c}^0)^{-2}z_{1c}z_{2c}^0\{E_1(1-z_{1c})+E_2^0(1-z_{2c}^0)\}^2 \times\ln\{2k\beta_c^2\,|\,T-T_c\,|\,(z_{2c}^{0-1}+1)[E_1(1-z_{1c})+E_2^0(1-z_{2c}^0)]\}+K. \tag{4.37}$$

Comparing this with (4.15) gives

$$\begin{aligned}K=&-4k\beta_c^2\pi^{-1}[(1-z_{1c})^{-2}(1-z_{2c}^0)^{-2}z_{1c}z_{2c}^0[E_1(1-z_{1c})+E_2^0(1-z_{2c}^0)]^2\\&\times\{-\ln 2(z_{2c}^{0-1}+1)[E_1(1-z_{1c})+E_2^0(1-z_{2c}^0)]+\ln\tfrac{1}{8}[E_1(z_{1c}+z_{1c}^{-1})+E_2^0(z_{2c}^0+z_{2c}^{0-1})]\}\\&+E_1^2 4z_{1c}^2z_{2c}^{02}(1-z_{2c}^0)^{-4}\,\mathrm{gd}\,2\beta_cE_1+2E_1E_2^0+E_2^{02}4z_{1c}^2z_{2c}^{02}(1-z_{1c})^{-4}\,\mathrm{gd}\,2\beta_cE_2^0]\\=&-4k\beta_c^2\pi^{-1}\{\tfrac{1}{2}[E_1^2\sinh 2\beta_cE_2^0+2E_1E_2^0+E_2^{02}\sinh 2\beta_cE_1]\\&\times[\ln\tfrac{1}{4}(E_1\coth 2\beta_cE_1+E_2^0\coth 2\beta_cE_2^0)-\ln 2\{(1+\coth\beta_cE_2^0)[E_1(1-\tanh\beta_cE_1)+E_2^0(1-\tanh\beta_cE_2^0)]\}]\\&+E_1^2\sinh^2 2\beta_cE_2^0\,\mathrm{gd}\,2\beta_cE_1+2E_1E_2^0+E_2^{02}\sinh^2 2\beta_cE_1\,\mathrm{gd}\,2\beta_cE_2^0\}.\end{aligned} \tag{4.38}$$

We therefore have as our final result that when $T-T_c=O(N^{-2})$

$$C_v^r\sim 2\pi^{-1}k\beta_c^2\{E_1^2\sinh 2\beta_cE_2^0+2E_1E_2^0+E_2^{02}\sinh 2\beta_cE_1\}\left\{\int_0^\infty d\phi\left[\frac{\partial^2}{\partial\delta^2}\ln K_\delta(\phi)-(\phi+1)^{-1}\right]+\ln N^2\right\}+K, \tag{4.39}$$

where δ is defined by (4.19) and K by (4.38). We note in particular that C_v^r is an even function of δ.

The specific heat (4.39) has been calculated for the random lattice characterized by $\mu(\lambda)$ given by (3.11).

---

[8] *Higher Transcendental Functions*, edited by A. Erdélyi (McGraw-Hill Book Co., New York, 1953), Vol. 2, p. 82.

[9] Reference 8, Vol. 2, p. 26.

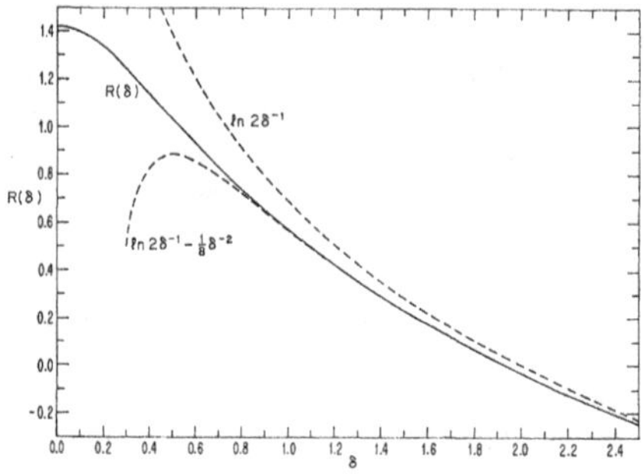

FIG. 2. The integral $R(\delta)$ as a function of δ.

The corresponding probability density $P(E_2)$ is

$$P(E_2)=2N(\tanh E_2{}^0\beta)^{-2N}\beta(\mathrm{sech}^2\beta E_2)(\tanh\beta E_2)^{2N-1}$$
$$\text{for}\quad 0\leq E_2\leq E_2{}^0. \quad (4.40)$$

This probability clearly depends on temperature. The probabilities envisioned in the original statement of the model were to be temperature-independent. Fortunately, because the only significant deviations of $C_v{}^r$ from $C_v{}^r$ occurs when $T-T_c=O(N^{-2})$, we see that, in the temperature range where (4.39) is valid,

$$P(E_2)=\beta_c(\mathrm{sech}^2\beta_c E_2)\,2N(\tanh E_2{}^0\beta_c)^{-2N}$$
$$\times(\tanh\beta_c E_2)^{2N-1}[1+O(N^{-1})]. \quad (4.41)$$

Therefore $P(E_2)$ as given by (4.41) differs, when $T-T_c=O(N^{-2})$, from a temperature-independent probability by a term of $O(N^{-1})$, which is negligible.

To justify calling the temperature given by the solution of (4.1) the critical temperature we must show that $C_v{}^r$ as given by (4.39) is not analytic at $\delta=0$. For this purpose we integrate the integral in (4.39) by parts once and consider

$$\int_0^\infty d\phi\left[\frac{\partial^2}{\partial\delta^2}\frac{\phi K_\delta'(\phi)}{K_\delta(\phi)}+(\phi+1)^{-1}\right]. \quad (4.42)$$

When $\delta\to 0$ the singular behavior of this function comes from the region near $\phi=0$. Thus we may obtain the most singular behavior of (4.42) if we integrate only from 0 to some small positive upper limit ϵ and expand for ϕ and δ near zero

$$K_\delta(\phi)\sim(2\delta)^{-1}[(\phi/2)^{-\delta}-(\phi/2)^{\delta}]. \quad (4.43)$$

Therefore we study

$$\int_0^\epsilon d\phi\,\delta(\phi^{-\delta}+\phi^{\delta})/(\phi^{-\delta}-\phi^{\delta})=\int_0^\epsilon d\phi\,2\delta(1-\phi^{2\delta})^{-1}-\delta\epsilon. \quad (4.44)$$

Now if $\epsilon<1$,

$$\int_0^\epsilon d\phi/\ln\phi$$

exists, so the analytic properties of (4.43) are the same as

$$I(\delta)=\int_0^1 d\phi\,[\delta(1-\varphi^{\delta})^{-1}+(\ln\phi)^{-1}]. \quad (4.45)$$

We have been able to replace ϵ by 1 in (4.45) because the integrand is regular at $\phi=1$. It is easily seen that $I(\delta)-I(-\delta)=\delta$ so we need consider only $\delta>0$. Then if we let

$$\phi=e^{-\xi/\delta}, \quad (4.46)$$

we find[10]

$$I(\delta)=\int_0^\infty d\xi\,e^{-\xi/\delta}[(1-e^{-\xi})^{-1}-\xi^{-1}]$$
$$-\ln\delta-\psi(\delta^{-1}), \quad (4.47)$$

where $\psi(z)=\Gamma'(z)/\Gamma(z)$. For small δ we may use the asymptotic expansion[11] for $\psi(\delta^{-1})$ to obtain the formal power series valid for both positive and negative δ,

$$I(\delta)=\tfrac{1}{2}\delta+\sum_{n=1}^{\infty}B_{2n}\delta^{2n}(2n)^{-1}. \quad (4.48)$$

Here B_{2n} are the Bernoulli numbers. Therefore, the most singular part of $C_v{}^r$ is an *infinitely differentiable* function of T even at $T=T_c$. However, because (4.48) diverges for all $\delta\neq0$, $C_v{}^r$ is *not* an analytic function of T at $T=T_c$. It is therefore correct to call T_c as determined from (4.1) the critical temperature.

Finally, it is useful to supplement these analytic considerations with a numerical evaluation of $C_v{}^r$ when $T-T_c=O(N^{-2})$. We have done this on an IBM 7044 computer and present the results in Fig. 2 and 3.

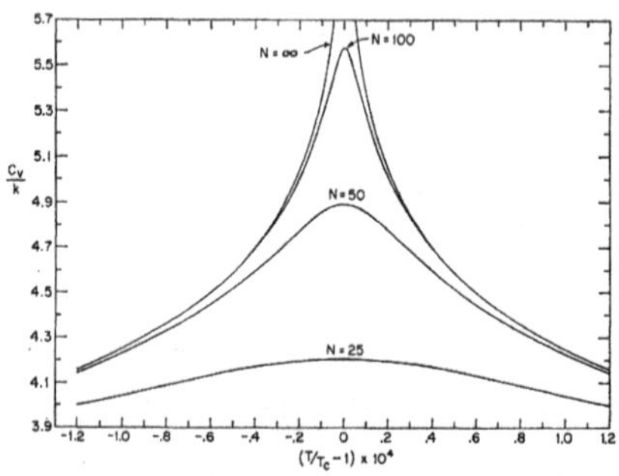

FIG. 3. Comparison of C_v of the pure Onsager lattice and $C_v{}^r$ for several values of N for the case $E_1=E_2{}^0$.

[10] Reference 8, Vol. 1, p. 18.
[11] Reference 8, Vol. 1, p. 47.

5. DISCUSSION

The principal result of this study is to exhibit explicitly a model with random impurities where, to order 1, the specific heat is an infinitely differentiable function of the temperature even at T_c. This behavior is qualitatively in beautiful agreement with the experimental findings previously quoted. We therefore feel justified in proposing[12] that random impurities can be the origin of the smooth specific heats observed experimentally.

On the other hand, this particular model of random impurities is obviously artificial. It is not realistic to assume that all vertical bonds $E_2(j)$ in a given row will be equal if we are allowing $E_2(j)$ to vary from row to row. A realistic model of random impurities should not have such a large amount of correlation between the random bonds. For this reason, we are extremely fortunate that the specific heat found in Sec. 4 is infinitely differentiable. If random impurities had increased the observed order of the phase transition to a higher but finite order, the question could very reasonably be asked if relaxation of this very stringent correlation requirement would further increase the observed order. However, even the limited amount of randomness allowed in this model has made the dominant contribution to all derivatives of the specific heat continuous, so no further increase of the observed order is possible. This very smooth specific heat is in distinct contrast to the result of Syozi's model.[13]

We are unable to ascertain which of our results are qualitatively dependent on the very special sort of randomness allowed in our model. For example, we find that if the impurities have a narrow width of order N^{-1}, there is an effect of order 1 on the specific heat only when $|T-T_c|=O(N^{-2})$. We do not know if this order of magnitude persists in general.

It must be emphasized that in this computation of the observed (order 1) specific heat we have in no way settled the question of the order of the phase transition. To answer this question, which admittedly is not very important physically, one must study the analytic behavior of all contributions to $C_v{}^r$. There are several reasons why this is not trivial. We mention two.

(1) The relationship of the differential equation (4.3) to the integral equation (3.28) must be studied in detail. The naive procedure of using more terms in the Taylor-series expansion of $U(\tau)$ leaves one with a differential equation of order larger than 1 and the problem of determining the boundary conditions must be resolved.

[12] B. M. McCoy and T. T. Wu, Phys. Rev. Letters, **21**, 549 (1968).

[13] I. Syozi, Progr. Theoret. Phys. (Kyoto) **34**, 189 (1965); I. Syozi and S. Miyazima, *ibid.* **36**, 1083 (1966); J. W. Essom and H. Garelick, Proc. Phys. Soc. (London) **92**, 136 (1967).

(2) Formally, it is possible to write down an exact iterative solution to the integral equation (3.28). This iterative solution is analytic only in the segments $B^{2n}<\eta<B^{2(n-1)}$. The approximation employed in Sec. 4 approximates the analytic function which this segmented solution approaches when $n\rightarrow\infty$. This is the important region for the order-1 contributions to $C_v{}^r$, but in general the effects (if any) of the segmented nature of the exact solution are not understood.

There are several ways in which the considerations of this paper may be generalized and extended.

(A) The present discussion has been concerned with the case in which E_2 is the random variable. However, the general formulation given in Sec. 2 applies equally well when E_1 (and even both E_1 and E_2) is random. A new feature which can occur when E_1 is random is that, as may be seen from (2.29), T_c may be zero even though $|E_1|$ is never small if $P(E_1)=P(-E_1)$.

(B) The specialization to the power-law distribution (3.11) has been made for clarity of presentation only and many of the above considerations may be generalized to any narrow distribution.

(C) It is also possible to consider distributions which do not have small widths. In particular, the power-law distribution (3.11) when N is not large may be analyzed in terms of a difference-differential equation.

(D) A more challenging problem is the study of spin-correlation functions. The average value of the nearest-neighbor correlation functions may be readily obtained if we combine the results of IV which express correlation functions in terms of C_n and D_n with the considerations of the present paper. However, the spin correlation functions themselves possess a probability distribution function. To study the moments of this distribution and also to study spin correlations other than nearest neighbor, it is necessary to consider joint stationary probability distributions. In particular for the second moment of the nearest-neighbor correlation-functions' probability function the relevant joint probability function satisfies a two-dimensional integral equation. This equation may be approximated by a partial differential equation in a manner analogous to that done in this paper.

(E) The most outstanding feature of this model that remains to be investigated is the behavior of the spontaneous magnetization near T_c. Without a doubt the eighth-root singularity found by Yang will be weakened by the presence of random impurities. The difficult question is, how much?

ACKNOWLEDGMENTS

The authors are deeply indebted to Professor C. N. Yang for the most helpful discussions. One of us (T.T.W) also wishes to thank Professor G. C. Rota and Professor H. McKean for fruitful conversations.

8. Reprinted article 236, listed in Chapter 9:

PHYSICAL REVIEW B VOLUME 13, NUMBER 1 1 JANUARY 1976

Spin-spin correlation functions for the two-dimensional Ising model: Exact theory in the scaling region

Tai Tsun Wu
Division of Engineering and Applied Physics, Harvard University, Cambridge, Massachusetts 02138

Barry M. McCoy*
Institute for Theoretical Physics, State University of New York, Stony Brook, New York 11794

Craig A. Tracy†§
Institute for Fundamental Studies, Department of Physics and Astronomy, University of Rochester, Rochester, New York 14627

Eytan Barouch‡
Department of Mathematics, Clarkson College of Technology, Potsdam, New York 13676
(Received 13 May 1975)

We compute exactly the spin-spin correlation functions $\langle\sigma_{0,0}\sigma_{M,N}\rangle$ for the two-dimensional Ising model on a square lattice in zero magnetic field for $T>T_c$ and $T<T_c$. We then analyze the correlation functions in the scaling limit $T\to T_c$, $M^2+N^2\to\infty$ such that $(M^2+N^2)^{1/2}(T-T_c)$ is fixed. In this scaling limit $\langle\sigma_{0,0}\sigma_{M,N}\rangle = R^{-1/4}F_\pm(t) + R^{-5/4}F_{1\pm}(t)+o(R^{-5/4})$, where t is the scaling variable R/ξ and $F_\pm(t)$ and $F_{1\pm}(t)$ are the scaling functions (ξ is the correlation length). We derive exact expressions for these scaling functions, in terms of a Painlevé function of the third kind and analyze both the small- and large-t behavior. A table of values for $F_\pm(t)$ (good to ten significant digits) is also given. As an application we compute the coefficients $C_{0\pm}$ and $C_{1\pm}$ in the expansion $k_BT\chi(T) = C_{0\pm}|1-T_c/T|^{-7/4} + C_{1\pm}|1-T_c/T|^{-3/4}+O(1)$ of the zero-field susceptibility $\chi(T)$ as $T\to T_c^\pm$.

I. INTRODUCTION

In this paper we calculate,[1] *in the scaling limit*, the spin-spin correlation function $\langle\sigma_{0,0}\sigma_{M,N}\rangle$ for the two-dimensional Ising model[2–5] in the absence of a magnetic field. By scaling limit we mean the limit

$$T\to T_c \tag{1.1}$$

and

$$M^2+N^2\to\infty\ , \tag{1.2}$$

such that

$$(M^2+N^2)^{1/2}(T-T_c) \tag{1.3}$$

is fixed. Here T_c is the critical temperature and the quantity in (1.3) may be either positive or negative (or zero).

In this limit the correlation function is expressible in terms of a function of one variable, which is a simple combination M, N, T, and the interaction energies. This function is known as the scaling function. Scaling functions have been found useful under many circumstances, and general properties of such scaling functions have been discussed by many authors[6–13] in the last ten years. However, the calculation reported here represents the first time that an *exact* calculation of any nontrivial scaling two-spin correlation function has ever been achieved.

As an application of our result we will obtain, for the zero-field magnetic susceptibility $\chi(T)$, all terms which are unbounded at T_c. More precisely, we will compute exactly the four numbers $C_{0\pm}$ and $C_{1\pm}$ in the expansion, valid as $T\to T_c$,

$$k_BT\chi(T)=C_{0\pm}|1-T_c/T|^{-7/4}+C_{1\pm}|1-T_c/T|^{-3/4}+O(1)\ . \tag{1.4}$$

Here k_B is Boltzmann's constant and the subscript + (−) means that $T\to T_c$ from above (below). Our results supercede the extensive numerical studies of these coefficients made during the last fifteen years by means of series expansions.[14]

II. SUMMARY OF RESULTS AND NOTATION

In view of the length and complexity of this paper we collect together our principal results and notation in this section so that they are readily accessible for easy reference.

A. Definitions and notations

The two-dimensional Ising model on a square lattice is specified by the interaction energy

$$\mathcal{E}=-E_1\sum_{j,k}\sigma_{j,k}\sigma_{j,k+1}-E_2\sum_{j,k}\sigma_{j,k}\sigma_{j+1,k}\ , \tag{2.1}$$

where the first (second) index of $\sigma_{j,k}$ specifies the row (column) of the lattice and $\sigma_{j,k}=\pm1$.

In this section we will use the following notation with regard to equation numbers. First, if an equation number is followed by an S, such as (2.2S), that equation applies only to the symmetrical case $E_1=E_2=E$. Second, in order to facilitate the location of the place in this paper where a particular result is derived, the equation number for a result is sometimes followed by a number in square brackets. Thus, for example, (2.9)-[3.66] means the result (2.9) is derived in the text shortly after (3.66).

We define

$$z_1 = \tanh(\beta E_1)\,, \quad z_2 = \tanh(\beta E_2)\,, \tag{2.2}$$

and for the symmetrical lattice,

$$z = z_1 = z_2 = \tanh(\beta E)\,, \tag{2.2S}$$

where $\beta = (k_B T)^{-1}$. At $T = T_c$

$$\sinh 2\beta_c E_1 \sinh 2\beta_c E_2 = 1 \tag{2.3}$$

or, equivalently,

$$z_{1c} z_{2c} + z_{1c} + z_{2c} - 1 = 0\,. \tag{2.4}$$

In particular, for the symmetrical lattice,

$$z_c = \sqrt{2} - 1\,. \tag{2.4S}$$

The scaled variable t is defined by

$$t = |z_1 z_2 + z_1 + z_2 - 1| \left(\frac{M^2}{z_2(1-z_1^2)} + \frac{N^2}{z_1(1-z_2^2)} \right)^{1/2}$$
$$= |z_1 z_2 + z_1 + z_2 - 1|\, [z_1 z_2 (1-z_1^2)(1-z_2^2)]^{-1/4} R\,, \tag{2.5}$$

where the appropriate radial measure of spatial distance is

$$R = \left[\left(\frac{z_1(1-z_2^2)}{z_2(1-z_1^2)} \right)^{1/2} M^2 + \left(\frac{z_2(1-z_1^2)}{z_1(1-z_2^2)} \right)^{1/2} N^2 \right]^{1/2}$$
$$= \left[\left(\frac{\sinh 2\beta E_1}{\sinh 2\beta E_2} \right)^{1/2} M^2 + \left(\frac{\sinh 2\beta E_2}{\sinh 2\beta E_1} \right)^{1/2} N^2 \right]^{1/2}. \tag{2.6}$$

For the symmetrical case one has simply

$$t = |z^2 + 2z - 1|\,[z(1-z^2)]^{-1/2} R \tag{2.5S}$$

and

$$R = (M^2 + N^2)^{1/2}\,. \tag{2.6S}$$

As $T \to T_c$ we find

$$z_1 z_2 + z_1 + z_2 - 1 \sim (\beta - \beta_c) 4 z_{1c} z_{2c} [E_1 (1 - z_{2c})^{-1} + E_2 (1 - z_{1c})^{-1}] \tag{2.7}$$

and

$$t/R = |\beta - \beta_c|\, 2\sqrt{2}\, (z_{1c} z_{2c})^{1/2} [E_1 (1 - z_{2c})^{-1} + E_2 (1 - z_{1c})^{-1}] + O((\beta - \beta_c)^2)$$
$$= |\beta - \beta_c|\, 2(\sinh 2\beta_c E_1 + \sinh 2\beta_c E_2)^{-1/2} \times (E_1 \coth 2\beta_c E_1 + E_2 \coth 2\beta_c E_2) + O((\beta - \beta_c)^2)\,. \tag{2.8}$$

For the symmetric lattice the behavior is

$$t/R = |\beta - \beta_c|\, 4E[1 - 2^{-1/2} E (\beta - \beta_c) + O((\beta - \beta_c)^2)]\,. \tag{2.8S}$$

B. Results for correlation functions

When $T < T_c$, one has

$$\langle \sigma_{0,0} \sigma_{M,N} \rangle = S_\infty^{<} \exp(-F_{<MN})\,, \qquad (2.9)\text{-}[3.66]$$

where

$$S_\infty^{<} = [1 - (\sinh 2\beta E_1 \sinh 2\beta E_2)^{-2}]^{1/4} \tag{2.10}$$

and

$$F_{<MN} = \sum_{n=1}^{\infty} F_{<MN}^{(2n)}\,, \tag{2.11}$$

$$F_{<MN}^{(2n)} = (-1)^n [2 z_2 (1 - z_1^2)]^{2n} (2n)^{-1} (2\pi)^{-4n} \times \int_{-\pi}^{\pi} d\phi_1 \int_{-\pi}^{\pi} d\phi_2 \cdots \int_{-\pi}^{\pi} d\phi_{4n} \times \prod_{j=1}^{2n} \left(\frac{e^{-iM\phi_{2j} - iN\phi_{2j}}}{\Delta(\phi_{2j-1}, \phi_{2j})} \, \frac{\sin\frac{1}{2}(\phi_{2j-1} - \phi_{2j+1})}{\sin\frac{1}{2}(\phi_{2j} + \phi_{2j+2})} \right). \qquad (2.12)\text{-}[3.84]$$

Here

$$\phi_{4n+1} \equiv \phi_1\,, \quad \phi_{4n+2} \equiv \phi_2\,;$$
$$\mathrm{Im}\phi_j < 0\,, \quad j = 1, 2, \ldots, 4n\,;$$

and

$$\Delta(\phi_{2j-1}, \phi_{2j}) = (1 + z_1^2)(1 + z_2^2) - 2 z_2 (1 - z_1^2) \cos\phi_{2j-1} - 2 z_1 (1 - z_2^2) \cos\phi_{2j}\,. \tag{2.13}$$

When $T > T_c$, the result is

$$\langle \sigma_{0,0} \sigma_{M,N} \rangle = S_\infty^{>} \left(\sum_{k=1}^{\infty} x_{>MN}^{(2k-1)} \right) \exp\left(- \sum_{n=1}^{\infty} F_{>MN}^{(2n)} \right), \qquad (2.14)\text{-}[4.87]$$

where

$$x_{>MN}^{(k)} = [-i 2 z_2 (1 - z_1^2)]^{k-1} (2\pi)^{-2k} \int_{-\pi}^{\pi} d\phi_1 \int_{-\pi}^{\pi} d\phi_2 \cdots \int_{-\pi}^{\pi} d\phi_{2k}$$
$$\times \prod_{j=1}^{k} \left(\frac{e^{-iM\phi_{2j-1} - iN\phi_{2j}}}{\Delta(\phi_{2j-1}, \phi_{2j})} \right) \prod_{j=1}^{k-1} \left(\frac{\sin\frac{1}{2}(\phi_{2j-1} - \phi_{2j+1})}{\sin\frac{1}{2}(\phi_{2j} + \phi_{2j+2})} \cos\tfrac{1}{2}(\phi_{2j} + \phi_{2j+2}) \cos\tfrac{1}{2}(\phi_{2j-1} + \phi_{2j+1}) \right) \tag{2.15}$$

[if $k = 1$, the second product is replaced by unity],

$$F_{>MN}^{(2n)} = (-1)^n [2 z_2 (1 - z_1^2)]^{2n} (2n)^{-1} (2\pi)^{-4n} \int_{-\pi}^{\pi} d\phi_1 \int_{-\pi}^{\pi} d\phi_2 \cdots \int_{-\pi}^{\pi} d\phi_{4n}$$

$$\times \prod_{j=1}^{2n} \frac{e^{-iM\phi_{2j-1}-iN\phi_{2j}}}{\Delta(\phi_{2j-1},\phi_{2j})} \frac{\sin\frac{1}{2}(\phi_{2j-1}-\phi_{2j+1})}{\sin\frac{1}{2}(\phi_{2j}+\phi_{2j+2})} \cos\tfrac{1}{2}(\phi_{2j}+\phi_{2j+2})\cos\tfrac{1}{2}(\phi_{2j-1}+\phi_{2j+1}) \tag{2.16}$$

[with $\phi_{4n+1}=\phi_1$ and $\phi_{4n+2}=\phi_2$], $\mathrm{Im}\phi_j<0$, and

$$S_\infty^{>} = [4z_1z_2(1-z_1^2)(1-z_2^2)]^{1/2}[(\sinh 2\beta E_1 \sinh 2\beta E_2)^{-2}-1]^{1/4}. \tag{2.17}$$

The above results hold for all T. From here on we concentrate on the scaling limit (1.1)–(1.3). Equivalently, this limit is

$$T\to T_c, \tag{2.18}$$

$$R\to\infty, \tag{2.19}$$

with

$$t \text{ fixed}. \tag{2.20}$$

In this limit we find that

$$\langle\sigma_{0,0}\sigma_{M,N}\rangle = |1-(\sinh 2\beta E_1 \sinh 2\beta E_2)^{-2}|^{1/4} \times[\hat{F}_\pm(t)+o(R^{-1})], \tag{2.21}$$

where $\hat{F}_\pm(t)$ depends upon E_1, E_2, M, and N only through the variable t, and the subscript + (−) means $T>T_c$ ($T<T_c$). The first factor may be expanded as a function of t/R, and we find

$$\langle\sigma_{0,0}\sigma_{M,N}\rangle = R^{-1/4}F_\pm(t)+R^{-5/4}F_{1\pm}(t)+o(R^{-5/4}), \tag{2.22}$$

where

$$F_\pm(t) = (2t)^{1/4}(\sinh 2\beta_c E_1 + \sinh 2\beta_c E_2)^{1/8}\hat{F}_\pm(t) \tag{2.23}$$

and

$$F_{1+}(t)/F_+(t) = -F_{1-}(t)/F_-(t) = -tR_1, \tag{2.24}$$

where

$$R_1 = (16\sqrt{2})^{-1}(z_{1c}z_{2c})^{-1/2}(z_{1c}+z_{2c})^{-1}[E_1(1-z_{2c})^{-1} + E_2(1-z_{1c})^{-1}]^{-1}[E_1(1-z_{2c})^{-1}(-5z_{2c}^2-3z_{1c}^2+6z_{2c} + 10z_{1c}-8)+E_2(1-z_{1c})^{-1}(-5z_{1c}^2-3z_{2c}^2 + 6z_{1c}+10z_{2c}-8)]. \tag{2.25}$$

For the symmetric lattice R_1 reduces to

$$R_1 = -2^{-3/2}. \tag{2.25S}$$

The ratio R_1 is plotted in Fig. 1 for fixed T_c as a function of the ratio E_1/E_2.

Furthermore, we find explicitly

$$\hat{F}_-(t) = \exp\left(-\sum_{n=1}^{\infty} f^{(2n)}(t)\right), \tag{2.26}$$

with

$$f^{(2n)}(t) = (-1)^n \pi^{-2n} n^{-1} \int_1^\infty dy_1 \cdots \int_1^\infty dy_{2n} \times \prod_{j=1}^{2n} \frac{e^{-ty_j}}{(y_j^2-1)^{1/2}(y_j+y_{j+1})} \prod_{j=1}^{n}(y_{2j}^2-1)$$

(2.27)-[3.128]

$[y_{2n+1}\equiv y_1]$

and also

$$\hat{F}_+(t) = G(t)\hat{F}_-(t), \tag{2.28}$$

where

$$G(t) = \sum_{k=0}^{\infty} g^{(2k+1)}(t), \tag{2.29}$$

with

$$g^{(2k+1)}(t) = (-1)^k \pi^{-2k-1}\int_1^\infty dy_1 \cdots \int_1^\infty dy_{2k+1} \times \prod_{j=1}^{2k+1}\frac{e^{-ty_j}}{(y_j^2-1)^{1/2}} \prod_{j=1}^{2k}(y_j+y_{j+1})^{-1} \prod_{j=1}^{k}(y_{2j}^2-1)$$

(2.30)-[4.150]

[for $k=0$ the last two products are replaced by unity].

As $t\to\infty$ we have

$$F_+(t) = (\sinh 2\beta_c E_1 + \sinh 2\beta_c E_2)^{1/8} \times(2t)^{1/4}\pi^{-1}K_0(t)+O(e^{-3t})$$

(2.31a)-[4.155]

and

$$F_-(t) = (\sinh 2\beta_c E_1 + \sinh 2\beta_c E_2)^{1/8}(2t)^{1/4} \times(1+\pi^{-2}\{t^2[K_1^2(t)-K_0^2(t)]-tK_0(t)K_1(t) + \tfrac{1}{2}K_0^2(t)\}+O(e^{-4t})).$$

(2.31b)-[3.151]

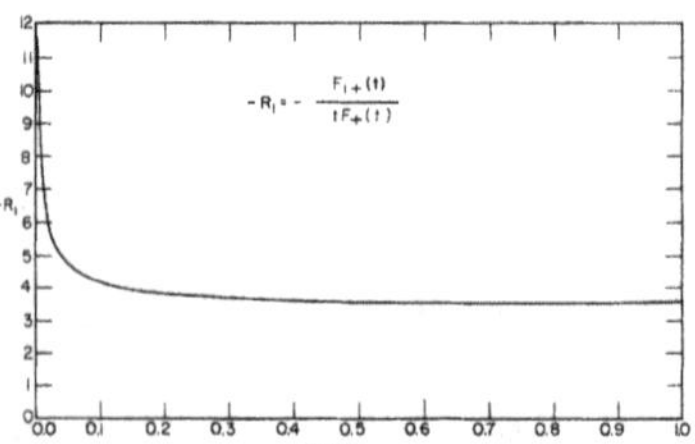

FIG. 1. Ratio $R_1=F_{1+}(t)/[tF_+(t)]$ as a function of $x=E_1/E_2$ for fixed T_c. For $x=1$, $R_1=-2^{-3/2}$.

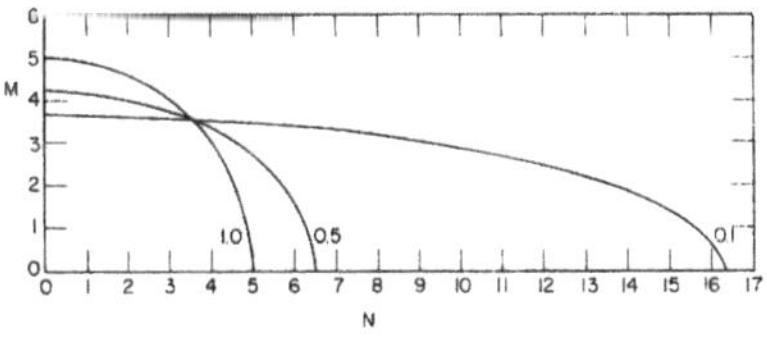

FIG. 2. Ellipses in the M, N plane along which $R^{-1/4}F_{\pm}(t)$ (for fixed t) is a constant. Three different ellipses corresponding to the ratios $E_2/E_1 = 1.0$, 0.5 and 0.1 are shown. Note that for $M = N$ the value of $R^{-1/4}F_{\pm}(t)$ is independent of the ratio E_1/E_2.

Conversely, as $t \to 0$ we obtain

$$F_{\pm}(t) = (\sinh 2\beta_c E_1 + \sinh 2\beta_c E_2)^{1/8} e^{1/4} \times 2^{1/12} A^{-3}[1 \pm \tfrac{1}{2}t\Omega + \tfrac{1}{16}t^2 \pm \tfrac{1}{32}t^3\Omega + \tfrac{1}{256}t^4(-\Omega^2 + \Omega + \tfrac{1}{8}) + O(t^5\Omega^4)] , \quad (2.32)\text{-}[5.53], [6.123]$$

where

$$\Omega = \ln(t/8) + \gamma_E , \quad (2.33)$$

$$\gamma_E = 0.577\,215\,664\,901\,53 \cdots \quad (2.34)$$

is Euler's constant, and

$$A = 1.282\,427\,129\,100\,62 \cdots \quad (2.35)$$

is Glaisher's constant.

In Fig. 2, we plot, for various values of the ratio E_2/E_1, the ellipses in the M, N plane along which $R^{-1/4}F_{\pm}(t)$ is constant for fixed t as $T \to T_c$. Note in particular that for $M = N$ the value of $R^{-1/4}F_{\pm}(t)$ is independent of the ratio E_1/E_2.

The functions $F_{\pm}(t)$ may also be expressed in terms of a Painlevé function (of the third kind)[33] which we denote by $\eta(0)$. This function satisfies the second-order nonlinear differential equation

$$\frac{d^2\eta}{d\theta^2} = \frac{1}{\eta}\left(\frac{d\eta}{d\theta}\right)^2 - \eta^{-1} + \eta^3 - \theta^{-1}\frac{d\eta}{d\theta} , \quad (2.36)$$

with the boundary conditions

$$\eta(\theta) = -\theta[\ln(\tfrac{1}{4}\theta) + \gamma_E] + O(\theta^5 \ln^3\theta) \quad (2.37)\text{-}[B8]$$

as $\theta \to 0$, and

$$\eta(\theta) = 1 - 2\pi^{-1}K_0(2\theta) + O(e^{-4\theta}) \quad (2.38)\text{-}[6.95]$$

as $\theta \to \infty$ [see Sec. VI for further discussion]. We then have

$$F_{\pm}(t) = 2^{-1/2}(\sinh 2\beta_c E_1 + \sinh 2\beta_c E_2)^{1/8}\theta^{1/4}[1 \mp \eta(\theta)]\eta^{-1/2}(\theta) \exp\left(\int_\theta^\infty dx\, \tfrac{1}{4}x\eta^{-2}(x)\{[1 - \eta^2(x)]^2 - [\eta'(x)]^2\}\right)$$

$$= 2^{-1/2}(\sinh 2\beta_c E_1 + \sinh 2\beta_c E_2)^{1/8}\theta^{1/4}[1 \mp \eta(\theta)] \exp\left(\int_\theta^\infty dx\, x \ln x[1 - \eta^2(x)] - h(\theta)\right) , \quad (2.39)\text{-}[6.67]$$

where

$$\theta = \tfrac{1}{2}t \quad (2.40)$$

and, in the last line, we have

$$h(\theta) = \left(\frac{\theta\eta'}{2\eta} + \frac{\theta^2}{4\eta^2}[(1 - \eta^2)^2 - \eta'^2]\right)\ln\theta . \quad (2.41)$$

The functions $F_{\pm}(t)$ for $E_1 = E_2$ are plotted in Fig. 3. The function $[F_-(t) - (\sinh 2\beta_c E_1 + \sinh 2\beta_c E_2)^{1/8} \times (2t)^{1/4}]$ for $E_1 = E_2$ is plotted in Fig. 4.

If one believes in some form of universality, then in the scaling limit (2.39) gives (aside from some lattice-dependent factors which are independent of t) the spin-spin correlation not only for the two-dimensional Ising model but also for a large class of two-dimensional models with short-range interactions.

C. Results for magnetic susceptibility

The magnetic susceptibility at $H = 0$,

$$\chi(T) = \left.\frac{\partial \mathfrak{M}(T, H)}{\partial H}\right|_{H=0} , \quad (2.42)$$

is related to $\langle \sigma_{0,0}\sigma_{M,N} \rangle$ by

$$\beta^{-1}\chi(T) = \sum_{M=-\infty}^{+\infty}\sum_{N=-\infty}^{+\infty}[\langle \sigma_{0,0}\sigma_{M,N} \rangle - \mathfrak{M}^2(T, 0)] , \quad (2.43)$$

where

$$\mathfrak{M}(T, 0) = (S_\infty^{<})^{1/2} \quad \text{if } T < T_c$$
$$= 0 \quad \text{if } T \geq T_c . \quad (2.44)$$

As $T \to T_c$ we find that

$$\beta^{-1}\chi(T) = C_{0\pm}|1 - T_c/T|^{-7/4} + C_{1\pm}|1 - T_c/T|^{-3/4} + O(1) , \quad (2.45)$$

where

$$C_{0+} = D\int_0^\infty d\theta\, \theta[1 - \eta(\theta)] \times \exp\left(\int_\theta^\infty dx[1 - \eta^2(x)]x \ln x - h(\theta)\right) , \quad (2.46\text{a})\text{-}[7.6]$$

$$C_{0-} = D\int_0^\infty d\theta\, \theta\left[[1 + \eta(\theta)] \times \exp\left(\int_\theta^\infty dx[1 - \eta^2(x)]x \ln x - h(\theta)\right) - 2\right] , \quad (2.46\text{b})\text{-}[7.17]$$

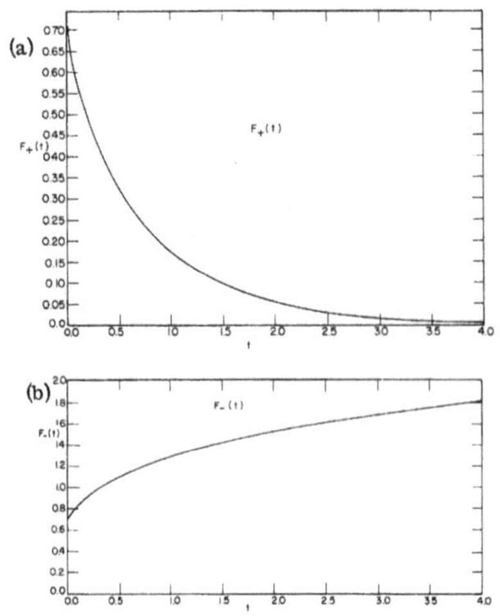

FIG. 3. (a) Scaling function $F_+(t)$ as a function of t for $E_1=E_2$. Note that $F_+(0)=0.703\,380\,157\,7\cdots$. (b) Scaling function $F_-(t)$ as a function of t for $E_1=E_2$. At $t=0$ $F_-(0)=F_+(0)$.

and

$$C_{1+}/C_{0+}=-C_{1-}/C_{0-}=-R_0\beta_c\ , \tag{2.47}$$

with

$$D=2^{-1/2}\pi(z_{1c}z_{2c})^{-1}(z_{1c}+z_{2c})^{1/4}\times\{\beta_c[E_1(1-z_{2c})^{-1}+E_2(1-z_{1c})^{-1}]\}^{-7/4}$$
$$=2^{1/2}\pi(\sinh 2\beta_c E_1+\sinh 2\beta_c E_2)\times\{\beta_c[E_1\coth 2\beta_c E_1+E_2\coth 2\beta_c E_2]\}^{-7/4} \tag{2.48}$$

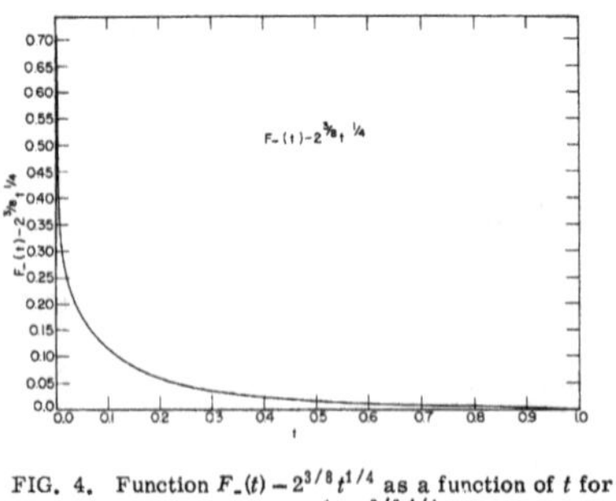

FIG. 4. Function $F_-(t)-2^{3/8}t^{1/4}$ as a function of t for $E_1=E_2$. Note that for $E_1=E_2$ $S_\infty^< \sim 2^{3/8}t^{1/4}$.

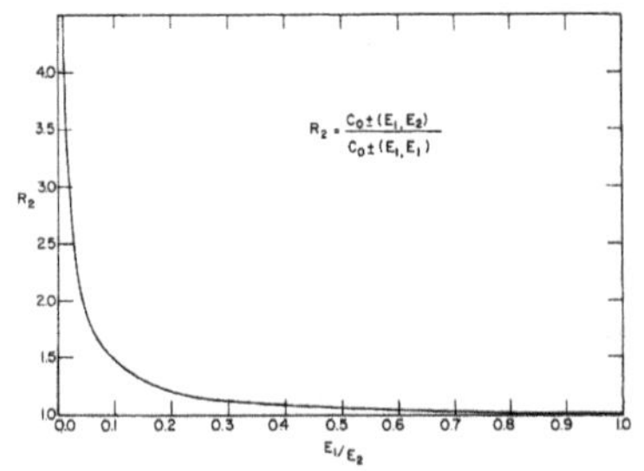

FIG. 5. Amplitude ratio $R_2=C_{0\pm}(E_1,E_2)/C_{0\pm}(E_1,E_1)$ as a function of the ratio E_1/E_2 for fixed T_c.

and

$$R_0=[E_1^2z_{2c}^2(1+6z_{1c}^2+z_{1c}^4)+E_2^2z_{1c}^2(1+6z_{2c}^2+z_{2c}^4)-8E_1E_2z_{1c}z_{2c}(z_{1c}+z_{2c})^2]\{8z_{1c}z_{2c}(z_{1c}+z_{2c})\times[E_1(1-z_{1c})+E_2(1-z_{2c})]\}^{-1}\ . \tag{2.49}$$

In the symmetric case we have

$$D=2^{5/8}\pi[\ln(1+\sqrt{2})]^{-7/4}\ , \tag{2.48S}$$

$$R_0=-\tfrac{1}{8}\sqrt{2}E\ . \tag{2.50S}$$

In Fig. 5 we plot

$$R_2=C_{0\pm}(E_1,E_2)/C_{0\pm}(E_1,E_1) \tag{2.51}$$

for fixed T_c as a function of E_1/E_2.

In Fig. 6 we plot the dimensionless quantity $R_0(E_1+E_2)^{-1}$ for fixed T_c as a function of E_1/E_2. Note that this quantity is $\frac{1}{2}$ at $E_1/E_2=0$ and $-\frac{1}{16}\sqrt{2}$ at $E_1=E_2$.

To compare these results for $C_{0\pm}$ and $C_{1\pm}$ with the available series-expansion results,[14] we specialize to the symmetric lattice and numerically evaluate $C_{0\pm}$ and $C_{1\pm}$ to find

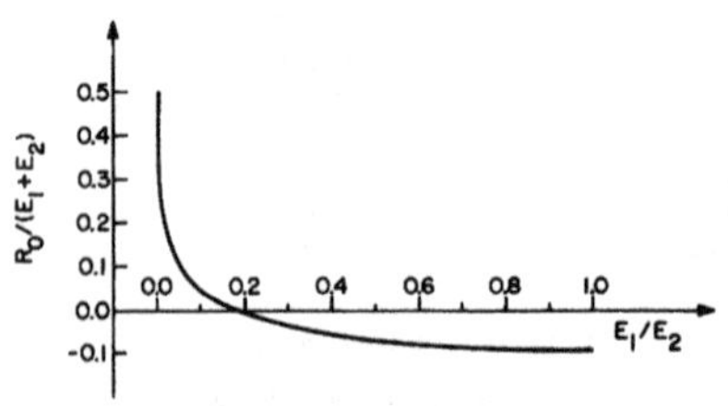

FIG. 6. Dimensionless quantity $R_0/(E_1+E_2)$ as a function of $x=E_1/E_2$. When $x=1$ the value is $-\frac{1}{16}\sqrt{2}$. As $x\to 0$ the function behaves as $\frac{1}{2}+3/(2\ln x)$. Curve crosses zero at ~ 0.1993.

$$C_{0+} = 0.962\,581\,732\,2\cdots,$$
$$C_{0-} = 0.025\,536\,971\,9\cdots,$$
$$C_{1+} = 0.074\,988\,153\,8\cdots, \quad (2.52S)$$
$$C_{1-} = -0.001\,989\,410\,7\cdots.$$

These are to be compared with the series-expansion estimates of Sykes, Gaunt, Roberts, and Wyles[14a] above T_c

$$C_{0+} = 0.962\,59 \pm 0.000\,03, \quad (2.53S)$$

$$C_{1+} \simeq 0.0742, \quad (2.54S)$$

and of Essam and Hunter[14b] below T_c

$$C_{0-} = 0.025\,68 \pm 0.000\,12. \quad (2.55S)$$

Some insight into the close agreement of the exact results (2.52S) and the series results (2.53S)–(2.55S) is obtained by expanding (2.26) as

$$\hat{F}_-(t) = 1 - f^{(2)}(t) - \{f^{(4)}(t) - \tfrac{1}{2}[f^{(2)}(t)]^2\} + \cdots, \quad (2.56)$$

and expanding (2.28) as

$$\hat{F}_+(t) = g^{(1)}(t) + [g^{(3)}(t) - g^{(1)}(t) f^{(2)}(t)] + (g^{(5)}(t) - g^{(3)}(t) f^{(2)}(t) - g^{(1)}(t)\{f^{(4)}(t) - \tfrac{1}{2}[f^{(2)}(t)]^2\}) + \cdots. \quad (2.57)$$

In (2.56), $f^{(2)}(t)$ behaves as e^{-2t} as $t\to\infty$, and the third term behaves as e^{-4t}. In (2.57), $g^{(1)}(t)$ behaves as e^{-t} as $t\to\infty$, the second term behaves as e^{-3t}, and the third term behaves as e^{-5t}. Here the phrase "behaves as e^{-at}" means that as $t\to\infty$ the quantity is asymptotically equal to e^{-at} times some power of t. Then if we define $C_{0+}^{(2n+1)}$ as the contribution to C_{0+} coming from the term in (2.57) behaving as $e^{-(2n+1)t}$ as $t\to\infty$, and $C_{0-}^{(2n)}$ as the contribution to C_{0-} coming from the terms in (2.56) behaving as e^{-2nt} as $t\to\infty$, we have

$$C_{0+} = \sum_{n=0}^{\infty} C_{0+}^{(2n+1)}, \quad (2.58a)$$

$$C_{0-} = \sum_{n=1}^{\infty} C_{0-}^{(2n)}. \quad (2.58b)$$

Numerically we find for $E_1 = E_2$

$$C_{0+}^{(1)} = [2\ln(1+\sqrt{2})]^{-7/4}\,2^{11/8} = 0.961\,797\,616\,76\cdots, \quad (2.59S)\text{-}[7.8]$$

$$C_{0+}^{(3)} = 0.000\,783\,348\,15\cdots, \quad (2.60S)$$

$$C_{0+}^{(5)} = 0.000\,000\,766\,5\cdots, \quad (2.61S)$$

and

$$C_{0-}^{(2)} = [2\ln(1+\sqrt{2})]^{-7/4}\,2^{11/8}(12\pi)^{-1} = 0.025\,512\,474\,16\cdots, \quad (2.62S)\text{-}[7.20]$$

$$C_{0-}^{(4)} = 0.000\,024\,476\cdots. \quad (2.63S)$$

These perturbation-expansion results are compared with the exact results (2.52S) and the series-expansion results (2.53S)–(2.55S) in Table I.

Finally a remark must be made concerning the $O(1)$ term of (1.4). This term arises from short-range contributions to the sum (2.43) and hence cannot be obtained from the scaling functions $F_\pm(t)$.

TABLE I. As $T\to T_c^\pm$ the zero-field susceptibility $\chi(T)$ behaves as $\beta^{-1}\chi(T) = C_{0\pm}|1-T_c/T|^{-7/4} + C_{1\pm}|1-T_c/T|^{-3/4} + O(1)$. The four numbers $C_{0\pm}$ and $C_{1\pm}$ for the symmetric lattice $E_1 = E_2$ are given. See the text for the definition of the numbers $C_{0\pm}^{(n)}$.

Quantity	Exact result	Series-expansion result	Perturbation-expansion approximation
C_{0+}	$0.962\,581\,732\,2\cdots$	$0.962\,59\pm0.000\,03$[a]	$C_{0+}^{(1)} = 0.961\,797\,616\,76\cdots$
			$C_{0+}^{(1)} + C_{0+}^{(3)} = 0.962\,580\,964\,91\cdots$
			$C_{0+}^{(1)} + C_{0+}^{(3)} + C_{0+}^{(5)} = 0.962\,581\,731\,4\cdots$
C_{0-}	$0.025\,536\,971\,9\cdots$	$0.025\,68\pm0.000\,12$[b]	$C_{0-}^{(2)} = 0.025\,512\,474\,16\cdots$
			$C_{0-}^{(2)} + C_{0-}^{(4)} = 0.025\,536\,950\cdots$
C_{0+}/C_{0-}	$37.693\,652\,01\cdots$		$C_{0+}^{(1)}/C_{0-}^{(2)} = 12\pi = 37.699\,111\,84$
			$(C_{0+}^{(1)} + C_{0+}^{(3)})/(C_{0-}^{(2)} + C_{0-}^{(4)}) = 37.693\,654\,3$
C_{1+}	$0.074\,988\,153\,8\cdots$	0.0742[a]	
C_{1-}	$-0.001\,989\,410\,7\cdots$		

[a]Reference 14a. [b]Reference 14b.

From the series-expansion results[14] this term is known to be nonzero.

D. Outline of main text

As a guide to the remainder of the paper, we present the following outline.

III. PERTURBATION EXPANSION FOR $\langle\sigma_{0,0}\sigma_{M,N}\rangle$ FOR $T<T_c$ AND LARGE (M^2+N^2)

A. Preliminary notation

B. Perturbation expansion, $T<T_c$

C. Recursion relations

D. Quantity f_{MN}

E. $\langle\sigma_{0,0}\sigma_{M,N}\rangle$ in terms of f_{MN}

F. $F^{(2n)}_{<MN}$ for arbitrary n

G. $\langle\sigma_{0,0}\sigma_{M,N}\rangle$ in scaling limit $(T<T_c)$

IV. PERTURBATION EXPANSION FOR $\langle\sigma_{0,0}\sigma_{M,N}\rangle$ FOR $T>T_c$ AND LARGE (M^2+N^2)

A. Perturbation-expansion formalism

B. Recursion relations

C. Quantity $\bar{f}_{MN}$

D. Quantity $\bar{F}_{MN}$

E. $x^{(k)}_{>MN}$ and $F^{(2n)}_{>MN}$

F. $\langle\sigma_{0,0}\sigma_{M,N}\rangle$ in scaling limit $(T>T_c)$

V. PERTURBATION EXPANSION FOR $\langle\sigma_{0,0}\sigma_{N,N}\rangle$ FOR LARGE N AND SMALL $N|T-T_c|$

A. Preliminary notation

B. Analysis of Δ in the scaling limit

C. Quantity $\mathrm{Tr}[(\Delta A_0^{-1})^n]$ in limit $n\to\infty$, $T\to T_c$, s fixed

D. Small-t expansion of $F_\pm(t)$

VI. SCALING FUNCTIONS $F_\pm(t)$ IN TERMS OF PAINLEVÉ FUNCTION OF THIRD KIND

A. Relating $x_0(N)$ to integral equations

B. $x_0(N)$ in terms of $\eta(\theta)$

C. $F_\pm(t)$ in terms of $\eta(\theta)$

D. $F_\pm(t)$ for $t\to 0$ and $t\to\infty$ as derived from $\eta(\theta)$ representation

VII. SUSCEPTIBILITY $\chi(T)$

A. Leading divergence of $\chi(T)$ as $T\to T_c^\pm$

B. $\chi(T)$ for $T<T_c$

C. $\chi(T)$ for $T>T_c$

APPENDIX A: NUMERICAL WORK

APPENDIX B: INTEGRAL EQUATION AND PAINLEVÉ FUNCTION $\eta(\theta)$

APPENDIX C: VARIOUS IDENTITIES FOR ISING MODEL

III. PERTURBATION EXPANSION FOR $\langle\sigma_{0,0}\sigma_{M,N}\rangle$ FOR $T<T_c$ AND LARGE (M^2+N^2)

Using the Pfaffian approach[15,16] to the two-dimensional Ising model, Cheng and Wu[17] (CW) developed a perturbation expansion for $\langle\sigma_{0,0}\sigma_{M,N}\rangle$ for $T<T_c$ (and $T>T_c$) and large M^2+N^2 which they then analyzed to leading order. Here we examine in this perturbation expansion scheme the nth-order terms. To facilitate the readability of this paper, we summarize all relevant results of CW that we will need. The interested reader should consult either CW or Chap. 12 of Ref. 5 whenever we say "it has been shown that . . ." for a demonstration of the stated result. In this way the reader is able to follow the derivations of this section (and Sec. IV) without being familiar with the results of CW.

A. Preliminary notation

We first define the following infinite matrices:

$$S_{mn}=(2\pi)^{-1}\int_{-\pi}^{\pi}d\phi\, e^{-i(n-m)\phi}S(e^{i\phi}), \tag{3.1}$$

$$V_{mn}=(2\pi)^{-1}\int_{-\pi}^{\pi}d\phi\, e^{-i(m+n)\phi}V(e^{i\phi}), \tag{3.2}$$

$$U_{mn}=(2\pi)^{-1}\int_{-\pi}^{\pi}d\phi\, e^{-i(m+n)\phi}U(e^{i\phi}), \tag{3.3}$$

$$T_{mn}=(2\pi)^{-1}\int_{-\pi}^{\pi}d\phi\, e^{-i(m+n)\phi}T(e^{i\phi}), \tag{3.4}$$

for $m,\ n=0,\ 1,\ 2,\ \ldots,$ where the generating functions $S(e^{i\phi})$, $V(e^{i\phi})$, $U(e^{i\phi})$, and $T(e^{i\phi})$ are given[18]

$$S(e^{i\phi})=[(1-\alpha_1e^{i\phi})(1-\alpha_2e^{-i\phi})\times(1-\alpha_1e^{-i\phi})^{-1}(1-\alpha_2e^{i\phi})^{-1}]^{1/2}, \tag{3.5}$$

$$V(e^{i\phi_2})=(1-z_1^2)(2\pi)^{-1}\times\int_{-\pi}^{\pi}d\phi_1\frac{e^{-iM\phi_1-iN\phi_2}}{\Delta(\phi_1,\phi_2)}e^{-2i\phi_2}\lambda(\phi_1,-\phi_2), \tag{3.6}$$

$$U(e^{i\phi_2})=-e^{-i\phi_2}(1-z_1^2)(2\pi)^{-1}\times\int_{-\pi}^{\pi}d\phi_1\frac{e^{-iM\phi_1-iN\phi_2}}{\Delta(\phi_1,\phi_2)}2iz_2\sin\phi_1, \tag{3.7}$$

and

$$T(e^{i\phi_2})=-(1-z_1^2)(2\pi)^{-1}\times\int_{-\pi}^{\pi}d\phi_1\frac{e^{-iM\phi_1-iN\phi_2}}{\Delta(\phi_1,\phi_2)}\lambda(\phi_1,\phi_2), \tag{3.8}$$

and with z_1 and z_2 given by (2.2), $\Delta(\phi_1, \phi_2)$ by (2.13),

$$\lambda(\phi_1, \phi_2) = 1 - z_2^2 - z_1(1 + z_2^2 + 2z_2\cos\phi_1)e^{-i\phi_2} \tag{3.9}$$

and

$$\begin{aligned} \alpha_1 &= z_1(1 - z_2)(1 + z_2)^{-1}, \\ \alpha_2 &= z_1^{-1}(1 - z_2)(1 + z_2)^{-1}. \end{aligned} \tag{3.10}$$

The square root in (3.5) is defined so that $S(e^{i\theta}) > 0$. It is convenient to define (following Refs. 15 and 17)

$$\begin{aligned} a &= (1 + z_1^2)(1 + z_2^2), \\ \gamma_1 &= 2z_2(1 - z_1^2), \end{aligned} \tag{3.11}$$

and

$$\gamma_2 = 2z_1(1 - z_2^2).$$

From (2.4) it follows that the $T = T_c$ condition can be expressed either as

$$\alpha_2 = 1 \tag{3.12}$$

or as

$$a - \gamma_1 - \gamma_2 = (z_1 z_2 + z_1 + z_2 - 1)^2 = 0. \tag{3.13}$$

We also note that for $T < T_c$ $(0 < \alpha_1 < \alpha_2 < 1)$ the index[19] of the generating function (3.5) is zero.

We define the matrices A, B, and C by

$$A = \begin{bmatrix} 0 & S & 0 & 0 \\ -S^T & 0 & 0 & 0 \\ 0 & 0 & 0 & -S \\ 0 & 0 & S^T & 0 \end{bmatrix}, \tag{3.14}$$

$$B = \begin{bmatrix} 0 & 0 & T & U \\ 0 & 0 & -U & V \\ -T & U & 0 & 0 \\ -U & -V & 0 & 0 \end{bmatrix}, \tag{3.15}$$

and

$$C = A + B, \tag{3.16}$$

with the entries in A and B being the operators S, V, U, and T defined by (3.1), (3.2), (3.3), and (3.4), respectively. The superscript "T" denotes the transpose operation.

B. Perturbation expansion, $T < T_c$

If we define for M, $N \geq 0$ the quantity

$$f_{MN} = \langle \sigma_{0,0}\sigma_{M,N+1}\rangle / \langle \sigma_{0,0}\sigma_{M,N}\rangle, \tag{3.17}$$

then CW have shown that

$$(f_{MN})^2 = x_{30}x'_{40} - x'_{30}x_{40}, \tag{3.18}$$

where x_{30}, x_{40}, x'_{30}, and x'_{40} are the zeroth component of the vectors x_3, x_4, x'_3, and x'_4, respectively, which are defined by

$$Cx = y_3 \quad \text{and} \quad Cx' = y_4, \tag{3.19}$$

where

$$x = \begin{bmatrix} x_1 \\ x_2 \\ x_3 \\ x_4 \end{bmatrix}, \quad x' = \begin{bmatrix} x'_1 \\ x'_2 \\ x'_3 \\ x'_4 \end{bmatrix}, \quad y_3 = \begin{bmatrix} 0 \\ 0 \\ \delta \\ 0 \end{bmatrix}, \quad y_4 = \begin{bmatrix} 0 \\ 0 \\ 0 \\ \delta \end{bmatrix}, \tag{3.20}$$

and δ is the unit vector in l_2 given by

$$\delta = \begin{bmatrix} 1 \\ 0 \\ 0 \\ \cdot \\ \cdot \\ \cdot \end{bmatrix}. \tag{3.21}$$

Since C is antisymmetric it follows that $x_{30} = x'_{40} = 0$ and $x'_{30} = -x_{40}$. Thus from (3.18)

$$f_{MN} = |x'_{30}|. \tag{3.22}$$

That x and x' exist is guaranteed by the fact that C^{-1} exists.[19]

The perturbation expansion of CW consists in expanding C^{-1} as

$$\begin{aligned} C^{-1} &= (1 + A^{-1}B)^{-1}A^{-1} \\ &= \sum_{n=0}^{\infty} (-1)^n (A^{-1}B)^n A^{-1}. \end{aligned} \tag{3.23}$$

This will be a valid expansion for $T < T_c$ and large $M^2 + N^2$ since the elements of S are order 1 while those of T, U, and V are exponentially small.

We can easily see that

$$A^{-1} = \begin{bmatrix} 0 & -(S^T)^{-1} & 0 & 0 \\ S^{-1} & 0 & 0 & 0 \\ 0 & 0 & 0 & (S^T)^{-1} \\ 0 & 0 & -S^{-1} & 0 \end{bmatrix} \tag{3.24}$$

and

$$A^{-1}B = \begin{bmatrix} 0 & 0 & (S^T)^{-1}U & -(S^T)^{-1}V \\ 0 & 0 & S^{-1}T & S^{-1}U \\ -(S^T)^{-1}U & -(S^T)^{-1}V & 0 & 0 \\ S^{-1}T & -S^{-1}U & 0 & 0 \end{bmatrix}. \tag{3.25}$$

It thus follows from (3.22)–(3.25) that

$$f_{MN} = 1 + \delta^T \sum_{n=1}^{\infty} [(A^{-1}B)^{2n}]_{33}(S^T)^{-1}\delta. \tag{3.26}$$

The subscripts refer to the matrix elements,

i.e., $(A^{-1}B)_{32}=-(S^T)^{-1}V$.

The Wiener-Hopf[19] method can be applied to obtain the matrix elements of S^{-1} by solving the equation

$$\sum_{l=0}^{\infty} S_{ml}(S^{-1})_{ln}=\delta_{m,n}\,. \tag{3.27}$$

The result is

$$(S^{-1})_{mn}=(2\pi i)^{-2}\oint d\xi\,\xi^{-(n+1)}P(\xi)$$
$$\times\oint d\xi'\,\frac{\xi'^m}{\xi'-\xi}\,Q(\xi'^{-1})\,, \tag{3.28}$$

where

$$P(\xi)=[(1-\alpha_2\xi)(1-\alpha_1\xi)^{-1}]^{1/2} \tag{3.29}$$

and

$$Q(\xi)=[(1-\alpha_1\xi)(1-\alpha_2\xi)^{-1}]^{1/2}\,. \tag{3.30}$$

The contours of integration in (3.28) are the unit circles, except that the one for ξ' is to be indented outward near $\xi'=\xi$. We note the P and Q functions have the property that

$$P(\xi)Q(\xi)=1\,. \tag{3.31}$$

C. Recursion relations

It is both useful and convenient to define the operators

$$P_1=U(S^T)^{-1}U+VS^{-1}T\,, \tag{3.32}$$

$$P_2=-VS^{-1}U+U(S^T)^{-1}V\,, \tag{3.33}$$

and

$$P_3=T(S^T)^{-1}U-US^{-1}T\,. \tag{3.34}$$

Then using (3.25) and (3.32)–(3.34) we see that

$$[(A^{-1}B)^{2k}]_{33}=-(S^T)^{-1}\{P_1[(A^{-1}B)^{2k-2}]_{33}$$
$$-P_2[(A^{-1}B)^{2k-2}]_{43}\} \tag{3.35}$$

and

$$[(A^{-1}B)^{2k}]_{43}=S^{-1}\{P_3[(A^{-1}B)^{2k-2}]_{33}$$
$$-P_1^T[(A^{-1}B)^{2k-2}]_{43}\} \tag{3.36}$$

for $k\geq 2$, and for $k=1$

$$[(A^{-1}B)^2]_{33}=-(S^T)^{-1}P_1 \tag{3.37}$$

and

$$[(A^{-1}B)^2]_{43}=S^{-1}P_3\,. \tag{3.38}$$

Equations (3.35)–(3.38) are the basic recursion relations. They will prove most useful if we restate these relations in generating function terms. Accordingly we define

$$S^{-1}(\xi,\xi')=\sum_{m,n=0}^{\infty}\xi^m(S^{-1})_{mn}(\xi')^n\,, \tag{3.39}$$

$$P_j(\xi,\xi')=\sum_{m,n=0}^{\infty}\xi^m(P_j)_{mn}(\xi')^n\,,\quad j=1,2,\text{ and }3, \tag{3.40}$$

and

$$R_{\alpha\beta}^{(2k)}(\xi,\xi')=\sum_{m,n=0}^{\infty}\xi^m\,\{[(A^{-1}B)^{2k}]_{\alpha\beta}(S^T)^{-1}\}_{mn}(\xi')^n\,,$$
$$\alpha=3,4\text{ and }\beta=3,4\,. \tag{3.41}$$

In (3.39), $S^{-1}(\xi,\xi')$ should not be confused with $[S(\xi)]^{-1}$. Using (3.28)–(3.30) we see that (3.39) becomes

$$S^{-1}(\xi,\xi')=Q(\xi)P(\xi')(1-\xi\xi')^{-1}\,. \tag{3.42}$$

From (3.32)–(3.34) and (3.42) the generating functions $P_j(\xi,\xi')$ of (3.40) are

$$P_1(1,2)=U(1)S^{-1}(\bar 2,\bar 1)U(2)+V(1)S^{-1}(\bar 1,\bar 2)T(2)\,, \tag{3.43}$$

$$P_2(1,2)=-V(1)S^{-1}(\bar 1,\bar 2)U(2)+U(1)S^{-1}(\bar 2,\bar 1)V(2)\,, \tag{3.44}$$

and

$$P_3(1,2)=T(1)S^{-1}(\bar 2,\bar 1)U(2)-U(1)S^{-1}(\bar 1,\bar 2)T(2)\,. \tag{3.45}$$

The notation $S^{-1}(1,2)$ means $S^{-1}(e^{i\phi_1},e^{i\phi_2})$, and $S^{-1}(\bar 1,\bar 2)$ means $S^{-1}(e^{-i\phi_1},e^{-i\phi_2})$. Likewise for the other functions $P_j(1,2)$, $j=1$, 2, and 3, and the functions $U(1)$, $V(1)$, etc. The steps leading to (3.42) are valid only if $|\xi|$ and $|\xi'|$ are less than 1. Hence when we write $S^{-1}(\bar 1,\bar 2)$ we must have $\mathrm{Im}\phi_1<0$, and $\mathrm{Im}\phi_2<0$, so that (3.43)–(3.45) have the restriction $\mathrm{Im}\phi_j<0$, $j=1$, 2.

Then in the notation (3.39)–(3.45), the recursion relations (3.35)–(3.38) become

$$R_{33}^{(2k)}(\xi,\xi')=(2\pi)^{-2}\int_{-\pi}^{\pi}\int_{-\pi}^{\pi}d(2)d(4)S^{-1}(\bar 2,\xi)$$
$$\times[-P_1(2,4)R_{33}^{(2k-2)}(\bar 4,\xi')$$
$$+P_2(2,4)R_{43}^{(2k-2)}(\bar 4,\xi')] \tag{3.46}$$

and

$$R_{43}^{(2k)}(\xi,\xi')=(2\pi)^{-2}\int_{-\pi}^{\pi}\int_{-\pi}^{\pi}d(2)d(4)S^{-1}(\xi,\bar 2)$$
$$\times[P_3(2,4)R_{33}^{(2k-2)}(\bar 4,\xi')$$
$$-P_1^T(2,4)R_{43}^{(2k-2)}(\bar 4,\xi')] \tag{3.47}$$

for $k\geq 2$, where $d(2)$ denotes $d\phi_2$, etc., and $P_1^T(2,4)$ is the generating function corresponding to the transpose operator P_1^T, i.e.,

$$P_1^T(1,2)=U(1)S^{-1}(\bar 1,\bar 2)U(2)+T(1)S^{-1}(\bar 2,\bar 1)V(2)\,. \tag{3.48}$$

For $k=1$ we have

$$R_{33}^{(2)}(\xi,\xi')=-(2\pi)^{-2}\int_{-\pi}^{\pi}\int_{-\pi}^{\pi}d(2)d(4)$$
$$\times S^{-1}(\bar 2,\xi)P_1(2,4)S^{-1}(\xi',\bar 4) \tag{3.49}$$

and

$$R_{43}^{(2)}(\xi,\xi')=(2\pi)^{-2}\int_{-\pi}^{\pi}\int_{-\pi}^{\pi}d(2)\,d(4)\times S^{-1}(\xi,\bar{2})P_3(2,4)S^{-1}(\xi',\bar{4})\,. \qquad (3.50)$$

To summarize we see from (3.26) and (3.41) with $\alpha=\beta=3$ that

$$f_{MN}=1+\sum_{n=1}^{\infty}R_{33}^{(2n)}(0,0)\,. \qquad (3.51)$$

D. Quantity f_{MN}

In this section we will show that all positive integers n and all $M,\ N\geq 0$ that

$$R_{33}^{(2n)}(0,0)=(2\gamma_1)^{2n}(2\pi)^{-4n}\int_{-\pi}^{\pi}\cdots\int_{-\pi}^{\pi}d\phi_1\cdots d\phi_{4n}\prod_{j=1}^{2n}\frac{e^{-iM\phi_{2j-1}-i(N+1)\phi_{2j}}}{\Delta(\phi_{2j-1},\phi_{2j})}$$
$$\times\cos\tfrac{1}{2}\phi_1\prod_{j=1}^{2n-1}\left(\frac{\sin\frac{1}{2}(\phi_{2j-1}-\phi_{2j+1})}{1-e^{-i(\phi_{2j}+\phi_{2j+2})}}\right)\sin\tfrac{1}{2}\phi_{4n-1}\,, \qquad (3.52)$$

where $\Delta(\phi_1,\phi_2)$ is defined by (2.13) and $\mathrm{Im}(\phi_j)<0$, $j=1,2,\ldots,4n$.

To establish (3.52) we compute $R_{33}^{(2n)}(\xi,\xi')$ and $R_{43}^{(2n)}(\xi,\xi')$ by induction. Once we have computed $R_{33}^{(2)}(\xi,\xi')$ and $R_{43}^{(2)}(\xi,\xi')$ we use the recursion relations (3.46) and (3.47) to establish the general case.

For $n=1$ we use (3.42)–(3.45) in (3.49) and (3.50) to find

$$R_{33}^{(2)}(\xi,\xi')=-(2\pi)^{-2}\int_{-\pi}^{\pi}\int_{-\pi}^{\pi}d(2)\,d(4)\,P(\xi)Q(\xi')\left[(1-\xi e^{-i\phi_2})(1-\xi' e^{-i\phi_4})(1-e^{-i(\phi_2+\phi_4)})\right]^{-1}$$
$$\times Q(\bar{2})\left[U(4)Q(\bar{4})P(\bar{2})U(2)+V(2)Q(\bar{2})P(\bar{4})T(4)\right]P(\bar{4}) \qquad (3.53)$$

and

$$R_{43}^{(2)}(\xi,\xi')=(2\pi)^{-2}\int_{-\pi}^{\pi}\int_{-\pi}^{\pi}d(2)\,d(4)\,Q(\xi)Q(\xi')\left[(1-\xi e^{-i\phi_2})(1-\xi' e^{-i\phi_4})(1-e^{-i(\phi_2+\phi_4)})\right]^{-1}$$
$$\times P(\bar{2})\left[T(2)Q(\bar{4})P(\bar{2})U(4)-U(2)P(\bar{4})Q(\bar{2})T(4)\right]P(\bar{4})\,. \qquad (3.54)$$

Since the values of the functions $T(e^{i\phi_2})$, $U(e^{i\phi_2})$, and $V(e^{i\phi_2})$ are determined by the residue at the pole coming from $\Delta(\phi_1,\phi_2)=0$, we may make use of the identity $\cos\phi_2=\gamma_2^{-1}(a-\gamma_1\cos\phi_1)$ when working with an integral over ϕ_1. This leads to the following identities:

$$T(2)[P(\bar{2})]^2=(2\pi)^{-1}\int_{-\pi}^{\pi}d\phi_1\frac{e^{-iM\phi_1-iN\phi_2}}{\Delta(\phi_1,\phi_2)}\left[-2z_2z_1^{-1}(1-z_1^2)e^{-i\phi_2}(1-\cos\phi_1)\right] \qquad (3.55)$$

and

$$V(2)[Q(\bar{2})]^2=(2\pi)^{-1}\int_{-\pi}^{\pi}d\phi_1\frac{e^{-iM\phi_1-iN\phi_2}}{\Delta(\phi_1,\phi_2)}\left[-2z_1z_2(1-z_1^2)e^{-i\phi_2}(1+\cos\phi_1)\right]\,. \qquad (3.56)$$

Making use of (3.31), (3.55), and (3.56) in (3.53) and (3.54) we obtain

$$R_{33}^{(2)}(\xi,\xi')=-(2\pi)^{-4}\int_{-\pi}^{\pi}\cdots\int_{-\pi}^{\pi}d(1)\cdots d(4)\frac{P(\xi)Q(\xi')}{(1-\xi e^{-i\phi_2})(1-\xi' e^{-i\phi_4})}$$
$$\times\prod_{j=1}^{2}\frac{e^{-iM\phi_{2j-1}-i(N+1)\phi_{2j}}}{\Delta(\phi_{2j-1},\phi_{2j})}(1-e^{-i(\phi_2+\phi_4)})^{-1}$$
$$\times\left[-(1-z_1^2)^2\,4z_2^2\sin\phi_1\sin\phi_3+4z_2^2(1-z_1^2)^2(1+\cos\phi_1)(1-\cos\phi_3)\right] \qquad (3.57a)$$

and

$$R_{43}^{(2)}(\xi,\xi')=(2\pi)^{-4}\int_{-\pi}^{\pi}\cdots\int_{-\pi}^{\pi}d(1)\cdots d(4)\frac{Q(\xi)Q(\xi')}{(1-\xi e^{-i\phi_2})(1-\xi' e^{-i\phi_4})}$$
$$\times\prod_{j=1}^{2}\frac{e^{-iM\phi_{2j-1}-i(N+1)\phi_{2j}}}{\Delta(\phi_{2j-1},\phi_{2j})}\,4iz_1^{-1}z_2^2(1-z_1^2)^2\left[(1-\cos\phi_1)\sin\phi_3-\sin\phi_1(1-\cos\phi_3)\right](1-e^{-i(\phi_2+\phi_4)})^{-1}\,. \qquad (3.57b)$$

Using the trigonometric identities

$$(1-\cos\phi_1)\sin\phi_3-\sin\phi_1(1-\cos\phi_3)=4\sin\tfrac{1}{2}\phi_1\sin\tfrac{1}{2}(\phi_1-\phi_3)\sin\tfrac{1}{2}\phi_3\,,$$
$$(1+\cos\phi_1)(1-\cos\phi_3)-\sin\phi_1\sin\phi_3=-4\cos\tfrac{1}{2}\phi_1\sin\tfrac{1}{2}(\phi_1-\phi_3)\sin\tfrac{1}{2}\phi_3 \qquad (3.58)$$

in (3. 57), we obtain the final form for $R_{33}^{(2)}(\xi,\xi')$ and $R_{43}^{(2)}(\xi,\xi')$:

$$R_{33}^{(2)}(\xi,\xi')=(2\gamma_1)^2(2\pi)^{-4}\int_{-\pi}^{\pi}\cdots\int_{-\pi}^{\pi}d(1)\cdots d(4)\frac{P(\xi)Q(\xi')}{(1-\xi e^{-i\phi_2})(1-\xi' e^{-i\phi_4})}$$
$$\times\prod_{j=1}^{2}\frac{e^{-iM\phi_{2j-1}-i(N+1)\phi_{2j}}}{\Delta(\phi_{2j-1},\phi_{2j})}(1-e^{-i(\phi_2+\phi_4)})^{-1}\cos\tfrac{1}{2}\phi_1\sin\tfrac{1}{2}(\phi_1-\phi_3)\sin\tfrac{1}{2}\phi_3 \qquad (3.59a)$$

and

$$R_{43}^{(2)}(\xi,\xi')=iz_1^{-1}(2\gamma_1)^2(2\pi)^{-4}\int_{-\pi}^{\pi}\cdots\int_{-\pi}^{\pi}d(1)\cdots d(4)\frac{Q(\xi)Q(\xi')}{(1-\xi e^{-i\phi_2})(1-\xi' e^{-i\phi_4})}$$
$$\times\prod_{j=1}^{2}\frac{e^{-iM\phi_{2j-1}-i(N+1)\phi_{2j}}}{\Delta(\phi_{2j-1},\phi_{2j})}(1-e^{-i(\phi_2+\phi_4)})^{-1}\sin\tfrac{1}{2}\phi_1\sin\tfrac{1}{2}(\phi_1-\phi_3)\sin\tfrac{1}{2}\phi_3\,. \qquad (3.59b)$$

We have used the definition (3. 11) of γ_1 in obtaining (3. 59). If we set $\xi=\xi'=0$ in (3. 59a) we obtain the result originally derived by CW.

For k any even integer, we now assume that

$$R_{33}^{(k)}(\xi,\xi')=(2\gamma_1)^k(2\pi)^{-2k}\int_{-\pi}^{\pi}\cdots\int_{-\pi}^{\pi}d(1)\cdots d(2k)\frac{P(\xi)Q(\xi')}{(1-\xi e^{-i\phi_2})(1-\xi' e^{-i\phi_{2k}})}$$
$$\times\prod_{j=1}^{k}\frac{e^{-iM\phi_{2j-1}-i(N+1)\phi_{2j}}}{\Delta(\phi_{2j-1},\phi_{2j})}\cos\tfrac{1}{2}\phi_1\prod_{j=1}^{k-1}\frac{\sin\frac{1}{2}(\phi_{2j-1}-\phi_{2j+1})}{1-e^{-i(\phi_{2j}+\phi_{2j+2})}}\sin\tfrac{1}{2}\phi_{2k-1} \qquad (3.60a)$$

and

$$R_{43}^{(k)}(\xi,\xi')=iz_1^{-1}(2\gamma_1)^k(2\pi)^{-2k}\int_{-\pi}^{\pi}\cdots\int_{-\pi}^{\pi}d(1)\cdots d(2k)\frac{Q(\xi)Q(\xi')}{(1-\xi e^{-i\phi_2})(1-\xi' e^{-i\phi_{2k}})}$$
$$\times\prod_{j=1}^{k}\frac{e^{-iM\phi_{2j-1}-i(N+1)\phi_{2j}}}{\Delta(\phi_{2j-1},\phi_{2j})}\sin\tfrac{1}{2}\phi_1\prod_{j=1}^{k-1}\frac{\sin\frac{1}{2}(\phi_{2j-1}-\phi_{2j+1})}{1-e^{-i(\phi_{2j}+\phi_{2j+2})}}\sin\tfrac{1}{2}\phi_{2k-1}\,. \qquad (3.60b)$$

We now proceed to demonstrate that these relations are true for k replaced by $k+2$. Using the recursion relations (3. 46) and (3. 47) and the induction hypothesis we have

$$R_{33}^{(k+2)}(\xi,\xi')=(2\pi)^{-2}\int_{-\pi}^{\pi}\int_{-\pi}^{\pi}d(2)d(4)S^{-1}(\bar{2},\xi)[-P_1(2,4)R_{33}^{(k)}(\bar{4},\xi')+P_2(2,4)R_{43}^{(k)}(\bar{4},\xi')]$$
$$=(2\pi)^{-2}\int_{-\pi}^{\pi}\int_{-\pi}^{\pi}d(2)d(4)\frac{P(\xi)Q(\bar{2})}{1-\xi e^{-i\phi_2}}\{-[U(4)S^{-1}(\bar{4},\bar{2})U(2)+V(2)S^{-1}(\bar{2},\bar{4})T(4)]$$
$$\times R_{33}^{(k)}(\bar{4},\xi')+[U(2)S^{-1}(\bar{4},\bar{2})V(4)-V(2)S^{-1}(\bar{2},\bar{4})U(4)]R_{43}^{(k)}(\bar{4},\xi')\}$$
$$=(2\pi)^{-2}\int_{-\pi}^{\pi}\int_{-\pi}^{\pi}d(2)d(4)P(\xi)(1-\xi e^{-i\phi_2})^{-1}(1-e^{-i(\phi_2+\phi_4)})^{-1}\{-[U(2)U(4)Q(\bar{4})$$
$$+V(2)[Q(\bar{2})]^2P(\bar{4})T(4)]R_{33}^{(k)}(\bar{4},\xi')+[U(2)Q(\bar{4})V(4)-V(2)[Q(\bar{2})]^2P(\bar{4})U(4)]R_{43}^{(k)}(\bar{4},\xi')\}$$
$$=(2\pi)^{-2-2k}(2\gamma_1)^k\int_{-\pi}^{\pi}\cdots\int_{-\pi}^{\pi}d(2)d(4)d(5)d(6)\cdots d(2k+4)\frac{P(\xi)Q(\xi')}{(1-\xi e^{-i\phi_2})(1-\xi' e^{-i\phi_{2k+4}})}$$
$$\times\prod_{j=3}^{k+2}\frac{e^{-iM\phi_{2j-1}-i(N+1)\phi_{2j}}}{\Delta(\phi_{2j-1},\phi_{2j})}\prod_{j=1}^{k+1}(1-e^{-i(\phi_{2j}+\phi_{2j+2})})^{-1}D(2,4,5)\prod_{j=3}^{k+1}\sin\tfrac{1}{2}(\phi_{2j-1}-\phi_{2j+1})\sin\tfrac{1}{2}\phi_{2k+3}\,, \qquad (3.61)$$

where in the last line

$$D(2,4,5)=-[U(2)U(4)+V(2)[Q(\bar{2})]^2T(4)[P(\bar{4})]^2]\cos\tfrac{1}{2}\phi_5$$
$$+[U(2)[Q(\bar{4})]^2V(4)-V(2)[Q(\bar{2})]^2U(4)](iz_1^{-1})\sin\tfrac{1}{2}\phi_5\,. \qquad (3.62)$$

Using (3. 7), (3. 55), and (3. 56) in (3. 62) we have

$$D(2,4,5)=(2\pi)^{-4}\int_{-\pi}^{\pi}\int_{-\pi}^{\pi}d(1)d(3)\frac{e^{-iM(\phi_1+\phi_3)-i(N+1)(\phi_2+\phi_4)}}{\Delta(\phi_1,\phi_2)\Delta(\phi_3,\phi_4)}$$
$$\times\{-[-4z_2^2(1-z_1^2)^2\sin\phi_1\sin\phi_3+4z_2^2(1-z_1^2)^2(1+\cos\phi_1)(1-\cos\phi_3)]\cos\tfrac{1}{2}\phi_5$$
$$+[2iz_2\sin\phi_1\,2z_1z_2(1+\cos\phi_3)-2z_1z_2(1+\cos\phi_1)2iz_2\sin\phi_3](iz_1^{-1})\sin\tfrac{1}{2}\phi_5\}, \tag{3.63a}$$

$$D(2,4,5)=(2\gamma_1)^2(2\pi)^{-4}\int_{-\pi}^{\pi}\int_{-\pi}^{\pi}d(1)d(3)\frac{e^{-iM(\phi_1+\phi_3)-i(N+1)(\phi_2+\phi_4)}}{\Delta(\phi_1,\phi_2)\Delta(\phi_3,\phi_4)}$$
$$\times[\cos\tfrac{1}{2}\phi_1\sin\tfrac{1}{2}(\phi_1-\phi_3)\sin\tfrac{1}{2}\phi_3\cos\tfrac{1}{2}\phi_5-\cos\tfrac{1}{2}\phi_1\sin\tfrac{1}{2}(\phi_1-\phi_3)\cos\tfrac{1}{2}\phi_3\sin\tfrac{1}{2}\phi_5], \tag{3.63b}$$

$$D(2,4,5)=(2\gamma_1)^2(2\pi)^{-4}\int_{-\pi}^{\pi}\int_{-\pi}^{\pi}d(1)d(3)\frac{e^{-iM(\phi_1+\phi_3)-i(N+1)(\phi_2+\phi_4)}}{\Delta(\phi_1,\phi_2)\Delta(\phi_3,\phi_4)}\cos\tfrac{1}{2}\phi_1\sin\tfrac{1}{2}(\phi_1-\phi_3)\sin\tfrac{1}{2}(\phi_3-\phi_5), \tag{3.64}$$

where we used the trigonometric identities (3.58) in going from (3.63a) to (3.63b). Substitution of the expression (3.64) for $D(2,4,5)$ into (3.61) gives (3.60a) with k replaced by $k+2$.

The induction proof for $R_{33}^{(k)}(\xi,\xi')$, k any even integer, is similar. Substituting $\xi=\xi'=0$ into (3.60a) and writing the even integer k as $2n$ we obtain (3.52). Equations (3.51) and (3.52) give us the perturbation expansion of the quantity f_{MN} defined by (3.17).

It will be convenient to define

$$f_{MN}^{(2k)}=R_{33}^{(2k)}(0,0), \tag{3.65}$$

where $R_{33}^{(2k)}(0,0)$ is given by (3.52).

E. $\langle\sigma_{0,0}\sigma_{M,N}\rangle$ in terms of f_{MN}

1. *Basic notation*

The correlation function $\langle\sigma_{0,0}\sigma_{M,N}\rangle$ can in view of (3.17) be expressed as

$$\langle\sigma_{0,0}\sigma_{M,N}\rangle=S_\infty^<\left(\prod_{k=N}^{\infty}f_{Mk}\right)^{-1}$$
$$=S_\infty^<\exp\left(-\sum_{k=N}^{\infty}\ln f_{Mk}\right), \tag{3.66}$$

where $S_\infty^<$ is given by (2.10). Recalling (3.51) and definition (3.65) we define $F_{<MN}$ by

$$F_{<MN}=\sum_{k=N}^{\infty}\ln\left(1+\sum_{n=1}^{\infty}f_{Mk}^{(2n)}\right). \tag{3.67}$$

Expanding the logarithm in (3.67) we obtain the expansion

$$F_{<MN}=\sum_{n=1}^{\infty}F_{<MN}^{(2n)}, \tag{3.68}$$

with

$$F_{<MN}^{(2)}=\sum_{k=N}^{\infty}f_{Mk}^{(2)}, \tag{3.69a}$$

$$F_{<MN}^{(4)}=\sum_{k=N}^{\infty}[f_{Mk}^{(4)}-\tfrac{1}{2}(f_{Mk}^{(2)})^2], \tag{3.69b}$$

$$F_{<MN}^{(6)}=\sum_{k=N}^{\infty}[f_{Mk}^{(6)}-f_{Mk}^{(2)}f_{Mk}^{(4)}+\tfrac{1}{3}(f_{Mk}^{(2)})^3], \tag{3.69c}$$

etc.

The remainder of this section and all of Sec. III F will be devoted to proving that $F_{<MN}^{(2n)}$ is given by (2.12). We first show that (2.12) is true for $n=1$ and $n=2$. This will then motivate the general case which is proved in Sec. III F.

2. $F_{<MN}^{(2)}$

Using (3.52) for $n=1$ [though first let $\phi_1\to\phi_3$ and $\phi_2\to\phi_4$ and add this to (3.52)], (3.65), and (3.69a) we have

$$F_{<MN}^{(2)}=\tfrac{1}{2}(2\gamma_1)^2(2\pi)^{-4}\int_{-\pi}^{\pi}d\phi_1\cdots\int_{-\pi}^{\pi}d\phi_4$$
$$\times\frac{e^{-iM(\phi_1+\phi_3)-i(N+1)(\phi_2+\phi_4)}}{\Delta(\phi_1,\phi_2)\Delta(\phi_3,\phi_4)}(1-e^{-i(\phi_2+\phi_4)})^{-2}$$
$$\times\sin\tfrac{1}{2}(\phi_1-\phi_3)\sin\tfrac{1}{2}(\phi_3-\phi_1)$$

(recall $\mathrm{Im}\phi_j<0$, $j=1,2,3,4$).

Multiplying numerator and denominator by $e^{i(\phi_2+\phi_4)}$ we have

$$F_{<MN}^{(2)}=(-1)\gamma_1^2\tfrac{1}{2}(2\pi)^{-4}\int_{-\pi}^{\pi}d\phi_1\cdots\int_{-\pi}^{\pi}d\phi_4$$
$$\times\prod_{j=1}^{2}\frac{e^{-iM\phi_{2j-1}-iN\phi_{2j}}}{\Delta(\phi_{2j-1},\phi_{2j})}\frac{\sin\tfrac{1}{2}(\phi_{2j-1}-\phi_{2j+1})}{\sin\tfrac{1}{2}(\phi_{2j}+\phi_{2j+2})}, \tag{3.70}$$

where $\phi_5\equiv\phi_1$ and $\phi_6\equiv\phi_2$. This is (2.12) for $n=1$.

3. $F_{<MN}^{(4)}$

Using (3.52) for $n=2$ and (3.65) we have

$$\sum_{k=N}^{\infty}f_{Mk}^{(4)}=(2\gamma_1)^4(2\pi)^{-8}\int_{-\pi}^{\pi}d\phi_1\cdots\int_{-\pi}^{\pi}d\phi_8\prod_{j=1}^{4}\frac{e^{-iM\phi_{2j-1}-i(N+1)\phi_{2j}}}{\Delta(\phi_{2j-1},\phi_{2j})}$$

$$\times\left(1-e^{-i(\phi_2+\phi_4+\phi_6+\phi_8)}\right)^{-1}\cos\tfrac{1}{2}\phi_1\prod_{j=1}^{3}\left(\frac{\sin\frac{1}{2}(\phi_{2j-1}-\phi_{2j+1})}{1-e^{-i(\phi_{2j}+\phi_{2j+2})}}\right)\sin\tfrac{1}{2}\phi_7\,. \tag{3.71}$$

Making the change of variables $\phi_1\to\phi_7$, $\phi_3\to\phi_5$, $\phi_5\to\phi_3$, $\phi_7\to\phi_1$ and similarly for ϕ_{2j} in (3.71) we obtain the equivalent expression

$$\sum_{k=N}^{\infty}f_{MK}^{(4)}=-(2\gamma_1)^4(2\pi)^{-8}\int_{-\pi}^{\pi}d\phi_1\cdots\int_{-\pi}^{\pi}d\phi_8\prod_{j=1}^{4}\left(\frac{e^{-iM\phi_{2j-1}-i(N+1)\phi_{2j}}}{\Delta(\phi_{2j-1},\phi_{2j})}\right)$$

$$\times\left(1-e^{-i(\phi_2+\phi_4+\phi_6+\phi_8)}\right)^{-1}\cos\tfrac{1}{2}\phi_7\prod_{j=1}^{3}\left(\frac{\sin\frac{1}{2}(\phi_{2j-1}-\phi_{2j+1})}{1-e^{-i(\phi_{2j}+\phi_{2j+2})}}\right)\sin\tfrac{1}{2}\phi_1\,. \tag{3.72}$$

Adding (3.71) and (3.72) and dividing by 2 we obtain

$$\sum_{k=N}^{\infty}f_{MK}^{(4)}=\tfrac{1}{2}(2\gamma_1)^4(2\pi)^{-8}\int_{-\pi}^{\pi}d\phi_1\cdots\int_{-\pi}^{\pi}d\phi_8\prod_{j=1}^{4}\left(\frac{e^{-iM\phi_{2j-1}-i(N+1)\phi_{2j}}}{\Delta(\phi_{2j-1},\phi_{2j})}\sin\tfrac{1}{2}(\phi_{2j-1}-\phi_{2j+1})\right)$$

$$\times\left(\prod_{j=1}^{3}\left(1-e^{-i(\phi_{2j}+\phi_{2j+2})}\right)^{-1}\right)\left(1-e^{-i(\phi_2+\phi_4+\phi_6+\phi_8)}\right)^{-1}, \tag{3.73}$$

with $\phi_{10}\equiv\phi_2$, $\phi_9\equiv\phi_1$.

We want to obtain the factor

$$\prod_{j=1}^{4}\left(1-e^{-i(\phi_{2j}+\phi_{2j+2})}\right)^{-1},$$

from the factor

$$\prod_{j=1}^{3}\left(1-e^{-i(\phi_{2j}+\phi_{2j+2})}\right)^{-1},$$

that occurs in (3.73). To accomplish this we cyclically permute the integration variables in (3.73). Under these cyclic permutations the only noninvariant factor in (3.73) is

$$\prod_{j=1}^{3}\left(1-e^{-i(\phi_{2j}+\phi_{2j+2})}\right)^{-1}.$$

Doing this (3.73) becomes

$$\sum_{k=N}^{\infty}f_{MK}^{(4)}=\tfrac{1}{2}(2\gamma_1)^4(2\pi)^{-8}\int_{-\pi}^{\pi}d\phi_1\cdots\int_{-\pi}^{\pi}d\phi_8\prod_{j=1}^{4}\left(\frac{e^{-iM\phi_{2j-1}-iN\phi_{2j}}}{\Delta(\phi_{2j-1},\phi_{2j})}\,\frac{\sin\frac{1}{2}(\phi_{2j-1}-\phi_{2j+1})}{1-e^{-i(\phi_{2j}+\phi_{2j+2})}}\right)A_4(\phi_2,\phi_4,\phi_6,\phi_8)\,, \tag{3.74}$$

where

$$A_4(\phi_2,\phi_4,\phi_6,\phi_8)=e^{-i(\phi_2+\phi_4+\phi_6+\phi_8)}\left(1-e^{-i(\phi_2+\phi_4+\phi_6+\phi_8)}\right)^{-1}\frac{1}{4}\left(4-\sum_{j=1}^{4}e^{-i(\phi_{2j}+\phi_{2j+2})}\right). \tag{3.75}$$

We can rewrite A_4 as

$$A_4(\phi_2,\phi_4,\phi_6,\phi_8)=\tfrac{1}{2}e^{-i(\phi_2+\phi_4+\phi_6+\phi_8)}+\tfrac{1}{4}e^{-i(\phi_2+\phi_4+\phi_6+\phi_8)}\left(1-e^{-i(\phi_2+\phi_4+\phi_6+\phi_8)}\right)^{-1}$$

$$\times\left(2+2e^{-i(\phi_2+\phi_4+\phi_6+\phi_8)}-\sum_{j=1}^{4}e^{-i(\phi_{2j}+\phi_{2j+2})}\right). \tag{3.76}$$

Using (3.76) in (3.74)

$$\sum_{k=N}^{\infty}f_{MK}^{(4)}=F_{<MN}^{(4)}+R_{MN}^{(4)}\,,$$

where $F_{<MN}^{(4)}$ is the right-hand side of (2.12) for $n=2$ and

$$R_{MN}^{(4)}=\tfrac{1}{2}(2\gamma_1)^4(2\pi)^{-8}\int_{-\pi}^{\pi}d\phi_1\cdots\int_{-\pi}^{\pi}d\phi_8\prod_{j=1}^{4}\left(\frac{e^{-iM\phi_{2j-1}-iN\phi_{2j}}}{\Delta(\phi_{2j-1},\phi_{2j})}\,\frac{\sin\frac{1}{2}(\phi_{2j-1}-\phi_{2j+1})}{1-e^{-i(\phi_{2j}+\phi_{2j+2})}}\right)$$

$$\times\left[\tfrac{1}{4}e^{-i(\phi_2+\phi_4+\phi_6+\phi_8)}(1-e^{-i(\phi_2+\phi_4+\phi_6+\phi_8)})^{-1}\left(2+2e^{-i(\phi_2+\phi_4+\phi_6+\phi_8)}-\sum_{j=1}^{4}e^{-i(\phi_{2j}+\phi_{2j+2})}\right)\right]. \tag{3.77}$$

Thus $F^{(4)}_{<MN}$ as defined by (3.69b) will be given by (2.12) for $n=2$ if we can show that

$$R^{(4)}_{MN}-\frac{1}{2}\sum_{k=N}^{\infty}(f^{(2)}_{Mk})^2=0. \tag{3.78}$$

Letting $\phi_2\leftrightarrow\phi_4$ and $\phi_1\leftrightarrow\phi_3$ in (3.52) for $n=1$, adding this to (3.52) for $n=1$ and summing we obtain

$$-\frac{1}{2}\sum_{k=N}^{\infty}(f^{(2)}_{Mk})^2=-\tfrac{1}{8}(2\gamma_1)^4(2\pi)^{-8}\int_{-\pi}^{\pi}d\phi_1\cdots\int_{-\pi}^{\pi}d\phi_8\left(\prod_{j=1}^{4}\frac{e^{-iM\phi_{2j-1}-iN\phi_{2j}}}{\Delta(\phi_{2j-1},\phi_{2j})}\right)$$
$$\times e^{-i(\phi_2+\cdots+\phi_8)}(1-e^{-i(\phi_2+\cdots+\phi_8)})^{-1}\sin^2\tfrac{1}{2}(\phi_1-\phi_3)\sin^2\tfrac{1}{2}(\phi_5-\phi_7)\,[(1-e^{-i(\phi_2+\phi_4)})(1-e^{-i(\phi_6+\phi_8)})]^{-1}. \tag{3.79}$$

In (3.79) the sine factors loop as $1\to3\to1$ and $5\to7\to5$ as compared with $1\to3\to5\to7\to1$ of (3.77). To compare $R^{(4)}_{MN}$ of (3.77) with (3.79) it is useful to make one loop in (3.79) rather than two separate loops. This is done by use of the identity

$$\sin\tfrac{1}{2}(\phi_a-\phi_b)\sin\tfrac{1}{2}(\phi_c-\phi_d)+\sin\tfrac{1}{2}(\phi_a-\phi_d)\sin\tfrac{1}{2}(\phi_b-\phi_c)=\sin\tfrac{1}{2}(\phi_a-\phi_c)\sin\tfrac{1}{2}(\phi_b-\phi_d). \tag{3.80}$$

Letting $a\to1$, $b\to5$, $c\to3$, and $d\to7$ in (3.80) and using this in (3.79) we obtain

$$-\frac{1}{2}\sum_{k=N}^{\infty}(f^{(2)}_{Mk})^2=-\tfrac{1}{8}(2\gamma_1)^4(2\pi)^{-8}\int_{-\pi}^{\pi}d\phi_1\cdots\int_{-\pi}^{\pi}d\phi_8\prod_{j=1}^{4}\frac{e^{-iM\phi_{2j-1}-iN\phi_{2j}}}{\Delta(\phi_{2j-1},\phi_{2j})}$$
$$\times e^{-i(\phi_2+\cdots+\phi_8)}(1-e^{-i(\phi_2+\cdots+\phi_8)})^{-1}[(1-e^{-i(\phi_2+\phi_4)})(1-e^{-i(\phi_6+\phi_8)})]^{-1}$$
$$\times[\sin\tfrac{1}{2}(\phi_1-\phi_5)\sin\tfrac{1}{2}(\phi_3-\phi_7)+\sin\tfrac{1}{2}(\phi_3-\phi_5)\sin\tfrac{1}{2}(\phi_7-\phi_1)]\sin\tfrac{1}{2}(\phi_1-\phi_3)\sin\tfrac{1}{2}(\phi_5-\phi_7). \tag{3.81}$$

One part of (3.81) has the correct $1\to3\to\cdots7\to1$ structure. The other part can be brought into this form by relabeling the integration variables ($5\leftrightarrow7$, $6\leftrightarrow8$). Doing this (3.81) becomes

$$-\frac{1}{2}\sum_{k=N}^{\infty}(f^{(2)}_{Mk})^2=-\tfrac{1}{4}(2\gamma_1)^4(2\pi)^{-8}\int_{-\pi}^{\pi}d\phi_1\cdots\int_{-\pi}^{\pi}d\phi_8\prod_{j=1}^{4}\frac{e^{-iM\phi_{2j-1}-iN\phi_{2j}}}{\Delta(\phi_{2j-1},\phi_{2j})}$$
$$\times\sin\tfrac{1}{2}(\phi_{2j-1}-\phi_{2j+1})e^{-i(\phi_2+\cdots+\phi_8)}[1-e^{-i(\phi_2+\cdots+\phi_8)}]^{-1}[(1-e^{-i(\phi_2+\phi_4)})(1-e^{-i(\phi_6+\phi_8)})]^{-1}. \tag{3.82}$$

Finally, if we add to (3.82) the relabeled version $2\to4\to6\to8\to2$ and $1\to3\to5\to7\to1$, and divide by 2, then

$$-\frac{1}{2}\sum_{k=N}^{\infty}(f^{(2)}_{Mk})^2=-\tfrac{1}{8}(2\gamma_1)^4(2\pi)^{-8}\int_{-\pi}^{\pi}d\phi_1\cdots\int_{-\pi}^{\pi}d\phi_8\prod_{j=1}^{4}\left(\frac{e^{-iM\phi_{2j-1}-iN\phi_{2j}}}{\Delta(\phi_{2j-1},\phi_{2j})}\right.$$
$$\left.\times\sin\tfrac{1}{2}(\phi_{2j-1}-\phi_{2j+1})(1-e^{-i(\phi_{2j}+\phi_{2j+2})})^{-1}\right)\left(2+2e^{-i(\phi_2+\phi_4+\phi_6+\phi_8)}-\sum_{j=1}^{4}e^{-i(\phi_{2j}+\phi_{2j+2})}\right)$$
$$\times e^{-i(\phi_2+\phi_4+\phi_6+\phi_8)}[1-e^{-i(\phi_2+\phi_4+\phi_6+\phi_8)}]^{-1}. \tag{3.83}$$

Comparing (3.83) and (3.77) we see that (3.78) holds; and hence $F^{(4)}_{<MN}$ is given by (2.12) for $n=2$.

F. $F^{(2n)}_{<MN}$ for arbitrary n

We now prove (2.12) for arbitrary n. We begin with the definition

$$\tilde F^{(2n)}_{<MN}=(-1)^n\gamma_1^{2n}(2n)^{-1}(2\pi)^{-4n}\int_{-\pi}^{\pi}d\phi_1\cdots\int_{-\pi}^{\pi}d\phi_{4n}$$
$$\times\prod_{j=1}^{2n}\left(\frac{e^{-iM\phi_{2j-1}-iN\phi_{2j}}}{\Delta(\phi_{2j-1},\phi_{2j})}\;\frac{\sin\frac{1}{2}(\phi_{2j-1}-\phi_{2j+1})}{\sin\frac{1}{2}(\phi_{2j}+\phi_{2j+2})}\right)$$
$$\times(1-e^{-i(\phi_2+\phi_4+\cdots+\phi_{4n})}). \tag{3.84}$$

Suppose we can prove that

$$\tilde F_{MN}\equiv\sum_{n=1}^{\infty}\tilde F^{(2n)}_{<MN}=\ln\left(1+\sum_{n=1}^{\infty}f^{(2n)}_{MN}\right), \tag{3.85}$$

where $\tilde F^{(2n)}_{<MN}$ is given by (3.84) and $f^{(2n)}_{MN}$ by (3.52). Then from (3.85) it follows that $F^{(2n)}_{<MN}$ is given by (2.12). To see this we note that by definition [see (3.66)–(3.68)] of $F^{(2n)}_{<MN}$

$$\langle\sigma_{0,0}\sigma_{M,N}\rangle=S^{<}_{\infty}\exp\left(-\sum_{n=1}^{\infty}F^{(2n)}_{<MN}\right), \tag{3.86}$$

and from (3.66) and the hypothesis (3.85)

$$\langle \sigma_{0,0}\sigma_{M,N}\rangle = S_\infty^< \exp\left(-\sum_{k=N}^{\infty} \ln\left(1+\sum_{n=1}^{\infty} f_{Mk}^{(2n)}\right)\right)$$

$$= S_\infty^< \exp\left(-\sum_{k=N}^{\infty}\sum_{n=1}^{\infty} \tilde{F}_{<Mk}^{(2n)}\right)$$

$$= S_\infty^< \exp\left(-\sum_{n=1}^{\infty}\sum_{k=N}^{\infty} \tilde{F}_{<Mk}^{(2n)}\right). \qquad (3.87)$$

Then from (3.86) and (3.87)

$$F_{<MN}^{(2n)} = \sum_{k=N}^{\infty} \tilde{F}_{<Mk}^{(2n)}. \qquad (3.88)$$

But doing the summation in (3.88) is trivial [see (3.84)]; and hence, $F_{<MN}^{(2n)}$ is given by (2.12). Hence if we establish (3.85) with $\tilde{F}_{<MN}^{(2n)}$ defined by (3.84) we have proved (2.12).

We define the generating functions

$$f(x) = \sum_{n=0}^{\infty} f_{MN}^{(2n)} x^n \qquad (3.89)$$

and

$$F(x) = \sum_{n=0}^{\infty} \tilde{F}_{<MN}^{(2n)} x^n, \qquad (3.90)$$

where $f_{MN}^{(0)} = 1$, $\tilde{F}_{<MN}^{(0)} = 0$, then (3.85) is a special case ($x = 1$) of

$$f(x) = e^{F(x)}. \qquad (3.91)$$

Differentiation of (3.91) gives

$$f'(x) = F'(x) f(x) \qquad (3.92)$$

or, equivalently,

$$n f_{MN}^{(2n)} = \sum_{l=1}^{n} l \tilde{F}_{<MN}^{(2l)} f_{MN}^{2(n-l)}. \qquad (3.93)$$

Therefore, (3.93) and the boundary conditions $f(0) = 1$, $F(0) = 0$ will imply (2.12). We now prove (3.93).

The variables in the identity

$$e^{-i(\phi_2+\phi_4+\phi_6+\cdots+\phi_{4n})} - 1 = (e^{-i(\phi_2+\phi_4+\phi_6+\cdots+\phi_{4n})} - e^{-i(\phi_6+\phi_8+\cdots+\phi_{4n})}) + (e^{-i(\phi_6+\phi_8+\cdots+\phi_{4n})} - e^{-i(\phi_{10}+\phi_{12}+\cdots+\phi_{4n})})$$
$$+ (e^{-i(\phi_{10}+\phi_{12}+\cdots+\phi_{4n})} - e^{-i(\phi_{14}+\phi_{16}+\cdots+\phi_{4n})}) + \cdots + (e^{-i(\phi_{4n-2}+\phi_{4n})} - 1) \qquad (3.94)$$

may be cyclically permuted and the identity (3.94) will continue to be valid *when* used in the integrand of (3.84) since the integrand in (3.84) is invariant under cyclic permutations of the integration variable labels.

Thus the identity

$$e^{-i(\phi_2+\phi_4+\phi_6+\cdots+\phi_{4n})} - 1 = -(1 - e^{-i(\phi_{4n}+\phi_2)})(e^{-i(\phi_4+\phi_6+\cdots+\phi_{4n-2})} + e^{-i(\phi_4+\phi_6+\cdots+\phi_{4n-6})}$$
$$+ e^{-i(\phi_4+\phi_6+\cdots+\phi_{4n-10})} + \cdots + 1) \qquad (3.95)$$

when substituted into (3.84) gives, with use of the identity

$$1 - e^{-i(\phi_{4n}+\phi_2)} = 2i \sin\tfrac{1}{2}(\phi_{4n}+\phi_2) e^{-1(\phi_{4n}+\phi_2)/2}, \qquad (3.96)$$

$$\tilde{F}_{<MN}^{(2n)} = (-1)^n \gamma_1^{2n} (2n)^{-1} (2\pi)^{-4n} \int_{-\pi}^{\pi} d\phi_1 \cdots \int_{-\pi}^{\pi} d\phi_{4n} \prod_{j=1}^{2n} \left\{ \frac{e^{-iM\phi_{2j-1}-iN\phi_{2j}}}{\Delta(\phi_{2j-1},\phi_{2j})} \sin\tfrac{1}{2}(\phi_{2j-1}-\phi_{2j+1}) \right\}$$
$$\times \prod_{j=1}^{2n-1} \frac{1}{\sin\frac{1}{2}(\phi_{2j}+\phi_{2j+2})} \, 2i e^{-i(\phi_{4n}+\phi_2)/2} \sum_{j=0}^{n-1} e^{-i(\phi_4+\phi_6+\cdots+\phi_{4n-2-4j})}, \qquad (3.97)$$

where the last term in the sum in (3.97) is to be interpreted as 1 if $n = 1$. From (3.52) and (3.65) we have

$$f_{MN}^{(2n)} = (2\gamma_1)^{2n} (2\pi)^{-4n} \int_{-\pi}^{\pi} d\phi_1 \cdots \int_{-\pi}^{\pi} d\phi_{4n} \prod_{j=1}^{2n} \left(\frac{e^{-iM\phi_{2j-1}-iN\phi_{2j}}}{\Delta(\phi_{2j-1},\phi_{2j})} \right)$$
$$\times \prod_{j=1}^{2n-1} \left(\frac{\sin\frac{1}{2}(\phi_{2j-1}-\phi_{2j+1})}{2i\sin\frac{1}{2}(\phi_{2j}+\phi_{2j+2})} \right) e^{-i(\phi_2+\phi_{4n})/2} \cos\tfrac{1}{2}\phi_1 \sin\tfrac{1}{2}\phi_{4n-1}. \qquad (3.98)$$

Letting $\phi_1 \to \phi_{4n-1}$, $\phi_2 \to \phi_{4n-2}$, ... in (3.98), adding the result to (3.98), and dividing by two we obtain,

$$f_{MN}^{(2n)} = \gamma_1^{2n} (2\pi)^{-4n} (-1)^n \int_{-\pi}^{\pi} d\phi_1 \cdots \int_{-\pi}^{\pi} d\phi_{4n} \prod_{j=1}^{2n} \left(\frac{e^{-iM\phi_{2j-1}-iN\phi_{2j}}}{\Delta(\phi_{2j-1},\phi_{2j})} \sin\tfrac{1}{2}(\phi_{2j-1}-\phi_{2j+1}) \right)$$
$$\times \prod_{j=1}^{2n-1} \left(\frac{1}{\sin\frac{1}{2}(\phi_{2j}+\phi_{2j+2})} \right) i e^{-i(\phi_2+\phi_{4n})/2}. \qquad (3.99)$$

Using (3.97) and (3.99) we find

$$\sum_{l=1}^{n-1} l\tilde{F}^{(2l)}_{<MN} f^{2(n-l)}_{MN} = (-1)^n \gamma_1^{2n} (2\pi)^{-4n} \sum_{l=1}^{n-1} \int_{-\pi}^{\pi} d\phi_1 \cdots \int_{-\pi}^{\pi} d\phi_{4l} \prod_{j=1}^{2l} \left(\frac{e^{-iM\phi_{2j-1} - iN\phi_{2j}}}{\Delta(\phi_{2j-1}, \phi_{2j})} \sin\tfrac{1}{2}(\phi_{2j-1} - \phi_{2j+1}) \right)$$
$$\times \prod_{j=1}^{2l-1} \left(\frac{1}{\sin\frac{1}{2}(\phi_{2j} + \phi_{2j+2})} \right) ie^{-i(\phi_2+\phi_{4l})/2} \sum_{j=0}^{l-1} e^{-i(\phi_4+\phi_6+\cdots+\phi_{4l-2-4j})}$$
$$\times \int_{-\pi}^{\pi} d\phi_{4l+1} \cdots \int_{-\pi}^{\pi} d\phi_{4n} \prod_{j=2l+1}^{2n} \frac{e^{-iM\phi_{2j-1} - iN\phi_{2j}}}{\Delta(\phi_{2j-1}, \phi_{2j})} \sin\tfrac{1}{2}(\phi_{2j-1} - \phi_{2j+1})$$
$$\times \prod_{j=2l+1}^{2n-1} \left(\frac{1}{\sin\frac{1}{2}(\phi_{2j} + \phi_{2j+2})} \right) ie^{-i(\phi_{4l+2}+\phi_{4n})} . \tag{3.100}$$

Using (3.80) ($a \to 4l-1$, $b \to 1$, $c \to 4n-1$, and $d \to 4l+1$) in (3.100), we find that one term has the correct $1 \to 3 \to 5 \to \cdots \to 4n-1 \to 1$ structure, and the other term can be brought to this form by relabeling variables [this is the same argument as that following (3.81)]. Using this new form of (3.100) and adding $n\tilde{F}^{(2n)}_{<MN}$ to this [see (3.97)] we obtain

$$n\tilde{F}^{(2n)}_{<MN} + \sum_{l=1}^{n-1} l\tilde{F}^{(2l)}_{<MN} f^{2(n-l)}_{MN} = i(-1)^n \gamma_1^{2n} (2\pi)^{-4n} \int_{-\pi}^{\pi} d\phi_1 \cdots \int_{-\pi}^{\pi} d\phi_{4n} \prod_{j=1}^{2n} \frac{e^{-iM\phi_{2j-1} - iN\phi_{2j}}}{\Delta(\phi_{2j-1}, \phi_{2j})} \sin\tfrac{1}{2}(\phi_{2j-1} - \phi_{2j+1})$$
$$\times \prod_{j=1}^{2n-1} \frac{1}{\sin\frac{1}{2}(\phi_{2j} + \phi_{2j+2})} \tilde{H}(\phi_2, \phi_4, \ldots, \phi_{4n}) , \tag{3.101}$$

where

$$\tilde{H}(\phi_2, \phi_4, \ldots, \phi_{4n}) = e^{-i(\phi_2+\phi_{4n})/2} \sum_{j=0}^{n-1} e^{-i(\phi_4+\phi_6+\cdots+\phi_{4n-2-4j})} + \sum_{l=1}^{n-1} e^{-i(\phi_2+\phi_{4l})/2}$$
$$\times \left(\sum_{k=0}^{l-1} e^{-i(\phi_4+\phi_6+\cdots+\phi_{4l-2-4k})} e^{-i(\phi_{4l+2}+\phi_{4n})/2} \left[e^{i(\phi_{4l}+\phi_{4l+2})/2} - e^{-i(\phi_{4l}+\phi_{4l+2})/2} \right] \right) . \tag{3.102}$$

If we can show that $\tilde{H}(\phi_2, \phi_4, \ldots, \phi_{4n})$ can be replaced by $ne^{-i(\phi_2+\phi_{4n})/2}$ in (3.101), we will have established (3.93) [see (3.99)].

We can rewrite (3.102) as

$$\tilde{H}(\phi_2, \phi_4, \ldots, \phi_{4n}) = e^{-i(\phi_2+\phi_{4n})/2} \left(\sum_{l=0}^{n-1} e^{-i(\phi_4+\phi_6+\cdots+\phi_{4n-2-4l})} + \sum_{l=1}^{n-1} (1 - e^{-i(\phi_{4l}+\phi_{4l+2})}) \right.$$
$$\left. \times \sum_{k=0}^{l-1} e^{-i(\phi_4+\phi_6+\cdots+\phi_{4l-2-4k})} \right) , \tag{3.103a}$$

$$\tilde{H}(\phi_2, \phi_4, \ldots, \phi_{4n}) = e^{-i(\phi_2+\phi_{4n})/2} H(\phi_2, \ldots, \phi_{4n}) , \tag{3.103b}$$

where Eq. (3.103b) defines the function $H(\phi_2, \ldots, \phi_{4n})$. Now

$$\sum_{l=1}^{n-1} \sum_{k=0}^{l-1} e^{-i(\phi_4+\phi_6+\cdots+\phi_{4l-2-4k})} = (n-1) + (n-2)e^{-i(\phi_4+\phi_6)} + (n-3)e^{-i(\phi_4+\phi_6+\phi_8+\phi_{10})} + \cdots + e^{-i(\phi_4+\phi_6+\cdots+\phi_{4n-6})} ,$$

so that

$$H(\phi_2, \ldots, \phi_{4n}) = R(\phi_2, \ldots, \phi_{4n}) + n , \tag{3.104}$$

where

$$R(\phi_2, \ldots, \phi_{4n}) = \sum_{l=0}^{n-2} (l+1) e^{-i(\phi_4+\phi_6+\cdots+\phi_{4n-2-4l})} - \sum_{l=0}^{n-1} e^{-i(\phi_{4l}+\phi_{4l+2})} \sum_{k=0}^{l-1} e^{-i(\phi_4+\phi_6+\cdots+\phi_{4l-2-4k})} . \tag{3.105}$$

We now show that the integral (3.101) with $\tilde{H}$ replaced by $e^{-i(\phi_2+\phi_{4n})/2}R$ is zero. From (3.104) it will then follow that (3.93) is true. If we multiply the numerator and denominator in (3.101) by $\sin\frac{1}{2}(\phi_2 + \phi_{4n})$, then the factor

$$\prod_{j=1}^{2n} \left(\frac{e^{-iM\phi_{2j-1} - iN\phi_{2j}}}{\Delta(\phi_{2j-1}, \phi_{2j})} \frac{\sin\frac{1}{2}(\phi_{2j-1} - \phi_{2j+1})}{\sin\frac{1}{2}(\phi_{2j} + \phi_{2j+2})} \right)$$

occurring in (3.101) is invariant under cyclic permutations of the variables $(\phi_2, \phi_4, \ldots, \phi_{4n})$. Hence the value of the integral (3.101) (with $\bar{H} \to e^{-i(\phi_2+\phi_{4n})/2}R$) is not affected if we cyclically permute the variables of the factor

$$R' \equiv e^{i(\phi_2+\phi_{4n})/2} - e^{-i(\phi_2+\phi_{4n})/2} e^{-i(\phi_2+\phi_{4n})/2} R. \tag{3.106}$$

From (3.105) we obtain

$$R' = A_1 - A_2 - A_3 + A_4, \tag{3.107}$$

where

$$A_1 = \sum_{l=0}^{n-2} (l+1) e^{-i(\phi_4+\phi_6+\cdots+\phi_{4n-2-4l})}, \tag{3.108}$$

$$A_2 = \sum_{l=1}^{n-1} e^{-i(\phi_{4l}+\phi_{4l+2})} \times \sum_{k=0}^{l-1} e^{-i(\phi_4+\phi_6+\cdots+\phi_{4l-2-4k})}, \tag{3.109}$$

$$A_3 = \sum_{l=0}^{n-2} (l+1) e^{-i(\phi_2+\phi_{4n})} e^{-i(\phi_4+\phi_6+\cdots+\phi_{4n-2-4l})}, \tag{3.110}$$

$$A_4 = \sum_{l=1}^{n-1} e^{-i(\phi_{4l}+\phi_{4l+2})} e^{-i(\phi_2+\phi_{4n})} \times \sum_{k=0}^{l-1} e^{-i(\phi_4+\phi_6+\cdots+\phi_{4l-2-4k})}. \tag{3.111}$$

In (3.108) we cyclically permute the variables so that each term in the sum A_1 starts at ϕ_2. That is,

$$A_1 = \sum_{l=0}^{n-2} (l+1) e^{-i(\phi_2+\phi_4+\cdots+\phi_{4n-4-4l})}. \tag{3.112}$$

Note that we use an equality sign in (3.112) only in the sense that A_1 [as defined by (3.108)] and the right-hand side of (3.112) yield the same value when integrated over in (3.101). We will continue to use this sense of equality for the remainder of the proof. We rewrite A_2 as follows:

$$A_2 = \sum_{j=2}^{n-1} e^{-i(\phi_2+\phi_4+\cdots+\phi_{4n-4j})} \times \sum_{l=0}^{j-2} e^{-i(\phi_{4n-4j+2+4l}+\phi_{4n-4j+4l+4})} + (n-1) e^{-i(\phi_2+\phi_4)}. \tag{3.113}$$

To see how (3.113) follows from (3.109) we note the following transformations: from the $l=n-1$ and $k=0$ term

$$e^{-i(\phi_{4n-4}+\phi_{4n-2})} e^{-i(\phi_4+\phi_6+\cdots+\phi_{4n-6})} \to e^{-i(\phi_2+\phi_4+\cdots+\phi_{4n-8})} e^{-i(\phi_{4n-6}+\phi_{4n-4})};$$

from the $l=n-2$, $k=0$ and $l=n-1$, $k=1$ terms

$$e^{-i(\phi_{4n-8}+\phi_{4n-6})} e^{-i(\phi_4+\phi_6+\cdots+\phi_{4n-10})} + e^{-i(\phi_{4n-4}+\phi_{4n-2})} e^{-i(\phi_4+\phi_6+\cdots+\phi_{4n-10})}$$
$$\to e^{-i(\phi_2+\phi_4+\cdots+\phi_{4n-12})} (e^{-i(\phi_{4n-10}+\phi_{4n-8})} + e^{-i(\phi_{4n-6}+\phi_{4n-4})})$$
$$= e^{-i(\phi_2+\phi_4+\cdots+\phi_{4n-12})} \sum_{l=0}^{1} e^{-i(\phi_{4n-12+2+4l}+\phi_{4n-12+4+4l})};$$

from the $l=n-3$, $k=0$, $l=n-2$, $k=1$, and $l=n-1$, $k=2$ terms

$$e^{-i(\phi_{4n-12}+\phi_{4n-10})} e^{-i(\phi_4+\phi_6+\cdots+\phi_{4n-14})} + e^{-i(\phi_{4n-8}+\phi_{4n-6})} e^{-i(\phi_4+\phi_6+\cdots+\phi_{4n-14})}$$
$$+ e^{-i(\phi_{4n-4}+\phi_{4n-2})} e^{-i(\phi_2+\phi_4+\cdots+\phi_{4n-14})} \to e^{-i(\phi_2+\phi_4+\cdots+\phi_{4n-16})} \sum_{l=0}^{2} e^{-i(\phi_{4n-16+2+4l}+\phi_{4n-16+4+4l})}, \text{ etc.}$$

The $(n-1)e^{-i(\phi_2+\phi_4)}$ term comes from the $l=1$, $k=0$; $l=2$, $k=1$; ...; $l=n-1$, $k=n-2$ terms.

We now make the following transformations on the terms in (3.110):

the $l=0$ term

$$\phi_2+\phi_{4n}+\phi_4+\phi_6+\cdots+\phi_{4n-2} \to \phi_2+\phi_4+\cdots+\phi_{4n},$$

the $l=1$ term

$$\phi_2+\phi_{4n}+\phi_4+\phi_6+\cdots+\phi_{4n-6} \to \phi_2+\phi_4+\cdots+\phi_{4n-4},$$

the $l=2$ term

$$\phi_2+\phi_{4n}+\phi_4+\phi_6+\cdots+\phi_{4n-10} \to \phi_2+\phi_4+\cdots+\phi_{4n-8},$$

etc. Thus A_3 becomes

$$A_3 = \sum_{l=0}^{n-2} (l+1) e^{-i(\phi_2+\phi_4+\cdots+\phi_{4n-4l})}, \tag{3.114a}$$

and hence

$$A_1 - A_3 = -\sum_{l=0}^{n-2} e^{-i(\phi_2+\phi_4+\cdots+\phi_{4n-l})} + (n-1) e^{-i(\phi_2+\phi_4)}. \tag{3.114b}$$

And finally for A_4 we can write [this is analogous to the transformations leading to (3.113)]

$$A_4 = \sum_{j=1}^{n-1} e^{-i(\phi_2+\phi_4+\cdots+\phi_{4n-4j})} \times \sum_{l=0}^{j-1} e^{-i(\phi_{4n-4j+2+4l}+\phi_{4n-4j+4+4l})}. \tag{3.115}$$

From (3.115) we can see that

$$A_4 = e^{-i(\phi_2+\phi_4+\cdots+\phi_{4n})} + 2e^{-i(\phi_2+\phi_4+\cdots+\phi_{4n-4})} + e^{-i(\phi_2+\phi_4+\cdots+\phi_{4n-8})} + \cdots + e^{-i(\phi_2+\phi_4+\phi_6+\phi_8)}$$
$$+ e^{-i(\phi_2+\phi_4+\cdots+\phi_{4n-12})}\sum_{l=1}^{2} e^{-i(\phi_{4n-12+2+4l}+\phi_{4n-12+4+4l})}$$
$$+ e^{-i(\phi_2+\phi_4+\cdots+\phi_{4n-16})}\sum_{l=1}^{3} e^{-i(\phi_{4n-16+2+4l}+\phi_{4n-16+4+4l})} + \cdots + e^{-i(\phi_2+\phi_4)}\sum_{l=1}^{n-2} e^{-i(\phi_{4l+8}+\phi_{4l+6})}, \qquad (3.116)$$

where we singled out the $j=1,\ l=0$; $j=2,\ l=0,\ 1$; $j=3,\ l=0$; ... ; $j=n-1,\ l=0$ terms to give the first terms in (3.116). Note that the $j=2,\ l=1$ term must be cyclically permuted to give form in (3.116).

From (3.113) and (3.116) it follows that (one must permute indices in certain terms to obtain cancellations)

$$-A_2+A_4 = \sum_{l=0}^{n-2} e^{-i(\phi_2+\phi_4+\cdots+\phi_{4n-4l})} - (n-1)e^{-i(\phi_2+\phi_4)}\ . \qquad (3.117)$$

Using (3.117) and (3.114b) in (3.107) it follows that

$$R' = 0\ . \qquad (3.118)$$

G. $\langle\sigma_{0,0}\sigma_{M,N}\rangle$ in scaling limit ($T<T_c$)

1. Analysis of $F^{(2)}_{<MN}$ in scaling limit

In this section we examine (2.12) in the scaling limit (1.1)–(1.3). By Cauchy's theorem we are allowed to expand the contours of integration in (2.12) out to the unit circle [recall the restriction $\mathrm{Im}\phi_j<0$ in (2.12)] provided that we do not integrate through any singularities. The only possible singularities are those coming from the factors $\sin\frac{1}{2}(\phi_{2j}+\phi_{2j+2})$ occurring in the denominator of (2.12) [$\Delta(\phi_{2j-1},\phi_{2j})\neq 0$ on the unit circle for $T<T_c$]. We first expand the ϕ_2 (and ϕ_1) integration to the unit circle. Now expand ϕ_4 (and ϕ_3) to the unit circle, except now at $\phi_4=-\phi_2$ we indent the ϕ_4 integration contour inward. Thus (2.12) can be interpreted as integration on the unit circle except at $\phi_{2j+2}=-\phi_{2j}$ the contour of integration for ϕ_{2j+2} is indented inward.

If the integrand in (2.12) is bounded, then the integral goes exponentially to zero in the limit $(M^2+N^2)\to\infty$. However at $T=T_c$ the denominator contains singular factors due to the vanishing of $\Delta(\phi_{2j-1},\phi_{2j})$ at $\phi_{2j-1}=\phi_{2j}=0$. Thus the leading contribution to the integral (2.12) for $T\to T_c^-$ comes from the behavior of the integrand around $\phi_j=0$, $j=1,2,\ldots,4n$. Thus to examine (2.12) in the scaling limit (1.1)–(1.3) we expand the integrand about $\phi_j=0$, $j=1,2,\ldots,4n$.

Since

$$a-\gamma_1\cos\phi_{2j-1}-\gamma_2\cos\phi_{2j} = (a-\gamma_1-\gamma_2)$$
$$\times[1+\tfrac{1}{2}\gamma_1(a-\gamma_1-\gamma_1)^{-1}\phi^2_{2j-1}+\tfrac{1}{2}(a-\gamma_1-\gamma_2)^{-1}\phi^2_{2j}$$
$$+O(\phi^4_{2j-1}(a-\gamma_1-\gamma_2)^{-1})+O(\phi^4_{2j}(a-\gamma_1-\gamma_2)^{-1})]\ , \qquad (3.119)$$

we define the scaled variables (or stretching variables)

$$x_j = [\tfrac{1}{2}\gamma_1(a-\gamma_1-\gamma_2)^{-1}]^{1/2}\phi_{2j-1}\ , \qquad (3.120a)$$

$$x_j = [z_2(1-z_1^2)]^{1/2}\,|z_1z_2+z_1+z_2-1|^{-1}\phi_{2j-1}\ ,$$
$$j=1,2,\ldots,2n \qquad (3.120b)$$

and

$$y_j = [\tfrac{1}{2}\gamma_2(a-\gamma_1-\gamma_2)^{-1}]^{1/2}\phi_{2j} \qquad (3.121a)$$

$$y_j = [z_1(1-z_2^2)]^{1/2}\,|z_1z_2+z_1+z_2-1|^{-1}\phi_{2j}\ ,$$
$$j=1,2,\ldots,2n\ . \qquad (3.121b)$$

The quantity $|z_1z_2+z_1+z_2-1|$ near T_c is given by (2.7). Then we have

$$\Delta(\phi_{2j-1},\phi_{2j}) = |z_1z_2+z_1+z_2-1|^2$$
$$\times[1+x_j^2+y_j^2+O(\phi^2_{2j-1}x_j^2)+O(\phi^2_{2j}y_j^2)]\ . \qquad (3.122)$$

Note that $O(\phi^2_{2j-1}x_j^2)=O(\phi^2_{2j-1})=O[(T-T_c)^2]$.

Using (3.120)–(3.122) we see that in the scaling limit (1.1)–(1.3)

$$\prod_{j=1}^{2n}\frac{d\phi_{2j-1}d\phi_{2j}}{\Delta(\phi_{2j-1},\phi_{2j})} \sim \left(\frac{4}{\gamma_1\gamma_2}\right)^n \prod_{j=1}^{2n}\frac{dx_j\,dy_j}{1+x_j^2+y_j^2}\ . \qquad (3.123)$$

$$e^{-iM(\phi_1+\cdots+\phi_{4n-1})-iN(\phi_2+\cdots+\phi_{4n})}$$
$$\sim e^{-i\bar{M}(x_1+\cdots+x_{2n})-i\bar{N}(y_1+\cdots+y_{2n})}\ , \qquad (3.124)$$

and

$$\prod_{j=1}^{2n}\frac{\sin\frac{1}{2}(\phi_{2j-1}-\phi_{2j+1})}{\sin\frac{1}{2}(\phi_{2j}+\phi_{2j+2})} \sim \left(\frac{\gamma_2}{\gamma_1}\right)^n \prod_{j=1}^{2n}\frac{x_j-x_{j+1}}{y_j+y_{j+1}}$$

$$(x_{2n+1}\equiv x_1 \text{ and } y_{2n+1}\equiv y_1)\ , \qquad (3.125)$$

where

$$\bar{M} = M(2/\gamma_1)^{1/2}\,|z_1z_2+z_1+z_2-1|\ , \qquad (3.126a)$$

$$\bar{M} = M[z_2(1-z_1^2)]^{-1/2}\,|z_1z_2+z_1+z_2-1| \qquad (3.126b)$$

and

$$\bar{N} = N(2/\gamma_2)^{1/2}\,|z_1z_2+z_1+z_2-1|\ , \qquad (3.127a)$$

$$\bar{N} = N[z_1(1-z_2^2)]^{-1/2}\,|z_1z_2+z_1+z_2-1|\ . \qquad (3.127b)$$

Also note that the correction terms in (3.123) and (3.125) are $O((T-T_c)^2)$.

Applying (3.123)–(3.127) to (2.12) we have

$$F^{(2n)}_{<MN} \sim (-1)^n 2^{-2n} \pi^{-4n} (2n)^{-1} \int_{-\infty}^{\infty} dx_1 \cdots \int_{-\infty}^{\infty} dx_{2n}$$

$$\times \int_{-\infty}^{\infty} dy_1 \cdots \int_{-\infty}^{\infty} dy_{2n}$$

$$\times \prod_{j=1}^{2n} \frac{e^{-i\bar{M}x_j - i\bar{N}y_j}}{1+x_j^2+y_j^2} \frac{x_j - x_{j+1}}{y_j + y_{j+1}} , \qquad (3.128)$$

where the contours of the y_{j+1} integration are to be indented downward at $-y_j$. The symbol "~" in (3.128) means that $F^{(2n)}_{<MN}$ is asymptotically equal to the right-hand side of (3.128) in the scaling limit (1.1)–(1.3). We denote the right-hand side of (3.128) by $f^{(2n)}(t)$; so that (3.128) becomes

$$F^{(2n)}_{<MN} \sim f^{(2n)}(t) \quad \text{(scaling limit)} , \qquad (3.129)$$

and the quantity t is defined by

$$t^2 = \bar{M}^2 + \bar{N}^2 , \qquad (3.130)$$

where $\bar{M}$ and $\bar{N}$ are given by (3.126) and (3.127), respectively. From (3.126) and (3.127) we see that the definition (3.130) of t is equivalent to (2.5). In terms of the interaction energies E_1 and E_2, the expression (2.5) for t can be written equivalently as

$$t^2 = \cosh\beta E_1 \cosh\beta E_2 (z_1 z_2)^{-1/2} (z_1 z_2 + z_1 + z_2 - 1)^2 [M^2(\sinh 2\beta E_1/\sinh 2\beta E_2)^{1/2} + N^2(\sinh 2\beta E_2/\sinh 2\beta E_1)^{1/2}] . \qquad (3.131)$$

Furthermore, as $T \to T_c$ we find

$$t \sim 4(\beta - \beta_c)(z_{1c} z_{2c})^{1/4} (\sinh\beta_c E_1 \sinh\beta_c E_2)^{1/2} [M^2(\sinh 2\beta_c E_1/\sinh 2\beta_c E_2)^{1/2} + N^2(\sinh 2\beta_c E_2/\sinh 2\beta_c E_1)^{1/2}]^{1/2} \qquad (3.132a)$$

$$t = (\beta - \beta_c)\, 4(z_{1c} z_{2c})^{1/4} (\sinh\beta_c E_1 \sinh\beta_c E_2)^{1/2} [E_1(1 - z_{2c})^{-1} + E_2(1 - z_{1c})^{-1}] (M^2 \sinh 2\beta_c E_1 + N^2 \sinh 2\beta_c E_2)^{1/2} . \qquad (3.132b)$$

We also define the quantity R by (2.6). Note that for the symmetric lattice $E_1 = E_2$, R is just $(M^2 + N^2)^{1/2}$ [see (2.65)].

2. "Rotational invariance" of $F^{(2n)}_{<MN}$ in scaling limit

In writing (3.129) we have implicitly assumed that (3.128) is only a function of $(\bar{M}^2 + \bar{N}^2)$ and not $\bar{M}$ and $\bar{N}$ separately. We now show this is the case. Defining the new integration variables

$$u_j = x_j \sin\theta + y_j \cos\theta ,$$
$$v_j = -x_j \cos\theta + y_j \sin\theta , \qquad (3.133)$$

for $j = 1, 2, \ldots, 2n$; and writing (which defines the angle θ)

$$\bar{M} = t \sin\theta , \quad \bar{N} = t\cos\theta , \qquad (3.134)$$

t given by (3.130), we see that (3.128) becomes

$$f^{(2n)}(t) = (-1)^n 2^{-2n} \pi^{-4n} (2n)^{-1} \int_{-\infty}^{\infty} \cdots \int_{-\infty}^{\infty} \prod_{j=1}^{2n} \left(\frac{du_j dv_j}{1+u_j^2+v_j^2} e^{-itu_j} \frac{(u_j - u_{j+1})\sin\theta - (v_j - v_{j+1})\cos\theta}{(u_j + u_{j+1})\cos\theta + (v_j + v_{j+1})\sin\theta} \right) , \qquad (3.135)$$

$u_{2n+1} \equiv u_1$ and $v_{2n+1} \equiv v_1$.

We now perform the v_j integrations $j = 1, 2, \ldots, 2n$ in (3.135) by closing the contour in the upper half-plane and evaluating the residue at $v_j = i(\sqrt{1+u_j^2})$. Hence (3.135) reduces to

$$f^{(2n)}(t) = (-1)^n 2^{-2n} \pi^{-2n} (2n)^{-1} \int_{-\infty}^{\infty} \cdots \int_{-\infty}^{\infty} \prod_{j=1}^{2n} \left(\frac{du_j}{(1+u_j^2)^{1/2}} e^{-itu_j} \frac{(u_j - u_{j+1})\sin\theta - i[(\sqrt{1+u_j^2}) - (\sqrt{1+u_{j+1}^2})]\cos\theta}{(u_j + u_{j+1})\cos\theta + i[(\sqrt{1+u_j^2}) + (\sqrt{1+u_{j+1}^2})]\sin\theta} \right) . \qquad (3.136)$$

We now show that the quantity

$$\frac{(u-v)\sin\theta - i[(\sqrt{1+u^2}) - (\sqrt{1+v^2})]\cos\theta}{(u+v)\cos\theta + i[(\sqrt{1+u^2}) + (\sqrt{1+v^2})]\sin\theta} \qquad (3.137)$$

is independent of the angle θ. To do this we introduce $u = \sinh\psi$ and $v = \sinh\chi$ into (3.137) obtaining

$$\frac{(\sinh\psi - \sinh\chi)\sin\theta - i(\cosh\psi - \cosh\chi)\cos\theta}{(\sinh\psi + \sinh\chi)\cos\theta + i(\cosh\psi + \cosh\chi)\sin\theta} . \qquad (3.138)$$

Using the hyperbolic addition formulas for $\cosh x \pm \cosh y$ and $\sinh x \pm \sinh y$ in (3.138) and rewriting the resulting expression in terms of the hyperbolic tangent function, (3.138) becomes simply

$$-i\tanh\tfrac{1}{2}(\psi-\chi)\,, \tag{3.139}$$

a quantity independent of the angle θ. Thus we may set $\theta=0$ in (3.137) to obtain the identity

$$\frac{(u-v)\sin\theta-i[(\sqrt{1+u^2})-(\sqrt{1+v^2})]\cos\theta}{(u+v)\cos\theta+i[(\sqrt{1+u^2})+(\sqrt{1+v^2})]\sin\theta} = -i\,\frac{(\sqrt{1+u^2})-(\sqrt{1+v^2})}{u+v} \tag{3.140}$$

Using (3.140) in (3.136), we obtain the explicitly "rotational invariant" form for $f^{(2n)}(t)$

$$f^{(2n)}(t)=2^{-2n}\pi^{-2n}(2n)^{-1}\int_{-\infty}^{\infty}\cdots\int_{-\infty}^{\infty}\prod_{j=1}^{2n}\left(\frac{du_j}{(1+u_j^2)^{1/2}}\times e^{-itu_j}\,\frac{(\sqrt{1+u_j^2})-(\sqrt{1+u_{j+1}^2})}{u_j+u_{j+1}}\right). \tag{3.141}$$

3. *Final integral form for $f^{(2n)}(t)$*

We can simplify (3.141) if we deform each u_j-integration contour into the lower half-plane so that we are now integrating on the loop about the branch point at $u_j=-i$. Letting $u_j=-iy_j$, we see that (3.141) will be of the form

$$f^{(2n)}(t)=2^{-2n}\pi^{-2n}(2n)^{-1}\int_1^{\infty}\cdots\int_1^{\infty}\prod_{j=1}^{2n}\left(\frac{e^{-ty_j}}{(y_j^2-1)^{1/2}}\,dy_j\right) \times P_n(y_1,y_2,\ldots,y_{2n})\,, \tag{3.142}$$

where $P_n(y_1,y_2,\ldots,y_{2n})$ is a sum of 2^{2n} terms coming from all possible combinations of plus and minus signs of the product term

$$\prod_{j=1}^{2n}\frac{(\sqrt{1+u_j^2})-(\sqrt{1+u_{j+1}^2})}{u_j+u_{j+1}}$$

in (3.141). This is so because of the square-root nature of the product term. Clearly the denominator (u_j+u_{j+1}) presents no difficulty; and so, we concentrate on the numerator where the signs of the square roots take on all possible plus and minus signs. We can represent a term in this sum as $(x_1-x_2)(x_2-x_3)\cdots(x_n-x_1)$. Summing over all possible sign changes means that we must evaluate the sum

$$S_{2n}=\sum_{(\epsilon_j=\pm1)}(\epsilon_1x_1-\epsilon_2x_2)\times(\epsilon_2x_2-\epsilon_3x_3)\cdots(\epsilon_{2n}x_{2n}-\epsilon_1x_1)\,. \tag{3.143}$$

The only terms in (3.143) which contribute are those quadratic in each of the ϵ_j. Thus we have

$$S_{2n}=(-1)^n2^{2n}(x_1^2x_3^2x_5^2\cdots x_{2n-1}^2+x_2^2x_4^2x_6^2\cdots x_{2n}^2)\,. \tag{3.144}$$

If we apply (3.144) to (3.142) we obtain[20] (2.27) [when one uses (3.144) in (3.142) the two terms resulting from (3.144) can be written as one term as we have done in (2.27)].

Since (2.27) is of the standard Laplace type, it is straightforward to analyze the large-t behavior. We find that for fixed n and $t\to\infty$

$$f^{(2n)}(t)=(-1)^n\pi^{-n}(2n)^{-1}2^{-3n+1}(e^{-2nt}/t^{2n}) \times[1-\tfrac{7}{4}nt^{-1}+\tfrac{1}{128}n(196n+272)t^{-2}+O(t^{-3})]\,. \tag{3.145}$$

For small t we have

$$f^{(2n)}(t)=C_{2n}(\ln t)^{2n}+O(\ln^{2n-1}t)\quad(t\to0)\,, \tag{3.146}$$

where C_{2n} is a constant.

4. *$f^{(2)}(t)$ in terms of Bessel functions*

The function $f^{(2)}(t)$ can be explicitly computed. From (2.27) we see that

$$\frac{d^2f^{(2)}(t)}{dt^2}=-\pi^{-2}\int_1^{\infty}dy_1 \times\int_1^{\infty}dy_2\,\frac{e^{-t(y_1+y_2)}}{[(y_1^2-1)(y_2^2-1)]^{1/2}}\,(y_2^2-1)\,. \tag{3.147}$$

Using the integral representation

$$K_0(t)=\int_1^{\infty}\frac{dx}{(x^2-1)^{1/2}}\,e^{-tx} \tag{3.148}$$

for the modified Bessel function $K_0(t)$, we see that (3.147) becomes

$$\frac{d^2f^{(2)}(t)}{dt^2}=\pi^{-2}[K_1'(t)K_0(t)+K_0^2(t)] =\pi^{-2}t^{-1}K_0(t)K_1(t)\,. \tag{3.149}$$

Differentiation verifies that

$$-f^{(2)}(t)=\pi^{-2}\{t^2[K_1^2(t)-K_0^2(t)] -tK_0(t)K_1(t)+\tfrac{1}{2}K_0^2(t)\}\,. \tag{3.150}$$

From (2.23), (2.26), (3.145), and (3.150) the result (2.32b) follows.

5. *$\hat{F}^{(2)}(t)$*

From (2.26) we see that we can write $\hat{F}_-(t)$ as

$$\hat{F}_-(t)=1+\sum_{n=1}^{\infty}\hat{F}_-^{(2n)}(t)\,, \tag{3.151}$$

where

$$\hat{F}_-^{(2)}(t)=-f^{(2)}(t)\quad[\text{recall (3.150)}]\,, \tag{3.152a}$$

$$\hat{F}_-^{(4)}(t)=-f^{(4)}(t)+(1/2!)[f^{(2)}(t)]^2\,, \tag{3.152b}$$

$$\hat{F}_-^{(6)}(t)=-f^{(6)}(t)+f^{(2)}(t)f^{(4)}(t)-(1/3!)[f^{(2)}(t)]^3\,, \tag{3.152c}$$

etc. From (3.145) we see that $\hat{F}_-^{(2n)}(t)$ is asymptotically equal to e^{-2nt} times some power of t as $t\to\infty$.

If one uses (3.145) in (3.152b) or (3.152c) in order to compute (for $t\to\infty$) the asymptotic behavior of $\hat{F}_-^{(4)}(t)$ or $\hat{F}_-^{(6)}(t)$, respectively, then one finds that not only does the leading term in (3.145) cancel but so do the two correction terms. This suggests we should examine (3.152) in more detail.

We discuss here only the case for $n=2$, i.e., $\hat{F}_-^{(4)}(t)$. Using (3.152b) and (2.27) we see that

$$\hat{F}_-^{(4)}(t)=\pi^{-4}\int_1^\infty dy_1\cdots\int_1^\infty dy_4\prod_{j=1}^4\left(\frac{e^{-ty_j}}{(y_j^2-1)^{1/2}}\right)(y_2^2-1)(y_4^2-1)$$
$$\times\{-\tfrac{1}{2}[(y_1+y_2)(y_2+y_3)(y_3+y_4)(y_4+y_1)]^{-1}+\tfrac{1}{4}[(y_1+y_2)^2(y_3+y_4)^2]^{-1}+\tfrac{1}{4}[(y_1+y_4)^2(y_2+y_3)^2]^{-1}\}$$
$$=4^{-1}\pi^{-4}\int_1^\infty dy_1\cdots\int_1^\infty dy_4\prod_{j=1}^4\left(\frac{e^{-ty_j}}{(y_j^2-1)^{1/2}}\right)(y_2^2-1)(y_4^2-1)\left(\frac{(y_2-y_4)(y_1-y_3)}{(y_1+y_2)(y_2+y_3)(y_3+y_4)(y_4+y_1)}\right)^2. \quad (3.153)$$

The cancellation that occurs in (3.152b) has been explicitly carried out in (3.153) in the sense that the leading term in the large-t expansion of (3.153) is nonvanishing. An elementary computation shows that as $t\to\infty$

$$\hat{F}_-^{(4)}(t)=3\,2^{-12}\pi^{-2}(e^{-4t}/t^8)[1-\tfrac{31}{2}t^{-1}+O(t^{-2})], \quad (3.154)$$

6. *Scaling functions F(t) and $F_1(t)$*

From the error estimates of Sec. IIIG1 it is clear that

$$\exp(-F_{\langle MN\rangle})=\hat{F}_-(t)+o(R^{-1})\ \text{(scaling limit)}. \quad (3.155)$$

The essential point is that the integrand in (2.12) has correction terms of order $(T-T_c)^2$ in the scaling limit since the variables ϕ_j appear only as sine and cosine factors. See, for instance, (3.119) and (3.122). Thus from (2.9) follows (2.21). To obtain $F_-(t)$ and $F_{1-}(t)$ we must expand $|1-(\sinh 2\beta E_1\times\sinh 2\beta E_2)^{-2}|^{1/4}$ to second order in the scaling limit.[21] The result of this expansion is given by (2.23), (2.24), and (2.25). Note that at $t=0$ it follows from (2.24) $F_{1-}(0)=0$, as it must to agree with the known[5,31] $T=T_c$ expansion of $\langle\sigma_{0,0}\sigma_{0,N}\rangle$ and $\langle\sigma_{0,0}\sigma_{N,N}\rangle$ for $N\to\infty$.

7. *Convergence proof*

It is straightforward to study the convergence of (2.26). Using the inequality

$$(y_j+y_{j+1})^{-1}\le(1+y_{j+1})^{-1}\quad\text{for } y_j, y_{j+1}\ge 1, \quad (3.156)$$

and the fact that the integrand in (2.27) is nonnegative, we have

$$|f^{(2n)}(t)|\le\pi^{-2n}(2n)^{-1}\int_1^\infty dy_1\cdots\int_1^\infty dy_{2n}\prod_{j=1}^{2n}\frac{e^{-ty_j}}{(y_j^2-1)^{1/2}}\prod_{j=1}^{n}\frac{y_{2j}^2-1}{(1+y_{2j})^2}$$
$$=\pi^{-2n}(2n)^{-1}\left(\int_1^\infty dy\frac{e^{-ty}}{(y^2-1)^{1/2}}\right)^n\left(\int_1^\infty dy'\frac{e^{-ty'}}{(y'^2-1)^{1/2}}\frac{y'-1}{y'+1}\right)^n. \quad (3.157)$$

We define t^* as the solution to the equation

$$\frac{1}{\pi^2}\int_1^\infty dy\frac{e^{-ty}}{(y^2-1)^{1/2}}\int_1^\infty dy'\frac{e^{-ty'}}{(y'^2-1)^{1/2}}\frac{y'-1}{y'+1}=1, \quad (3.158)$$

and we find numerically that

$$t^*=0.03302. \quad (3.159)$$

Then for all $t>t^*$ the sum

$$\sum_{n=1}^\infty f^{(2n)}(t)$$

converges. Clearly this is only a sufficient condition.

IV. PERTURBATION EXPANSION FOR $\langle\sigma_{0,0}\sigma_{M,N}\rangle$ FOR $T>T_c$ AND LARGE M^2+N^2

A. Perturbation-expansion formalism

The index of the function $S(e^{i\phi})$ defined by (3.5) is not zero for $T>T_c$. Consequently we modify the Toeplitz determinant of Sec. III so that we work with zero index generating functions. So following CW we define the following "barred" quantities[22]:

$$\bar{S}_{mn}=(2\pi)^{-1}\int_{-\pi}^{\pi}d\phi\, e^{-i(n-m)\phi}\bar{S}(e^{i\phi}), \quad (4.1)$$

$$\bar{V}_{mn}=(2\pi)^{-1}\int_{-\pi}^{\pi}d\phi\, e^{-i(m+n)\phi}\bar{V}(e^{i\phi}), \quad (4.2)$$

$$\bar{U}_{mn}=(2\pi)^{-1}\int_{-\pi}^{\pi}d\phi\, e^{-i(m+n)\phi}\bar{U}(e^{i\phi}), \quad (4.3)$$

for m, $n=0, 1, 2, \ldots$, where the generating functions $\bar{S}(e^{i\phi})$, $\bar{V}(e^{i\phi})$, and $\bar{U}(e^{i\phi})$ are given by

$$\bar{S}(e^{i\phi}) = e^{-i\phi}S(e^{i\phi}) = [(1-\alpha_1 e^{i\phi})(1-\alpha_2^{-1}e^{i\phi}) \times(1-\alpha_1 e^{-i\phi})^{-1}(1-\alpha_2^{-1}e^{-i\phi})^{-1}]^{1/2}, \quad (4.4)$$

$$\bar{V}(e^{i\phi_2}) = e^{2i\phi_2}V(e^{i\phi_2}) = (1-z_1^2)(2\pi)^{-1}\int_{-\pi}^{\pi} d\phi_1 \frac{e^{-iM\phi_1 - iN\phi_2}}{\Delta(\phi_1,\phi_2)} \times \lambda(\phi_1, -\phi_2), \quad (4.5)$$

$$\bar{U}(e^{i\phi_2}) = e^{i\phi_2}U(e^{i\phi_2}) = -(1-z_1^2)(2\pi)^{-1}\int_{-\pi}^{\pi} d\phi_1 \frac{e^{-iM\phi_1 - iN\phi_2}}{\Delta(\phi_1,\phi_2)} \times 2iz_2\sin\phi_1, \quad (4.6)$$

with $S(e^{i\phi})$, $V(e^{i\phi})$, and $U(e^{i\phi})$ given by (3.5)–(3.7), respectively, $\Delta(\phi_1, \phi_2)$ and $\lambda(\phi_1, \phi_2)$ given by (2.13) and (3.9), respectively.

We now introduce the barred version of the matrices A, B, and C of Sec. III:

$$\bar{A} = \begin{bmatrix} 0 & \bar{S} & 0 & 0 \\ -\bar{S}^T & 0 & 0 & 0 \\ 0 & 0 & 0 & -\bar{S} \\ 0 & 0 & \bar{S}^T & 0 \end{bmatrix}, \quad (4.7)$$

$$\bar{B} = \begin{bmatrix} 0 & 0 & T & \bar{U} \\ 0 & 0 & -\bar{U} & \bar{V} \\ -T & \bar{U} & 0 & 0 \\ -\bar{U} & -\bar{V} & 0 & 0 \end{bmatrix}, \quad (4.8)$$

and

$$\bar{C} = \bar{A} + \bar{B} \quad (4.9)$$

The operators $\bar{S}$, $\bar{V}$, and $\bar{U}$ appearing in (4.7) and (4.8) are defined by the matrices (4.1)–(4.3), respectively. The operator T is the same as in Sec. III and is defined by (3.4).

Then it is known[5,17] that for arbitrary $T > T_c$ and arbitrary M, $N \ge 0$

$$\langle \sigma_{0,0}\sigma_{M,N}\rangle = [r(M,N)]^{1/2}\det(\bar{C}), \quad (4.10)$$

with

$$r(M,N) = x_{20}''x_{40}' - x_{20}'x_{40}'', \quad (4.11)$$

and $\bar{C}$ defined by (4.9). Furthermore x_{20}'', x_{40}'', x_{20}', and x_{40}' are the zeroth components of the vectors x_2'', x_4'', x_2', and x_4', respectively, which are solutions to the equations

$$\bar{C}x'' = y_2 \quad \text{and} \quad \bar{C}x' = y_4, \quad (4.12a)$$

with

$$x'' = \begin{pmatrix} x_1'' \\ x_2'' \\ x_3'' \\ x_4'' \end{pmatrix}, \quad y_2 = \begin{pmatrix} 0 \\ \delta \\ 0 \\ 0 \end{pmatrix}, \quad (4.12b)$$

and x' and y_4 of the form (3.20). Since $\bar{S}(e^{i\phi})$ is of index zero for $T > T_c$, the methods of Sec. III can be applied to compute $\det(\bar{C})$ in (4.10). That is to say, if we define in analogy with (3.17)

$$\bar{f}_{MN} = \frac{\det(\bar{C}_{M,N+1})}{\det(\bar{C}_{M,N})}, \quad (4.13)$$

where we made the dependence of $\bar{C}$ on M and N explicit by writing $\bar{C}_{M,N}$, then

$$(\bar{f}_{MN})^2 = x_{30}x_{40}' - x_{30}'x_{40}, \quad (4.14a)$$

where

$$\bar{C}x = y_3, \quad (4.14b)$$

x and y_3 are of the form (3.20), and x_{30} and x_{40} (x_{30}' and x_{40}') are the zeroth components of the vectors x_3 and x_4 (x_3' and x_4') that are solutions to (4.14b) [(4.12)].

Thus in view of the results of Sec. III we may write

$$\langle \sigma_{0,0}\sigma_{M,N}\rangle = \bar{S}_\infty[r(M,N)]^{1/2}\exp(-\bar{F}_{MN}), \quad (4.15)$$

with

$$\bar{F}_{MN} = \sum_{k=N}^{\infty} \ln \bar{f}_{Mk}, \quad (4.16)$$

$$\bar{f}_{MN} = 1 + \delta^T \sum_{n=1}^{\infty} [(\bar{A}^{-1}\bar{B})^{2n}]_{33}(\bar{S}^T)^{-1}\delta, \quad (4.17)$$

$$\bar{S}_\infty = \lim_{N\to\infty} \det(\bar{C}_{M,N}), \quad (4.18)$$

and the quantity δ is defined by (3.21). As shown by CW[23]

$$\bar{S}_\infty = (1-z_2^2)^{-1}(\gamma_1\gamma_2)^{1/2} \times[(\sinh 2\beta E_1 \sinh 2\beta E_2)^{-2} - 1]^{1/4}, \quad (4.19)$$

$$\bar{S}_\infty = (1-z_2^2)^{-1}S_\infty^>, \quad (4.20)$$

where we used the definition (2.17) of $S_\infty^>$ in going from (4.19) to (4.20).

Using the fact that $\bar{C}$ is antisymmetric we obtain from (4.11) and (4.12) the result

$$r(M,N) = (x_{20}')^2. \quad (4.21)$$

We can compute x_{20}' from (4.12) by writing $\bar{C}^{-1}$ as

$$\bar{C}^{-1} = \sum_{n=0}^{\infty} (-1)^n(\bar{A}^{-1}\bar{B})^n\bar{A}^{-1}, \quad (4.22)$$

so that x_{20}' can be written perturbatively as

$$-x_{20}' = \delta^T \sum_{n=0}^{\infty} [(\bar{A}^{-1}\bar{B})^{2n+1}]_{23}(\bar{S}^T)^{-1}\delta. \quad (4.23)$$

Note that in (4.23) only odd powers of $(\bar{A}^{-1}\bar{B})$ contribute to x'_{20}. This should be compared with (3.26), the perturbation expansion for f_{MN}, where only even powers of $(A^{-1}B)$ contribute.

The method of Wiener-Hopf[19] can be applied to obtain the matrix elements of $\bar{S}^{-1}$ which are needed in (4.23). The result is

$$(\bar{S}^{-1})_{mn}=(2\pi i)^{-2}\oint d\xi^{-(n+1)}\bar{P}(\xi)\oint d\xi'\,\frac{\xi'^{m}}{\xi'-\xi}\,\bar{Q}(\xi'^{-1}), \tag{4.24}$$

where now $\bar{P}(\xi)$ and $\bar{Q}(\xi)$ are given by[24]

$$\bar{P}(\xi)=[(1-\alpha_1\xi)(1-\alpha_2^{-1}\xi)]^{-1/2} \tag{4.25}$$

and

$$\bar{Q}(\xi)=[(1-\alpha_1\xi)(1-\alpha_2^{-1}\xi)]^{1/2}, \tag{4.26}$$

with the property that

$$\bar{P}(\xi)\bar{Q}(\xi)=1. \tag{4.27}$$

The contours of integration in (4.24) are the unit circles, except that the one for ξ' is to be indented outward near $\xi'=\xi$.

We now turn to the problem of computing the $(2n+1)$th term, $n=0, 1, 2, \ldots$ in (4.23). As in Sec. III we introduce certain recursion relations which when solved will give us $\delta^T[(\bar{A}^{-1}\bar{B})^{2n+1}]_{23}(\bar{S}^T)^{-1}\delta$ of (4.23) for $n=0, 1, 2, \ldots$.

B. Recursion relations

We define the operators $\bar{P}_1$, $\bar{P}_2$, and $\bar{P}_3$ as the "barred" version of (3.32)–(3.34). That is,

$$\bar{P}_1=\bar{U}(\bar{S}^T)^{-1}\bar{U}+\bar{V}\bar{S}^{-1}T, \tag{4.28}$$

$$\bar{P}_2=-\bar{V}\bar{S}^{-1}\bar{U}+\bar{U}(\bar{S}^T)^{-1}\bar{V}, \tag{4.29}$$

and

$$\bar{P}_3=\bar{T}(\bar{S}^T)^{-1}\bar{U}-\bar{U}\bar{S}^{-1}\bar{T}. \tag{4.30}$$

Then our basic recursion relations for $T>T_c$ are (for $k\geq 1$)

$$[(\bar{A}^{-1}\bar{B})^{2k+1}]_{23}=-\bar{S}^{-1}\{\bar{P}_3[(\bar{A}^{-1}\bar{B})^{2k-1}]_{13}+\bar{P}_1^T[(\bar{A}^{-1}\bar{B})^{2k-1}]_{23}\} \tag{4.31}$$

and

$$[(\bar{A}^{-1}\bar{B})^{2k+1}]_{13}=-(\bar{S}^T)^{-1}\{\bar{P}_1[(\bar{A}^{-1}\bar{B})^{2k-1}]_{13}+\bar{P}_2[(\bar{A}^{-1}\bar{B})^{2k-1}]_{23}\}. \tag{4.32}$$

For $k=0$ $(\bar{A}^{-1}\bar{B})_{13}$ and $(\bar{A}^{-1}\bar{B})_{23}$ are given below [(4.92) and (4.93)].

We introduce the generating functions

$$\bar{S}^{-1}(\xi,\xi')=\sum_{m,n=0}^{\infty}\xi^m(\bar{S}^{-1})_{mn}(\xi')^n, \tag{4.33}$$

$$\bar{S}^{-1}(\xi,\xi')=\bar{Q}(\xi)\bar{P}(\xi')(1-\xi\xi')^{-1}, \tag{4.34}$$

and

$$\bar{R}^{(k)}_{\alpha\beta}(\xi,\xi')=\sum_{m,n=0}^{\infty}\xi^m\{[(\bar{A}^{-1}\bar{B})^k]_{\alpha\beta}(\bar{S}^T)^{-1}\}_{mn}\xi'^n. \tag{4.35}$$

In going from (4.33) to (4.34) we made use of (4.24)–(4.26). Using (4.35) we see that (4.31) and (4.32) can be written equivalently as

$$\bar{R}^{(2k+1)}_{23}(\xi,\xi')=-(2\pi)^{-2}\int_{-\pi}^{\pi}\int_{-\pi}^{\pi}d(2)d(4)\bar{S}^{-1}(\xi,\bar{2})\times\{\bar{P}_3(2,4)\bar{R}^{(2k-1)}_{13}(\bar{4},\xi')+\bar{P}_1^T(2,4)\bar{R}^{(2k-1)}_{23}(\bar{4},\xi')\} \tag{4.36}$$

and

$$\bar{R}^{(2k+1)}_{13}(\xi,\xi')=-(2\pi)^{-2}\int\int_{\pi}^{\pi}d(2)d(4)\bar{S}^{-1}(2,\xi)\times\{\bar{P}_1(2,4)\bar{R}^{(2k-1)}_{13}(\bar{4},\xi')+\bar{P}_2(2,4)\bar{R}^{(2k-1)}_{23}(\bar{4},\xi')\}, \tag{4.37}$$

where $k=1, 2, \ldots$, and

$$\bar{P}_1(1,2)=\bar{U}(2)\bar{S}^{-1}(\bar{2},\bar{1})\bar{U}(1)+\bar{V}(1)\bar{S}^{-1}(\bar{1},\bar{2})T(2), \tag{4.38}$$

$$\bar{P}_2(1,2)=-\bar{V}(1)\bar{S}^{-1}(\bar{1},\bar{2})\bar{U}(2)-\bar{U}(1)\bar{S}^{-1}(\bar{2},\bar{1})\bar{V}(2), \tag{4.39}$$

$$\bar{P}_3(1,2)=T(1)\bar{S}^{-1}(\bar{2},\bar{1})\bar{U}(2)-\bar{U}(1)\bar{S}^{-1}(\bar{1},\bar{2})T(2). \tag{4.40}$$

As in Sec. III the notation $\bar{S}^{-1}(1,2)$ means $\bar{S}^{-1}(e^{i\phi_1}, e^{i\phi_2})$ and $\bar{S}^{-1}(\bar{1},\bar{2})$ means $\bar{S}^{-1}(e^{-i\phi_1}, e^{-i\phi_2})$. Likewise for the functions $\bar{P}_j(1,2)$ and $j=1$, 2, and 3, and the functions $\bar{U}(1)$, $\bar{V}(1)$, etc.

In view of (4.23) and (4.36) we have

$$-x'_{20}=\sum_{k=0}^{\infty}\bar{R}^{(2k+1)}_{23}(0,0). \tag{4.41a}$$

and we define

$$\bar{x}^{(2k+1)}_{MN}=-(1-z_2^2)^{-1}\bar{R}^{(2k+1)}_{23}(0,0),\quad k=0,1,2,\ldots. \tag{4.41b}$$

For $k=0$ we have

$$(\bar{A}^{-1}\bar{B})_{13}=(\bar{S}^T)^{-1}\bar{U} \tag{4.42}$$

and

$$(\bar{A}^{-1}\bar{B})_{23}=\bar{S}^{-1}T, \tag{4.43}$$

so that $\bar{R}^{(1)}_{13}(\xi,\xi')$ and $\bar{R}^{(1)}_{23}(\xi,\xi')$ are given by

$$\bar{R}^{(1)}_{13}(\xi,\xi')=(2\pi)^{-1}\int_{-\pi}^{\pi}d(2)\bar{S}^{-1}(\bar{2},\xi)\bar{U}(2)\bar{S}^{-1}(\xi',2), \tag{4.44}$$

$$\bar{R}^{(1)}_{13}(\xi,\xi')=-i\gamma_1\bar{P}(\xi)\bar{Q}(\xi')(2\pi)^{-2}\times\int_{-\pi}^{\pi}\int_{-\pi}^{\pi}d\phi_1\,d\phi_2\,\frac{e^{-iM\phi_1-iN\phi_2}}{\Delta(\phi_1,\phi_2)}\times[(1-\xi e^{-i\phi_2})(1-\xi'e^{-i\phi_2})]^{-1}\sin\phi_1, \tag{4.45}$$

and

$$\overline{R}_{23}^{(1)}(\xi,\xi')=(2\pi)^{-1}\int_{-\pi}^{\pi}d(2)\overline{S}^{-1}(\xi,\overline{2})T(2)\overline{S}^{-1}(\xi',\overline{2}), \tag{4.46}$$

$$\overline{R}_{23}^{(1)}(\xi,\xi')=-(1-z_2^2)\overline{Q}(\xi)\overline{Q}(\xi')(2\pi)^{-2}\times\int_{-\pi}^{\pi}\int_{-\pi}^{\pi}d\phi_1\,d\phi_2\frac{e^{-iM\phi_1-iN\phi_2}}{\Delta(\phi_1,\phi_2)}\times[(1-\xi e^{-i\phi_2})(1-\xi' e^{-i\phi_2})]^{-1}. \tag{4.47}$$

We used (4.25)–(4.27) and (4.34) in going from (4.44) to (4.45). To obtain (4.47) from (4.46) one uses the identity

$$T(e^{i\phi_2})[\overline{P}(e^{-i\phi_2})]^2=-(1-z_2^2)(2\pi)^{-1}\times\int_{-\pi}^{\pi}d\phi_1\frac{e^{-iM\phi_1-iN\phi_2}}{\Delta(\phi_1,\phi_2)}. \tag{4.48}$$

This identity is derived in a manner analogous to (3.55). The barred version of (3.56) is the identity

$$V(e^{i\phi_2})[\overline{Q}(e^{-i\phi_2})]^2=-\gamma_1^2(1-z_2^2)^{-1}(2\pi)^{-1}\times\int_{-\pi}^{\pi}d\phi_1\frac{e^{-iM\phi_1-iN\phi_2}}{\Delta(\phi_1,\phi_2)}\sin^2\phi_1. \tag{4.49}$$

Hence the first term in (4.41), $\overline{x}_{MN}^{(1)}$, is given by the $\xi=\xi'=0$ value of (4.47) [recall definition (4.41b)],

$$\overline{x}_{MN}^{(1)}=(2\pi)^{-2}\int_{-\pi}^{\pi}\int_{-\pi}^{\pi}d\phi_1\,d\phi_2\frac{e^{-iM\phi_1-iN\phi_2}}{\Delta(\phi_1,\phi_2)}. \tag{4.50}$$

Equation (4.50) was first derived by CW.

Since the general structure of $\overline{R}_{13}^{(2k+1)}(\xi,\xi')$ and $\overline{R}_{23}^{(2k+1)}(\xi,\xi')$ is not revealed by the $k=0$ case, we proceed to the $k=1$ case. Making use of (4.34), (4.38)–(4.40), (4.45), and (4.47) in (4.36) for $k=1$ we obtain

$$\overline{R}_{23}^{(3)}(\xi,\xi')=-\overline{Q}(\xi)\overline{Q}(\xi')(2\pi)^{-4}\int_{-\pi}^{\pi}\cdots\int_{-\pi}^{\pi}d(2)d(4)d(5)d(6)\frac{e^{-M\phi_5-iN\phi_6}}{\Delta(\phi_5,\phi_6)}[(1-\xi e^{-i\phi_2})(1-\xi' e^{-i\phi_6})]^{-1}\prod_{j=1}^{2}(1-e^{-i(\phi_{2j}+\phi_{2j+2})})^{-1}$$
$$\times(\{[\overline{P}(\overline{2})]^2T(2)\overline{U}(4)-\overline{U}(2)[\overline{P}(4)]^2T(4)\}(-i\gamma_1\sin\phi_5)+\{\overline{U}(2)\overline{U}(4)+[\overline{P}(\overline{2})]^2T(2)[\overline{Q}(\overline{4})]^2$$
$$\times\overline{V}(4)\}[-(1-z_2^2)]). \tag{4.51}$$

Now use the definition (4.6) of $\overline{U}(2)$ and identities (4.48)–(4.49) in (4.51) to obtain

$$\overline{R}_{23}^{(3)}(\xi,\xi')=-(1-z_2^2)\gamma_1^2\overline{Q}(\xi)\overline{Q}(\xi')(2\pi)^{-6}\int_{-\pi}^{\pi}\cdots\int_{-\pi}^{\pi}d(1)\cdots d(6)\prod_{j=1}^{3}\frac{e^{-iM\phi_{2j-1}-iN\phi_{2j}}}{\Delta(\phi_{2j-1},\phi_{2j})}$$
$$\times[(1-\xi e^{-i\phi_2})(1-\xi' e^{-i\phi_6})]^{-1}\left(\prod_{j=1}^{2}(1-e^{-i(\phi_{2j}+\phi_{2j+2})})^{-1}\right)(\sin\phi_1-\sin\phi_3)(\sin\phi_3-\sin\phi_5). \tag{4.52}$$

Similarly,

$$\overline{R}_{13}^{(3)}(\xi,\xi')=-\overline{P}(\xi)\overline{Q}(\xi')(2\pi)^{-4}\int_{-\pi}^{\pi}\cdots\int_{-\pi}^{\pi}d(2)d(4)d(5)d(6)\frac{e^{-iM\phi_5-iN\phi_6}}{\Delta(\phi_5,\phi_6)}[(1-\xi e^{-i\phi_2})(1-\xi' e^{-i\phi_6})]^{-1}\prod_{j=1}^{2}(1-e^{-i(\phi_{2j}+\phi_{2j+2})})^{-1}$$
$$(\{\overline{U}(4)\overline{U}(2)+\overline{V}(2)[\overline{Q}(\overline{2})]^2T(4)[\overline{P}(\overline{4})]^2\}(-i\gamma_1\sin\phi_5)+\{-[\overline{Q}(\overline{2})]^2\overline{V}(2)\overline{U}(4)+\overline{U}(2)[\overline{Q}(\overline{4})]^2\overline{V}(4)\}[-(1-z_2^2)])$$
$$=-i\gamma_1^3\overline{P}(\xi)\overline{Q}(\xi')(2\pi)^{-6}\int_{-\pi}^{\pi}\cdots\int_{-\pi}^{\pi}d(1)\cdots d(6)\prod_{j=1}^{3}\frac{e^{-iM\phi_{2j-1}-iN\phi_{2j}}}{\Delta(\phi_{2j+1},\phi_{2j})}[(1-\xi e^{-i\phi_2})(1-\xi' e^{-i\phi_6})]^{-1}$$
$$\times\prod_{j=1}^{2}(1-e^{-i(\phi_{2j}+\phi_{2j+2})})^{-1}\sin\phi_1(\sin\phi_1-\sin\phi_3)(\sin\phi_3-\sin\phi_5). \tag{4.53}$$

The general structure of $\overline{R}_{23}^{(2k+1)}(\xi,\xi')$ and $\overline{R}_{13}^{(2k+1)}(\xi,\xi')$ is now clear from (4.52) and (4.53), respectively. That is we have for $k\geq 1$

$$\overline{R}_{13}^{(2k+1)}(\xi,\xi')=-i\gamma_1^{2k+1}\overline{P}(\xi)\overline{Q}(\xi')(2\pi)^{-4k-2}\int_{-\pi}^{\pi}\cdots\int_{-\pi}^{\pi}d\phi_1\cdots d\phi_{4k+2}\prod_{j=1}^{2k+1}\frac{e^{-iM\phi_{2j-1}-iN\phi_{2j}}}{\Delta(\phi_{2j-1},\phi_{2j})}[(1-\xi e^{-i\phi_2})(1-\xi' e^{-i\phi_{4k+2}})]^{-1}$$
$$\times\prod_{j=1}^{2k}(1-e^{-i(\phi_{2j}+\phi_{2j+2})})^{-1}\sin\phi_1\prod_{j=1}^{2k}(\sin\phi_{2j-1}-\sin\phi_{2j+1}) \tag{4.54}$$

and

$$\overline{R}_{23}^{(2k+1)}(\xi,\xi')=-(1-z_2^2)\gamma_1^{2k}\overline{Q}(\xi)\overline{Q}(\xi')(2\pi)^{-4k-2}\int_{-\pi}^{\pi}\cdots\int_{-\pi}^{\pi}d\phi_1\cdots d\phi_{4k+2}$$

$$\times\prod_{j=1}^{2k+1}\frac{e^{-iM\phi_{2j-1}-iN\phi_{2j}}}{\Delta(\phi_{2j-1},\phi_{2j})}[(1-\xi e^{-i\phi_2})(1-\xi' e^{-i\phi_{4k+2}})]^{-1}\prod_{j=1}^{2k}(1-e^{-i(\phi_{2j}+\phi_{2j+2})})^{-1}\prod_{j=1}^{2k}(\sin\phi_{2j-1}-\sin\phi_{2j+1})\,. \tag{4.55}$$

For $k=1$ (4.54) and (4.55) have been established. The method of proof for the general case is to assume (4.54) and (4.55) hold for $2k-1$ and show that the truth of (4.54) and (4.55) follows for $2k+1$. To do this we use the induction hypothesis in the recursion relations (4.36) and (4.37). Then we make use of (4.38)–(4.40) and identities (4.48) and (4.49) to simplify the resulting expression. The details of this argument is just repeating the $k=1$ case but this time carrying along the extra factor $(\sin\phi_5-\sin\phi_7)\cdots(\sin\phi_{4k-1}-\sin\phi_{4k+1})$. Thus we spare the reader and consider (4.54) and (4.55) established.

From (4.41) and (4.55) we have for $k\geq 1$

$$\bar{x}_{MN}^{(2k+1)}=\gamma_1^{2k}(2\pi)^{-4k-2}\int_{-\pi}^{\pi}d\phi_1\cdots\int_{-\pi}^{\pi}d\phi_{4k+2}\prod_{j=1}^{2k+1}\left(\frac{e^{-iM\phi_{2j-1}-iN\phi_{2j}}}{\Delta(\phi_{2j-1},\phi_{2j})}\right)\prod_{j=1}^{2k}\left(\frac{\sin\phi_{2j-1}-\sin\phi_{2j+1}}{1-e^{-i(\phi_{2j}+\phi_{2j+2})}}\right), \tag{4.56a}$$

$$\bar{x}_{MN}^{(2k+1)}=(-i\gamma_1)^{2k}(2\pi)^{-4k-2}\int_{-\pi}^{\pi}d\phi_1\cdots\int_{-\pi}^{\pi}d\phi_{4k+2}\prod_{j=1}^{2k+1}\left(\frac{e^{-iM\phi_{2j-1}-iN\phi_{2j}}}{\Delta(\phi_{2j-1},\phi_{2j})}\right)\prod_{j=1}^{2k}\left(\frac{\sin\frac{1}{2}(\phi_{2j-1}-\phi_{2j+1})\cos\frac{1}{2}(\phi_{2j-1}-\phi_{2j+1})}{\sin\frac{1}{2}(\phi_{2j}+\phi_{2j+2})}\times e^{i(\phi_{2j}+\phi_{2j+2})/2}\right). \tag{4.56b}$$

The quantity $\bar{x}_{MN}^{(1)}$ is given by (4.50). Equation (4.56) provides a perturbation expansion for x'_{20} [see (4.41)] for arbitrary $T>T_c$ and arbitrary $M, N\geq 0$. As in Sec. IV, the condition $\mathrm{Im}\phi_j<0$, $j=1, 2, \ldots, 4k+2$ is understood in (4.56).

C. Quantity $\bar{f}_{MN}$

In this section we compute the quantity $\bar{f}_{MN}^{(2k)}$ [defined by (4.13)] for arbitrary $T>T_c$ and $M, N\geq 0$.

If we define

$$\bar{f}_{MN}^{(2k)}=\delta^T[(\bar{A}^{-1}\bar{B})^{2k}]_{33}(\bar{S}^T)^{-1}\delta\,, \tag{4.57}$$

then the perturbation expansion (4.17) becomes

$$\bar{f}_{MN}=1+\sum_{k=1}^{\infty}\bar{f}_{MN}^{(2k)}\,. \tag{4.58}$$

From (4.35) and (4.57) we have

$$\bar{f}_{MN}^{(2k)}=\bar{R}_{33}^{(2k)}(0,0)\,. \tag{4.59}$$

We can compute $\bar{R}_{33}^{(2k)}(\xi,\xi')$ and $\bar{R}_{43}^{(2k)}(\xi,\xi)$ by using for $k\geq 2$ the "barred" version of (3.46) and (3.47), respectively. For $k=1$ $\bar{R}_{33}^{(2)}(\xi,\xi')$ and $\bar{R}_{43}^{(2)}(\xi,\xi')$ are the "barred" version of (3.49) and (3.50), respectively. Thus, for instance,

$$\bar{R}_{33}^{(2)}(\xi,\xi')=-(2\pi)^{-2}\int_{-\pi}^{\pi}\int_{-\pi}^{\pi}d(2)\,d(4)\,\bar{S}^{-1}(\bar{2},\xi)\,\bar{P}_1(2,4)\,\bar{S}^{-1}(\xi',\bar{4})\,, \tag{4.60}$$

$$\bar{R}_{33}^{(2)}(\xi,\xi')=-(2\pi)^{-2}\int_{-\pi}^{\pi}\int_{-\pi}^{\pi}d(2)\,d(4)\,\bar{P}(\xi)\,\bar{Q}(\xi')[(1-\xi e^{-i\phi_2})(1-e^{-i(\phi_2+\phi_4)})(1-\xi' e^{-i\phi_4})]^{-1}\times\{\bar{U}(2)\,\bar{U}(4)+\bar{V}(2)[\bar{Q}(\bar{2})]^2\,T(4)[\bar{P}(\bar{4})]^2\}\,, \tag{4.61}$$

where we used (4.27), (4.34), and (4.38) in going from (4.60) to (4.61). Using the definition (4.6) of $\bar{U}(2)$ and the identities (4.48) and (4.49), (4.61) becomes

$$\bar{R}_{33}^{(2)}(\xi,\xi')=-\gamma_1^2(2\pi)^{-4}\int_{-\pi}^{\pi}d\phi_1\cdots\int_{-\pi}^{\pi}d\phi_4\prod_{j=1}^{2}\left(\frac{e^{-iM\phi_{2j-1}-iN\phi_{2j}}}{\Delta(\phi_{2j-1},\phi_{2j})}\right)\bar{P}(\xi)\,\bar{Q}(\xi')\times[(1-\xi e^{-i\phi_2})(1-e^{-i(\phi_2+\phi_4)})(1-\xi' e^{-i\phi_4})]^{-1}\sin\phi_1(\sin\phi_1-\sin\phi_3)\,, \tag{4.62}$$

where $\mathrm{Im}\phi_j<0$, $j=1, 2, 3$, and 4.

Likewise from the "barred" version of (3.50) we obtain, with the help of (4.27), (4.34), and (4.40),

$$\bar{R}_{43}^{(2)}(\xi,\xi')=(2\pi)^{-2}\int_{-\pi}^{\pi}\int_{-\pi}^{\pi}d(2)\,d(4)\,\bar{S}^{-1}(\xi,\bar{2})\,\bar{P}_3(2,4)\,\bar{S}^{-1}(\xi',\bar{4})\,, \tag{4.63}$$

$$\bar{R}_{43}^{(2)}(\xi,\xi')=(2\pi)^{-2}\int_{-\pi}^{\pi}\int_{-\pi}^{\pi}d(2)\,d(4)\,\bar{Q}(\xi)\,\bar{Q}(\xi')[(1-\xi e^{-i\phi_2})(1-e^{-i(\phi_2+\phi_4)})(1-\xi' e^{-i\phi_4})]^{-1}\times\{T(2)[\bar{P}(\bar{2})]^2\,\bar{U}(4)-\bar{U}(2)\,T(4)[\bar{P}(\bar{4})]^2\}\,. \tag{4.64}$$

Using (4.6), (4.48), and (4.49) in (4.64) we obtain

$$\bar{R}_{43}^{(2)}(\xi,\xi')=-i\gamma_1(1-z_2^2)(2\pi)^{-4}\int_{-\pi}^{\pi}d\phi_1\cdots\int_{-\pi}^{\pi}d\phi_4\prod_{j=1}^{2}\left(\frac{e^{-iM\phi_{2j-1}-iN\phi_{2j}}}{\Delta(\phi_{2j-1},\phi_{2j})}\right)\bar{Q}(\xi)\,\bar{Q}(\xi')$$
$$\times[(1-\xi e^{-i\phi_2})(1-e^{-i(\phi_2+\phi_4)})(1-\xi' e^{-i\phi_4})]^{-1}(\sin\phi_1-\sin\phi_3)\,, \tag{4.65}$$

where $\mathrm{Im}\phi_j<0$, $j=1, 2, 3$, and 4.

For arbitrary $n=1, 2, 3, \ldots$ we now establish that

$$\bar{R}_{33}^{(2n)}(\xi,\xi')=-\gamma_1^{2n}(2\pi)^{-4n}\int_{-\pi}^{\pi}d\phi_1\cdots\int_{-\pi}^{\pi}d\phi_{4n}\prod_{j=1}^{2n}\left(\frac{e^{-iM\phi_{2j-1}-iN\phi_{2j}}}{\Delta(\phi_{2j-1},\phi_{2j})}\right)\bar{P}(\xi)\,\bar{Q}(\xi')$$
$$\times[(1-\xi e^{-i\phi_2})(1-\xi' e^{-i\phi_{4n}})]^{-1}\prod_{j=1}^{2n-1}\left(\frac{\sin\phi_{2j-1}-\sin\phi_{2j+1}}{(1-e^{-i(\phi_{2j}+\phi_{2j+2})})}\right)\sin\phi_1 \tag{4.66}$$

and

$$\bar{R}_{43}^{(2n)}(\xi,\xi')=-i\gamma_1^{2n-1}(1-z_2^2)(2\pi)^{-4n}\int_{-\pi}^{\pi}d\phi_1\cdots\int_{-\pi}^{\pi}d\phi_{4n}\prod_{j=1}^{2n}\left(\frac{e^{-iM\phi_{2j-1}-iN\phi_{2j}}}{\Delta(\phi_{2j-1},\phi_{2j})}\right)\bar{Q}(\xi)\,\bar{Q}(\xi')$$
$$\times[(1-\xi e^{-i\phi_2})(1-\xi' e^{-i\phi_{4n}})]^{-1}\prod_{j=1}^{2n-1}\left(\frac{\sin\phi_{2j-1}-\sin\phi_{2j+1}}{(1-e^{-i(\phi_{2j}+\phi_{2j+2})})}\right)\,. \tag{4.67}$$

We will assume that (4.66) and (4.67) hold for $n=k-1$ and then show by use of the "barred" version of (3.46) and (3.47) that (4.66) and (4.67) hold for $n=k$. Since (4.62) and (4.65) establish the case $k=1$, we will then have shown (4.66) and (4.67) for arbitrary n.

Using (4.66) and (4.67) for $n=k-1$ in the barred version of (3.46) we obtain after using (4.38)–(4.40) and (4.34)

$$\bar{R}_{33}^{(2k)}(\xi,\xi')=(2\pi)^{-2}\int_{-\pi}^{\pi}d\phi_2\int_{-\pi}^{\pi}d\phi_4[(1-\xi e^{-i\phi_2})(1-e^{-i(\phi_2+\phi_4)})]^{-1}\bar{P}(\xi)\,(-\{\bar{U}(2)\,\bar{U}(4)\,\bar{Q}(\bar{4})$$
$$+\bar{V}(2)\,[\bar{Q}(\bar{2})]^2\,T(4)\,\bar{P}(\bar{4})\}\bar{R}_{33}^{(2k-2)}(\bar{4},\xi')+\{\bar{V}(2)[\bar{Q}(\bar{2})]^2\bar{U}(4)\bar{P}(\bar{4})+\bar{U}(2)\bar{V}(4)\bar{Q}(\bar{4})\}\bar{R}_{43}^{(2k-2)}(\bar{4},\bar{\xi}'))\,, \tag{4.68}$$

$$\bar{R}_{33}^{(2k)}(\xi,\xi')=(2\pi)^{-4k+2}\int_{-\pi}^{\pi}d\phi_2\int_{-\pi}^{\pi}d\phi_4\int_{-\pi}^{\pi}d\phi_5\cdots\int_{-\pi}^{\pi}d\phi_{4k}\prod_{j=3}^{2k}\left(\frac{e^{-iM\phi_{2j-1}-iN\phi_{2j}}}{\Delta(\phi_{2j-1},\phi_{2j})}\right)\bar{P}(\xi)\,\bar{Q}(\xi')$$
$$\times[(1-\xi e^{-i\phi_2})(1-\xi' e^{-i\phi_{4k}})]^{-1}\prod_{j=1}^{2k-1}(1-e^{-i(\phi_{2j}+\phi_{2j+2})})^{-1}\prod_{j=3}^{2k-1}(\sin\phi_{2j-1}-\sin\phi_{2j+1})\,(-\{\bar{U}(2)\,\bar{U}(4)$$
$$+\bar{V}(2)[\bar{Q}(\bar{2})]^2T(4)\,[\bar{P}(\bar{4})]^2\}(-\gamma_1^{2k-2}\sin\phi_5)+\{-\bar{V}(2)[\bar{Q}(\bar{2})]^2\bar{U}(4)+\bar{U}(2)\,\bar{V}(4)[\bar{Q}(4)]^2\}[-i\gamma_1^{2k-3}(1-z_2^2)])\,. \tag{4.69}$$

Using (4.6), (4.48), and (4.49), (4.69) becomes just (4.66) with n replaced by k. The "barred" version of (3.47) establishes (4.67) in a similar manner.

From (4.59) and (4.66)

$$\bar{f}_{MN}^{(2k)}=-\gamma_1^{2k}(2\pi)^{-4k}\int_{-\pi}^{\pi}d\phi_1\cdots\int_{-\pi}^{\pi}d\phi_{4k}\prod_{j=1}^{2k}\left(\frac{e^{-iM\phi_{2j-1}-iN\phi_{2j}}}{\Delta(\phi_{2j-1},\phi_{2j})}\right)\prod_{j=1}^{2k-1}\left(\frac{\sin\phi_{2j-1}-\sin\phi_{2j+1}}{1-e^{-i(\phi_{2j}+\phi_{2j+2})}}\right)\sin\phi_1\,. \tag{4.70}$$

D. Quantity $\bar{F}_{MN}$

From (4.16), (4.17), (4.57), and (4.58) we have

$$\bar{F}_{MN}=\sum_{k=N}^{\infty}\ln\left(1+\sum_{n=1}^{\infty}\bar{f}_{Mk}^{(2n)}\right), \tag{4.71}$$

where $\bar{f}_{Mk}^{(2n)}$ is given by (4.70). Equation (4.71) should be compared with (3.67), the analog below T_c. Defining [in analogy with (3.68)] $\bar{F}_{MN}^{(2n)}$

$$\bar{F}_{MN}=\sum_{n=1}^{\infty}\bar{F}_{MN}^{(2n)}\,, \tag{4.72}$$

where

$$\bar{F}_{MN}^{(2)}=\sum_{k=N}^{\infty}\bar{f}_{Mk}^{(2)}\,, \tag{4.73a}$$

$$\bar{F}_{MN}^{(4)}=\sum_{k=N}^{\infty}[\bar{f}_{Mk}^{(4)}-\tfrac{1}{2}(\bar{f}_{Mk}^{(2)})^2]\,, \tag{4.73b}$$

etc., we will show that

$$\bar{F}_{MN}^{(2n)}=(-1)^n\gamma_1^{2n}(2n)^{-1}(2\pi)^{-4n}\int_{-\pi}^{\pi}d\phi_1\cdots\int_{-\pi}^{\pi}d\phi_{4n}$$
$$\times\prod_{j=1}^{2n}\frac{e^{-iM\phi_{2j-1}-iN\phi_{2j}}}{\Delta(\phi_{2j-1},\phi_{2j})}\,\frac{\sin\frac{1}{2}(\phi_{2j-1}-\phi_{2j+1})}{\sin\frac{1}{2}(\phi_{2j}+\phi_{2j+2})}$$
$$\times e^{i\phi_{2j}}\cos\tfrac{1}{2}(\phi_{2j-1}+\phi_{2j+1})\,, \tag{4.74}$$

with $\phi_{4n+1}\equiv\phi_1$, $\phi_{4n+2}\equiv\phi_2$.

Equation (4.74) should be compared with (2.12).

1. $\bar{F}^{(2)}_{MN}$

Using (4.70) for $k=1$ in (4.73a) we have

$$\bar{F}^{(2)}_{MN}=-\gamma_1^2(2\pi)^{-4}\int_{-\pi}^{\pi}d\phi_1\cdots\int_{-\pi}^{\pi}d\phi_4\times\prod_{j=1}^{2}\left(\frac{e^{-iM\phi_{2j-1}-iN\phi_{2j}}}{\Delta(\phi_{2j-1},\phi_{2j})}\right)\frac{\sin\phi_1-\sin\phi_3}{(1-e^{-i(\phi_2+\phi_4)})^2}\sin\phi_1\,. \tag{4.75}$$

Letting $\phi_1\rightarrow\phi_3$ and $\phi_2\rightarrow\phi_4$ in (4.75), adding this to (4.75) and dividing by 2, gives

$$\bar{F}^{(2)}_{MN}=\tfrac{1}{2}\gamma_1^2(2\pi)^{-4}\int_{-\pi}^{\pi}d\phi_1\cdots\int_{-\pi}^{\pi}d\phi_4\times\prod_{j=1}^{2}\left(\frac{e^{-iM\phi_{2j-1}-iN\phi_{2j}}}{\Delta(\phi_{2j-1},\phi_{2j})}\frac{\sin\phi_{2j-1}-\sin\phi_{2j+1}}{(1-e^{-i(\phi_{2j}+\phi_{2j+2})})}\right), \tag{4.76}$$

with $\phi_5\equiv\phi_1$, $\phi_6\equiv\phi_2$.

We can rewrite (4.76) as

$$\bar{F}^{(2)}_{MN}=-\gamma_1^2\tfrac{1}{2}(2\pi)^{-4}\int_{-\pi}^{\pi}d\phi_1\cdots\int_{-\pi}^{\pi}d\phi_4\times\prod_{j=1}^{2}\left(\frac{e^{-iM\phi_{2j-1}-iN\phi_{2j}}}{\Delta(\phi_{2j-1},\phi_{2j})}\frac{\sin\frac{1}{2}(\phi_{2j-1}-\phi_{2j+1})}{\sin\frac{1}{2}(\phi_{2j}+\phi_{2j+2})}\times e^{i\phi_{2j}}\cos\tfrac{1}{2}(\phi_{2j-1}+\phi_{2j+1})\right), \tag{4.77}$$

where we used the identity $\sin x-\sin y=\frac{1}{2}\sin\frac{1}{2}(x-y)\cos\frac{1}{2}(x+y)$.

This is just (4.74) for $n=1$.

2. $\bar{F}^{(2n)}_{MN}$ for arbitrary n

We define

$$\bar{F}^{(2n)}_{MN}=(-1)^n\gamma_1^{2n}(2n)^{-1}(2\pi)^{-4n}\int_{-\pi}^{\pi}d\phi_1\cdots\int_{-\pi}^{\pi}d\phi_{4n}\times\prod_{j=1}^{2n}\left(\frac{e^{-iM\phi_{2j-1}-iN\phi_{2j}}}{\Delta(\phi_{2j-1},\phi_{2j})}\frac{\sin\frac{1}{2}(\phi_{2j-1}-\phi_{2j+1})}{\sin\frac{1}{2}(\phi_{2j}+\phi_{2j+2})}\times e^{i\phi_{2j}}\cos\tfrac{1}{2}(\phi_{2j-1}+\phi_{2j+1})\right)(1-e^{-i(\phi_2+\phi_4+\cdots+\phi_{4n})}) \tag{4.78}$$

$$\bar{f}(x)=\sum_{n=0}^{\infty}\bar{f}^{(2n)}_{MN}x^{2n}, \tag{4.79}$$

and

$$\bar{F}(x)=\sum_{n=0}^{\infty}\bar{F}^{(2n)}_{MN}x^{2n}, \tag{4.80}$$

with $\bar{f}^{(0)}_{MN}=1$, $\bar{F}^{(0)}_{MN}=0$. Then as in Sec. III F we see that (4.74) will be true if we can prove that

$$n\bar{f}^{(2n)}_{MN}=\sum_{b=1}^{n}l\,\bar{F}^{(2n)}_{MN}\bar{f}^{2(n-l)}_{MN}\,. \tag{4.81}$$

In (4.70), if we let $\phi_1\rightarrow\phi_{4n-1}$, $\phi_2\rightarrow\phi_{4n}$, $\phi_3\rightarrow\phi_{4n-3}$, $\cdots$, add this to (4.70), and divide by two, we obtain

$$\bar{f}^{(2k)}_{MN}=2^{-1}\gamma_1^{2k}(2\pi)^{-4k}\int_{-\pi}^{\pi}d\phi_1\cdots\int_{-\pi}^{\pi}d\phi_{4k}\prod_{j=1}^{2k}\left(\frac{e^{-iM\phi_{2j-1}-iN\phi_{2j}}}{\Delta(\phi_{2j-1},\phi_{2j})}(\sin\phi_{2j-1}-\sin\phi_{2j+1})\right)\prod_{j=1}^{2k-1}(1-e^{-i(\phi_{2j}+\phi_{2j+2})})^{-1}\,. \tag{4.82}$$

Furthermore, (4.82) can be written as

$$\bar{f}^{(2k)}_{MN}=\gamma_1^{2k}(2\pi)^{-4k}(-1)^k\int_{-\pi}^{\pi}d\phi_1\cdots\int_{-\pi}^{\pi}d\phi_{4k}\prod_{j=1}^{2k}\left(\frac{e^{-iM\phi_{2j-1}-iN\phi_{2j}}}{\Delta(\phi_{2j-1},\phi_{2j})}\sin\tfrac{1}{2}(\phi_{2j-1}-\phi_{2j+1})\cos\tfrac{1}{2}(\phi_{2j-1}+\phi_{2j+1})e^{i\phi_{2j}}\right)\times\prod_{j=1}^{2k-1}\left(\frac{1}{\sin\frac{1}{2}(\phi_{2j}+\phi_{2j+2})}\right)ie^{-(\phi_2+\phi_{4k})/2}\,. \tag{4.83}$$

Upon comparing (4.78) with (3.84) and (4.83) with (3.99) we see that the proof of (4.81) is the same as (3.93). Hence (4.74) is proved.

Thus we have shown that for $T>T_c$ and $M,N\geqslant 0$

$$\langle\sigma_{0,0}\sigma_{M,N}\rangle=S_\infty^{>}\sum_{k=1}^{\infty}\bar{x}^{(2k-1)}_{MN}\exp\left(-\sum_{n=1}^{\infty}\bar{F}^{(2n)}_{MN}\right), \tag{4.84}$$

where $\bar{x}^{(2k+1)}_{MN}$ is given by (4.56) for $k\geq 1$, $\bar{x}^{(1)}_{MN}$ by (4.50), $\bar{F}^{(2n)}_{MN}$ by (4.74), and $S_\infty^{>}$ by (4.19) and (4.20).

E. $x^{(k)}_{MN}$ and $F^{(2n)}_{>MN}$

In (2.12) we may make use of the identity, valid in the integrand of (2.12),

$$\gamma_1\sin\tfrac{1}{2}(\phi_{2j-1}-\phi_{2j+1})\sin\tfrac{1}{2}(\phi_{2j-1}+\phi_{2j+1})+\gamma_2\sin\tfrac{1}{2}(\phi_{2j}-\phi_{2j+2})\sin\tfrac{1}{2}(\phi_{2j}+\phi_{2j+2})=0 \tag{4.85}$$

[see discussion following (3.54)] to rewrite (2.12)

$$F^{(2n)}_{<MN}=[4z_1z_2(1-z_1^2)(1-z_2^2)]^n(2n)^{-1}(2\pi)^{-4n}\times\int_{-\pi}^{\pi}d\phi_1\cdots\int_{-\pi}^{\pi}d\phi_{4n}\prod_{j=1}^{2n}\left(\frac{e^{-iM\phi_{2j-1}-iN\phi_{2j}}}{\Delta(\phi_{2j-1},\phi_{2j})}\right)\times\left(\frac{\sin\frac{1}{2}(\phi_{2j-1}-\phi_{2j+1})\sin\frac{1}{2}(\phi_{2j}-\phi_{2j+2})}{\sin\frac{1}{2}(\phi_{2j}+\phi_{2j+2})\sin\frac{1}{2}(\phi_{2j-1}+\phi_{2j+1})}\right)^{1/2}. \tag{4.86}$$

From (4.86) it is clear that $F_{<MN}^{(2n)}$ is invariant under the interchange $E_1 \leftrightarrow E_2$ and $M \leftrightarrow N$. And hence from (2.9)–(2.12), it is clear that $\langle\sigma_{0,0}\sigma_{M,N}\rangle$ for $T < T_c$ is invariant under the interchange $E_1 \leftrightarrow E_2$ and $M \leftrightarrow N$, as it must be.

For $T > T_c$, $\langle\sigma_{0,0}\sigma_{M,N}\rangle$ must still be invariant under $E_1 \leftrightarrow E_2$ and $M \leftrightarrow N$. However, an examination of $\bar{x}_{MN}^{(2k+1)}$ [see (4.56b)] and $\bar{F}_{MN}^{(2n)}$ [see (4.74)] that occur in (4.84) shows that $\bar{x}_{MN}^{(2k+1)}$ for $k \geq 1$ and $\bar{F}_{MN}^{(2n)}$, $n \geq 1$, are not *individually* invariant under $E_1 \leftrightarrow E_2$ and $M \leftrightarrow N$. Note that $\bar{x}_{MN}^{(1)}$ is invariant under $E_1 \leftrightarrow E_2$, $M \leftrightarrow N$. This suggests that (4.84) is not the best representation for $\langle\sigma_{0,0}\sigma_{M,N}\rangle$ $(T > T_c)$. In this section we show that

$$\sum_{k=1}^{\infty} \bar{x}_{MN}^{(2k-1)} \exp\left(-\sum_{n=1}^{\infty} \bar{F}_{MN}^{(2n)}\right) = \sum_{k=1}^{\infty} x_{>MN}^{(2k-1)} \exp\left(-\sum_{n=1}^{\infty} F_{>MN}^{(2n)}\right), \tag{4.87}$$

where $x_{>MN}^{(2k-1)}$ is given by (2.15) and $F_{>MN}^{(2n)}$ by (2.16). Using the identity (4.85) it is easy to show that $x_{>MN}^{(2n)}$ and $F_{>MN}^{(2n)}$ are individually invariant under $E_1 \leftrightarrow E_2$ and $M \leftrightarrow N$. Also note that

$$\bar{x}_{MN}^{(1)} = x_{>MN}^{(1)}. \tag{4.88}$$

1. *Special case*

If one examines the expression (4.56b) for $\bar{x}_{MN}^{(2k+1)}$, then one notices that if $e^{i(\phi_{2j}+\phi_{2j+2})/2}$ were replaced $\cos\frac{1}{2}(\phi_{2j}+\phi_{2j+2})$, then the resulting expression would be invariant under $E_1 \leftrightarrow E_2$, $M \leftrightarrow N$. A similar remark holds for $\bar{F}_{MN}^{(2n)}$. Hence we guess that

$$\bar{x}_{MN}^{(3)} - \bar{x}_{MN}^{(1)}\bar{F}_{MN}^{(2)} = x_{>MN}^{(3)} - x_{>MN}^{(1)}F_{>MN}^{(2)}, \tag{4.89}$$

where $x_{>MN}^{(3)}$ is defined by the $k=3$ case of (2.15), and $F_{>MN}^{(2)}$ the $n=1$ case of (2.16).

Thus we have

$$\bar{F}_{MN}^{(2)} - F_{>MN}^{(2)} = +\tfrac{1}{2}\gamma_1^2(2\pi)^{-4}\int_{-\pi}^{\pi}d\phi_1\cdots\int_{-\pi}^{\pi}d\phi_4 \prod_{j=1}^{2}\frac{e^{-iM\phi_{2j-1}-iN\phi_{2j}}}{\Delta(\phi_{2j-1},\phi_{2j})}\left(\frac{\sin\frac{1}{2}(\phi_1-\phi_3)}{\sin\frac{1}{2}(\phi_2+\phi_4)}\cos\tfrac{1}{2}(\phi_1+\phi_3)\right)^2 \times\left[e^{i(\phi_2+\phi_4)} - \cos^2\tfrac{1}{2}(\phi_2+\phi_4)\right]. \tag{4.90}$$

Now

$$e^{i(\phi_2+\phi_4)} - \cos^2\tfrac{1}{2}(\phi_2+\phi_4) = i\sin\tfrac{1}{2}(\phi_2+\phi_4)\left[e^{i(\phi_2+\phi_4)/2} + \cos\tfrac{1}{2}(\phi_2+\phi_4)\right],, \tag{4.91}$$

so that

$$\bar{F}_{MN}^{(2)} - F_{>MN}^{(2)} = \tfrac{1}{2}\gamma_1^2(2\pi)^{-4}\int_{-\pi}^{\pi}d\phi_1\cdots\int_{-\pi}^{\pi}d\phi_4 \prod_{j=1}^{2}\frac{e^{-iM\phi_{2j-1}-iN\phi_{2j}}}{\Delta(\phi_{2j-1},\phi_{2j})}\, i\sin^2\tfrac{1}{2}(\phi_1-\phi_3)\cos^2\tfrac{1}{2}(\phi_1+\phi_3)\,G(\phi_2,\phi_4), \tag{4.92}$$

where

$$G(\phi_2,\phi_4) = \left[e^{i(\phi_2+\phi_4)/2} + \cos\tfrac{1}{2}(\phi_2+\phi_4)\right]/\sin\tfrac{1}{2}(\phi_2+\phi_4). \tag{4.93}$$

Now from (4.56b) and (2.15)

$$\bar{x}_{MN}^{(3)} - x_{>MN}^{(3)} = -\gamma_1^2(2\pi)^{-6}\int_{-\pi}^{\pi}d\phi_1\cdots\int_{-\pi}^{\pi}d\phi_6 \prod_{j=1}^{3}\frac{e^{-iM\phi_{2j-1}-iN\phi_{2j}}}{\Delta(\phi_{2j-1},\phi_{2j})}\prod_{j=1}^{2}\frac{\sin\frac{1}{2}(\phi_{2j-1}-\phi_{2j+1})\cos\frac{1}{2}(\phi_{2j-1}+\phi_{2j+1})}{\sin\frac{1}{2}(\phi_{2j}+\phi_{2j+2})} \times\left[e^{i(\phi_2/2+\phi_4+\phi_6/2)} - \cos\tfrac{1}{2}(\phi_2+\phi_4)\cos\tfrac{1}{2}(\phi_4+\phi_6)\right]. \tag{4.94}$$

Writing

$$e^{i(\phi_2/2+\phi_4+\phi_6/2)} - \cos\tfrac{1}{2}(\phi_2+\phi_4)\cos\tfrac{1}{2}(\phi_4+\phi_6) = i\left[e^{i(\phi_2+\phi_4)/2}\sin\tfrac{1}{2}(\phi_4+\phi_6) + \sin\tfrac{1}{2}(\phi_2+\phi_4)\cos\tfrac{1}{2}(\phi_4+\phi_6)\right] \tag{4.95}$$

in (4.94), changing the integration variable labels by $1 \leftrightarrow 5$ and $2 \leftrightarrow 6$, adding this to (4.94), and dividing by two gives

$$\bar{x}_{MN}^{(3)} - x_{>MN}^{(3)} = -\tfrac{1}{2}i\gamma_1^2(2\pi)^{-6}\int_{-\pi}^{\pi}d\phi_1\cdots\int_{-\pi}^{\pi}d\phi_6 \prod_{j=1}^{3}\left(\frac{e^{-iM\phi_{2j-1}-iN\phi_{2j}}}{\Delta(\phi_{2j-1},\phi_{2j})}\right)\prod_{j=1}^{2}\left[\sin\tfrac{1}{2}(\phi_{2j-1}-\phi_{2j+1})\cos\tfrac{1}{2}(\phi_{2j-1}+\phi_{2j+1})\right] \times\left[G(\phi_4,\phi_6)+G(\phi_2,\phi_4)\right], \tag{4.96}$$

where we used the definition (4.93). We can rewrite (4.96)

$$\bar{x}_{MN}^{(3)} - x_{>MN}^{(3)} = -\tfrac{1}{8}i\gamma_1^2(2\pi)^{-6}\int_{-\pi}^{\pi}d\phi_1\cdots\int_{-\pi}^{\pi}d\phi_6 \prod_{j=1}^{3}\left(\frac{e^{-iM\phi_{2j-1}-iN\phi_{2j}}}{\Delta(\phi_{2j-1},\phi_{2j})}\right)(\sin\phi_1-\sin\phi_3)(\sin\phi_3-\sin\phi_5) \times\left[G(\phi_4,\phi_6)+G(\phi_2,\phi_4)\right]. \tag{4.97}$$

Now consider the three cyclic permutations of the integration variable labels in (4.97). That part of the integrand of (4.97) that is not invariant becomes

$$(\sin\phi_1-\sin\phi_3)(\sin\phi_3-\sin\phi_5)[G(\phi_2,\phi_4)+G(\phi_4,\phi_6)]+(\sin\phi_3-\sin\phi_5)(\sin\phi_5-\sin\phi_1)[G(\phi_4,\phi_6)+G(\phi_6,\phi_2)]$$
$$+(\sin\phi_5-\sin\phi_1)(\sin\phi_1-\sin\phi_3)[G(\phi_6,\phi_2)+G(\phi_2,\phi_4)]=-G(\phi_2,\phi_4)(\sin\phi_1-\sin\phi_3)^2$$
$$-G(\phi_4,\phi_6)(\sin\phi_3-\sin\phi_5)^2-G(\phi_6,\phi_2)(\sin\phi_5-\sin\phi_1)^2. \tag{4.98}$$

Hence (4.97) becomes

$$\bar{x}^{(3)}_{MN}-x^{(3)}_{>MN}=\tfrac{1}{8}i\gamma_1(2\pi)^{-6}\int_{-\pi}^{\pi}d\phi_1\cdots\int_{-\pi}^{\pi}d\phi_6\prod_{j=1}^{3}\left(\frac{e^{-iM\phi_{2j-1}-iN\phi_{2j}}}{\Delta(\phi_{2j-1},\phi_{2j})}\right)(\sin\phi_1-\sin\phi_3)^2G(\phi_2,\phi_4). \tag{4.99}$$

Comparing (4.99) with (4.92), and recalling (4.50) and (4.88), we see that (4.89) follows.

2. *The general case*

Equations (4.88) and (4.89) have established (4.87) to second order. Here we show that (4.87) holds to every order.

We define

$$x(z)=\sum_{k=0}^{\infty}x^{(k)}z^k,\quad \bar{x}(z)=\sum_{k=0}^{\infty}\bar{x}^{(k)}z^k, \tag{4.100a}$$

where

$$x^{(k)}=\begin{cases}x^{(k)}_{>MN} & \text{for } k \text{ odd}\\ 0 & \text{for } k \text{ even}\end{cases} \tag{4.100b}$$

and

$$\bar{x}^{(k)}=\begin{cases}\bar{x}^{(k)}_{MN} & \text{for } k \text{ odd}\\ 0 & \text{for } k \text{ even.}\end{cases} \tag{4.100c}$$

Also we define

$$F(z)=\sum_{n=0}^{\infty}F^{(n)}z^n,\quad \bar{F}(z)=\sum_{n=0}^{\infty}\bar{F}^{(n)}z^n, \tag{4.101a}$$

where

$$F^{(n)}=\begin{cases}F^{(n)}_{>MN} & \text{for } n \text{ even}\\ 0 & \text{for } n \text{ odd}\end{cases} \tag{4.101b}$$

and

$$\bar{F}^{(n)}=\begin{cases}\bar{F}^{(n)}_{MN} & \text{for } n \text{ even}\\ 0 & \text{for } n \text{ odd,}\end{cases} \tag{4.101c}$$

and finally,

$$G(z)\equiv\bar{F}(z)-F(z), \tag{4.102a}$$

$$G^{(n)}\equiv\bar{F}^{(n)}-F^{(n)} \tag{4.102b}$$

[where this $G(z)$ is not to be confused with that of (2.29)].

Then (4.87) is the $z=1$ case of

$$\bar{x}(z)\exp[-\bar{F}(z)]=x(z)\exp[-F(z)] \tag{4.103}$$

or, equivalently,

$$x(z)=\bar{x}(z)\exp[-G(z)]. \tag{4.104}$$

Differentiating (4.104) and multiplying the resulting equation by $\bar{x}(z)$ yields

$$x'(z)\bar{x}(z)=\bar{x}'(z)x(z)-\bar{x}(z)x(z)G'(z). \tag{4.105}$$

Suppose that we can prove (4.105), then (4.87) must follow since (4.105) can be written

$$\frac{x'(z)}{x(z)}=\frac{\bar{x}'(z)}{\bar{x}(z)}-G'(z), \tag{4.106}$$

which upon integrating gives

$$\ln x(z)=\ln\bar{x}(z)-G(z)+(\text{const}) \tag{4.107}$$

or

$$x(z)=[\bar{x}(z)e^{-G(z)}](\text{const}). \tag{4.108}$$

Now let z approach zero in (4.108).

For $z\to 0$ we have

$$x(z)\sim zx^{(1)},$$
$$\bar{x}(z)\sim z\bar{x}^{(1)},$$
$$G(z)\sim z^2G^{(2)}, \tag{4.109}$$

so that in the limit $z\to 0$ (4.108) becomes

$$x^{(1)}=\bar{x}^{(1)}(\text{const}). \tag{4.110}$$

Recalling (4.88) we see that the constant in (4.110) must be equal to one; and hence, from (4.105) follows (4.104).

From (4.100)

$$\begin{aligned}x'(z)\bar{x}(z)-\bar{x}'(z)x(z)&=\sum_{L=0}^{\infty}z^L\sum_{k=0}^{L}[(k+1)x^{(k+1)}\bar{x}^{(L-k)}-(L-k+1)x^{(k)}\bar{x}^{(L-k+1)}]\\&=\sum_{L=0}^{\infty}z^L\left((L+1)x^{(L+1)}\bar{x}^{(0)}-(L+1)x^{(0)}\bar{x}^{(L+1)}+\sum_{k=0}^{L-1}(2k-L-1)x^{(k)}\bar{x}^{(L-k+1)}\right)\\&=\sum_{L=0}^{\infty}z^L\sum_{k=0}^{L-1}(2k-L-1)x^{(k)}\bar{x}^{(L-k+1)}.\end{aligned} \tag{4.111}$$

From (4.100) (4.102)

$$x(z)\bar{x}(z)G'(z)=\sum_{k=0}^{\infty}x^{(k)}z^k\sum_{L=0}^{\infty}\bar{x}^{(L)}z^L\sum_{m=0}^{\infty}G^{(m+1)}(m+1)z^m$$

$$=\sum_{k=0}^{\infty}x^{(k)}z^k\sum_{L=0}^{\infty}z^L\sum_{m=0}^{L}\bar{x}^{(m)}G^{(L-m+1)}(L-m+1)$$

$$=\sum_{L=0}^{\infty}z^L\sum_{k=0}^{L}x^{(k)}\sum_{m=0}^{L-k}\bar{x}^{(m)}G^{(L-m-k+1)}(L-m-k+1)\,. \tag{4.112}$$

Thus from (4.105), (4.111), and (4.112)

$$\sum_{k=1}^{L}(2k-L-1)x^{(k)}\bar{x}^{(L-k+1)}$$

$$=-\sum_{k=0}^{L}x^{(k)}\sum_{m=0}^{L-k}(L-k-m+1)G^{(L-m-k+1)}\bar{x}^{(m)}$$

$$=-\sum_{k=1}^{L-1}x^{(k)}\sum_{m=1}^{L-k}\bar{x}^{(m)}G^{(L-m-k+1)}(L-m-k+1)\,, \tag{4.113}$$

where we used $\bar{x}^{(0)}=x^{(0)}=0$. If we establish (4.113), the identity (4.87) follows.

We define for all positive integers l

$$x^{(l)}=(-\tfrac{1}{2}i\lambda_1)^{l-1}(2\pi)^{-2l}\int_{-\pi}^{\pi}d\phi_1\cdots\int_{-\pi}^{\pi}d\phi_{2l}\prod_{j=1}^{l}\frac{e^{-iM\phi_{2j-1}-iN\phi_{2j}}}{\Delta(\phi_{2j-1},\phi_{2j})}\prod_{j=1}^{l-1}\left((\sin\phi_{2j-1}-\sin\phi_{2j+1})\frac{\cos\frac{1}{2}(\phi_{2j}+\phi_{2j+2})}{\sin\frac{1}{2}(\phi_{2j}+\phi_{2j+2})}\right), \tag{4.114}$$

$$\bar{x}^{(l)}=(-\tfrac{1}{2}i\gamma)^{l-1}(2\pi)^{-2l}\int_{-\pi}^{\pi}d\phi_1\cdots\int_{-\pi}^{\pi}d\phi_{2l}\prod_{j=1}^{l}\frac{e^{-iM\phi_{2j-1}-iN\phi_{2j}}}{\Delta(\phi_{2j-1},\phi_{2j})}\prod_{j=1}^{l-1}\left(\frac{\sin\phi_{2j-1}-\sin\phi_{2j+1}}{\sin\frac{1}{2}(\phi_{2j}+\phi_{2j+2})}e^{i(\phi_{2j}+\phi_{2j+2})/2}\right), \tag{4.115}$$

and

$$G^{(l)}=\bar{F}^{(l)}-F^{(l)}=(-\tfrac{1}{2}i\gamma_1)^{l-1}l^{-1}(2\pi)^{-2l}\int_{-\pi}^{\pi}d\phi_1\cdots\int_{-\pi}^{\pi}d\phi_{2l}\prod_{j=1}^{l}\left(\frac{e^{-iM\phi_{2j-1}-iN\phi_{2j}}}{\Delta(\phi_{2j-1},\phi_{2j})}\frac{(\sin\phi_{2j-1}-\sin\phi_{2j+1})}{\sin\frac{1}{2}(\phi_{2j}+\phi_{2j+2})}\right)$$
$$\times\left(\prod_{j=1}^{l}e^{i(\phi_{2j}+\phi_{2j+2})/2}-\prod_{j=1}^{l}\cos\tfrac{1}{2}(\phi_{2j}+\phi_{2j+2})\right)\,. \tag{4.116}$$

If we establish (4.113) with the definitions (4.114)–(4.116) of $x^{(l)}$, $\bar{x}^{(l)}$, and $G^{(l)}$ [as compared with (4.100)–(4.102)], then (4.113) will also follow for (4.100)–(4.102).

Before we analyze the sum (4.113), it is useful to cast $G^{(l)}$ into a different form. As $G^{(l)}$ is represented in (4.116) the factor $\sin\frac{1}{2}(\phi_{2l}+\phi_2)$ appears in the denominator. This factor does not appear in $x^{(l)}$ or $\bar{x}^{(l)}$; and so, it is advantageous to eliminate this term in $G^{(l)}$. To do this we begin by noting

$$e^{ix/2}-\cos\tfrac{1}{2}x=i\sin\tfrac{1}{2}x \tag{4.117}$$

to show that

$$\prod_{j=1}^{l}e^{i(\phi_{2j}+\phi_{2j+2})/2}-\prod_{j=1}^{l}\cos\tfrac{1}{2}(\phi_{2j}+\phi_{2j+2})=i\sin\tfrac{1}{2}(\phi_2+\phi_4)\prod_{j=2}^{l}e^{i(\phi_{2j}+\phi_{2j+2})/2}+i\cos\tfrac{1}{2}(\phi_2+\phi_4)\sin\tfrac{1}{2}(\phi_4+\phi_6)$$
$$\times\prod_{j=3}^{l}e^{i(\phi_{2j}+\phi_{2j+2})/2}+i\cos\tfrac{1}{2}(\phi_2+\phi_4)\cos\tfrac{1}{2}(\phi_4+\phi_6)\sin\tfrac{1}{2}(\phi_6+\phi_8)\prod_{j=4}^{l}e^{i(\phi_{2j}+\phi_{2j+2})/2}$$
$$+\cdots+i\cos\tfrac{1}{2}(\phi_2+\phi_4)\cos\tfrac{1}{2}(\phi_4+\phi_6)\cdots\cos\tfrac{1}{2}(\phi_{2l-2}+\phi_{2l})\sin\tfrac{1}{2}(\phi_{2l}+\phi_2)\,. \tag{4.118}$$

The first product term in (4.116) is invariant under cyclic permutations of the even and odd integration variable labels ($\phi_1\to\phi_3\to\cdots\to\phi_{2l-1}\to\phi_1$ and $\phi_2\to\phi_4\to\cdots\to\phi_{2l}\to\phi_2$). Thus we may cyclically permute the labels of the right-hand side of (4.118), and (4.118) will continue to remain an identity if used in the intergrand of (4.116). So we cyclically permute each term in (4.118) so that each term contains a factor of $\sin\frac{1}{2}(\phi_{2l}+\phi_2)$. In this manner we obtain

$$G^{(l)}=i(-\tfrac{1}{2}i\gamma_1)^{l}l^{-1}(2\pi)^{-2l}\int_{-\pi}^{\pi}d\phi_1\cdots\int_{-\pi}^{\pi}d\phi_{2l}\prod_{j=1}^{l}\frac{e^{-iM\phi_{2j-1}-iN\phi_{2j}}}{\Delta(\phi_{2j-1},\phi_{2j})}(\sin\phi_{2j-1}-\sin\phi_{2j+1})$$
$$\times\prod_{j=1}^{l-1}\frac{1}{\sin\frac{1}{2}(\phi_{2j}+\phi_{2j+2})}\left(\sum_{k=1}^{l}\prod_{j=k}^{l-1}\cos\tfrac{1}{2}(\phi_{2j}+\phi_{2j+2})\prod_{j=1}^{k-1}e^{i(\phi_{2j}+\phi_{2j+2})/2}\right), \tag{4.119}$$

where the products

$$\prod_{j=k}^{l-1}\cos\tfrac{1}{2}(\phi_{2j}+\phi_{2j+2})\quad\text{and}\quad\prod_{j=1}^{k-1}e^{i(\phi_{2j}+\phi_{2j+2})/2}$$

are to be interpreted as unity for $j=k>l-1$ and $j=1>k-1$, respectively.

We examine the structure of the product

$$\bar{x}^{(l)}G^{(m)}x^{(n)}\,, \tag{4.120}$$

where we denote the integration variables associated with $\bar{x}^{(l)}$, $G^{(m)}$, $x^{(n)}$ by ϕ, ϕ', and ϕ'', respectively. The term $(\sin\phi'_{2m-1}-\sin\phi'_1)$ occurring in (4.119) we write as follows:

$$\sin\phi'_{2m-1}-\sin\phi'_1=(\sin\phi''_{2n-1}-\sin\phi'_1)+(\sin\phi'_{2m-1}-\sin\phi_1)+(\sin\phi_1-\sin\phi''_{2n-1})\,. \tag{4.121}$$

We now define

$$\tilde{G}^{(m)}=i(-\tfrac{1}{2}i\gamma_1)^m m^{-1}(2\pi)^{-2m}\int_{-\pi}^{\pi}d\phi_1\cdots\int_{-\pi}^{\pi}d\phi_{2m}\prod_{j=1}^{m}\left(\frac{e^{-iM\phi_{2j-1}-iN\phi_{2j}}}{\Delta(\phi_{2j-1},\phi_{2j})}\right)$$
$$\times\prod_{j=1}^{m-1}\left(\frac{\sin\phi_{2j-1}-\sin\phi_{2j+1}}{\sin\frac{1}{2}(\phi_{2j}+\phi_{2j+2})}\right)\left(\sum_{k=1}^{m}\prod_{j=k}^{m-1}\cos\tfrac{1}{2}(\phi_{2j}+\phi_{2j+2})\prod_{j=1}^{k-1}e^{i(\phi_{2j}+\phi_{2j+2})/2}\right), \tag{4.122}$$

where we follow the same convention concerning the products [see discussion following (4.119)]. Now if we use (4.121) in (4.119), then the single term (4.120) becomes the sum of three terms; and the factor $(\sin\phi''_{2n-1}-\sin\phi'_1)$ naturally combines $\tilde{G}^{(m)}$ and $x^{(n)}$, the factor $(\sin\phi'_{2m-1}-\sin\phi_1)$ combines $\tilde{G}^{(m)}$ and $\bar{x}^{(l)}$; and the factor $(\sin\phi_1-\sin\phi''_{2n-1})$ combines $x^{(n)}$ and $\bar{x}^{(l)}$.

Thus we have

$$m\bar{x}^{(l)}G^{(m)}x^{(n)}=S^{(1)}_{lmn}+S^{(2)}_{lmn}+S^{(3)}_{lmn}\,, \tag{4.123}$$

where

$$S^{(1)}_{lmn}=\bar{x}^{(l)}i(-\tfrac{1}{2}i\gamma_1)^{m+n-1}(2\pi)^{-2m-2n}\int_{-\pi}^{\pi}d\phi_1\cdots\int_{-\pi}^{\pi}d\phi_{2(m+n)}\prod_{j=1}^{m+n}\left(\frac{e^{-iM\phi_{2j-1}-iN\phi_{2j}}}{\Delta(\phi_{2j-1},\phi_{2j})}\right)\prod_{j=1}^{m+n-1}\left(\frac{\sin\phi_{2j-1}-\sin\phi_{2j+1}}{\sin\frac{1}{2}(\phi_{2j}+\phi_{2j+2})}\right)$$
$$\times\prod_{j=1}^{n-1}\cos\tfrac{1}{2}(\phi_{2j}+\phi_{2j+2})\sin\tfrac{1}{2}(\phi_{2n}+\phi_{2n+2})\sum_{k=n+1}^{n+m}\prod_{j=k}^{n+m-1}\cos\tfrac{1}{2}(\phi_{2j}+\phi_{2j+2})\prod_{j=n+1}^{k-1}e^{i(\phi_{2j}+\phi_{2j+2})/2}\,, \tag{4.124}$$

$$S^{(2)}_{lmn}=x^{(n)}i(-\tfrac{1}{2}i\gamma_1)^{l+m-1}(2\pi)^{-2l-2m}\int_{-\pi}^{\pi}d\phi_1\cdots\int_{-\pi}^{\pi}d\phi_{2(l+n)}\prod_{j=1}^{l+n}\left(\frac{e^{-iM\phi_{2j-1}-iN\phi_{2j}}}{\Delta(\phi_{2j-1},\phi_{2j})}\right)\prod_{j=1}^{l+m-1}\left(\frac{\sin\phi_{2j-1}-\sin\phi_{2j+1}}{\sin\frac{1}{2}(\phi_{2j}+\phi_{2j+2})}\right)$$
$$\times\left(\sum_{k=1}^{m}\prod_{j=k}^{m-1}\cos\tfrac{1}{2}(\phi_{2j}+\phi_{2j+2})\prod_{j=1}^{k-1}e^{i(\phi_{2j}+\phi_{2j+2})/2}\right)\sin\tfrac{1}{2}(\phi_{2m}+\phi_{2m+2})\prod_{j=m+1}^{m+l-1}e^{i(\phi_{2j}+\phi_{2j+2})/2}\,, \tag{4.125}$$

and

$$S^{(3)}_{lmn}=-m\tilde{G}^{(m)}(-\tfrac{1}{2}i\gamma_1)^{l+n-2}(2\pi)^{-2(l+n)}\int_{-\pi}^{\pi}d\phi_1\cdots\int_{-\pi}^{\pi}d\phi_{2(l+n)}\prod_{j=1}^{l+n}\left(\frac{e^{-iM\phi_{2j-1}-iN\phi_{2j}}}{\Delta(\phi_{2j-1},\phi_{2j})}\right)\prod_{j=1}^{l+n-1}\left(\frac{\sin\phi_{2j-1}-\sin\phi_{2j+1}}{\sin\frac{1}{2}(\phi_{2j}+\phi_{2j+2})}\right)$$
$$\times\left(\prod_{j=1}^{l-1}\cos\tfrac{1}{2}(\phi_{2j}+\phi_{2j+2})\right)\sin\tfrac{1}{2}(\phi_{2n}+\phi_{2n+2})\prod_{j=l+1}^{l+n-1}e^{i(\phi_{2j}+\phi_{2j+2})/2}\,. \tag{4.126}$$

Letting

$$m=L-l-n+1\,, \tag{4.127}$$

we have from (4.123)–(4.126)

$$\sum_{l=1}^{L-1}\sum_{n=1}^{L-l}\bar{x}^{(l)}(L-l-n+1)G^{(L-l-n+1)}x^{(n)}=S_1+S_2+S_3\,, \tag{4.128}$$

where

$$S_j=\sum_{l=1}^{L-1}\sum_{n=1}^{L-l}S^{(j)}_{lmn}\,,\quad j=1,\ 2,\ \text{and } 3\,, \tag{4.129}$$

and m is given by (4.127).

In computing the sum S_1 we need the following result:

$$\sum_{n=1}^{L-l}\left(\prod_{j=1}^{n-1}A_j\right)(B_n-A_n)\sum_{k=n+1}^{L-l+1}\prod_{j=k}^{L-l}A_j\prod_{j=n+1}^{k-1}B_j=\sum_{n=1}^{L-l}\left(\prod_{j=1}^{n}B_j\prod_{j=n+1}^{L-l}A_j\right)-(L-l)\prod_{j=1}^{L-l}A_j\,, \tag{4.130}$$

where $\{A_j\}$ and $\{B_j\}$ denote sequences of real numbers, and where we follow our product convention [see discussion following (4.119)]. To prove (4.130) we write the left-hand side of (4.130), which we denote by I, as follows:

$$I=B_1\sum_{k=2}^{L-l+1}\prod_{j=k}^{L-l}A_j\prod_{j=2}^{k-1}B_j+\sum_{n=2}^{L-l}\Big(\prod_{j=1}^{n-1}A_j\Big)B_n\sum_{k=n+1}^{L-l+1}\prod_{j=k}^{L-l}A_j\prod_{j=n+1}^{k-1}B_j-\sum_{n=1}^{L-l}\prod_{j=1}^{n}A_j\sum_{k=n+1}^{L-l+1}\prod_{j=k}^{L-l}A_j\prod_{j=n+1}^{k-1}D_j$$

$$=\sum_{k=2}^{L-l+1}\prod_{j=1}^{k-1}B_j\prod_{j=k}^{L-l}A_j-\prod_{j=1}^{L-l}A_j+\sum_{n=2}^{L-l}\Big(\prod_{j=1}^{n-1}A_j\Big)B_n\sum_{k=n+1}^{L-l+1}\prod_{j=k}^{L-l}A_j\prod_{j=n+1}^{k-1}B_j-\sum_{n=1}^{L-l-1}\Big(\prod_{j=1}^{n}A_j\Big)\sum_{k=n+1}^{L-l+1}\Big(\prod_{j=k}^{L-l}A_j\Big)\prod_{j=n+1}^{k-1}B_j\,. \tag{4.131}$$

The last term in (4.131) can be written as

$$\sum_{n=1}^{L-l-1}\Big(\prod_{j=1}^{n}A_j\Big)\sum_{k=n+1}^{L-l+1}\Big(\prod_{j=k}^{L-l}A_j\Big)\prod_{j=n+1}^{k-1}B_j=\sum_{n=1}^{L-l-1}\prod_{j=1}^{L-l}A_j+\sum_{n=1}^{L-l-1}\Big(\prod_{j=1}^{n}A_j\Big)\sum_{k=n+2}^{L-l+1}\Big(\prod_{j=k}^{L-l}A_j\Big)\prod_{j=n+1}^{k-1}B_j$$

$$=(L-l-1)\prod_{j=1}^{L-l}A_j+\sum_{n=1}^{L-l+1}\Big(\prod_{j=1}^{n}A_j\Big)\sum_{k=n+2}^{L-l}\Big(\prod_{j=k}^{L-l}A_j\Big)\prod_{j=n+1}^{k-1}B_j\,,$$

so that

$$I=\sum_{k=2}^{L-l+1}\prod_{j=1}^{k-1}B_j\prod_{j=k}^{L-l}A_j-(L-l)\prod_{j=1}^{L-l}A_j+\sum_{n=2}^{L-l}\Big(\prod_{j=1}^{n-1}A_j\Big)B_n\sum_{k=n+1}^{L-l+1}\prod_{j=k}^{L-l}A_j\prod_{j=n+1}^{k-1}B_j-\sum_{n=1}^{L-l-1}\prod_{j=1}^{n}A_j\sum_{k=n+2}^{L-l+1}\prod_{j=k}^{L-l}A_j\prod_{j=n+1}^{k-1}B_j. \tag{4.132}$$

That last two terms in (4.132) cancel [let $n\to n-1$ in the third term of (4.132)], and we thus see that (4.130) follows.

Using (4.117) on the $\sin\frac{1}{2}(\phi_{2n}+\phi_{2n+2})$ term in (4.124) we see that (4.129) for $j=1$ becomes upon using (4.130) [$A_j\to\cos\frac{1}{2}(\phi_{2j}+\phi_{2j+2})$ and $B_j\to e^{i(\phi_{2j}+\phi_{2j+2})/2}$]

$$S_1=\sum_{l=1}^{L-1}\bar{x}^{(l)}(-\tfrac{1}{2}i\gamma_1)^{L-l}(2\pi)^{-2(L-l+1)}\int_{-\pi}^{\pi}d\phi_1\cdots\int_{-\pi}^{\pi}d\phi_{2(L-l+1)}\prod_{j=1}^{L-l+1}\left(\frac{e^{-iM\phi_{2j-1}-iN\phi_{2j}}}{\Delta(\phi_{2j-1},\phi_{2j})}\right)\prod_{j=1}^{L-l}\left(\frac{\sin\phi_{2j-1}-\sin\phi_{2j+1}}{\sin\frac{1}{2}(\phi_{2j}+\phi_{2j+2})}\right)$$

$$\times\left[\sum_{n=1}^{L-l}\Big(\prod_{j=1}^{n}e^{i(\phi_{2j}+\phi_{2j+2})/2}\prod_{j=n+1}^{L-l}\cos\tfrac{1}{2}(\phi_{2j}+\phi_{2j+2})\Big)-(L-l)\prod_{j=1}^{L-l}\cos\tfrac{1}{2}(\phi_{2j}+\phi_{2j+2})\right]. \tag{4.133}$$

Using the definitions (4.115) and (4.119) we see that S_1 can be written as

$$S_1=\sum_{l=1}^{L-1}[\bar{x}^{(l)}(L-l+1)(\tfrac{1}{2}\gamma_1)^{-1}\tilde{G}^{(L-l+1)}-\bar{x}^{(l)}x^{(L-l+1)}-(L-l)\bar{x}^{(l)}x^{(L-l+1)}]$$

$$=\sum_{l=1}^{L-1}[-(L-l+1)\bar{x}^{(l)}x^{(L-l+1)}+(L-l+1)(\tfrac{1}{2}\gamma_1)^{-1}\bar{x}^{(l)}\tilde{G}^{(L-l+1)}] \tag{4.134}$$

To compute S_2 we need the identity

$$\sum_{l=1}^{L-n}\Big(\prod_{j=L-n+2}^{L-n}B_j\Big)(B_{L-l-n+1}-A_{L-l-n+1})\sum_{k=1}^{L-l-n+1}\Big[\prod_{j=k}^{L-l-n}A_j\prod_{j=1}^{k-1}B_j\Big]=\sum_{l=1}^{L-n}\Big(\prod_{j=l+1}^{L-n}B_j\Big)(B_l-A_l)\sum_{k=1}^{l}\Big[\prod_{j=k}^{l-1}A_j\prod_{j=1}^{k-1}B_j\Big] \tag{4.135a}$$

$$=(L-n)\prod_{j=1}^{L-n}B_j-\sum_{k=1}^{L-1}\prod_{j=k}^{L-n}A_j\prod_{j=1}^{k-1}B_j\,. \tag{4.135b}$$

The first step in (4.135) is just letting $l'=L-l-n+1$, and the proof from (4.135a) to (4.135b) is similar to (4.130).

To compute S_2 we first interchange sums

$$\sum_{l=1}^{L-1}\sum_{n=1}^{L-l}=\sum_{n=1}^{L-1}\sum_{l=1}^{L-n}\,, \tag{4.136}$$

use the identity (4.117) on the term $\sin\frac{1}{2}(\phi_{2(L-l-n+1)}+\phi_{2(L-l-n+1)+2})$ [recall (4.127)], and thus from (4.135) we obtain

$$S_2=\sum_{n=1}^{L-1}x^{(n)}(-\tfrac{1}{2}i\gamma_1)^{L-n}(2\pi)^{-2(L-n+1)}\int_{-\pi}^{\pi}d\phi_1\cdots\int_{-\pi}^{\pi}d\phi_{2(L-n+1)}\prod_{j=1}^{L-n+1}\frac{e^{-iM\phi_{2j-1}-iN\phi_{2j}}}{\Delta(\phi_{2j-1},\phi_{2j})}\prod_{j=1}^{L-n}\left(\frac{\sin\phi_{2j-1}-\sin\phi_{2j+1}}{\sin\frac{1}{2}(\phi_{2j}+\phi_{2j+2})}\right)$$

$$\times\left((L-n)\prod_{j=1}^{L-n}e^{i(\phi_{2j}+\phi_{2j+2})/2}-\sum_{k=1}^{L-n}\prod_{j=k}^{L-n}\cos\tfrac{1}{2}(\phi_{2j}+\phi_{2j+2})\prod_{j=1}^{k-1}e^{i(\phi_{2j}+\phi_{2j+2})/2}\right)$$

$$= \sum_{n=1}^{L-1} [x^{(n)}(L-n)\bar{x}^{(L-n+1)} - x^{(n)}(\tfrac{1}{2}\gamma_1)^{-1}(L-n+1)\tilde{G}^{(L-n+1)} + x^{(n)}\bar{x}^{(L-n+1)}]$$

$$= \sum_{n=1}^{L-1} [(L-n+1)x^{(n)}\bar{x}^{(L-n+1)} - x^{(n)}(\tfrac{1}{2}\gamma_1)^{-1}(L-n+1)\tilde{G}^{(L-n+1)}]. \quad (4.137)$$

Letting $s = l+n$ the sum in (4.129) can be written as

$$S_2 = \sum_{s=2}^{L} \sum_{n=1}^{s-1} S_{lmn}^{(3)},$$

so that from (4.126) [and using (4.117)]

$$S_3 = \sum_{s=2}^{L} (L-s+1)\, i\tilde{G}^{(L-s+1)}(-\tfrac{1}{2}i\gamma_1)^{s-2}(2\pi)^{-2s}\int_{-\pi}^{\pi} d\phi_1 \cdots \int_{-\pi}^{\pi} d\phi_{2s} \prod_{j=1}^{s} \frac{e^{-iM\phi_{2j-1}-iN\phi_{2j}}}{\Delta(\phi_{2j-1},\phi_{2j})} \prod_{j=1}^{s-1}\left(\frac{\sin\phi_{2j-1}-\sin\phi_{2j+1}}{\sin\frac{1}{2}(\phi_{2j}+\phi_{2j+2})}\right)$$

$$\times \sum_{n=1}^{s-1}\left[\left(\prod_{j=1}^{n-1}\cos\tfrac{1}{2}(\phi_{2j}+\phi_{2j+2})\right)\left(e^{i(\phi_{2n}+\phi_{2n+2})/2} - \cos\tfrac{1}{2}(\phi_{2n}+\phi_{2n+2})\right)\prod_{j=n+1}^{s-1} e^{i(\phi_{2j}+\phi_{2j+2})/2}\right]. \quad (4.139)$$

Now

$$\sum_{n=1}^{s-1}\left(\prod_{j=1}^{n-1}A_j\right)(B_n - A_n)\prod_{j=n+1}^{s-1}B_j = \prod_{j=1}^{s-1}B_j - \prod_{j=1}^{s-1}A_j, \quad (4.140)$$

so that (4.139) becomes

$$S_3 = \sum_{s=2}^{L} (L-s+1)\, i\,\tilde{G}^{(L-s+1)}(\bar{x}^{(s)} - x^{(s)})(-\tfrac{1}{2}i\gamma_1)^{-1} \quad (4.141a)$$

$$= \sum_{s=1}^{L} (L-s+1)\, i\,\tilde{G}^{(L-s+1)}(\bar{x}^{(s)} - x^{(s)})(-\tfrac{1}{2}i\gamma_1)^{-1}, \quad (4.141b)$$

where we used the fact that $\bar{x}^{(1)} = x^{(1)}$ [recall (4.88)] in going from (4.141a) to (4.141b).

Using (4.134), (4.137), (4.141b), in (4.128) we obtain

$$-\sum_{l=1}^{L-1}\sum_{n=1}^{L-l} \bar{x}^{(l)}(L-l-n+1)G^{(L-l-n+1)}x^{(n)}$$

$$= \sum_{l=1}^{L-1} [(L-l+1)\bar{x}^{(l)}x^{(L-l+1)} - (L-l+1)x^{(l)}\bar{x}^{(L-l+1)}]$$

$$+ \tilde{G}^{(1)}(\bar{x}^{(L)} - x^{(L)})(\tfrac{1}{2}\gamma_1)^{-1}. \quad (4.142)$$

Now [recall (4.122)]

$$(\tfrac{1}{2}\gamma_1)^{-1}\tilde{G}^{(1)}$$

$$= (2\pi)^{-2}\int_{-\pi}^{\pi} d\phi_1 \int_{-\pi}^{\pi} d\phi_2 \frac{e^{-iM\phi_1 - iN\phi_2}}{\Delta(\phi_1,\phi_2)} = x^{(1)} = \bar{x}^{(1)} \quad (4.143)$$

so that

$$\tilde{G}^{(1)}(\bar{x}^{(L)} - x^{(L)})(\tfrac{1}{2}\gamma_1)^{-1} = x^{(1)}\bar{x}^{(L)} - \bar{x}^{(1)}x^{(L)}. \quad (4.144)$$

Using (4.144) in (4.142) we have

$$-\sum_{l=1}^{L-1}\sum_{n=1}^{L-l} \bar{x}^{(l)}(L-l-n+1)G^{(L-l-n+1)}x^{(n)}$$

$$= \sum_{l=1}^{L} [(L-l+1)\bar{x}^{(l)}\bar{x}^{(L-l+1)} - (L-l+1)x^{(l)}\bar{x}^{(L-l+1)}]$$

$$= \sum_{l=1}^{L} (2l-L-1)\bar{x}^{(L-l+1)}x^{(n)}. \quad (4.145)$$

Thus (4.113) is proved.

F. $\langle\sigma_{0,0}\sigma_{M,N}\rangle$ in scaling limit ($T > T_c$)

From the analysis of Sec. III G it is clear that in the scaling limit the quantities $x_{>MN}^{(k)}$ [see (2.15)] and $F_{>MN}^{(2n)}$ [see (2.16)] become

$$x_{>MN}^{(2k+1)} \sim (-1)^k(\gamma_1\gamma_2)^{-1/2}(2\pi^2)^{-2k-1}$$

$$\times \int_{-\infty}^{\infty}\cdots\int_{-\infty}^{\infty}\prod_{j=1}^{2k+1}\left(\frac{dx_j\,dy_j}{(1+x_j^2+y_j^2)}e^{-i\bar{M}x_j - i\bar{N}y_j}\right)$$

$$\times \prod_{j=1}^{2k}\left(\frac{x_j - x_{j+1}}{y_j + y_{j+1}}\right) \quad \text{(scaling limit)} \quad (4.146)$$

for $k \geq 1$, and $k = 0$

$$x_{>MN}^{(1)} \sim (2\pi^2)^{-1}(\gamma_1\gamma_2)^{-1/2}$$

$$\times \int_{-\infty}^{\infty}\int_{-\infty}^{\infty}\frac{dx_1\,dy_1}{(1+x_1^2+y_1^2)}e^{-i\bar{M}x_1 - i\bar{N}y_1} \quad \text{(scaling limit)} \quad (4.147)$$

and

$$F_{>MN}^{(2n)} \sim F_{<MN}^{(2n)} \sim f^{(2n)}(t) \quad \text{(scaling limit)}, \quad (4.148)$$

where we have used the notation of Sec. III G.

As in Section III G the quantity (4.146) can be simplified by first defining the new integration variables u_j and v_j [see (3.133)], and then performing the v_j integrations to obtain

$$x_{>MN}^{(2k+1)} \sim (\gamma_1\gamma_2)^{-1/2}g^{(2k+1)}(t) \quad \text{(scaling limit)}, \quad (4.149)$$

where

$$g^{(2k+1)}(t) = (-1)^k (2\pi)^{-2k-1} \times \int_{-\infty}^{\infty} \cdots \int_{-\infty}^{\infty} \prod_{j=1}^{2k+1} \frac{du_j e^{-itu_j}}{(1+u_j^2)^{1/2}} \times \prod_{j=1}^{2k} \frac{(1+u_j^2)^{1/2} - (1+u_{j+1}^2)^{1/2}}{u_j + u_{j+1}} \qquad (4.150a)$$

for $k \geq 1$, and for $k = 0$

$$g^{(1)}(t) = (2\pi)^{-1} \int_{-\infty}^{+\infty} du_1 \frac{e^{-itu_1}}{(1+u_1^2)^{1/2}} \; . \qquad (4.150b)$$

The final simplification occurs when we deform the contours of integration for u_j [in (4.150) $\mathrm{Im} u_j < 0$, $j = 1, 2, \ldots, 2k+1$] to a loop enclosing the branch point at $u_j = -i$, $j = 1, 2, \ldots, 2k+1$. The problem then is to determine how the second product in (4.149) sums up. The relevant identity is

$$\sum_{\{\epsilon_j = \pm 1\}} (\epsilon_1 x_1 - \epsilon_2 x_2)(\epsilon_2 x_2 - \epsilon_3 x_3) \cdots (\epsilon_{2n-2} x_{2n-2} - \epsilon_{2n-1} x_{2n-1}) = (-1)^{n-1} 2^{2n-1} x_2^2 x_4^2 \cdots x_{2n-2}^2 \; . \qquad (4.151)$$

Thus (4.150) becomes (2.30) for $k \geq 1$, and for $k = 0$

$$g^{(1)}(t) = \frac{1}{\pi} \int_1^{\infty} \frac{dy}{(y^2-1)^{1/2}} e^{-ty} = \frac{1}{\pi} K_0(t) \; , \qquad (4.152)$$

where $K_0(t)$ is the modified Bessel function. Hence in the scaling limit (2.14) becomes

$$\langle \sigma_{0,0} \sigma_{M,N} \rangle \sim \left| 1 - (\sinh 2\beta E_1 \sinh 2\beta E_2)^{-2} \right|^{1/4} \hat{F}_+(t), \qquad (4.153)$$

where $\hat{F}_+(t)$ is given by (2.28).

Furthermore, the error estimate $o(R^{-1})$ in (2.28) follows in the same manner as in Sec. III G [see discussion following (3.155)]. It should be pointed out that if one examines the representation (4.84) for $\langle \sigma_{0,0} \sigma_{M,N} \rangle$ (for $T > T_c$) in the scaling limit, then the error estimate $o(R^{-1})$ in (2.28) does *not* follow from (4.84). It is only after we proved (4.87) could we obtain the error estimate $o(R^{-1})$.

The large-t behavior of $g^{(2k+1)}(t)$ follows from (2.30) by standard techniques. We find for k fixed and $t \to \infty$

$$g^{(2k+1)}(t) = (-1)^k \pi^{-k-1/2} 2^{-3k-1/2} [e^{-(2k+1)t} / t^{2k+1/2}] \times [1 - \tfrac{1}{8}(14k+1) t^{-1} + O(t^{-2})] \; . \qquad (4.154)$$

Hence we expand $\hat{F}_+(t)$ as

$$\hat{F}_+(t) = \sum_{k=0}^{\infty} \hat{F}_+^{(2k+1)}(t) \; , \qquad (4.155)$$

where from (2.28), (2.29), and (3.162)

$$\hat{F}_+^{(1)}(t) = g^{(1)}(t) = \pi^{-1} K_0(t) \; , \qquad (4.156a)$$

$$\hat{F}_+^{(3)}(t) = g^{(3)}(t) + g^{(1)}(t) \hat{F}_-^{(2)}(t) = g^{(3)}(t) - g^{(1)}(t) f^{(2)}(t) \; , \qquad (4.156b)$$

$$\hat{F}_+^{(5)}(t) = g^{(5)}(t) + g^{(3)}(t) \hat{F}_-^{(2)}(t) + g^{(1)}(t) \hat{F}_-^{(4)}(t) = g^{(5)}(t) - g^{(3)}(t) f^{(2)}(t) + g^{(1)}(t) \times \{ -f^{(4)}(t) + \tfrac{1}{2} [f^{(2)}(t)]^2 \} \; , \qquad (4.156c)$$

etc., and the functions $f^{(2n)}(t)$ are given by (2.27). From (3.145) and (4.154) it follows that $\hat{F}_+^{(2k+1)}(t)$ for fixed k and $t \to \infty$ is asymptotically equal to $e^{-(2k+1)t}$ times some power of t. If one uses (4.154) in (4.156b) or (4.156c) [and using (3.145)] one finds not only the leading term in these expansions cancels, but the next correction term.

Of course the large-t expansion of $\hat{F}_+^{(1)}(t)$ is known, and the first few terms are

$$\hat{F}_+^{(1)}(t) = (2\pi)^{-1/2} \frac{e^{-t}}{t^{1/2}} \left(1 - \frac{1}{8t} + \frac{9}{128 t^2} + O(t^{-3}) \right) \quad (t \to \infty). \qquad (4.157)$$

We examine $\hat{F}_+^{(3)}(t)$. From (4.156b) we obtain using (2.27) and (2.30)

$$\hat{F}_+^{(3)}(t) = \pi^{-3} \int_1^{\infty} dy_1 \int_1^{\infty} dy_2 \int_1^{\infty} dy_3 \times \prod_{j=1}^{3} \frac{e^{-ty_j}}{(y_j^2-1)^{1/2}} \left(-\frac{y_2^2-1}{(y_1+y_2)(y_2+y_3)} + \frac{y_2^2-1}{(y_1+y_2)^2} \right)$$

$$= \pi^{-3} \int_1^{\infty} dy_1 \int_1^{\infty} dy_2 \int_1^{\infty} dy_3 \times \prod_{j=1}^{3} \left(\frac{e^{-ty_j}}{(y_j^2-1)^{1/2}} \right) (y_2^2 - 1) \frac{y_3 - y_1}{(y_1+y_2)^2 (y_2+y_3)} \; . \qquad (4.158)$$

Letting $y_1 \rightleftarrows y_3$ in (4.158), adding the resulting expression to (4.158), and dividing by 2 we obtain

$$\hat{F}_+^{(3)}(t) = 2^{-1} \pi^{-3} \int_1^{\infty} dy_1 \int_1^{\infty} dy_2 \int_1^{\infty} dy_3 \times \left(\prod_{j=1}^{3} \frac{e^{-ty_j}}{(y_j^2-1)^{1/2}} \right) (y_2^2 - 1) \left(\frac{y_1 - y_3}{(y_1+y_2)(y_2+y_3)} \right)^2 . \qquad (4.159)$$

From (4.159) it is straightforward to obtain for $t \to \infty$

$$\hat{F}_+^{(3)}(t) = 2^{-13/2} \pi^{-3/2} \times (e^{-3t} / t^{9/2}) [1 - \tfrac{51}{8} t^{-1} + O(t^{-2})] \; . \qquad (4.160)$$

V. PERTURBATION EXPANSION FOR $\langle \sigma_{0,0} \sigma_{N,N} \rangle$ FOR LARGE N AND SMALL $N|T-T_c|$

A. Preliminary notation

For the diagonal case $M = N$ the variables t and R of (2.5) and (2.6), respectively, reduce to

$$t = |z_1 z_2 + z_1 + z_2 - 1| \left(\frac{1}{z_1(1-z_2^2)} + \frac{1}{z_2(1-z_1^2)} \right)^{1/2} N$$

$$= |z_1 z_2 + z_1 + z_2 - 1|$$

$$\times[z_1z_2(1-z_1^2)(1-z_2^2)]^{-1/4}R \qquad (5.1)$$

and

$$R=\left[\left(\frac{z_1(1-z_2^2)}{z_2(1-z_1^2)}\right)^{1/2}+\left(\frac{z_2(1-z_1^2)}{z_1(1-z_2^2)}\right)^{1/2}\right]^{1/2}N$$

$$=\left[\left(\frac{\sinh 2\beta E_1}{\sinh 2\beta E_2}\right)^{1/2}+\left(\frac{\sinh 2\beta E_2}{\sinh 2\beta E_1}\right)^{1/2}\right]^{1/2}N. \qquad (5.2)$$

Note that for the symmetric case $R=\sqrt{2}N$. We have established that in the scaling limit (2.18)–(2.20)

$$\langle\sigma_{0,0}\sigma_{M,N}\rangle=R^{-1/4}F_{\pm}(t)+R^{-5/4}F_{1_\pm}(t)+o(R^{-5/4}), \qquad (5.3)$$

where for arbitrary M and N, t and R are given by (2.5) and (2.6), respectively, and $F_{1_\pm}(t)$ is related to $F_{\pm}(t)$ by (2.24). Now if we compute the diagonal correlation function $\langle\sigma_{0,0}\sigma_{N,N}\rangle$ in the scaling limit (2.18)–(2.20) where we identify t by (5.1) and R by (5.2), then any property we establish for $F_{\pm}(t)$ holds in general for the $M\neq N$ case if we merely use the definitions (2.5) and (2.6) for t and R, respectively. Hence from now on we restrict ourselves to the diagonal correlation function $\langle\sigma_{0,0}\sigma_{N,N}\rangle$ with no loss in generality in the scaling limit (2.18)–(2.20).

It is known[5,25] that the diagonal correlation function can be expressed as a Toeplitz determinant

$$\langle\sigma_{0,0}\sigma_{N,N}\rangle=\det(a_{m-n})_{m,n=0}^{N-1}, \qquad (5.4)$$

with

$$a_n=(2\pi)^{-1}\int_{-\pi}^{\pi}e^{-in\theta}W(e^{i\theta})\,d\theta, \qquad (5.5a)$$

$$W(\xi)=[(1-k_0\xi^{-1})/(1-k_0\xi)]^{1/2}, \qquad (5.5b)$$

and

$$k_0=(\sinh 2\beta E_1\sinh 2\beta E_2)^{-1} \qquad (5.6)$$

The square root in (5.5) is defined so that $W(e^{i\theta})$ is positive at $\theta=\pi$.

At $T=T_c$ ($k_0=1$), a_n reduces to

$$a_n=a_n^{(0)}\quad(T=T_c), \qquad (5.7)$$

with

$$a_n^{(0)}=\pi^{-1}(n+\tfrac{1}{2})^{-1}. \qquad (5.8)$$

The matrix A_0 with elements $(A_0)_{m,n}=a_{m-n}^{(0)}$ is a Cauchy matrix, and so the evaluation of $\det(a_{m-n}^{(0)})_{m,n=0}^{N-1}$ is particularly simple.[26,27] One finds for $N\to\infty$

$$\langle\sigma_{0,0}\sigma_{N,N}\rangle|_{T=T_c}=(e^{1/4}2^{1/12}A^{-3}/N^{1/4})\times[1-\tfrac{1}{64}N^{-2}+O(N^{-4})], \qquad (5.9)$$

where A is Glaisher's constant[28] [see (2.35)]. From (5.2), (5.3), and (5.9) it follows

$$F_{\pm}(0)=(\sinh 2\beta_cE_1+\sinh 2\beta_cE_2)^{1/8}e^{1/4}2^{1/12}A^{-3}\equiv F(0). \qquad (5.10)$$

For $E_1=E_2$ we have

$$F(0)=2^{1/8}e^{1/4}2^{1/12}A^{-3}=0.703\,380\,157\,687\,723\cdots. \qquad (5.10S)$$

Denoting by $\tilde{A}$ the matrix (a_{m-n}), a_m given by (5.5), we see from (5.4) that

$$\langle\sigma_{0,0}\sigma_{N,N}\rangle=\det(\tilde{A})=\det(A_0+\Delta)=\det(A_0)\det(1+\Delta A_0^{-1}), \qquad (5.11)$$

with

$$\Delta=\tilde{A}-A_0. \qquad (5.12)$$

For T close to T_c we expect Δ to be "small." Thus the expansion

$$\det(1+\Delta A_0^{-1})=\exp\left(\sum_{n=1}^{\infty}\frac{(-1)^{n-1}}{n}\operatorname{Tr}[(\Delta A_0^{-1})^n]\right) \qquad (5.13)$$

should be (at least) asymptotic as $T\to T_c$.

This observation that (5.13) should provide a means of computing $\langle\sigma_{0,0}\sigma_{N,N}\rangle$ in the scaling limit (1.1)–(1.3) was first made by Ryazanov[29] and Vaks, Larkin, and Ovchinnikov.[30] They examined (5.13) to leading order and showed (in our notation) that

$$F_+(t)=F(0)[1+\tfrac{1}{2}t\ln t+O(t)]\quad(t\to 0), \qquad (5.14)$$

where $F(0)$ is given by (5.10). However their evaluation of the constant "$\frac{1}{2}$" multiplying the $t\ln t$ term is in error (these authors obtain $\frac{1}{4}$).

In this section we will use (5.13) to develop a systematic perturbation expansion for $F_{\pm}(t)$ in the limit $t\to 0$.

B. Analysis of Δ in scaling limit

The matrix Δ defined by (5.12) has elements

$$\Delta_{m,n}=\Delta_{m-n}=a_{m-n}-a_{m-n}^{(0)}, \qquad (5.15)$$

with a_m ($a_m^{(0)}$) given by (5.5) [(5.8)]. We analyze Δ_n in the limit $T\to T_c^{\pm}$, $n\to\infty$ such that $|T-T_c|n$ is fixed. We consider the $T\to T_c^+$ limit first. We have from (5.5)

$$a_n=(2\pi i)^{-1}\oint d\xi\,\xi^{-n-1}\left(\frac{1-k_0\xi^{-1}}{1-k_0\xi}\right)^{1/2}. \qquad (5.16)$$

For $n\geq 0$ we deform the contour in (5.16) outward looping the branch point at $\xi=k_0$ ($k_0>1$ for $T>T_c$) to obtain

$$a_n=\pi^{-1}\int_{k_0}^{\infty}dx\,x^{-n-1}\left(\frac{x-k_0}{x(k_0x-1)}\right)^{1/2}. \qquad (5.17)$$

For $n<0$ we contract the contour of integration in (5.16) about the branch points 0 and k_0^{-1} to obtain

$$a_{-n}=-\pi^{-1}\int_0^{k_0^{-1}}dx\,x^{n-1}\left(\frac{k_0-x}{x(1-k_0x)}\right)^{1/2}. \qquad (5.18)$$

Letting $x = k_0 y$ in (5.17) we have

$$a_n = \frac{k_0^{-n}}{\pi}\int_1^\infty dy\, y^{-n-1}\left(\frac{y-1}{y(k_0^2 y-1)}\right)^{1/2}, \quad (5.19)$$

and with the final change of variables

$$y = 1 + \tfrac{1}{2}(1-k_0^{-2})(z-1), \quad (5.20)$$

(5.19) becomes ($T > T_c$, $n \geq 0$)

$$a_n = (k_0^{-n}/\pi)\,\tfrac{1}{2}(1-k_0^{-2}) \times \int_1^\infty dz\left(\frac{z-1}{z+1}\right)^{1/2}[1+\tfrac{1}{2}(1-k_0^{-2})(z-1)]^{-n-1/2}. \quad (5.21)$$

From Appendix C we have

$$\tfrac{1}{2}(1-k_0^{-2}) \sim 2|\beta-\beta_c|(z_{1c}+z_{2c}) \times[E_1(1-z_{2c})^{-1}+E_2(1-z_{1c})^{-1}]. \quad (5.22)$$

Defining

$$s = |z_1 z_2 + z_1 + z_2 - 1| \times\{[z_1(1-z_2^2)]^{-1}+[z_2(1-z_1^2)]^{-1}\}n, \quad (5.23)$$

and using (2.4) and (2.7) we see from (5.22) and (5.23) that

$$\tfrac{1}{2}(1-k_0^{-2}) \sim s/n. \quad (5.24)$$

Using (5.24) in (5.21) we see that

$$a_n \sim \tfrac{1}{2}(1-k_0^{-2})\,\pi^{-1}\int_1^\infty \frac{dz}{(z^2-1)^{1/2}}(z-1)\,e^{-sz} \quad (5.25)$$

in the limit $n\to\infty$, $T\to T_c^+$ such that s as defined by (5.23) is fixed. Evaluating (5.25) we obtain

$$a_n \sim \tfrac{1}{2}(1-k_0^{-2})\pi^{-1}[K_1(s)-K_0(s)]. \quad (5.26)$$

Similar transformations on (5.18) result in

$$a_{-n} \sim \tfrac{1}{2}(1-k_0^{-2})\pi^{-1}[-K_1(s)-K_0(s)] \quad (5.27)$$

for $n\to+\infty$, $T\to T_c^+$, s fixed.

Combining (5.26) and (5.27) with (5.15) and (5.8) we obtain for Δ_n

$$\Delta_n \sim \tfrac{1}{2}(1-k_0^{-2})\pi^{-1}\Delta_+(s) \quad (n\to\pm\infty,\ T\to T_c^+,\ s \text{ fixed}), \quad (5.28)$$

with

$$\Delta_+(s) = \theta(s)K_1(|s|) - K_0(|s|) - 1/s, \quad (5.29)$$

$$\Theta(s) = \begin{cases} +1 & \text{if } s \geq 0 \\ -1 & \text{if } s < 0. \end{cases} \quad (5.30)$$

The analysis of a_n in the limit $n\to\pm\infty$, $T\to T_c^-$ and s fixed is quite similar, the result is

$$\Delta_n \sim \tfrac{1}{2}(1-k_0^{-2})\pi^{-1}\Delta_-(s) \quad (n\to\pm\infty,\ T\to T_c^-,\ s \text{ fixed}), \quad (5.31)$$

$$\Delta_-(s) = \Theta(s)K_1(|s|) + K_0(|s|) - 1/s, \quad (5.32)$$

and $\Theta(s)$ is given by (5.30).

C. Quantity $\mathrm{Tr}[(\Delta A_0^{-1})^n]$ in the limit $n\to\infty$, $T\to T$, s fixed

For simplicity we first analyze the case $n=1$. To compute $\mathrm{Tr}(\Delta A_0^{-1})$ we must know the matrix elements of A_0^{-1}. Since A_0 is a Cauchy matrix, the inverse matrix elements can be computed. This has been done by Wu[31] who finds

$$(A_0^{-1})_{lk} = \frac{f(k)g(l)}{\pi(k-l+\frac{1}{2})}, \quad k,\ l = 0,\ 1,\ 2,\ \ldots, N-1, \quad (5.33)$$

with

$$f(k) = \frac{\Gamma(N-k+\frac{1}{2})\Gamma(k+\frac{3}{2})}{\Gamma(N-k+1)\Gamma(k+1)}, \quad (5.34)$$

$$g(l) = \frac{\Gamma(N-l+\frac{3}{2})\Gamma(l+\frac{1}{2})}{\Gamma(N-l+1)\Gamma(l+1)}, \quad (5.35)$$

and $\Gamma(x)$ the gamma function. Thus

$$\mathrm{Tr}(\Delta A_0^{-1}) = \sum_{k,l=0}^{N-1}\frac{f(k)g(l)}{\pi(k-l+\frac{1}{2})}\Delta_{k-l}, \quad (5.36)$$

with Δ_k given by (5.15). Equations (5.13) and (5.36) are exact with no approximations having been made. We wish to extract from (5.36) the leading term in the limit $N\to\infty$, $T\to T_c^\pm$ such that t is fixed.

Letting $s=\frac{1}{2}(1-k_0^{-2})k$ and $s'=\frac{1}{2}(1-k_0^{-2})l$ we have, approximating the sum in (5.36) by an integral,

$$\mathrm{Tr}(\Delta A_0^{-1}) = \pi^{-2}\int_0^t ds\int_0^t ds'(s-s'+\tfrac{1}{2}\kappa_d)^{-1} \times f(s/\kappa_d)g(s'/\kappa_d)\Delta_\pm(s-s') + E, \quad (5.37)$$

where

$$\kappa_d \equiv \tfrac{1}{2}(1-k_0^{-2}), \quad (5.38)$$

and E is the error in approximating the sum by an integral. Using Stirling's formula we see that as $\kappa_d\to 0$

$$f(s/\kappa_d)g(s'/\kappa_d) = \left(\frac{s(t-s')}{s'(t-s)}\right)^{1/2} + O(\kappa_d). \quad (5.39)$$

In the limit $N\to\infty$, $T\to T_c^\pm$ such that t is fixed, the error E in (5.37) is negligible to leading order and we have

$$\mathrm{Tr}(\Delta A_0^{-1}) \sim \pi^{-2}\,P\int_0^t ds\,P\int_0^t ds'\frac{1}{s-s'}\left[\frac{s(t-s')}{s'(t-s)}\right]^{1/2} \times \Delta_\pm(s-s'). \quad (5.40)$$

Changing variables we have

$$\mathrm{Tr}(\Delta A_0^{-1}) \sim \frac{t}{\pi^2}P\int_0^1 ds\,P\int_0^1 du'\left(\frac{u(1-u')}{u'(1-u)}\right)^{1/2} \times\frac{\Delta\pm[t(u-u')]}{u-u'} \quad (T\to T_c^\pm,\ N\to\infty,\ t \text{ fixed}). \quad (5.41)$$

The symbol $P\int$ denotes the Cauchy principal value.

Note that although (5.39) is correct, its use in (5.37) is a nonuniform approximation. Clearly s/κ_d is not large when s is in the range $(0, \kappa_d)$. If the integrand were integrable (nonsingular), supplying the estimates that would allow the use of (5.39) in (5.37) would be relatively straightforward. However, the singular factor prevents such a straightforward analysis. Heuristically, the singular part of the denominator goes like $O(\kappa_d^{-1})$ in the range $(0, \kappa_d)$ if we interpret the integrals as Cauchy principal-value integrals. The rest of the integrand is integrable and so the double integral goes like $O(\kappa_d^2)$. Hence the error involved in using (5.39) in (5.37) is $O(\kappa_d)$ as $\kappa_d \to 0$.

The result (5.41) can be easily generalized to $\mathrm{Tr}[(\Delta A_0^{-1})^n]$. We find for $N\to\infty$, $T\to T_c^{\pm}$ such that t is fixed

$$\mathrm{Tr}[(\Delta A_0^{-1})^n] \sim t^n \pi^{-2n} P\int_0^1 du_1 \cdots P\int_0^1 du_{2n} \times \prod_{j=1}^{n} \frac{\Delta_{\pm}[t(u_{2j-1}-u_{2j})]}{u_{2j+1}-u_{2j}} \times \left(\frac{u_{2j+1}(1-u_{2j})}{u_{2j}(1-u_{2j+1})}\right)^{1/2}, \tag{5.42}$$

with $u_{2n+1}=u_1$. We define the right-hand side of (5.42) to be $\Lambda_n^{\pm}(t)$.

D. Small-t expansion of $F_{\pm}(t)$

From (5.9), (5.10), (5.11), and (5.13) we see that to compute the scaling functions $F_{\pm}(t)$ we need (5.13) in the scaling limit. Recalling the definition of $\Lambda_n^{\pm}(t)$ [(5.42)] we have

$$F_{\pm}(t) = F(0)\exp\left(\sum_{n=1}^{\infty} \frac{(-1)^{n-1}}{n}\Lambda_n^{\pm}(t)\right), \tag{5.43}$$

with $F(0)$ given by (5.10). From (5.29), (5.32), and (5.42) we see that

$$\Lambda_n^{\pm}(t) = O(t^n \ln^n t) \quad (t\to 0), \tag{5.44}$$

so (5.43) provides a small-t expansion of $F_{\pm}(t)$. Thus the program is to expand the functions $\Delta_{\pm}[t(u_{2j-1}-u_{2j})]$ appearing in the definition of $\Lambda_n^{\pm}(t)$ [we do this by using the small-t expansion of $K_0(t)$ and $K_1(t)$], evaluate the resulting integrals, and hence obtain $\Lambda_n^{\pm}(t)$. From (5.29) and (5.32) we see that

$$\Delta_-(s) = -\Delta_+(-s). \tag{5.45}$$

Hence from (5.42) and (5.45) it becomes clear that the small-t expansion of $\Lambda_n^-(t)$ can be obtained from the small-t expansion of $\Lambda_n^+(t)$ by replacing all factors of t in the small-t expansion of $\Lambda_n^+(t)$ by $-t$ where we interpret all logarithmic factors $\ln t$ in $\Lambda_n^+(t)$ as $\ln|t|$. This prescription carries over to $F_{\pm}(t)$. Hence we compute the small-t expansion of $F_+(t)$ and apply this prescription to obtain the small-t expansion of $F_-(t)$.

The general structure of $\Lambda_n^+(t)$ is clear from (5.29), (5.42), and the small-x expansions of $K_0(x)$ and $K_1(x)$:

$$\Lambda_n^+(t) = \sum_{k=n}^{\infty}\sum_{l=0}^{n} C_{kl}^{(n)} t^k (\ln t)^l, \tag{5.46}$$

where $C_{kl}^{(n)}$ are constants, and are (in general) expressed as $2n$-dimensional integrals.

Using the expansion for $s\to 0$

$$\Delta_+(s) = \ln(s/2) + \gamma_E + (s/2)(\ln s + \gamma_E - \tfrac{1}{2}) + (s/2)^2(\ln(s/2) + \gamma_E - 1) + \cdots \tag{5.47}$$

in (5.42) for $n=1$ [recall (5.29)] we see that, for example,

$$C_{11}^{(1)} = \pi^{-2}\int_0^1 du\, P\int_0^1 du' \frac{1}{u-u'}\left(\frac{u(1-u')}{u'(1-u)}\right)^{1/2} = \frac{1}{2} \tag{5.48}$$

and

$$C_{10}^{(1)} = (\gamma_E - \ln 2)C_{11}^{(1)} + \pi^{-2}\int_0^1 du\, P\int_0^1 du' \times \frac{1}{u-u'}\left(\frac{u(1-u')}{u'(1-u)}\right)^{1/2}\ln|u-u'| = \tfrac{1}{2}(\gamma_E - 3\ln 2). \tag{5.49}$$

From (5.48) and (5.49) we can already conclude that

$$\Lambda_1^+(t) = \tfrac{1}{2}t\ln t + \tfrac{1}{2}t(\gamma_E - 3\ln 2) + O(t^2\ln^2 t) \tag{5.50a}$$

$$= \tfrac{1}{2}t\Omega + O(t^2\ln^2 t), \tag{5.50b}$$

where Ω is defined by (2.33). From (5.50), (5.44), and (5.43) we have

$$F_{\pm}(t) = F(0)[1 \pm \tfrac{1}{2}t\Omega + O(t^2\ln^2 t)], \tag{5.51a}$$

where we used the prescription of replacing $t\to -t$ to obtain the small-t expansion of $F_-(t)$ from the small-t expansion of $F_+(t)$. Using (5.50a) we can write (5.51) as

$$F_{\pm}(t) = F(0)[1 \pm \tfrac{1}{2}t\ln t \pm \tfrac{1}{2}t(\gamma_E - 3\ln 2) + O(t^2\ln^2 t)]. \tag{5.51b}$$

The coefficient $\tfrac{1}{2}(\gamma_E - 3\ln 2)$ has the numerical value

$$\tfrac{1}{2}(\gamma_E - 3\ln 2) = -0.751\,112\,938\,389\,14\cdots. \tag{5.52}$$

To obtain the higher-order terms in (5.51) we must keep the higher-order terms in $\Lambda_1^+(t)$ and include the contributions coming from $\Lambda_2^+(t)$, $\Lambda_3^+(t)$, ... In principle, this is straightforward, though by order t^3 it becomes rather messy.

Rather than work through each order in detail we present the results of our computation in Table II. Certain integrals that appear frequently are given in Table III. By using Table II we can conclude that for $t\to 0$

$$\ln[F_{\pm}(t)/F(0)] = \pm\tfrac{1}{2}t\Omega - \tfrac{1}{8}t^2\Omega^2 + \tfrac{1}{16}t^2$$

TABLE II. Coefficients $C_{kl}^{(n)}$ in the small-t expansion of $\Lambda_n^{+}(t)$. All relevant coefficients to compute $F_{\pm}(t)$ up to and including the term $t^3 \ln t$ are presented. See Eqs. (5.43) and (5.46) in text.

$n=1$	$C_{11}^{(1)}=\frac{1}{2}$
	$C_{10}^{(1)}=\frac{1}{2}(\gamma_E-3\ln 2)$
	$C_{21}^{(1)}=\frac{1}{8}$
	$C_{20}^{(1)}=\frac{1}{8}(\gamma_E-3\ln 2)$
	$C_{31}^{(1)}=\frac{1}{32}$
$n=2$	$C_{22}^{(2)}=\frac{1}{4}$
	$C_{21}^{(2)}=\frac{1}{2}(\gamma_E-3\ln 2)+\frac{1}{4}$
	$C_{20}^{(2)}=\frac{1}{4}(\gamma_E-3\ln 2)^2+\frac{1}{4}(\gamma_E-3\ln 2)-\frac{1}{8}$
	$C_{32}^{(2)}=\frac{1}{8}$
	$C_{31}^{(2)}=\frac{1}{4}(\gamma_E-3\ln 2)+\frac{1}{16}$
$n=3$	$C_{31}^{(3)}=\frac{1}{8}$
	$C_{33}^{(3)}=\frac{3}{8}(\gamma_E-3\ln 2)+\frac{3}{16}$
	$C_{31}^{(3)}=\frac{3}{8}(\gamma_E-3\ln 2)^2+\frac{3}{8}(\gamma_E-3\ln 2)$

$$\pm\tfrac{1}{24}t^3\Omega^3 \pm \tfrac{1}{64}t^3\Omega + O(t^4\ln^4 t)\,. \tag{5.53}$$

Actually not enough information is given in Table II to evaluate the coefficient of the t^3 term in (5.53). Strictly from Table II one can compute up to and including the $t^3\ln t$ term in (5.53). The t^3 plus all t^4 terms will be computed by a more efficient method in Sec. VI. We should emphasize it is not necessary to use the techniques of Sec. VI to establish these higher-order terms, it is just easier. For this reason we computed only through $t^3 \ln t$ by the techniques of this section.

It is easy to see that

$$C_{nn}^{(n)}=2^{-n} \tag{5.54}$$

since the $2n$-dimensional integral for this coefficient factorizes. Thus,

$$\Lambda_n^{+}(t)=(\tfrac{1}{2}t\ln t)^n+O(t^n\ln^{n-1}t)\,, \tag{5.55}$$

and hence

$$\sum_{n=1}^{\infty}\frac{(-1)^{n-1}}{n}\Lambda_n^{+}(t)=\ln(1+\tfrac{1}{2}t\ln t)+O\Big(\sum_n t^n\ln^{n-1}t\Big)\,. \tag{5.56}$$

Using (5.56) in (5.43) we see

$$F_{+}(t)=F(0)\left[1+\tfrac{1}{2}t\ln t+O(t)\right]. \tag{5.57}$$

The point being that all $t^n(\ln t)^n$ terms in the exponential add up to give the $\frac{1}{2}t\ln t$ term in $F_{+}(t)$. Hence we cannot have a $t^n(\ln t)^n$ term in the small-t expansion of $F_{\pm}(t)$. At most we could have is $t^n(\ln t)^{n-1}$.

In Fig. 7 we plot $[1+\frac{1}{2}t\Omega]$ and $F_{+}(t)/F(0)$ as a function of t. In Fig. 8 we plot $[1-\frac{1}{2}t\Omega]$ and $F_{-}(t)/F(0)$ as a function of t.

We conclude this section by evaluating one of the nontrivial integrals given in Table III. Let

$$F(x,y)=\pi^{-1}P\int_0^1\frac{du}{u-x}\left(\frac{1-u}{u}\right)^{1/2}\ln|u-y|\,, \tag{5.58}$$

with $0<x<1$ and $0<y<1$. It is not difficult to show that

$$F(x,0)=\lim_{y\to 0^+}F(x,y)\,.$$

Thus we can write

$$F(x,y)-F(x,0)=\pi^{-1}P\int_0^1\frac{du}{u-x}\left(\frac{1-u}{u}\right)^{1/2}\ln\left|\frac{u-y}{u}\right| \tag{5.59}$$

To evaluate (5.59) consider $\int_C f(z)\,dz$ where

$$f(z)=\frac{1}{z-x}\left(\frac{z-1}{z}\right)^{1/2}\ln\left(\frac{z-y}{z}\right), \tag{5.60}$$

and the contour C is shown in Fig. 9. Thus we see that

$$F(x,y)=F(x,0)-\pi\left(\frac{1-x}{x}\right)^{1/2} \quad \text{for } x<y\,,$$

$$=F(x,0)-\frac{\pi}{2}\left(\frac{1-x}{x}\right)^{1/2} \quad \text{for } x=y\,,$$

and

$$=F(x,0) \quad \text{for } x>y\,. \tag{5.61}$$

It is not difficult to see that

$$\lim_{x\to 1^-}F(x,0)=F(1,0)$$

$$=-\frac{1}{\pi}\int_0^1 du\ln u\,[u(1-u)]^{-1/2}$$

$$=2\ln 2\,, \tag{5.62}$$

so that

$$F(x,0)-F(1,0)=(1-x)P\int_0^1\frac{du}{u-x}\times\ln u\,[u(1-u)]^{-1/2}\,. \tag{5.63}$$

Letting

$$\ln u=\int_1^u\frac{dt}{t}$$

$$=(u-1)\int_0^1\frac{d\xi}{1+\xi(u-1)}\,, \tag{5.64}$$

in (5.63) we see that

$$F(x,0)-F(1,0)=-\int_0^1 d\xi\,P\int_0^1\frac{du}{u-x}\times\left(\frac{1-u}{u}\right)^{1/2}\frac{1}{1-\xi(1-u)}\,, \tag{5.65}$$

TABLE III. Some of the definite integrals needed to compute the coefficients $C_{kl}^{(n)}$. Variables x and y are always restricted to the range $0<x<1$ and $0<y<1$.

1. $$\pi^{-1}\int_0^1 du\, u^n \left(\frac{u}{1-u}\right)^{1/2} = 2^{-(2m+2)} \binom{2n+2}{n+1}$$

2. $$\pi^{-1}\int_0^1 du\, u^n [u(1-u)]^{1/2} = 2^{-(2n+2)}(n+2)^{-1} \binom{2n+1}{n+1}$$

3. $$\pi^{-1}\int_0^1 du \left(\frac{1-u}{u}\right)^{1/2} = \tfrac{1}{2}$$

4. $$\pi^{-1}\int_0^1 du\, [u(1-u)]^{-1/2} = 1$$

5. $$\pi^{-1}\int_0^1 du\, [u(1-u)]^{-1/2} \ln|u-x| = -2\ln 2$$

6. $$\pi^{-1}\int_0^1 du \left(\frac{u}{1-u}\right)^{1/2} \ln|u-x| = -x+\tfrac{1}{2}-\ln 2$$

7. $$\pi^{-1}\int_0^1 du \left(\frac{1-u}{u}\right)^{1/2} \ln|u-x| = x-\tfrac{1}{2}-\ln 2$$

8. $$\pi^{-1}\int_0^1 du\, [u(1-u)]^{1/2} \ln|u-x| = \tfrac{1}{2}x^2-\tfrac{1}{2}x+\tfrac{1}{16}-\tfrac{1}{4}\ln 2$$

9. $$\pi^{-1}\int_0^1 du\, u[u(1-u)]^{1/2} \ln|u-x| = -\tfrac{1}{3}x^3+\tfrac{1}{4}x^2+\tfrac{1}{8}x+\tfrac{5}{96}-\tfrac{1}{8}\ln 2$$

10. $$\pi^{-1} P\int_0^1 \frac{du}{u-x} \left(\frac{1-u}{u}\right)^{1/2} = -1$$

11. $$\pi^{-1} P\int_0^1 \frac{du}{u-x} \left(\frac{u}{1-u}\right)^{1/2} = +1$$

12. $$\pi^{-1} P\int_0^1 \frac{du}{u-x}\, u \left(\frac{1-u}{u}\right)^{1/2} = -x+\tfrac{1}{2}$$

13. $$\pi^{-1} P\int_0^1 \frac{du}{u-x}\, u \left(\frac{u}{1-u}\right)^{1/2} = x+\tfrac{1}{2}$$

14. $$\pi^{-1} P\int_0^1 \frac{du}{u-x}\, u[u(1-u)]^{1/2} = -x^2+\tfrac{1}{2}x-\tfrac{1}{8}$$

15. $$\pi^{-1} P\int_0^1 \frac{du}{u-x}\, u^2[u(1-u)]^{1/2} = -x^3+\tfrac{1}{2}x^2+\tfrac{1}{8}x+\tfrac{1}{16}$$

16. $$\pi^{-1} P\int_0^1 \frac{du}{u-x} \left(\frac{1-u}{u}\right)^{1/2} \ln|u-y| = \begin{cases} 2\ln 2 + 2\left(\frac{1-x}{x}\right)^{1/2} \arccos(x^{1/2}), & x>y \\ 2\ln 2 + 2\left(\frac{1-x}{x}\right)^{1/2} \arccos(x^{1/2}) - \frac{\pi}{2}\left(\frac{1-x}{x}\right)^{1/2}, & x=y \\ 2\ln 2 + 2\left(\frac{1-x}{x}\right)^{1/2} \arccos(x^{1/2}) - \pi\left(\frac{1-x}{x}\right)^{1/2}, & x<y\,. \end{cases}$$

where the interchange of the ξ and u integration can be rigorously justified.[32] The inner integral in (5.65) is straightforward to compute, and so after the ξ integration we obtain

$$F(x,0) = 2\ln 2 + 2\left(\frac{1-x}{x}\right)^{1/2} \arccos(x^{1/2})\,. \quad (5.66)$$

Equations (5.61) and (5.66) give $F(x,y)$.

VI. SCALING FUNCTIONS $F_\pm(t)$ IN TERMS OF PAINLEVÉ FUNCTION OF THIRD KIND

In this section we unify the two expansions of the Secs. III-V by demonstrating that $F_\pm(t)$ may be expressed in terms of the Painlevé function of the

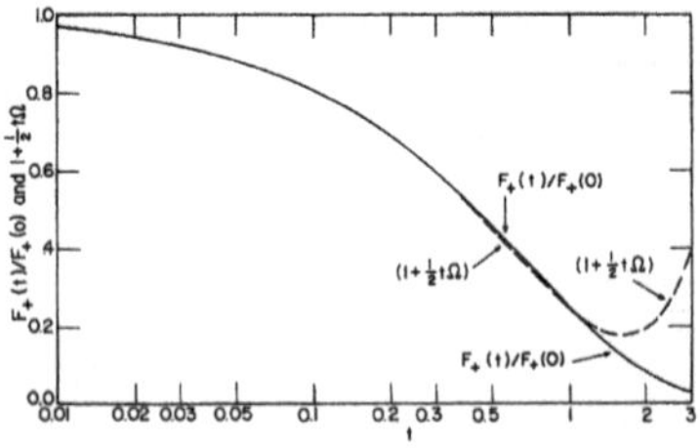

FIG. 7. Functions $(1+\frac{1}{2}t\Omega)$, $\Omega=\ln(\frac{1}{8}t)+\gamma_E$, γ_E = Euler's constant, and $F_+(t)/F_+(0)$ as a function of t.

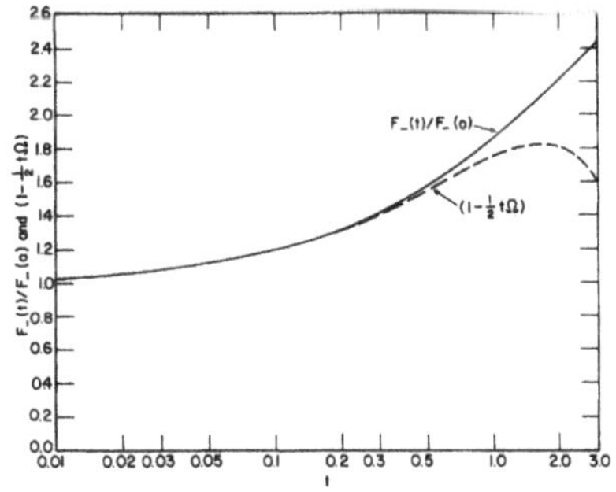

FIG. 8. Functions $(1-\frac{1}{2}t\Omega)$, $\Omega=\ln(\frac{1}{8}t)+\gamma_E$, γ_E = Euler's constant, and $F_-(t)/F_-(0)$ as a function of t.

third kind defined by (2. 36)–(2. 38).[33]

The Painlevé function arises when we are lead from considering the ratio

$$x_0(N)=\langle\sigma_{0,0}\sigma_{N-1,N-1}\rangle/\langle\sigma_{0,0}\sigma_{N,N}\rangle \tag{6. 1}$$

in the scaling limit (1. 1)–(1. 3) to integral equations of the form

$$\int_0^t K_0(|s-s'|)\,x(s')\,ds'=y(s)\ , \tag{6. 2}$$

where $K_0(x)$ is the modified Bessel function and $y(s)$ is taken as a known function. This integral equation has been the subject of much investigation in scattering theory where it arises in the context of electromagnetic scattering from a strip.

Myers[34,35] has expressed the solution to (6. 2) [for particular $y(s)$'s] in terms of the Painlevé function $\eta(\theta)$. Therefore when our analysis leads to equations of the form (6.2), we can take over the work of Myers and apply it directly to our problem. Since this work is not well known in the statistical mechanics literature, we have summarized the relevant results of Myers's thesis in Appendix B. A reading of Appendix B is ultimately necessary for an understanding of the computations of this section.

A. Relating $x_0(N)$ to integral equations

We first define

$$S_N=\langle\sigma_{0,0}\sigma_{N,N}\rangle \tag{6. 3}$$

and recall that S_N is then given by (5. 4). Consider the sum equation

$$\sum_{m=0}^{N-1} a_{n-m}\,x_m=\delta_{n,0}\ . \tag{6. 4}$$

Then by Cramer's rule and (5. 4) we have $x_0(N)$ being given by (6. 1). The quantities a_n in (6. 4) are defined by (5. 5).

We introduce the auxiliary generating function

$$\overline{W}(\xi)=(1-k_0\xi^{-1})^{-1}\,W(\xi)\ , \tag{6. 5}$$

where $W(\xi)$ and k_0 are given by (5. 5b) and (5. 6), respectively, and define the auxiliary coefficients

$$\overline{a}_n=(2\pi)^{-1}\int_{-\pi}^{\pi} e^{-in\theta}\,\overline{W}(e^{i\theta})\,d\theta\ . \tag{6. 6}$$

If we define for $0\le n\le N$

$$y_n=\sum_{m=0}^{N-1}\overline{a}_{n-m}\,x_m\ , \tag{6. 7}$$

then from (6. 4)–(6. 6) it follows that

$$y_n-k_0\,y_{n+1}=\delta_{n,0} \tag{6. 8}$$

for $0\le n\le N$. Equation (6. 8) has the solution

$$y_n=\delta_{n,0}+\kappa\,k_0^{-n} \tag{6. 9}$$

with κ a constant to be determined.

If we define

$$x_N=0\ , \tag{6. 10}$$

then (6. 7) is equivalent to the system of $(N+1)$ linear equations

$$\sum_{m=0}^{N}\overline{a}_{n-m}\,x_m=\delta_{n,0}+\kappa\,k_0^{-n}\ . \tag{6. 11}$$

Let $\overline{x}_m$ and $\overline{\overline{x}}_m$ be the solutions of

$$\sum_{m=0}^{N}\overline{a}_{n-m}\,\overline{x}_m=\delta_{n,0} \tag{6. 12}$$

and

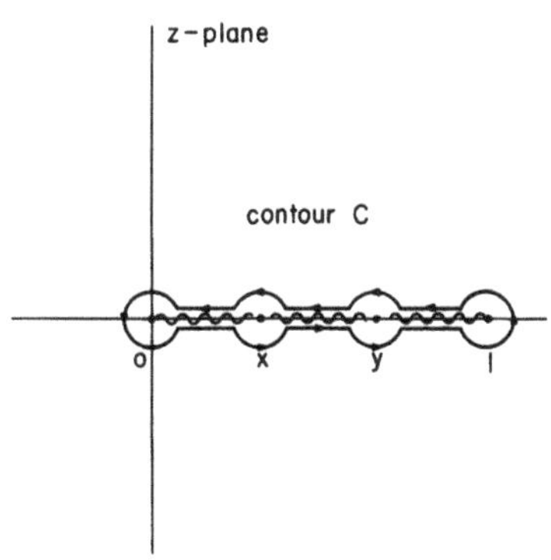

FIG. 9. Contour of integration C.

$$\sum_{m=0}^{N} \bar{a}_{n-m}\, \tilde{x}_m = k_0^{-n} \,, \qquad (6.13)$$

respectively, with $0 \le n \le N$. Then

$$\kappa = -\bar{x}_N / \tilde{x}_N \,, \qquad (6.14)$$

and so

$$x_m = \bar{x}_m - \tilde{x}_m \bar{x}_N / \tilde{x}_N \,. \qquad (6.15)$$

In the scaling limit we have in a manner similar to that of (5.16)–(5.26)

$$\bar{a}_n \sim \pi^{-1} K_0(s) \quad (n \to \infty,\ T \to T_c^{\pm},\ s \text{ fixed}) \,, \qquad (6.16)$$

with s being defined by (5.23). Given (6.12), (6.13), and (6.16) it is suggestive to consider the pair of integral equations

$$\int_0^t K_0(|s-s'|)\, \bar{x}(s';Z)\, ds' = Z^{1/2} e^{-Zs} \qquad (6.17)$$

and

$$\int_0^t K_0(|s-s'|)\, \tilde{x}_{\pm}(s')\, ds' = e^{\pm s} \,, \qquad (6.18)$$

with t defined by (5.1) [the $M=N$ case of (2.5)], and Z a large positive parameter so that $Z^{1/2}e^{-Zs}$ mocks the Kronecker δ function $\delta_{0,n}$. Equations (6.17) and (6.18) are, in some sense to be made precise, the continuum analog of (6.12) and (6.13), respectively. For $T > T_c$ ($k_0 > 1$) we associate $\tilde{x}_-(s)$ of (6.18) with (6.13), and for $T < T_c$ ($k_0 < 1$) we associate $\tilde{x}_+(s)$ with (6.13).

We now examine the pair of *infinite* equations

$$\sum_{m=0}^{\infty} \bar{a}_{n-m}\, \hat{x}_m = \hat{y}_n \qquad (6.19)$$

and

$$\int_0^{\infty} K_0(|s-s'|)\, \hat{x}(s')\, ds' = \hat{y}(s) \,, \qquad (6.20)$$

where

$$\hat{y}_n = \hat{y}(s) \qquad (6.21)$$

for s given by (5.23), and we assume that $\hat{y}$ is slowly varying in the sense that $\hat{y}_n - \hat{y}_{nH}$ is small. We can apply the standard methods of Wiener and Hopf[19] to (6.19) and (6.20). Writing

$$[\bar{W}(\xi)]^{-1} = P(\xi)\, Q(\xi^{-1}) \,, \qquad (6.22)$$

with

$$P(\xi) = \begin{cases} (1-k_0\xi)^{1/2} & \text{for } T < T_c \\ (\xi - k_0)^{1/2} & \text{for } T > T_c \end{cases} \qquad (6.23)$$

and

$$Q(\xi) = \begin{cases} (1-k_0\xi)^{1/2} & \text{for } T < T_c \\ (\xi - k_0)^{1/2} & \text{for } T > T_c \,, \end{cases} \qquad (6.24)$$

it follows that

$$X(\xi) = P(\xi)\,[Q(\xi^{-1})\,Y(\xi)]_+ \,, \qquad (6.25)$$

where

$$X(\xi) = \sum_{n=0}^{\infty} \hat{x}_n \xi^n \,, \qquad (6.26a)$$

$$Y(\xi) = \sum_{n=0}^{\infty} \hat{y}_n \xi^n \,, \qquad (6.26b)$$

and $[\cdots]_+$ denotes the plus part of a function.[36]

Using (6.23)–(6.26) we have

$$\hat{x}_0 = X(0) = (2\pi)^{-1} \int_{-\pi}^{\pi} d\theta\, (1 - k_0 e^{-i\theta})^{1/2}\, Y(e^{i\theta}) \,. \qquad (6.27)$$

Defining

$$\zeta = \theta (1-k_0)^{-1} \,, \qquad (6.28)$$

(6.27) becomes

$$\hat{x}_0 = (2\pi)^{-1} \int_{-\pi(1-k_0)^{-1}}^{\pi(1-k_0)^{-1}} d\zeta\, (1-k_0) \times (1 - k_0 e^{-i\zeta(1-k_0)})^{1/2}\, Y(e^{i\zeta(1-k_0)}) \,, \qquad (6.29)$$

which to leading order in the scaling limit is

$$\hat{x}_0 \sim (2\pi)^{-1} \int_{-\infty}^{+\infty} d\zeta\, (1-k_0)^{3/2} (1+i\zeta)^{1/2}\, Y(e^{i\zeta(1-k_0)}) \,. \qquad (6.30)$$

From the Poisson sum formula it follows that

$$\mathcal{Y}(\zeta) \sim (1-k_0)\, Y(e^{i\zeta(1-k_0)}) \,, \qquad (6.31)$$

where

$$\mathcal{Y}(\zeta) = \int_0^{\infty} e^{i\zeta t}\, \hat{y}(t)\, dt \,, \qquad (6.32)$$

and it is assumed that $\hat{y}_n = \hat{y}(s)$ for $s = n(1-k_0)$. Hence (6.30) becomes

$$\hat{x}_0 \sim (1-k_0)^{1/2} \int_{-\infty}^{+\infty} d\zeta\, (1+i\zeta)^{1/2} \mathcal{Y}(\zeta) \,. \qquad (6.33)$$

Applying the Wiener-Hopf analysis to (6.20) we obtain

$$\chi(\zeta) = \pi^{-1} (1-i\zeta)^{1/2} [(1+i\zeta)^{1/2} \mathcal{Y}(\zeta)]_+ \,, \qquad (6.34)$$

where

$$\chi(\zeta) = \int_0^{\infty} e^{i\zeta t}\, \hat{x}(t)\, dt \,. \qquad (6.35)$$

Using the fact that

$$\lim_{s\to 0} s^{1/2} \hat{x}(s) = \lim_{\zeta\to\infty} e^{-i\pi/4} (\zeta/\pi)^{1/2} \chi(\zeta) \,, \qquad (6.36)$$

and (6.34) we see that (6.33) can be written

$$\hat{x}_0 \sim \pi^{3/2} (1-k_0)^{1/2} \lim_{s\to 0} s^{1/2} \hat{x}(s) \,, \qquad (6.37)$$

where to summarize $\tilde{x}_0$ is the $m=0$ solution to (6.19) and $\hat{x}(s)$ is the solution to (6.20).

We now apply this result to (6.13) and (6.18). Let us define

$$\hat{y}(s)=\int_0^t ds' K_0(|s-s'|)\tilde{x}_\pm(s') \tag{6.38}$$

for all $s \geq 0$. Then by (6.18) $\hat{y}(s)=e^{\pm s}$ for $0 \leq s \leq t$. Similarly, let us put

$$y_n'=\sum_{m=0}^{N} \bar{a}_{n-m}\tilde{x}_m \tag{6.39}$$

for all $n \geq 0$. Then by (6.13) $y_n'=\hat{y}_n$ for $0 \leq n \leq N$ with $\hat{y}_n=\hat{y}(s)$ for $s=n(1-k_0)$. Although this does not hold for $n>N$, the difference between y_n and y_n' is small and negligible. We can therefore apply (6.37) to obtain

$$\tilde{x}_0 \sim \pi^{3/2}(1-k_0)^{1/2}\lim_{s\to 0} s^{1/2}\tilde{x}_\pm(s). \tag{6.40}$$

Similarly,

$$\tilde{x}_N \sim \pi^{3/2}(1-k_0)^{1/2}\lim_{s\to t}(t-s)^{1/2}\tilde{x}_\pm(s). \tag{6.41}$$

Equation (6.37) cannot be applied to (6.12) and (6.17) as $\delta_{n,0}$ is not a slowly varying function. If we examine the sum equation

$$\sum_{m=0}^{\infty} \bar{a}_{n-m}\bar{x}_m^{(\infty)}=\delta_{n,0}, \tag{6.42}$$

it is straightforward to show

$$\bar{x}_0^{(\infty)}=\begin{cases} 1 & \text{for } k_0<1 \\ k_0 & \text{for } k_0>1, \end{cases} \tag{6.43}$$

and for $n\to\infty$

$$\bar{x}_n^{(\infty)}\sim\begin{cases} -\tfrac{1}{2}\pi^{-1/2}k_0^{n}n^{-3/2} & \text{for } k_0<1 \\ -\tfrac{1}{2}\pi^{-1/2}k_0^{-n+1}n^{-3/2} & \text{for } k_0>1. \end{cases} \tag{6.44}$$

Likewise for the integral equation

$$\int_0^\infty K_0(|s-s'|)\bar{x}^{(\infty)}(s';Z)\,ds'=Z^{1/2}e^{-Zs}, \tag{6.45}$$

we can show that

$$\bar{x}^{(\infty)}(s;Z)\sim Z^{1/2}(Z+1)^{1/2}\pi^{-3/2}s^{-1/2} \tag{6.46}$$

for $sZ \ll 1$, and

$$\bar{x}^{(\infty)}(s;Z)\sim -\tfrac{1}{2}\pi^{-3/2}e^{-s}s^{-3/2}(1+Z^{-1})^{1/2} \tag{6.47}$$

for $sZ \gg 1$.

Now let

$$y''(s;Z)=\int_t^\infty ds' K_0(|s-s'|)\bar{x}^{(\infty)}(s';Z) \tag{6.48}$$

and

$$\hat{y}_n'''=\sum_{m=N+1}^{\infty}\bar{a}_{n-m}x_m^{(\infty)}. \tag{6.49}$$

Then it follows from (6.46)–(6.49) that

$$\hat{y}_n'''\sim(1-k_0)^{1/2}\lim_{Z\to\infty}\hat{y}''(s;Z) \tag{6.50}$$

for $s=n(1-k_0)$ and $0 \leq n \leq N$.

If we consider for $0 \leq n \leq N$ the sum equation

$$\sum_{m=0}^{N}\bar{a}_{n-m}(\bar{x}_m-\bar{x}_m^{(\infty)})=\hat{y}_n''', \tag{6.51a}$$

and the corresponding integral equation

$$\int_0^t ds' K_0(|s-s'|)[\bar{x}(s';Z)-\bar{x}^{(\infty)}(s';Z)]=y''(s;Z) \tag{6.51b}$$

then (6.37) can be applied to yield

$$\bar{x}_0(Z)-\bar{x}_0^{(\infty)}(Z)\sim\pi^{3/2}(1-k_0)\lim_{s\to 0}[s^{1/2}\bar{x}(s;Z) - s^{1/2}\bar{x}^{(\infty)}(s;Z)], \tag{6.52}$$

and very similarly

$$\bar{x}_N-\bar{x}_N^{(\infty)}\sim\pi^{3/2}(1-k_0)\lim_{s\to t}(t-s)^{1/2}[\bar{x}(s;Z) - \bar{x}^{(\infty)}(s;Z)]. \tag{6.53}$$

Combining all these results we obtain

$$x_0\sim\max(1,k_0)+\pi(1-k_0)\lim_{Z\to\infty}\Big(-Z^{1/2}(Z+1)^{1/2}\pi^{-1} + \lim_{s\to 0}(\pi s)^{1/2}\bar{x}(s;Z)-\pi^{1/2}\lim_{s\to t}\frac{\bar{x}(s;Z)}{\tilde{x}_\pm(s)}\lim_{s\to 0}s^{1/2}\tilde{x}_\pm(s)\Big). \tag{6.54}$$

B. $x_0(N)$ in terms of $\eta(\theta)$

The evaluation of the limits in (6.54) is straightforward if we make use of the results of Myers.[34] Using (B2), (B3), and (B7) of Appendix B we have

$$\lim_{s\to 0}s^{1/2}\tilde{x}_+(s)=\theta^{-1/2}e^{\theta}f_0(1-\eta(\theta)), \tag{6.55a}$$

$$\lim_{s\to 0}s^{1/2}\tilde{x}_-(s)=\theta^{-1/2}e^{-\theta}f_0(1+\eta(\theta)), \tag{6.55b}$$

$$\lim_{s\to t}(t-s)^{1/2}\tilde{x}_+(s)=\theta^{-1/2}e^{\theta}f_0(1+\eta(\theta)), \tag{6.55c}$$

$$\lim_{s\to t}(t-s)^{1/2}\tilde{x}_-(s)=\theta^{-1/2}e^{-\theta}f_0(1-\eta(\theta)), \tag{6.55d}$$

$$\lim_{s\to 0}s^{1/2}\bar{x}(s;Z)=-\theta^{-1/2}Z^{1/2}\times e^{-Z\theta}f_0(\lambda_1-\eta(\theta)\lambda_2), \tag{6.55e}$$

$$\lim_{s\to t}(t-s)^{1/2}\bar{x}(s;Z)=-\theta^{-1/2}Z^{1/2}\times e^{-Z\theta}f_0(\lambda_1+\eta(\theta)\lambda_2), \tag{6.55f}$$

with $\theta=\tfrac{1}{2}t$. The notation follows that of Appendix B. From (B13) we have

$$\lambda_1=\lambda_1(\theta,Z)=M(\theta)[\bar{f}(Z,\theta)-Z\bar{g}(Z,\theta)], \tag{6.56a}$$

$$\lambda_2 = \lambda_2(\theta, Z) = M(\theta)[\bar{g}(Z,\theta) - Z\bar{f}(Z,\theta)] , \quad (6.56b)$$

with the bar operation defined by (B13c).

Equations (6.55) allow one to evaluate the $s \to 0$ and $s \to t$ limits in (6.54). What remains is the $Z \to \infty$ limit. To evaluate this we must know the large-Z behavior of $\lambda_1(\theta, Z)$ and $\lambda_2(\theta, Z)$. From the power-series expansions (B7a) and (B7b) of $f(t)$ and $g(t)$ we can calculate the large-Z behavior of $\bar{f}(Z,\theta)$ and $g(Z,\theta)$.

Thus for $Z \to \infty$ we find

$$\bar{f}(Z,\theta) = \frac{e^{Z\theta}}{Z^{1/2}} \frac{f_0\sqrt{\pi}}{\theta^{1/2}} \left(1 + \frac{f_1}{2Z\theta} + \frac{3f_2}{4Z^2\theta^2} + \cdots\right) \quad (6.57a)$$

and

$$\bar{g}(Z,\theta) = -\frac{e^{Z\theta}}{Z^{1/2}} \frac{f_0\eta(\theta)\pi^{1/2}}{\theta^{1/2}} \times \left(1 + \frac{g_1}{2Z\theta} + \frac{3g_2}{4Z^2\theta^2} + \cdots\right) , \quad (6.57b)$$

with f_1, f_2, g_1, and g_2 given by (B7). From (6.57) we can determine the large-Z behavior of (6.56), and hence (6.55e) and (6.55f). We find that

$$\bar{x}_0(Z) = \max(1, k_0) + \pi(1-k_0)\{-[2M(\theta) \times f_0^2\pi\eta(\theta)\theta^{-1} + \pi^{-1}]Z - M(\theta)f_0^2\pi\theta^{-1}[1 + f_0\eta(\theta)/2\theta + g_1\eta(\theta)/2\theta + \eta^2(\theta)] - \tfrac{1}{2}\pi^{-1} + O(1/Z)\} \quad (6.58)$$

and

$$\bar{x}_N = -(1-k_0)f_0^2M(\theta)\pi^2\theta^{-1} \times [1 - f_1\eta(\theta)/2\theta + g_1\eta(\theta)/2\theta - \eta^2(\theta)] . \quad (6.59)$$

In Appendix B we show that

$$2\eta(\theta)f_0^2M(\theta)\pi^2\theta^{-1} = -1 . \quad (6.60)$$

Thus the coefficient of the term proportional to Z in (6.48) vanishes, as it must if the $Z \to \infty$ limit is to exist.

Using (6.60) in (6.58) and (6.59), it follows from (6.15) [or (6.54)] that

$$x_0 = k_0 + |1-k_0| \left(-\frac{3}{2} + \frac{[f_1 - \eta(\theta)g_1]}{2\theta[1-\eta(\theta)]}\right) \quad (T > T_c) , \quad (6.61)$$

$$x_0 = 1 + |1-k_0| \left(\frac{1}{2} + \frac{[f_1 + \eta(\theta)g_1]}{2\theta[1+\eta(\theta)]}\right) \quad (T < T_c) , \quad (6.62)$$

where we used the fact that

$$\frac{\bar{x}_0}{\bar{x}_N} = \begin{cases} (1+\eta)/(1-\eta) & \text{for } T > T_c , \quad (6.63a) \\ (1-\eta)/(1+\eta) & \text{for } T < T_c , \quad (6.63b) \end{cases}$$

which follows from (6.40), (6.41), and (6.55).

It is convenient to define the functions $H_\pm(\theta)$ by

$$H_+(\theta) = -\tfrac{3}{2} + (f_1 - \eta g_1)/2\theta(1-\eta) \quad (6.64a)$$

and

$$H_-(\theta) = \tfrac{1}{2} + (f_1 + \eta g_1)/2\theta(1+\eta) , \quad (6.64b)$$

so that (6.61) and (6.62) become

$$x_0 \sim \max(1, k_0) + |1-k_0| H_\pm(\theta)$$

$$(T \to T_c^\pm,\ N \to \infty,\ t \text{ fixed}) \quad (6.65)$$

Using (B7c) and (B7d) one can show that

$$H_+(\theta) = 1 + \frac{\eta'}{4\eta}\frac{1+\eta}{1-\eta} - \theta\frac{(\eta')^2}{8\eta^2} + \frac{\theta(1-\eta^2)^2}{8\eta^2} \quad (6.66a)$$

and

$$H_-(\theta) = \frac{\eta'}{4\eta}\frac{1-\eta}{1+\eta} - \theta\frac{(\eta')^2}{8\eta^2} + \frac{\theta(1-\eta^2)^2}{8\eta^2} , \quad (6.66b)$$

where the definition (B5a) was used.

C. $F_\pm(t)$ in terms of $\eta(\theta)$

Equation (6.65) gives $x_0(N)$ in the scaling limit. Using this and (6.1) we now construct S_N in the scaling limit. We do this first for $T < T_c$. Choose an N_0, $N_0 \gg N$. Then we can write

$$S_N = S_{N_0}\prod_{n=N+1}^{N_0} x_0(n) = S_{N_0}\exp\sum_{n=N+1}^{N_0}\ln x_0(n) . \quad (6.67)$$

Since we are interested in the limit $T \to T_c$ ($k_0 \to 1$), we can write [in view of (6.65)] to leading order

$$\ln x_0(n) \sim |1-k_0| H_-(\theta) . \quad (6.68)$$

Then again to leading order

$$\sum_{n=N+1}^{N_0}\ln x_0(n) \sim 2\int_{t/2}^{N_0|1-k_0|/2} H_-(\theta)\,d\theta . \quad (6.69)$$

The large-θ behavior of $H_-(\theta)$ follows from that of $\eta(\theta)$. Thus using (2.38) in (6.66b) we find

$$H_-(\theta) = O(e^{-4\theta}) , \quad \theta \to \infty . \quad (6.70)$$

For N_0 large, S_{N_0} becomes independent (again to leading order) of N_0 since it approaches the spontaneous magnetization squared $S_\infty^<$ [see (2.10)]. From (6.69) and (6.70) we see that

$$\exp\left(\sum_{n=N+1}^{\infty}\ln x_0(n)\right) \sim \exp\left(2\int_{t/2}^{\infty} d\theta' H_-(\theta')\right) . \quad (6.71)$$

From (2.10), (5.23), and (5.24) we can conclude that

$$S_\infty^< \sim (2t/N)^{1/4} , \quad (6.72)$$

where t is given by (5.1). Using (6.71) and (6.72) we conclude that

$$S_N^< \sim N^{-1/4}(2t)^{1/4}\exp\left(2\int_{t/2}^{\infty} d\theta' H_-(\theta')\right) . \quad (6.73)$$

In terms of the variable R given (for the diagonal case) by (5.2), (6.73) becomes

$$S_N^< \sim R^{-1/4}(\sinh 2\beta_c E_1 + \sinh 2\beta_c E_2)^{1/8} \times (2t)^{1/4} \exp\left(2\int_{t/2}^{\infty} d\theta'\, H_-(\theta')\right) \quad (T\to T_c^-,\ R\to\infty,\ t \text{ fixed}) . \tag{6.74}$$

We now consider the case $T > T_c$. Again choose $N_0 \gg N$. For $T > T_c$ we have from (6.65)

$$\ln x_0(N) \sim |k_0 - 1| + |k_0 - 1| H_+(\theta) . \tag{6.75}$$

So (6.67) in this case becomes

$$S_N^> \sim S_{N_0} \frac{(k_0)^{N_0}}{(k_0)^N} \exp\left(2\int_{t/2}^{N_0|k_0-1|/2} H_+(\theta')\, d\theta'\right). \tag{6.76}$$

For $\theta\to\infty$ we can show using (2.38)

$$H_+(\theta) = 1/4\theta + O(\theta^{-2}) . \tag{6.77}$$

Using (6.77) in (6.76)

$$S_N^> \sim S_{N_0} k_0^{N_0} k_0^{-N} \exp\left[2\int_{t/2}^{N_0|k_0-1|/2} \left(H_+(\theta') - \frac{1}{4\theta'}\right) d\theta' + \frac{1}{2}\int_{t/2}^{N_0|k_0-1|/2} \frac{d\theta'}{\theta'}\right] . \tag{6.78}$$

Integrating the second term in (6.78) we obtain

$$S_N^> \sim S_{N_0} k_0^{N_0} k_0^{-N} (N_0/N)^{1/2} \times \exp\left[2\int_{t/2}^{N_0|k_0-1|/2} \left(H_+(\theta') - \frac{1}{4\theta'}\right) d\theta'\right] . \tag{6.79}$$

For fixed $T > T_c$ and $N_0 \to \infty$ it is known[37] that

$$S_{N_0} = [k_0^{-N}/(\pi N_0)^{1/2}](1 - k_0^{-2})^{-1/4} + O(k_0^{-N}/N_0^{3/2}) . \tag{6.80}$$

Using (6.80) in (6.79) we obtain (note that the N_0 dependence cancels)

$$S_N^> \sim k_0^{-N}(\pi N)^{-1/2}(1 - k_0^{-2})^{-1/4} \times \exp\left[2\int_{t/2}^{\infty} \left(H_+(\theta') - \frac{1}{4\theta'}\right) d\theta'\right], \tag{6.81}$$

where we have replaced the upper limit $\frac{1}{2}N_0|k_0 - 1|$ in (6.79) by ∞ (valid to leading order). Again using (2.10), (5.23), and (5.24) we conclude that

$$S_N^> \sim \frac{\pi^{-1/2} e^{-t}(2t)^{-1/4}}{N^{1/4}} \exp\left[2\int_{t/2}^{\infty}\left(H_+(\theta') - \frac{1}{4\theta'}\right) d\theta'\right], \tag{6.82}$$

or in terms of R

$$S_N^> \sim R^{-1/4}(\sinh 2\beta_c E_1 + \sinh 2\beta_c E_2)^{1/8} \pi^{-1/2} e^{-t} \times (2t)^{-1/4} \exp\left[2\int_{t/2}^{\infty}\left(H_+(\theta') - \frac{1}{4\theta'}\right) d\theta'\right] . \tag{6.83}$$

We can simplify (6.74) and (6.83) by using

$$\int_\theta^\infty \frac{\eta'(s)}{2\eta(s)} \frac{1-\eta(s)}{1+\eta(s)}\, ds = \frac{1}{2}\int_{\eta(\theta)}^{1} \frac{d\eta}{\eta} \frac{1-\eta}{1+\eta} = \ln[1 + \eta(\theta)] - \ln 2 - \tfrac{1}{2}\ln\eta(\theta). \tag{6.84}$$

Recalling the definition (6.66b) of $H_-(\theta)$ we see that (6.84) implies that (6.74) can be written as

$$S_N^< = \frac{F_-(t)}{R^{1/4}} + o(R^{-1/4}) , \tag{6.85a}$$

$$F_-(t) = (\sinh 2\beta_c E_1 + \sinh 2\beta_c E_2)^{1/8} \times \frac{1+\eta(\theta)}{2[\sqrt{\eta(\theta)}]} (4\theta)^{1/4} \exp\left(\int_\theta^\infty ds\, \tilde{F}(s)\right) , \tag{6.85b}$$

with

$$\tilde{F}(s) = [s/4\eta^2(s)]\{[1 - \eta^2(s)]^2 - [\eta'(s)]^2\} \tag{6.85c}$$

and

$$\theta = \tfrac{1}{2}t . \tag{6.85d}$$

To simplify (6.83) we note from (6.66) the identity

$$H_+(\theta) = -1 + \eta'(\theta)/[1 - \eta^2(\theta)] + H_-(\theta) . \tag{6.86}$$

Substituting (6.86) into (6.83) and making use of (6.74) we obtain

$$S_N^> = \pi^{-1/2} e^{-t} (2t)^{-1/2} \times \exp\left[2\int_{t/2}^{\infty}\left(-1 + \frac{\eta'(s)}{1-\eta^2(s)} - \frac{1}{4s}\right) ds\right] S_N^< . \tag{6.87}$$

The integral in (6.87) can be evaluated:

$$\int_{t/2}^{\infty}\left(-1 + \frac{\eta'(s)}{1-\eta^2(s)} - \frac{1}{4s}\right) ds = \lim_{L\to\infty}\left[\int_{t/2}^{L}\left(-1 + \frac{\eta'}{1-\eta^2} - \frac{1}{4s}\right) ds\right]$$
$$= \lim_{L\to\infty}\left\{-L + \tfrac{1}{2}t + \tfrac{1}{2}\ln\frac{1-\eta(t/2)}{1+\eta(t/2)} - \tfrac{1}{2}\ln[1 - \eta(L)] + \tfrac{1}{2}\ln[1 + \eta(L)] - \tfrac{1}{4}\ln L + \tfrac{1}{4}\ln t/2\right\} . \tag{6.88}$$

Using $\ln[1 + \eta(L)] \sim \ln 2$ and $\ln[1 - \eta(L)] \sim -2L - \frac{1}{2}\ln\pi - \frac{1}{2}\ln L$ for $L\to\infty$ in (6.88)

$$2\int_{t/2}^{\infty}\left(-1 + \frac{\eta'}{1-\eta^2} - \frac{1}{4s}\right) ds = t + \ln\left(\frac{1-\eta(t/2)}{1+\eta(t/2)}\right) + \tfrac{1}{2}\ln\pi + \tfrac{1}{2}\ln(t/2) + \ln 2. \tag{6.89}$$

Using (6.89) in (6.87) we obtain

$$S_N^> = \frac{1-\eta(t/2)}{1+\eta(t/2)} S_N^< \quad (T\to T_c,\ N\to\infty,\ t \text{ fixed}). \tag{6.90}$$

Thus (6.90) implies that (6.83) can be written

$$S_N^> = F_+(t)/R^{1/4} + o(R^{-1/4}), \quad F_+(t) = (\sinh 2\beta_c E_1 + \sinh 2\beta_c E_2)^{1/8} \tag{6.91a}$$

$$\times \frac{1-\eta(\theta)}{2[\eta(\theta)]^{1/2}}(4\theta)^{1/4}\exp\left(\int_\theta^\infty ds\,\bar{F}(s)\right), \tag{6.91b}$$

and $\bar{F}(s)$ is given by (6.85c) and $\theta = \frac{1}{2}t$. Another way of stating (6.90) is

$$\frac{F_+(t)}{F_-(t)} = \frac{1-\eta(t/2)}{1+\eta(t/2)}\ . \tag{6.92}$$

Since $\eta(\theta)\to 0$ as $\theta\to 0$ [see (2.37) or Appendix B] we see from (6.92) that

$$\lim_{t\to 0}\frac{F_+(t)}{F_-(t)} = 1. \tag{6.93}$$

That (6.93) is the case is clear from the small-t perturbation methods of Sec. V. From Secs. III and IV we obtained the representation

$$F_+(t)/F_-(t) = G(t), \tag{6.94}$$

where $G(t)$ is given by (2.29). From (6.94) and (6.92) we obtain

$$\frac{1-\eta(t/2)}{1+\eta(t/2)} = G(t) = \sum_{k=0}^{\infty} g^{(2k+1)}(t), \tag{6.95}$$

with $g^{(2k+1)}(t)$ given by (2.30) and from (6.93) $\lim_{t\to 0} G(t) = 1$.

We have numerically solved for the function $\eta(\theta)$ [by solving the nonlinear differential equation for $\eta(\theta)$, see (2.36)–(2.38); and for more details, the reader should consult Appendices A and B], and from this we have obtained $F_\pm(t)$ to ten significant figures. Tables of $F_\pm(t)$ along with all other numerical work can be found in Appendix A. It is from this numerical work that we obtained Fig. 3.

We can write (6.85b) and (6.91b) in an alternative form that is useful in analyzing the small-t limit. For large t, (6.85b) and (6.91b) prove to be more convenient. We first note the identity

$$\int_y^\infty ds\, s[\eta(s)]^{-2}[1-\eta^2(s)]^2$$

$$= -y^2[2\eta^2(y)]^{-1}\{[1-\eta^2(y)]^2 - [\eta'(y)]^2\}. \tag{6.96}$$

Then from (6.85) and (6.96) we have

$$\int_\theta^\infty dy\,\bar{F}(y) = -\frac{1}{2}\int_\theta^\infty dy\int_y^\infty ds\,\frac{s}{\eta^2}(1-\eta^2)^2$$

$$= -\frac{1}{2}\int_\theta^\infty ds\,\frac{s}{\eta^2}(1-\eta^2)^2(\ln s - \ln\theta)$$

$$= -(\theta^2/4\eta)[(1-\eta^2)^2 - \eta'^2]\ln\theta$$

$$-\frac{1}{2}\int_\theta^\infty ds\, s\ln s\frac{1}{\eta^2}(1-\eta^2)^2. \tag{6.97}$$

The second term in (6.97) can be written

$$-\frac{1}{2}\int_\theta^\infty ds\, s\ln s(1/\eta^2)(1-\eta^2)^2$$

$$= \int_\theta^\infty ds\, s\ln s[1-\eta^2(s)]$$

$$+\frac{1}{2}\int_\theta^\infty ds\, s\ln s(\eta^2-\eta^{-2})\ . \tag{6.98}$$

If we make use of the identity

$$\frac{1}{s}\frac{d}{ds}\left(s\frac{\eta'}{\eta}\right) = \eta^2 - \eta^{-2}, \tag{6.99}$$

the second term in (6.98) becomes

$$\frac{1}{2}\int_\theta^\infty ds\, s\ln s(\eta^2-\eta^{-2})$$

$$= \frac{1}{2}\int_\theta^\infty ds\ln s\frac{d}{ds}\left(s\frac{\eta'}{\eta}\right)$$

$$= -\frac{\theta}{2}\ln\theta\frac{\eta'(\theta)}{\eta(\theta)} + \frac{1}{2}\ln\eta(\theta). \tag{6.100}$$

Using (6.100) in (6.98) we obtain for (6.97)

$$\int_\theta^\infty dy\,\bar{F}(y) = \int_\theta^\infty ds\, s\ln s[1-\eta^2(s)]$$

$$+\tfrac{1}{2}\ln\eta(\theta) - h(\theta), \tag{6.101}$$

where $h(\theta)$ is given by (2.41). Note that to derive (6.99) one must use (2.36). Substituting (6.101) into (6.85b) we obtain (2.39) [to obtain $F_+(t)$ we use (6.92)].

Using (2.37) one can show as $\theta\to 0$

$$\exp[-h(\theta)] = \theta^{-1/4} + o(1). \tag{6.102}$$

Combining (6.102) and (2.39) we see

$$\lim_{t\to 0} F_\pm(t) = 2^{-1/2}(\sinh 2\beta_c E_1 + \sinh 2\beta_c E_2)^{1/8}$$

$$\times\exp\int_0^\infty dx\, x\ln x[1-\eta^2(x)]. \tag{6.103}$$

From (5.10) we see that if (6.103) is to agree with the known value of $F_\pm(0)$ we must have the identity

$$\int_0^\infty dx\, x\ln x(1-\eta^2) = \tfrac{1}{4} + \tfrac{7}{12}\ln 2 - 3\ln A. \tag{6.104}$$

The importance of (6.104) is that the preceding analysis (Sec. VI A in particular) was to a large degree based upon insight gained from the large-t behavior of $F_\pm(t)$. Identity (6.104) shows in some sense that "nothing was left out" in the asymptotic estimates of Sec. VI A.

We have numerically solved (2.36) subject to the boundary conditions (2.37) and (2.38), and then numerically integrated the left-hand side of (6.104). The details of this numerical work can be found in Appendix A. Here we give the results:

$$\int_0^\infty ds\, s\ln s(1-\eta^2) = -0.919\,275\,757\,747\,071\times 10^{-2}$$

$$\text{(numerical work).} \tag{6.105}$$

The right-hand side of (6.104) is known to be

$$\tfrac{1}{4}+\tfrac{7}{12}\ln 2-3\ln A=-0.919\,275\,757\,747\,172\cdots\times 10^{-2}. \tag{6.106}$$

Thus we have verified (6.104) to 13 significant digits.

Dyson[38] has pointed out that Glaisher's constant A can be expressed in terms of the Riemann ζ function. To show this we use the representation[39]

$$\ln A=\tfrac{1}{12}+\lim_{N\to\infty}\left(\sum_{n=1}^{N-1} n\ln n-\tfrac{1}{2}(N^2-N+\tfrac{1}{6})\ln N+\tfrac{1}{4}N^2\right), \tag{6.107}$$

and the relation valid for $\mathrm{Re}\, s>-3$

$$\zeta(s)=\lim_{n\to\infty}\left(\sum_{m=1}^{n} m^{-s}-\frac{n^{1-s}}{1-s}-\tfrac{1}{2}n^{-s}+\tfrac{1}{12}sn^{-s-1}\right), \tag{6.108}$$

where $\zeta(s)$ is the Riemann ζ function.[40] Differentiation of (6.108) with respect to s and comparison with (6.107) shows that

$$\ln A=-\zeta'(-1). \tag{6.109}$$

Thus the identity (6.104) that remains to be *proved* is

$$\int_0^\infty dx\, x\ln x[1-\eta^2(x)]=\tfrac{1}{4}+\tfrac{7}{12}\ln 2+3\zeta'(-1). \tag{6.110}$$

D. $F_\pm(t)$ for $t\to 0$ and $t\to\infty$ as derived from $\eta(\theta)$ representation

From (2.23), (6.85b), and (6.91b) we have

$$F_\pm(t)=\frac{1\mp\eta(\theta)}{2[\eta(\theta)]^{1/2}}\exp\left(\int_\theta^\infty ds\,\bar{F}(s)\right). \tag{6.111}$$

In Secs. III and IV we computed the large-t expansion of $\hat{F}_\pm(t)$. Here using (6.111) we derive (2.31a) and (2.31b).

For $t\to\infty$ we have

$$\eta(t/2)=1-(2/\pi)K_0(t)+(2/\pi^2)K_0^2(t)+O(e^{-3t}). \tag{6.112}$$

Using this in (6.85) we can show that as $t\to\infty$

$$\int_{t/2}^\infty dx\,\frac{x}{4\eta^2}\,[(1-\eta^2)^2-\eta'^2]$$
$$=\frac{t^2}{\pi^2}[K_1^2(t)-K_0^2(t)-t^{-1}K_0(t)K_1(t)]+O(e^{-3t}), \tag{6.113}$$

where we used the indefinite integrals[41]

$$\int xK_0^2(x)\,dx=\tfrac{1}{2}x^2[K_0^2(x)-K_1^2(x)], \tag{6.114}$$

$$\int xK_1^2(x)\,dx=\tfrac{1}{2}x^2[K_1^2(x)-K_0^2(x)-(2/x)K_0(x)K_1(x)].$$

Substituting (6.112) and (6.113) into (6.111) we obtain

$$\hat{F}_+(t)=\pi^{-1}K_0(t)+O(e^{-3t}), \tag{6.115}$$

$$\hat{F}_-(t)=\pi^{-2}\{t^2[K_1^2(t)-K_0^2(t)]$$
$$-tK_0(t)K_1(t)+\tfrac{1}{2}K_0^2(t)\}+O(e^{-4t}). \tag{6.116}$$

To determine the small-t behavior of $F_\pm(t)$ from the Painlevé function representation, it is convenient to start with (2.39). In Appendix B we show that

$$\eta(\theta)=-\theta\Omega-\tfrac{1}{128}\theta^5(8\Omega^3-8\Omega^2+4\Omega-1)+O(\theta^9\Omega^5),\quad \theta\to 0, \tag{6.117}$$

where Ω is given by (2.33). Using (6.117) and (6.104) we have

$$\int_\theta^\infty ds\, s\ln s[1-\eta^2(s)]=\tfrac{1}{4}+\tfrac{7}{12}\ln 2-3\ln A$$
$$-\int_0^\theta ds\, s\ln[1-\eta^2(s)] \tag{6.118a}$$
$$=\tfrac{1}{4}+\tfrac{7}{12}\ln 2-3\ln A$$
$$-[I_{11}-I_{33}-2(\gamma_E-2\ln 2)I_{32}$$
$$-(\gamma_E-2\ln 2)^2 I_{31}+O(\theta^8\ln^5\theta)], \tag{6.118b}$$

with

$$I_{k,l}=\int_0^\theta x^k(\ln x)^l\,dx. \tag{6.119}$$

The integrals (6.119) are standard and we obtain for $t\to 0$

$$\int_\theta^\infty ds\, s\ln s[1-\eta^2(s)]=\tfrac{1}{4}+\tfrac{7}{12}\ln 2-3\ln A$$
$$-\{\tfrac{1}{2}\theta^2\ln\theta-\tfrac{1}{4}\theta^2-\tfrac{1}{4}\theta^4[\ln^3\theta-(\tfrac{3}{4}-2\lambda)\ln^2\theta$$
$$+(\tfrac{3}{8}-\lambda+\lambda^2)\ln\theta-(\tfrac{3}{32}-\tfrac{1}{4}\lambda+\tfrac{1}{4}\lambda^2)]\}+O(\theta^8\ln^5\theta), \tag{6.120}$$

with

$$\lambda=\gamma_E-2\ln 2. \tag{6.121}$$

Using (6.117) in (2.41) we obtain for $\theta\to 0$

$$h(\theta)=\ln\theta[\tfrac{1}{4}-\tfrac{1}{2}\theta^2+\tfrac{1}{32}\theta^4(8\Omega^2-4\Omega+1)+\cdots]. \tag{6.122}$$

Using (6.120) and (6.122) in (2.39) we obtain after expanding the exponential

$$F_\pm(t)=F(0)[1\pm\tfrac{1}{2}t\Omega+\tfrac{1}{16}t^2\pm\tfrac{1}{32}t^3\Omega$$
$$+\tfrac{1}{256}t^4(-\Omega^2+\Omega+\tfrac{1}{8})+O(t^5\Omega^4)], \tag{6.123}$$

where (to summarize)

$$F(0)=(\sinh 2\beta_c E_1+\sinh 2\beta_c E_2)^{1/8}e^{1/4}2^{1/12}A^{-3},$$
$$\Omega=\ln(t/8)+\gamma_E, \tag{6.124}$$
$$\gamma_E=\text{Euler's constant}=0.577\,215\,664\cdots.$$

For the symmetric lattice the value of $F(0)$ is given by (5.10S).

In Sec. V we computed $F_\pm(t)$ for $t\to 0$ through $t^3\ln t$, and as we see from (2.32) the two methods agree (as they must, of course). The error term $O(t^5\Omega^4)$ in (6.123) follows from the remarks following (5.57). It may well happen that the coefficient of this term vanishes, but without explicit computation we know that the correction term cannot exceed $t^5(\ln t)^4$. One should also note that the method using the Painlevé function representation (2.39) is much easier to carry out than is the perturbation scheme given in Sec. V.

VII. SUSCEPTIBILITY $\chi(T)$

A. Leading divergence of $\chi(T)$ as $T\to T_c^{\pm}$

The zero-field susceptibility $\chi(T)$ defined by (2.42) is related to $\langle\sigma_{0,0}\sigma_{M,N}\rangle$ by (2.43). As $T\to T_c^{\pm}$, $\chi(T)$ diverges, and it is this leading divergence we compute in this section. Since it is the large-M^2+N^2 behavior that determines the singularity of $\chi(T)$ [any finite sum of terms in (2.43) is finite and continuous at T_c], we may start the sum in (2.43) at any large, but finite, N_0 and M_0. Hence to leading order we write

$$\beta^{-1}\chi(T)\sim 4\sum_{N=N_0}^{\infty}\sum_{M=M_0}^{\infty}[\langle\sigma_{0,0}\sigma_{M,N}\rangle-\mathfrak{M}^2(T,0)]\,. \tag{7.1}$$

We consider the $T\to T_c^{+}$ case first. Recalling the definitions of $\bar{M}$ and $\bar{N}$ [see (3.126) and (3.127)] we may be leading order approximate (7.1) by an integral,

$$\beta^{-1}\chi(T)\sim 4(\tfrac14\gamma_1\gamma_2)^{1/2}(z_1z_2+z_1+z_2-1)^{-2}\times\int_0^\infty d\bar{N}\int_0^\infty d\bar{M}\,\frac{F_+(t)}{R^{1/4}}\,, \tag{7.2}$$

where we have used (2.22). Recalling that $t^2=\bar{M}^2+\bar{N}^2$ we can write (7.2) as a single integral [use (2.5) also] and obtain

$$\beta^{-1}\chi(T)\sim 2\pi\,|z_1z_2+z_1+z_2-1|^{-7/4}(\tfrac14\gamma_1\gamma_2)^{7/16}\times\int_0^\infty dt\,t^{3/4}F_+(t)\,. \tag{7.3}$$

Expanding the factors, multiplying the integral in (7.3) (using the various identities in Appendix C) we obtain, for $T\to T_c^{+}$,

$$\beta^{-1}\chi(T)\sim|1-T_c/T|^{-7/4}D'\int_0^\infty dt\,t^{3/4}F_+(t)=C_{0+}|1-T_c/T|^{-7/4}\,, \tag{7.4}$$

where

$$D'=2^{-13/8}\pi\{\beta_c[E_1(1-z_{2c})^{-1}+E_2(1-z_{1c})^{-1}]\}^{-7/4}(z_{1c}z_{2c})^{-7/8}\,. \tag{7.5}$$

If we use (2.39) we can write the integral appearing in (7.4) as

$$\int_0^\infty dt\,t^{3/4}F_+(t)=(\sinh 2\beta_cE_1+\sinh 2\beta_cE_2)^{1/8}2^{5/4}\int_0^\infty d\theta\,\theta[1-\eta(\theta)]\exp\left(\int_\theta^\infty dx\,x\ln x(1-\eta^2)-h(\theta)\right)\,. \tag{7.6}$$

Combining (7.4)–(7.6) we obtain (2.46a). To obtain (2.48) from (7.5) and (7.6) the various identities in Appendix C must be used. Using (2.23) in (7.4) we also obtain for C_{0+} the expression

$$C_{0+}=D'2^{1/4}(\sinh 2\beta_cE_1+\sinh 2\beta_cE_2)^{1/8}\times\int_0^\infty dt\,t\hat{F}_+(t)\,. \tag{7.7}$$

If we use (4.154) we can compute C_{0+} perturbatively as

$$C_{0+}=\sum_{n=1}^{\infty}C_{0+}^{(2n-1)}\,, \tag{7.8}$$

with

$$C_{0+}^{(2n-1)}=D'2^{1/4}(\sinh 2\beta_cE_1+\sinh 2\beta_cE_2)^{1/8}\times\int_0^\infty dt\,t\hat{F}_+^{(2n-1)}(t) \tag{7.9a}$$

$$=2^{-3/2}(z_{1c}+z_{2c})^{1/4}(z_{1c}z_{2c})^{-1}\times\{\beta_c[E_1(1-z_{2c})^{-1}+E_2(1-z_{1c})^{-1}]\}^{-7/4}I_{2n-1}^{+}, \tag{7.9b}$$

where we define

$$I_{2n-1}^{+}=\pi\int_0^\infty dt\,t\hat{F}_+^{(2n-1)}(t)\,dt\,, \tag{7.10}$$

and $\hat{F}_+^{(2n-1)}(t)$ is given by (4.156). From the analysis of the large-t behavior of $\hat{F}_+^{(2n-1)}(t)$ in Sec. IV we can conclude that $C_{0+}^{(2n-1)}$ represents the contribution to C_{0+} coming from that part of $\hat{F}_+(t)$ that behaves as $e^{-(2n-1)t}$ (times some power of t) for $t\to\infty$. Stated slightly differently, we have $C_{0+}^{(1)}$ being the contribution to C_{0+} coming from the Ornstein-Zernike pole, and $C_{0+}^{(2n-1)}$, $n\geq 2$, the contribution to C_{0+} coming from the $(2n-1)$-particle branch cut. This latter interpretation will become more evident when we examine the k-dependent susceptibility $\chi(\vec{k},T)$ in the scaling limit $k\to 0$, $T\to T_c$ such that $k|T-T_c|^{-1}$ is fixed.[42]

From (4.156a) and (7.10) for $n=1$ we can obtain[43]

$$I_1^{+}=\int_0^\infty dt\,tK_0(t)=1\,. \tag{7.11}$$

Using (4.159) we obtain for I_3^{+} the expression

$$I_3^+ = 2^{-1}\pi^{-2}\int_1^\infty dy_1\int_1^\infty dy_2\int_1^\infty dy_3\left(\frac{y_2^2-1}{(y_1^2-1)(y_3^2-1)}\right)^{1/2}$$
$$\times\left(\frac{y_1-y_3}{(y_1+y_2)(y_2+y_3)(y_1+y_2+y_3)}\right)^2 . \qquad (7.12)$$

Numerically evaluating this we find

$$I_3^+ = 0.814\,462\,565\,5\ldots\times 10^{-3} . \qquad (7.13)$$

Using (4.156c) we also numerically study I_5^+ to obtain

$$I_5^+ = 0.7969\times 10^{-6} . \qquad (7.14)$$

For the symmetric lattice these results have already been summarized in (2.59S)-(2.61S) and Table I.

We now consider (7.1) for $T\to T_c^-$. In this case instead of (7.1) we have

$$\beta^{-1}\chi(T)\sim 4(\tfrac14\gamma_1\gamma_2)^{1/2}\,|z_1z_2+z_1+z_2-1|^{-2}$$
$$\times\int_0^\infty d\bar N\int_0^\infty d\bar M\,[F_-(t)/R^{1/4}-(z_{1c}z_{2c})^{-1/8}$$
$$\times 2^{1/8}(z_{1c}+z_{2c})^{1/4}t^{1/4}/R^{1/4}] , \qquad (7.15)$$

where we used

$$S_\infty^< \sim (z_{1c}z_{2c})^{-1/8}[\sqrt2\,(z_{1c}+z_{2c})\,t/R]^{1/4}$$

(see Appendix C). Expanding the factors multiplying the integral, and using (2.5) we obtain

$$\beta^{-1}\chi(T)\sim |1-T_c/T|^{-7/4}D'\int_0^\infty dt\,[t^{3/4}F_-(t)$$
$$-(\sinh 2\beta_cE_1+\sinh 2\beta_cE_2)^{1/8}\,2^{1/4}t] , \qquad (7.16)$$

with D' given by (7.5). Using (2.39) we can write the integral appearing in (7.16) as

$$\int_0^\infty dt\,[t^{3/4}F_-(t)-(\sinh 2\beta_cE_1+\sinh 2\beta_cE_2)^{1/8}2^{1/4}t]$$
$$=(\sinh 2\beta_cE_1+\sinh 2\beta_cE_2)^{1/8}2^{5/4}\int_0^\infty d\theta\,\theta\Big[[1+\eta(\theta)]$$
$$\times\exp\left(\int_\theta^\infty dx\,x\ln x(1-\eta^2)-h(\theta)\right)-2\Big] . \qquad (7.17)$$

From (7.16) and (7.17) follows (2.46b).

Using (2.23) in (7.16) we obtain for C_{0-} the expression

$$C_{0-}=D'2^{1/4}(\sinh 2\beta_cE_1+\sinh 2\beta_cE_2)^{1/8}$$
$$\times\int_0^\infty dt\,t[\hat F_-(t)-1] . \qquad (7.18)$$

If we use (3.151) we can compute C_{0-} perturbatively as

$$C_{0-}=\sum_{n=1}^\infty C_{0-}^{(2n)} , \qquad (7.19)$$

with

$$C_{0-}^{(2n)}=D'2^{1/4}(\sinh 2\beta_cE_1+\sinh 2\beta_cE_2)^{1/8}$$
$$\times\int_0^\infty dt\,t\hat F_-^{(2n)}(t)=2^{1/2}(z_{1c}+z_{2c})^{1/4}(z_{1c}z_{2c})^{-1}$$
$$\times\{\beta_c[E_1(1-z_{2c})^{-1}+E_2(1-z_{1c})^{-1}]\}^{7/4}I_{2n}^- , \qquad (7.20)$$

where we define

$$I_{2n}^-=\pi\int_0^\infty dt\,t\hat F_-^{(2n)}(t) , \qquad (7.21)$$

and $\hat F_-^{(2n)}(t)$ is given by (3.152).

From (3.150), (3.152), and (7.21) for $n=1$ we obtain

$$I_2^-=\frac1\pi\int_0^\infty dt\,\{t^3[K_1^2(t)-K_0^2(t)]$$
$$-t^2K_0(t)K_1(t)+\tfrac12\,tK_0^2(t)\}$$
$$=(12\pi)^{-1}=0.026\,525\,823\,8\ldots , \qquad (7.22)$$

where we have used[44]

$$\int_0^\infty K_\mu(t)K_\nu(t)t^\rho\,dt=\frac{2^{\rho-2}}{\Gamma(1+\rho)}\Gamma[\tfrac12(1+\nu+\mu+\rho)]\Gamma[\tfrac12(1+\nu-\mu+\rho)]\Gamma[\tfrac12(1-\nu+\mu+\rho)]\Gamma[\tfrac12(1-\nu-\mu+\rho)] .$$

Using (3.153) and (7.21) for $n=2$ gives

$$I_4^-=(4\pi^3)^{-1}\int_1^\infty dy_1\cdots\int_1^\infty dy_4\left(\frac{(y_2^2-1)(y_4^2-1)}{(y_1^2-1)(y_3^2-1)}\right)^{1/2}\left(\frac{(y_2-y_4)(y_1-y_3)}{(y_1+y_2)(y_2+y_3)(y_3+y_4)(y_4+y_1)(y_1+y_2+y_3+y_4)}\right)^2 . \qquad (7.23)$$

Evaluating this numerically we find

$$I_4^-=0.254\,48\times 10^{-4} . \qquad (7.24)$$

B. $\chi(T)$ for $T<T_c$

Using (2.9) and (2.11) in (2.43) we see that for $T<T_c$ we can expand $\chi(T)$ as

$$\beta^{-1}\chi(T)=\sum_{n=1}^\infty \chi_<^{(2n)}(T)\quad (T<T_c) , \qquad (7.25)$$

where

$$\chi_<^{(2)}(T)=-S_\infty^<\sum_{M=-\infty}^{+\infty}\sum_{N=-\infty}^{+\infty}F_{<MN}^{(2)} , \qquad (7.26a)$$

$$\chi^{(4)}(T) = -S_\infty^< \sum_{M=-\infty}^{+\infty} \sum_{N=-\infty}^{+\infty} [F_{<MN}^{(4)} - \tfrac{1}{2}(F_{<MN}^{(2)})^2], \tag{7.26b}$$

$$\chi^{(6)}(T) = -S_\infty^< \sum_{M=-\infty}^{+\infty} \sum_{N=-\infty}^{+\infty} [F_{<MN}^{(6)} - F_{<MN}^{(2)} F_{<MN}^{(4)} + (1/3!)(F_{<MN}^{(2)})^3], \tag{7.26c}$$

etc. and with $F_{<MN}^{(2n)}$ given by (2.12). In the Sec. VII A we analyzed the leading divergence of $\chi(T)$ as $T \to T_c^\pm$. Here we analyze the next diverging term, viz., $|1 - T_c/T|^{-3/4}$. A general term in the expansion (7.26) of $\chi_<^{(2n)}(T)$ is of the form

$$\sum_{M=-\infty}^{+\infty} \sum_{N=-\infty}^{+\infty} F_{<MN}^{(2\alpha_1)} \cdots F_{<MN}^{(2\alpha_p)}, \tag{7.27a}$$

with

$$\alpha_1 + \alpha_2 + \cdots + \alpha_p = n, \quad p \le n. \tag{7.27b}$$

For $\mathrm{Im}\,\phi < 0$ we have

$$\sum_{l=-\infty}^{+\infty} e^{-i|l|\phi} = \frac{1 + e^{-i\phi}}{1 - e^{-i\phi}} = \cot\tfrac{1}{2}\phi. \tag{7.28}$$

Using (7.28) when we substitute (2.12) into (7.27a) we see that (7.3a) is of the form

$$(-1)^n [2z_2(1 - z_1^2)]^{2n} (2^p \alpha_1 \cdots \alpha_p)^{-1} (2\pi)^{-4n} \times \int_{-\pi}^{\pi} d\phi_1 \cdots \int_{-\pi}^{\pi} d\phi_{4n} \cot\tfrac{1}{2}(\phi_1 + \phi_3 + \cdots + \phi_{4n-1}) \times \cot\tfrac{1}{2}(\phi_2 + \phi_4 + \cdots + \phi_{4n}) S_{\alpha_1} S_{\alpha_2} \cdots S_{\alpha_p}, \tag{7.29}$$

where

$$S_{\alpha_1} = S_{\alpha_1}(\phi_1 \cdots \phi_{4\alpha_1})$$

$$= \prod_{j=1}^{2\alpha_1} \frac{1}{\Delta(\phi_{2j-1}, \phi_{2j})} \frac{\sin\frac{1}{2}(\phi_{2j-1} - \phi_{2j+1})}{\sin\frac{1}{2}(\phi_{2j} + \phi_{2j+2})}. \tag{7.30}$$

We now analyze (7.29) for $T \to T_c^-$. From arguments identical to those in Sec. III G we see that the factor $S_{\alpha_1} \cdots S_{\alpha_p} d\phi_1 \cdots d\phi_{4n}$ leads to an $O(1)$ quantity with a correction term of order $(\beta - \beta_c)^2$. The quantities $\cot\frac{1}{2}(\phi_1 + \cdots + \phi_{4n-1})$ and $\cot\frac{1}{2}(\phi_2 + \cdots + \phi_{4n})$ in (7.29) diverge as $(\beta - \beta_c)^{-2}$ with a correction term of order $(\beta - \beta_c)^2$. Hence as $T \to T_c^-$ we can expect (7.27a) to have a divergence of order $(\beta - \beta_c)^{-2}$ with a correction term smaller than order $(\beta - \beta_c)^{-1}$. That is (7.27a) will diverge like $|1 - T_c/T|^{-7/4}$ but will not have a term of the form $|1 - T_c/T|^{-3/4}$. The quantity $S_\infty^<$ behaves as

$$S_\infty^< \sim \mathrm{const}(\beta - \beta_c)^{1/4} [1 + R_0\beta_c(\beta/\beta_c - 1) + \cdots](\beta - \beta_c), \tag{7.31}$$

where the constant R_0 is given by (2.49). Using (7.31) and the fact that (7.27a) has no divergence of order $(\beta - \beta_c)^{-1}$ we must conclude that $\chi_<^{(2n)}(T)$ has two diverging terms, viz., $(\beta - \beta_c)^{-7/4}$ and $(\beta - \beta_c)^{-3/4}$, where the second diverging term, i.e., $(\beta - \beta_c)^{-3/4}$, comes from the *leading* divergence of (7.27a) times the second term in (7.31). Thus if we write, as $T \to T_c^-$,

$$\chi_<^{(2n)}(T) = C_{0-}^{(2n)} |1 - T_c/T|^{-7/4} + C_{1-}^{(2n)} |1 - T/T_c|^{-3/4} + O(1), \tag{7.32}$$

then $C_{1-}^{(2n)}$ is just $C_{0-}^{(2n)}$ times the constant $R_0\beta_c$. From (7.25) we can therefore conclude that C_{1-} is related to C_{0-} by (2.47). In the previous section we derived an *exact* formula [see (2.46b)] for C_{0-}; and hence, by (2.47) we have an exact formula for C_{1-}. For the symmetric lattice these results have been summarized in Table I.

Using (2.12) for $n = 1$ in (7.26a) we obtain

$$\beta^{-1}\chi_<^{(2)}(T) = S_\infty^< \frac{\gamma_1\gamma_2}{2(2\pi)^4} \int_{-\pi}^{\pi} d\phi_1 \cdots \int_{-\pi}^{\pi} d\phi_4 [\Delta(\phi_1, \phi_2)\Delta(\phi_3, \phi_4)]^{-1} \frac{1 + e^{-i(\phi_1+\phi_3)}}{1 - e^{-i(\phi_1+\phi_3)}} \frac{1 + e^{-i(\phi_2+\phi_4)}}{1 - e^{-i(\phi_2+\phi_4)}} \times \frac{(e^{i\phi_1} - e^{i\phi_3})(e^{i\phi_2} - e^{i\phi_4})}{(e^{i(\phi_1+\phi_3)} - 1)(e^{i(\phi_2+\phi_4)} - 1)}, \tag{7.33}$$

where in (7.33) we have expanded the contours of integration to the unit circle except that $\phi_1(\phi_2)$ is to be indented inward at $-\phi_3(-\phi_4)$. Noting that the integrand in (7.33) is odd under the transformation $\phi_{1,3} \to -\phi_{1,3}$ $(\phi_{2,4} \to -\phi_{2,4})$, we see that the contribution from the $\phi_1(\phi_2)$ integration comes from the residue of the pole at $-\phi_3(-\phi_4)$. Hence

$$\beta^{-1}\chi_<^{(2)}(T) = \tfrac{1}{8} S_\infty^< \gamma_1\gamma_2 (2\pi i)^{-2} \oint \frac{d\xi_3}{\xi_3} \oint \frac{d\xi_2}{\xi_2} \frac{\partial}{\partial \xi_1}\left(\frac{(\xi_1 + \xi_3^{-1})(\xi_3^{-1} - \xi_1^{-1})}{\Delta(\xi_1, \xi_2)}\right)\Bigg|_{\xi_1 = \xi_3^{-1}} \frac{\partial}{\partial \xi_4}\left(\frac{(\xi_4 + \xi_2^{-1})(\xi_4^{-1} - \xi_2^{-1})}{\Delta(\xi_3, \xi_4)}\right)\Bigg|_{\xi_4 = \xi_2^{-1}}. \tag{7.34}$$

Performing the indicated differentiations in (7.34) gives

$$\beta^{-1}\chi_<^{(2)}(T) = -\tfrac{1}{8} S_\infty^< \frac{\gamma_1\gamma_2}{(2\pi)^2} \int_{-\pi}^{\pi} d\phi_1 \int_{-\pi}^{\pi} d\phi_2 \left(\frac{2\cos\phi_1}{\Delta(\phi_1, \phi_2)} - \frac{4\gamma_1 \sin^2\phi_1}{\Delta^2(\phi_1, \phi_2)}\right)\left(\frac{2\cos\phi_2}{\Delta(\phi_1, \phi_2)} - \frac{4\gamma_2 \sin^2\phi_2}{\Delta^2(\phi_1, \phi_2)}\right). \tag{7.35}$$

Using the identity

$$\frac{\partial}{\partial\phi_1}\frac{\partial}{\partial\phi_2}\Delta^{-2}(\phi_1,\phi_2)=6\gamma_1\gamma_2\frac{\sin\phi_1\sin\phi_2}{\Delta^4(\phi_1,\phi_2)}, \quad (7.36)$$

we have

$$\gamma_1\gamma_2\int_{-\pi}^{\pi}\int_{-\pi}^{\pi}d\phi_1 d\phi_2\frac{\sin^2\phi_1\sin^2\phi_2}{\Delta^4(\phi_1,\phi_2)}=\frac{1}{6}\int_{-\pi}^{\pi}\int_{-\pi}^{\pi}d\phi_1 d\phi_2\frac{\cos\phi_1\cos\phi_2}{\Delta^2(\phi_1,\phi_2)}. \quad (7.37)$$

Using this and similar identities for the $\sin^2\phi_1\cos\phi_2$ and $\sin^2\phi_2\cos\phi_1$ terms, (7.35) becomes

$$\beta^{-1}\chi_<^{(2)}(T)=\tfrac{1}{6}S_\infty^<\frac{\gamma_1\gamma_2}{(2\pi)^2}\int_{-\pi}^{\pi}d\phi_1\times\int_{-\pi}^{\pi}d\phi_2\frac{\cos\phi_1\cos\phi_2}{\Delta^2(\phi_1,\phi_2)}. \quad (7.38)$$

In Fig. 10 we plot $\beta^{-1}\chi_<^{(2)}(T)$ for two values of E_1/E_2 as a function of $(1-T/T_c)$.

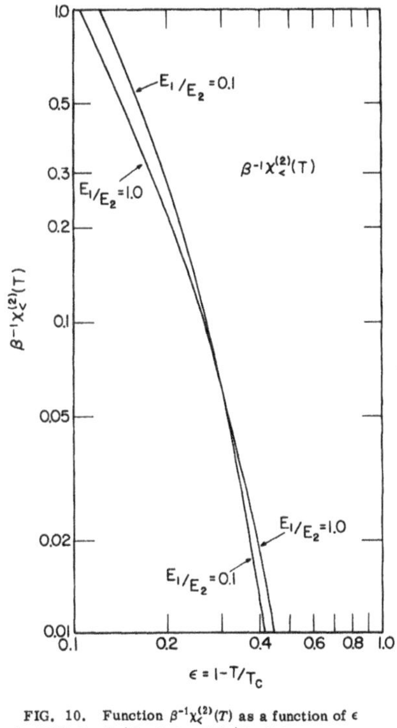

FIG. 10. Function $\beta^{-1}\chi_<^{(2)}(T)$ as a function of $\epsilon=1-T/T_c$ for $E_1/E_2=1.0$ and $E_1/E_2=0.1$.

TABLE IV. Low-temperature series expansion of $\beta^{-1}\chi(T)$ [see (7.40)] is compared with the low-temperature series expansion $\beta^{-1}\chi_<^{(2)}(T)$ [see (7.42)]. The coefficients h_n are taken from Ref. 45.

n	h_n	$h_n^{(2)}$
1	8	8
2	60	60
3	416	416
4	2791	2791
5	18 296	18 296
6	118 016	118 012
7	752 008	751 944
8	4 746 341	4 745 661
9	29 727 472	29 721 472
10		184 968 932

For the symmetric lattice the integrals in (7.38) can be expressed in terms of the complete elliptic integrals $E(k)$ and $K(k)$. The result is

$$\beta^{-1}\chi_<^{(2)}(T)=\frac{S_\infty^<}{6\pi}\left[\frac{a^2-2\gamma^2}{a^2-4\gamma^2}E\left(\frac{2\gamma}{a}\right)-K\left(\frac{2\gamma}{a}\right)\right], \quad (7.39)$$

where $\gamma=\gamma_1=\gamma_2$, and so as $T\to T_c$ we have $2\gamma/a\to 1$. Expanding this about $T=T_c$ we see that after the $|1-T_c/T|^{-7/4}$ and $|1-T_c/T|^{-3/4}$ terms, there is a term proportional to $|1-T_c/T|^{1/4}\ln|1-T_c/T|$.

It is instructive to compare the low-temperature expansion of $\beta^{-1}\chi_<^{(2)}(T)$ with the low-temperature expansion of $\beta^{-1}\chi(T)$. Following Essam and Fisher[45] we write (see also Sykes, Gaunt, Martin, Mattingly, and Essam[45b]) for the isotropic lattice

$$\beta^{-1}\chi(T)=4u^2\left(1+\sum_{n=1}^{\infty}h_n u^n\right), \quad (7.40)$$

with

$$u=e^{-4\beta E}. \quad (7.41)$$

In Ref. 45b the quantities h_n are computed for $n=1, 2, \ldots, 9$. We now make an analogous low-temperature expansion of $\beta^{-1}\chi_<^{(2)}(T)$:

$$\beta^{-1}\chi_<^{(2)}(T)=4u^2\left(1+\sum_{n=1}^{\infty}h_n^{(2)}u^n\right). \quad (7.42)$$

In computing $h_n^{(2)}$ we expanded the denominator in (7.38) and integrated term by term. Our results are displayed in Table IV. One sees that $h_n=h_n^{(2)}$ for $n=1, 2, \ldots, 5$; $h_6-h_6^{(2)}=4$, and $h_7-h_7^{(2)}=64$. Thus the low-temperature expansion of $\beta^{-1}\chi(T)$ is dominated in the early terms by contributions to $\beta^{-1}\chi(T)$ coming from the two-particle cut, i.e., $\beta^{-1}\chi_<^{(2)}(T)$. From (2.12) we see that the earliest the $2n$th branch cut can contribute to the coefficients h_k is for $k\geq n$.

C. $\chi(T)$ for $T>T_c$

Using (2.14)–(2.16) in (2.43) we see that for $T>T_c$ we can expand $\chi(T)$ as

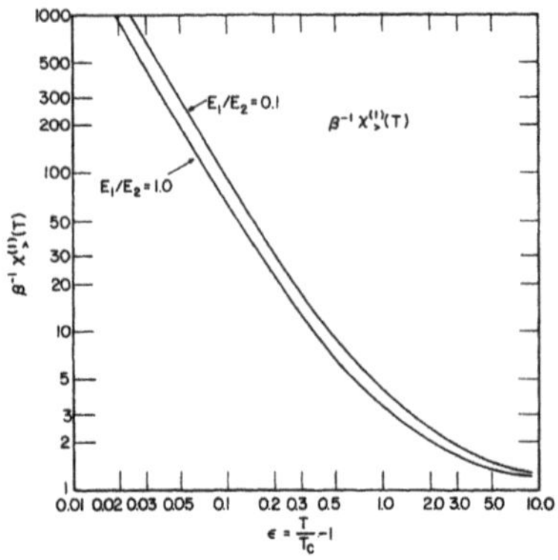

FIG. 11. Function $\beta^{-1}\chi_>^{(1)}(T)$ as a function of $\epsilon = T/T_c - 1$ for $E_1/E_2 = 1.0$ and $E_1/E_2 = 0.1$.

$$\beta^{-1}\chi(T) = \sum_{n=1}^{\infty} \chi_>^{(2n-1)}(T) , \qquad (7.43)$$

where

$$\chi_>^{(1)}(T) = S_\infty^> \sum_{M=-\infty}^{\infty} \sum_{N=-\infty}^{+\infty} x_{>MN}^{(1)} , \qquad (7.44a)$$

$$\chi_>^{(3)}(T) = S_\infty^> \sum_{M=-\infty}^{\infty} \sum_{N=-\infty}^{\infty} [x_{>MN}^{(3)} - x_{>MN}^{(1)} F_{>MN}^{(2)}] , \qquad (7.44b)$$

$$\chi_>^{(5)}(T) = S_\infty^> \sum_{M=-\infty}^{\infty} \sum_{N=-\infty}^{\infty} [x_{>MN}^{(5)} - F_{>MN}^{(2)} x_{>MN}^{(3)} - x_{>MN}^{(1)} F_{>MN}^{(4)} + \tfrac{1}{2} x_{>MN}^{(1)} (F_{>MN}^{(2)})^2] , \qquad (7.44c)$$

etc.

The arguments needed to relate C_{0+} and C_{1+} [see (2.47)] are similar to those given in Sec. VIIB relating C_{0-} and C_{1-}. Hence we consider (2.47) established.

Using (2.15) for $k=1$ and (7.44a) we have

$$\beta^{-1}\chi_>^{(1)}(T) = S_\infty^> (a - \gamma_1 - \gamma_2)^{-1} , \qquad (7.45)$$

which for the symmetric lattice is

$$\beta^{-1}\chi_>^{(1)}(T) = (1 - z^2)(1 - 4z^2 - 10z^4 - 4z^6 + z^8)^{1/4} \times (1 - 2z - z^2)^{-2} , \qquad (7.45a)$$

with $z = \tanh\beta E$. In Fig. 11 we plot $\beta^{-1}\chi_>^{(1)}(T)$ for two values of E_1/E_2 as a function of $(T/T_c - 1)$.

It is instructive to compare the high-temperature expansion of $\beta^{-1}\chi(T)$ with the high-temperature expansion of $\beta^{-1}\chi_>^{(1)}(T)$. We write for the symmetric lattice

$$\beta^{-1}\chi(T) = \sum_{n=0}^{\infty} \chi_n z^n \qquad (7.46)$$

and

$$\beta^{-1}\chi_>^{(1)}(T) = \sum_{n=0}^{\infty} \chi_n^{(1)} z^n . \qquad (7.47)$$

The coefficients χ_n in (7.46) have been computed for $0 \le n \le 21$ by Sykes *et al.*[14] In Table V we compare χ_n and $\chi_n^{(1)}$. We see that the first seven terms and the ninth term are identical. Clearly the lowest-lying excitation dominates the low-order coefficients in the high-temperature series expansion of $\beta^{-1}\chi(T)$. From the fact that $x_{>MN}^{(k)}$ in (2.15) is proportional to z^{k-1}, we see that the earliest the effect of the kth branch cut can have on the expansion (7.46) is at the $(k-1)$th order. From Table V we see that this is a poor bound.

ACKNOWLEDGMENTS

The authors express their gratitude to Professor C. N. Yang, Professor E. W. Montroll, Professor M. Blume, Professor A. Toomre, Dr. H. Au-Yang, Dr. J. M. Myers, and Professor M. E. Fisher for many stimulating discussions.

APPENDIX A: NUMERICAL WORK

We solved (2.36)–(2.38) by a standard Runge-Kutta method. The technique consisted in using both the large-θ and small-θ expansions of $\eta(\theta)$ and $\eta'(\theta)$ to obtain the initial conditions. For instance, by using all terms through order θ^{13} in the small-θ expansion of $\eta(\theta)$ [and the correspond-

TABLE V. High-temperature series expansion of $\beta^{-1}\chi(T)$ [see (7.46)] is compared with the high-temperature series expansion of $\beta^{-1}\chi_>^{(1)}(T)$ [see (7.47)]. The coefficients χ_n are taken from Ref. 14a.

n	χ_n	$\chi_n^{(1)}$	$\chi_n - \chi_n^{(1)}$
0	1	1	0
1	4	4	0
2	12	12	0
3	36	36	0
4	100	100	0
5	276	276	0
6	740	740	0
7	1 972	1 972	0
8	5172	5168	4
9	13 492	13 492	0
10	34 876	34 844	32
11	89 764	89 748	16
12	229 628	229 420	208
13	585 508	585 316	192
14	1 486 308	1 484 980	1328
15	3 763 460	3 761 860	1600
16	9 497 380	9 488 960	8420
17	23 918 708	23 906 884	11 824
18	60 080 156	60 027 340	52 816
19	150 660 388	150 577 892	82 568
20	377 009 300	376 680 820	328 480
21	942 105 604	941 551 252	554 352

ing expansion for $\eta'(\theta)$] we started the forward integration at $\theta_{start}=0.013$ with the initial values obtained by evaluating these small-θ expansions at $\theta=\theta_{start}$. For large we solved for

$$\epsilon(\theta)=1-\eta(\theta) \tag{A1}$$

by starting the backward integration at $\theta=10$. For initial conditions we used

$$\epsilon(\theta)=(2/\pi)K_0(2\theta)-(2/\pi^2)K_0^2(2\theta), \tag{A2}$$

$$\epsilon'(\theta)=-(4/\pi)K_1(2\theta)+(8/\pi^2)K_0(2\theta)K_1(2\theta) \tag{A3}$$

for $\theta=10$. We found that the forward and backward integration programs had an optimal overlap at $\theta=0.8$. The two values of $\eta(\eta')$ agreed to 1(3) digit(s) in the 13th place. In Table VI we present our numerical solution to (2.36)–(2.38). For $\theta \geq 0.9$ we display $\epsilon(\theta)$ rather than $\eta(\theta)$. In Fig. 12 we plot $\eta(\theta)$ and $\eta'(\theta)$ as a function of θ.

Once we obtained accurate $\eta(\theta)$ and $\eta'(\theta)$, (2.39) could be numerically integrated to give the scaling functions $F_\pm(t)$. In Table VII we give for the symmetric lattice the scaling functions $F_\pm(t)$. To obtain $F_\pm(t)$ for arbitrary E_1 and E_2 the entry in Table VII should be multiplied by $2^{-1/8}(\sinh 2\beta_c E_1+\sinh 2\beta_c E_2)^{1/8}$. For $E_1=E_2$ this quantity is one, as it must be. The susceptibility coefficients $C_{0\pm}$ were obtained by numerically integrating (2.36).

In Fig. 13 we plot the function $I(\theta)$ defined by

$$I(\theta)=\int_\theta^\infty dx\, x \ln x[1-\eta^2(x)]. \tag{A4}$$

APPENDIX B: INTEGRAL EQUATION AND PAINLEVÉ FUNCTION $\eta(\theta)$

In this appendix we discuss integral equations of the form

$$\int_a^b K_0(\theta|x-t|)\,\psi(t)\,dt=\phi(x), \tag{B1}$$

where a and b are finite real numbers, $a<x<b$, and $K_0(x)$ is the modified Bessel function of zeroth order. The cases of one or both of the end points infinite are solved by standard methods.

Equation (B1) has been the subject of considerable investigation as it arises in the context of scattering from a thin strip. Since the geometry of the strip is a degenerate case of an elliptic cylinder, it is not surprising that the solution to (B1) can be expressed as infinite series of Mathieu functions. Unfortunately the representation of a function as an infinite series of Mathieu functions is cumbersome at best.

Progress toward the solution of (B1) was made by Latta[46] when he showed that solutions to

$$\Gamma f=\cosh\theta t, \tag{B2a}$$

$$\Gamma g=\sinh\theta t, \tag{B2b}$$

where we define the operator Γ by

$$(\Gamma\phi)(t)=\int_{-1}^{1}K_0(\theta|t-s|)\,\phi(s)\,ds, \tag{B3}$$

satisfy a pair of linear first order coupled differential equations. Latta did not determine the coefficients in this set of differential equations (these coefficients will be a function of θ). This Myers[34,35] did by imposing the boundary conditions that are implicit in (B2) on the set of differential equations. These results can be summarized by a sequence of theorems.[47]

Theorem 1 (Latta-Myers)

If we define the operator Γ by (B3) and if $f(t)$ and $g(t)$ are the solution to (B2), then we have

$$(1-t^2)\frac{df}{dt}-(\tfrac{1}{2}-\rho)tf=[\eta^{-1}(\rho+\tfrac{1}{2})+\theta(1-t^2)]g, \tag{B4a}$$

$$(1-t^2)\frac{dg}{dt}-(\tfrac{1}{2}+\rho)tg=[\eta(\tfrac{1}{2}-\rho)+\theta(1-t^2)]f, \tag{B4b}$$

where

$$\rho=\theta[2\eta(\theta)]^{-1}\left(1-\eta^2(\theta)-\frac{d\eta}{d\theta}\right) \tag{B5a}$$

and

$$\eta(\theta)=\lim_{t\to 1}\frac{g(t)}{f(t)}. \tag{B5b}$$

Theorem 2 (Myers)

The function $\eta(\theta)$, (B5b), satisfies a Painlevé equation of third kind[33]

$$\frac{d^2\eta}{d\theta^2}=\eta^{-1}(\theta)\left(\frac{d\eta}{d\theta}\right)^2-\eta^{-1}(\theta)+\eta^3(\theta)-\theta^{-1}\frac{d\eta}{d\theta}. \tag{B6a}$$

Equation (B6a) is subject to the boundary conditions

$$\eta(\theta)=-\theta(\ln\tfrac{1}{4}\theta+\gamma_E)+O(\theta^5\ln^3\theta) \tag{B6b}$$

as $\theta\to 0$, and

$$\eta(\theta)=1-(2/\pi)K_0(2\theta)+O(e^{-4\theta}) \tag{B6c}$$

as $\theta\to\infty$. In (B6b) γ_E is Euler's constant.

In Sec VI we will need expansions of $f(t)$ and $g(t)$ about the singular point $t=1$. Since this point is a regular singular point, we can easily find from (B4)

$$f(t,\theta)=\frac{f_0(\theta)}{(1-t)^{1/2}}\left(1+\sum_{n=1}^{\infty}f_n(\theta)(1-t)^n\right), \tag{B7a}$$

$$g(t,\theta)=\frac{f_0(\theta)\,\eta(\theta)}{(1-t)^{1/2}}\left(1+\sum_{n=1}^{\infty}g_n(\theta)(1-t)^n\right), \tag{B7b}$$

with

$$\left.\begin{aligned}f_1&=-\tfrac{1}{2}(\tfrac{3}{2}+\rho)(\rho+2\theta\eta)-\tfrac{1}{2}(\tfrac{1}{2}+\rho)(\rho-2\theta/\eta),\\ f_2&=\tfrac{1}{24}f_1[(\tfrac{7}{2}+\rho)(\tfrac{1}{2}-\rho)+(\tfrac{1}{4}-\rho^2)+(4\theta/\eta)(\tfrac{1}{2}+\rho)]\end{aligned}\right\} \tag{B7c}$$

TABLE VI. Painlevé function $\eta(\theta)$ and its first derivative $\eta'(\theta)$ are given for $0 \le \theta \le 8$. For $0 \le \theta \le 0.8$ the entry in the second column is $\eta(\theta)$, and for $\theta \ge 0.9$ the entry in the second column is $\epsilon(\theta) = 1 - \eta(\theta)$. The entry in the third column is for $0 \le \theta \le 8$. If an entry is followed by a number enclosed in a parenthesis, e.g., $(-n)$, then this entry should be multiplied by 10^{-n}. Thus, for instance, the value of $\eta(\theta)$ at $\theta = 0.001$ is $0.771\,683\,397\,5 \times 10^{-2}$ and the value of $\eta'(\theta)$ at $\theta = 0.001$ is $0.671\,683\,397\,5 \times 10^{1}$.

θ	$\eta(\theta)$	$\eta'(\theta)$
0.0	0	∞
0.001	0.771 683 397 5 (−2)	0.671 683 397 5 (1)
0.002	0.140 473 735 9 (−1)	0.602 368 679 6 (1)
0.003	0.198 546 650 6 (−1)	0.561 822 169 4 (1)
0.004	0.253 221 584 8 (−1)	0.533 053 963 6 (1)
0.005	0.305 369 803 7 (−1)	0.510 739 611 0 (1)
0.006	0.355 504 471 6 (−1)	0.492 507 459 6 (1)
0.007	0.403 964 670 2 (−1)	0.477 092 398 1 (1)
0.008	0.450 991 399 1 (−1)	0.463 739 268 0 (1)
0.009	0.496 764 853 2 (−1)	0.451 960 976 9 (1)
0.010	0.541 424 900 1 (−1)	0.441 424 941 7 (1)
0.011	0.585 083 275 5 (−1)	0.431 893 944 5 (1)
0.012	0.627 831 306 0 (−1)	0.423 192 832 8 (1)
0.013	0.669 745 038 8 (−1)	0.415 188 593 8 (1)
0.014	0.710 888 784 0 (−1)	0.407 777 834 8 (1)
0.015	0.751 317 638 9 (−1)	0.400 878 592 9 (1)
0.016	0.791 079 337 4 (−1)	0.394 424 793 9 (1)
0.017	0.830 215 634 1 (−1)	0.388 362 393 4 (1)
0.018	0.868 763 362 4 (−1)	0.382 646 622 8 (1)
0.019	0.906 755 256 6 (−1)	0.377 239 981 4 (1)
0.02	0.944 220 600 3 (−1)	0.372 110 743 4 (1)
0.03	0.129 469 251 6	0.331 565 894 0 (1)
0.04	0.161 118 716 6	0.302 801 233 6 (1)
0.05	0.190 241 946 6	0.280 493 065 3 (1)
0.06	0.217 352 406 9	0.262 270 476 5 (1)
0.07	0.242 789 553 0	0.246 869 068 7 (1)
0.08	0.266 794 827 5	0.233 534 360 9 (1)
0.09	0.289 548 934 9	0.221 779 899 8 (1)
0.10	0.311 192 313 4	0.211 273 700 4 (1)
0.11	0.331 837 327 7	0.201 779 101 5 (1)
0.12	0.351 575 986 6	0.193 121 465 1 (1)
0.13	0.370 485 069 2	0.185 168 252 4 (1)
0.14	0.388 629 662 5	0.177 816 503 9 (1)
0.15	0.406 065 682 5	0.170 984 651 0 (1)
0.16	0.422 841 718 8	0.164 606 975 8 (1)
0.17	0.439 000 415 9	0.158 629 753 6 (1)
0.18	0.454 579 527 2	0.153 008 501 0 (1)
0.19	0.469 612 733 7	0.147 705 968 5 (1)
0.20	0.484 130 289 6	0.142 690 649 9 (1)
0.21	0.498 159 537 0	0.137 935 655 4 (1)
0.22	0.511 725 323 5	0.133 417 848 0 (1)
0.23	0.524 850 341 5	0.129 117 171 6 (1)
0.24	0.537 555 409 7	0.125 016 123 9 (1)
0.25	0.549 859 706 3	0.121 099 335 3 (1)
0.26	0.561 780 964 6	0.117 353 232 4 (1)
0.27	0.573 335 639 2	0.113 765 764 5 (1)
0.28	0.584 539 045 8	0.110 326 179 9 (1)
0.29	0.595 405 482 3	0.107 024 841 9 (1)
0.30	0.605 948 332 3	0.103 853 076 6 (1)
0.31	0.616 180 154 6	0.100 803 044 9 (1)

TABLE VI. (*Continued*)

θ	$\eta(\theta)$ [$\epsilon(\theta)$]	$\eta'(\theta)$
0.32	0.626 112 761 3	0.978 676 356 8
0.33	0.635 757 286 1	0.950 403 755 5
0.34	0.645 124 243 7	0.923 153 515 9
0.35	0.654 223 583 2	0.896 871 460 6
0.36	0.663 064 733 8	0.871 507 800 5
0.37	0.671 656 647 2	0.847 016 650 6
0.38	0.680 007 833 5	0.823 355 610 6
0.39	0.688 126 395 2	0.800 485 401 0
0.40	0.696 020 055 8	0.778 369 546 2
0.41	0.703 696 187 2	0.756 974 0964
0.42	0.711 161 833 3	0.736 267 384 6
0.43	0.718 423 731 8	0.716 219 812 1
0.44	0.725 488 334 1	0.696 803 659 0
0.45	0.732 361 822 9	0.677 992 916 7
0.46	0.739 050 128 8	0.659 763 138 8
0.47	0.745 558 945 3	0.642 091 308 7
0.48	0.751 893 742 3	0.624 955 721 0
0.49	0.758 059 778 7	0.608 335 875 5
0.50	0.764 062 114 2	0.592 212 382 7
0.52	0.775 594 986 9	0.561 381 942 2
0.54	0.786 529 235 3	0.532 328 441 8
0.56	0.796 899 149 2	0.504 929 014 3
0.58	0.806 736 677 3	0.479 072 130 5
0.60	0.816 071 638 5	0.454 656 146 0
0.62	0.824 931 907 6	0.431 588 090 3
0.64	0.833 343 578 0	0.409 782 647 8
0.66	0.841 331 106 1	0.389 161 295 9
0.68	0.848 917 439 5	0.369 651 571 6
0.70	0.856 124 131 9	0.351 186 442 4
0.72	0.862 971 445 9	0.333 703 764 2
0.74	0.869 478 446 4	0.317 145 811 6
0.76	0.875 663 084 3	0.301 458 867 9
0.80	0.887 131 959 7	0.272 501 075 4
0.90	0.887 641 364 4 (−1)	0.212 220 235 0
1.0	0.699 637 109 1 (−1)	0.165 787 696 0
1.1	0.552 571 023 1 (−1)	0.129 868 027 9
1.2	0.437 230 351 6 (−1)	0.101 980 718 6
1.3	0.346 558 553 8 (−1)	0.802 614 032 9 (−1)
1.4	0.275 125 892 1 (−1)	0.632 981 360 9 (−1)
1.5	0.218 738 225 1 (−1)	0.500 153 640 0 (−1)
1.6	0.174 144 827 7 (−1)	0.395 898 923 7 (−1)
1.7	0.138 818 467 9 (−1)	0.313 891 168 8 (−1)
1.8	0.110 788 873 6 (−1)	0.249 251 486 7 (−1)
1.9	0.885 160 017 3 (−2)	0.198 204 800 8 (−1)
2.0	0.707 931 889 0 (−2)	0.157 821 321 1 (−1)
2.2	0.454 093 067 1 (−2)	0.100 424 345 9 (−1)
2.4	0.292 242 112 9 (−2)	0.641 765 253 0 (−2)
2.6	0.188 627 153 6 (−2)	0.411 658 230 3 (−2)
2.8	0.122 060 623 8 (−2)	0.264 920 767 5 (−2)
3.0	0.791 637 912 4 (−3)	0.170 977 737 3 (−2)
3.2	0.514 453 726 3 (−3)	0.110 626 523 1 (−2)
3.4	0.334 919 280 2 (−3)	0.717 379 331 8 (−3)
3.6	0.218 387 124 9 (−3)	0.466 123 080 2 (−3)
3.8	0.142 605 824 2 (−3)	0.303 404 672 3 (−3)
4.0	0.932 417 997 9 (−4)	0.197 803 779 6 (−3)
4.2	0.610 372 497 6 (−4)	0.129 142 509 9 (−3)
4.4	0.399 985 887 5 (−4)	0.844 242 044 1 (−4)
4.6	0.262 373 474 2 (−4)	0.552 556 064 1 (−4)
4.8	0.172 260 488 2 (−4)	0.362 036 489 6 (−4)
5.0	0.113 190 751 6 (−4)	0.237 440 870 6 (−4)

TADLE VI. (*Continued*)

θ	$\epsilon(\theta)$	$\eta'(\theta)$
5.2	0.744 334 697 2 (−5)	0.155 865 837 5 (−4)
5.4	0.489 816 220 4 (−5)	0.102 401 878 6 (−4)
5.6	0.322 540 129 8 (−5)	0.673 285 185 8 (−5)
5.8	0.212 520 291 2 (−5)	0.442 996 113 4 (−5)
6.0	0.140 108 798 2 (−5)	0.291 667 890 5 (−5)
6.2	0.924 193 229 1 (−6)	0.192 152 267 1 (−5)
6.4	0.609 927 237 4 (−6)	0.126 663 901 9 (−5)
6.6	0.402 715 142 8 (−6)	0.835 400 302 5 (−6)
6.8	0.266 017 712 7 (−6)	0.551 259 710 0 (−6)
7.0	0.175 794 311 1 (−6)	0.363 935 553 2 (−6)
7.2	0.116 217 268 8 (−6)	0.240 373 856 2 (−6)
7.4	0.768 597 265 2 (−7)	0.158 830 325 0 (−6)
7.6	0.508 488 099 3 (−7)	0.104 991 203 4 (−6)
7.8	0.336 518 281 0 (−7)	0.694 282 730 5 (−7)
8.0	0.222 779 463 2 (−7)	0.459 277 498 4 (−7)

$$+\tfrac{1}{24}g_1[-(\tfrac{7}{2}+\rho)(\tfrac{1}{2}+\rho)$$

$$-4\eta\theta(\tfrac{7}{2}+\rho)-(\tfrac{1}{2}+\rho)^2]-\tfrac{1}{12}\rho(2+\rho), \qquad \text{(B7d)}$$

$$g_1=\tfrac{1}{2}(\tfrac{3}{2}-\rho)(\rho-2\theta/\eta)+\tfrac{1}{2}(\rho+2\theta\eta)(\tfrac{1}{2}-\rho), \qquad \text{(B7e)}$$

$$g_2=-\tfrac{1}{24}(2-5\rho+2\rho^2)f_1+\tfrac{1}{24}(2+3\rho-2\rho^2)g_1$$

$$-(\theta/6\eta)(\tfrac{7}{2}-\rho)f_1+\tfrac{1}{6}\eta\theta(\tfrac{1}{2}-\rho)g_1+\tfrac{1}{12}\rho(2-\rho), \qquad \text{(B7f)}$$

and ρ given by (B5c). Since $f(t)[g(t)]$ is an even (odd) function of t, this fact could be incorporated into the differential Eqs. (B5a) and (B5b). This will simplify things somewhat, but for our purposes (B7) is good enough.

It is useful to have higher-order terms in the expansion (B6b) of $\eta(\theta)$ for small θ. Writing

$$\eta(\theta)=\sum_{n=0}^{\infty}\theta^{4n+1}\rho_n, \qquad \text{(B8a)}$$

with

$$\rho_n=\sum_{k=0}^{2n+1}b_{n,k}\Omega^k \qquad \text{(B8b)}$$

and

$$\Omega=\ln\tfrac{1}{4}\theta+\gamma_E, \qquad \text{(B8c)}$$

and substituting (B8a) into (B6a) determines the coefficients $b_{n,k}$. Myers[34] did this through order θ^5. We have computed through order θ^{13}. Our results are given in Table VIII.

The large-θ expansion of $\eta(\theta)$ is given by

$$\eta(\theta)=1-(2/\pi)K_0(2\theta)+(2/\pi^2)K_0^2(2\theta)+O(e^{-6\theta}). \qquad \text{(B9)}$$

To compute the higher-order terms makes use of (6.95).

Myers considers another function $\chi(t;\,z)$ that is defined by the equation

$$\Gamma\chi=e^{-Z\theta t}. \qquad \text{(B10)}$$

His results are summarized in the following.

Theorem 3 (Myers)

If we denote by $\chi(t;\,Z)$ the solution to (B10), then

$$\chi(t;\,Z)=\theta(Z^2-1)^{1/2}\psi(t;\,Z)$$

$$-\lambda_1(\theta,Z)f(t,\theta)-\lambda_2(\theta,Z)g(t,\theta), \qquad \text{(B11)}$$

where f and g are solutions to (B2a) and (B2b), respectively. The function $\psi(t;\,Z)$ is the solution to the differential equation

$$\frac{\partial\psi}{\partial t}+Z\theta\psi(t,Z)$$

$$=(Z^2-1)^{1/2}[(\lambda_2+Z\lambda_1)f+(\lambda_1+Z\lambda_2)g], \qquad \text{(B12)}$$

subject to the boundary conditions $\psi(\pm 1)=0$. The functions $\lambda_1(\theta,Z)$ and $\lambda_2(\theta,Z)$ are determined by

$$\lambda_1(\theta,Z)=M(\theta)[\bar{f}(Z,\theta)-Z\bar{g}(Z,\theta)], \qquad \text{(B13a)}$$

$$\lambda_2(\theta,Z)=M(\theta)[\bar{g}(Z,\theta)-Z\bar{f}(Z,\theta)], \qquad \text{(B13b)}$$

where the bar operation is defined by

$$\bar{h}(\xi,\theta)=\int_{-1}^{1}e^{-t\xi\theta}h(t)\,dt. \qquad \text{(B13c)}$$

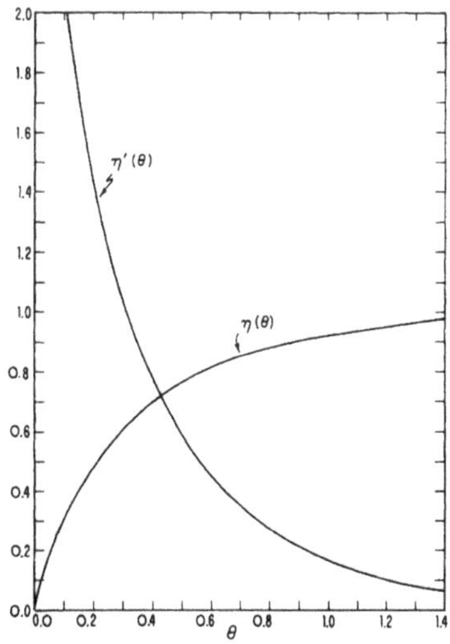

FIG. 12. Painlevé function $\eta(\theta)$ and its first derivative $\eta'(\theta)$.

TABLE VII. Scaling functions $F_{\pm}(t)$ for the symmetric lattice are given for $0 \le t \le 8.0$. To obtain the scaling functions $F_{\pm}(t)$ for arbitrary E_1 and E_2 the entry for the symmetric lattice should be multiplied by $(\sinh 2\beta_c E_1 + \sinh 2\beta_c E_2)^{1/8} 2^{-1/8}$. If an entry is followed by a number enclosed in parenthesis, e.g., $(-n)$, then this entry should be multiplied by 10^{-n}. Thus, for instance, the value of $F_+(t)$ at $t=2$ is 0.055 910 659 65.

t	$F_+(t)$	$F_-(t)$
0.0	0.703 380 157 7	0.703 380 157 7
0.002	0.697 952 464 3	0.708 808 202 8
0.004	0.693 500 207 3	0.713 261 514 8
0.006	0.689 416 331 3	0.717 347 149 0
0.008	0.685 571 795 6	0.721 194 145 7
0.010	0.681 905 312 4	0.724 863 792 9
0.012	0.678 380 781 7	0.728 392 189 9
0.014	0.674 974 348 7	0.731 803 191 2
0.016	0.671 669 057 7	0.735 113 752 4
0.018	0.668 452 229 7	0.738 336 551 9
0.020	0.665 314 022 6	0.741 481 431 4
0.022	0.662 246 573 1	0.744 556 253 7
0.024	0.659 243 453 5	0.747 567 446 0
0.026	0.656 299 311 1	0.750 520 360 6
0.028	0.653 409 618 5	0.754 319 524 3
0.030	0.650 570 496 6	0.756 268 815 6
0.032	0.647 778 584 3	0.759 071 595 1
0.034	0.645 030 941 1	0.761 830 802 6
0.036	0.642 324 972 7	0.764 549 031 8
0.038	0.639 658 373 2	0.767 228 588 0
0.040	0.637 029 079 5	0.769 871 533 6
0.042	0.634 435 234 6	0.772 479 725 0
0.044	0.631 875 158 6	0.775 054 841 3
0.046	0.629 347 323 9	0.777 598 409 4
0.048	0.626 850 335 8	0.780 111 823 3
0.050	0.624 382 915 7	0.782 596 360 8
0.06	0.612 451 118 3	0.794 624 092 9
0.07	0.601 114 045 8	0.806 074 271 5
0.08	0.590 286 569 4	0.817 031 918 0
0.09	0.579 904 960 0	0.827 560 647 1
0.10	0.569 919 685 0	0.837 709 872 0
0.11	0.560 291 117 3	0.847 519 095 0
0.12	0.550 986 818 7	0.857 020 625 6
0.13	0.541 979 733 9	0.866 241 386 3
0.14	0.533 246 945 0	0.875 204 159 0
0.15	0.524 768 782 4	0.883 928 474 7
0.16	0.516 528 173 8	0.892 431 264 1
0.17	0.508 510 157 5	0.900 727 345 5
0.18	0.500 701 509 1	0.908 829 797 6
0.19	0.493 090 452 5	0.916 750 249 3
0.20	0.485 666 431 6	0.924 499 108 3
0.21	0.478 419 927 1	0.932 085 743 8
0.22	0.471 342 309 2	0.939 518 634 6
0.23	0.464 425 716 5	0.946 805 490 6
0.24	0.457 662 955 8	0.953 953 352 0
0.25	0.451 047 419 7	0.960 968 673 5
0.26	0.444 573 015 7	0.967 857 393 8
0.27	0.438 234 108 0	0.974 624 995 0
0.28	0.432 025 466 3	0.981 276 553 1
0.29	0.425 942 223 2	0.987 816 781 5
0.30	0.419 979 836 6	0.994 250 068 1
0.31	0.414 134 057 8	1.000 580 507 4

TABLE VII. (*Continued*)

t	$F_+(t)$	$F_-(t)$
0.32	0.408 400 902 9	1.006 811 929 4
0.33	0.402 776 628 6	1.012 947 923 6
0.34	0.397 257 710 4	1.018 991 861 4
0.35	0.391 840 823 4	1.024 946 914 2
0.36	0.386 522 825 8	1.030 816 071 7
0.37	0.381 300 743 3	1.036 602 155 7
0.38	0.376 171 756 2	1.042 307 834 6
0.39	0.371 133 187 0	1.047 935 634 8
0.40	0.366 182 489 8	1.053 487 951 9
0.41	0.361 317 240 8	1.058 967 060 2
0.42	0.356 535 129 0	1.064 375 121 5
0.43	0.351 833 948 7	1.069 714 193 5
0.44	0.347 211 592 3	1.074 986 236 2
0.45	0.342 666 043 6	1.080 193 119 4
0.46	0.338 195 371 8	1.085 336 628 2
0.47	0.333 797 726 2	1.090 418 468 2
0.48	0.329 471 331 0	1.095 440 271 3
0.49	0.325 214 481 2	1.100 403 599 5
0.50	0.321 025 537 6	1.105 309 949 8
0.52	0.312 845 122 0	1.114 957 400 1
0.54	0.304 918 157 7	1.124 393 430 9
0.56	0.297 233 559 8	1.133 628 025 4
0.58	0.289 780 997 1	1.142 670 429 8
0.60	0.282 550 816 7	1.151 529 228 3
0.62	0.275 533 978 4	1.160 212 409 8
0.64	0.268 721 997 6	1.168 727 424 6
0.66	0.262 106 895 2	1.177 081 235 1
0.68	0.255 681 153 3	1.185 280 359 7
0.70	0.249 437 676 4	1.193 330 912 2
0.72	0.243 369 756 7	1.201 238 636 3
0.74	0.237 471 043 1	1.209 008 936 4
0.76	0.231 735 514 1	1.216 646 905 0
0.78	0.226 157 452 4	1.224 157 347 8
0.80	0.220 731 423 5	1.231 544 805 3
0.82	0.215 452 254 8	1.238 813 573 1
0.84	0.210 315 018 1	1.245 967 720 0
0.86	0.205 315 012 9	1.253 011 103 9
0.88	0.200 447 751 2	1.259 947 387 1
0.90	0.195 708 944 2	1.266 780 049 4
0.92	0.191 094 489 7	1.273 512 400 9
0.94	0.186 600 460 8	1.280 147 592 6
0.96	0.182 223 095 1	1.286 688 627 0
0.98	0.177 958 785 4	1.293 138 367 8
1.0	0.173 804 070 4	1.299 499 548 1
1.1	0.154 562 530 2	1.330 064 530 4
1.2	0.137 612 980 7	1.358 762 886 4
1.3	0.122 647 434 5	1.385 828 041 9
1.4	0.109 407 490 3	1.411 452 008 3
1.5	0.976 743 235 9 (−1)	1.435 794 682 2
1.6	0.872 612 080 4 (−1)	1.458 990 641 1
1.8	0.697 756 688 9 (−1)	1.502 383 351 4
2.0	0.559 106 596 5 (−1)	1.542 365 330 2
2.2	0.448 789 760 4 (−1)	1.579 490 531 5
2.4	0.360 772 704 3 (−1)	1.614 186 500 3
2.6	0.290 386 372 2 (−1)	1.646 789 986 5
2.8	0.233 990 436 6 (−1)	1.677 570 899 7
3.0	0.188 729 732 3 (−1)	1.706 748 894 3
3.2	0.152 354 133 3 (−1)	1.734 505 137 3

TABLE VII. (*Continued*)

t	$F_+(t)$	$F_-(t)$
3.4	0.123 083 337 1 (−1)	1.760 990 838 3
3.6	0.995 041 383 9 (−2)	1.786 333 548 5
3.8	0.804 916 302 5 (−2)	1.810 641 891 1
4.0	0.651 482 814 1 (−2)	1.834 009 166 4
4.2	0.527 565 272 9 (−2)	1.856 516 137 5
4.4	0.427 416 774 3 (−2)	1.878 233 208 9
4.6	0.346 427 618 9 (−2)	1.899 222 150 6
4.8	0.280 895 291 0 (−2)	1.919 537 475 6
5.0	0.227 842 371 4 (−2)	1.939 227 552 4
5.2	0.184 871 985 9 (−2)	1.958 335 510 2
5.4	0.150 052 769 5 (−2)	1.976 899 982 4
5.6	0.121 827 121 4 (−2)	1.994 955 721 8
5.8	0.989 378 990 (−3)	2.012 534 112 5
6.0	0.803 697 447 (−3)	2.029 663 600 6
6.2	0.653 020 574 (−3)	2.046 370 055 7
6.4	0.530 712 468 (−3)	2.062 677 079 8
6.6	0.431 404 022 (−3)	2.078 606 269 4
6.8	0.350 748 937 (−3)	2.094 177 440 9
7.0	0.285 227 267 (−3)	2.109 408 825 0
7.2	0.231 987 098 (−3)	2.124 317 235 0
7.4	0.188 716 856 (−3)	2.138 918 212 4
7.6	0.153 542 243 (−3)	2.153 226 155 0
7.8	0.124 943 002 (−3)	2.167 254 428 7
8.0	0.101 685 644 (−3)	2.181 015 465 5

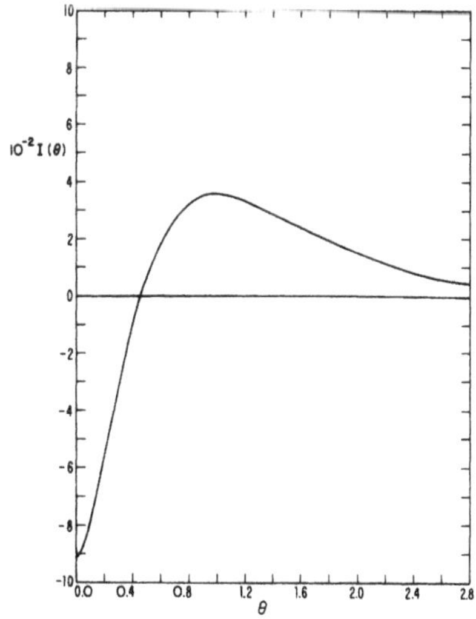

FIG. 13. Quantity $I(\theta)=\int_\theta^\infty x\ln x[1-\eta^2(x)]dx$ as a function of θ.

And finally,

$$\frac{\partial f(t,\theta)}{\partial\theta}-\theta^{-1}(\tfrac{1}{2}+\rho)f(t,\theta)-tg(t,\theta)=0, \qquad \text{(B13d)}$$

$$\frac{\partial\lambda_1(\theta,Z)}{\partial\theta}+(\lambda_1+Z\lambda_2)\,\eta(\theta)=0, \qquad \text{(B13e)}$$

and

$$\frac{\partial\lambda_2(\theta,Z)}{\partial\theta}+(\lambda_2+Z\lambda_1)\,\eta^{-1}(\theta)=0. \qquad \text{(B13f)}$$

We have the result

Theorem:

$$f_0^2(\theta)\,M(\theta)\,\eta(\theta)=-\theta/(2\pi^2). \qquad \text{(B14)}$$

To prove this we multiply (B13d) by $(1-t)^{1/2}$ and take the limit $t\to1$. We obtain

$$\frac{\partial f_0}{\partial\theta}-\frac{(\frac{1}{2}+\rho)}{\theta}f_0-f_0\eta(\theta)=0, \qquad \text{(B15)}$$

which implies

$$f_0^2(\theta)=(\text{const})\exp\left(2\int^\theta[\eta+(\tfrac{1}{2}+\rho)\theta^{-1}]\,d\theta\right) \qquad \text{(B16)}$$

for $Z=0$ (B13e) implies

$$\lambda_1(\theta)=(\text{const})\exp\left(-\int^\theta\eta(\theta)\,d\theta\right). \qquad \text{(B17)}$$

Multiplying (B13d) by $e^{-i\xi t}$ and integrating over t from -1 to $+1$ we obtain for the special case $\xi=0$

$$\frac{\partial}{\partial\theta}\bar f(0,\theta)-(\tfrac{1}{2}+\rho)\,\theta^{-1}\bar f(0,\theta)-\int_{-1}^{1}xg(x,\theta)\,dx=0. \qquad \text{(B18)}$$

One can show, (Myers does)

TABLE VIII. Coefficients $b_{n,k}$ in the small-θ expansion of $\eta(\theta)$. See Eq. (B8) for the definition of these coefficients.

n	k	$b_{n,k}$
0	0	0
0	1	-1
1	0	-2^{-7}
1	1	2^{-5}
1	2	-2^{-4}
1	3	2^{-4}
2	0	-145×2^{-20}
2	1	145×2^{-17}
2	2	-129×2^{-15}
2	3	121×2^{-14}
2	4	-2^{-7}
2	5	2^{-8}
3	0	$-28195\times3^{-5}\times2^{-25}$
3	1	$25603\times3^{-4}\times2^{-23}$
3	2	$-82729\times3^{-3}\times2^{-24}$
3	3	$9539\times3^{-2}\times2^{-21}$
3	4	-477×2^{-19}
3	5	275×2^{-18}
3	6	-3×2^{-12}
3	7	2^{-12}

$$\frac{d}{dt}(g-\eta tf)+[\eta(\tfrac{1}{2}+\rho)-\theta]f+\eta\theta tg=0\,. \qquad \text{(B19)}$$

Applying the bar operation to (B19) we have for $\xi=0$

$$[\eta(\tfrac{1}{2}+\rho)-\theta]\bar{f}(0,\theta)=-\theta\eta\int_{-1}^{1}xg(x,\theta)\,dx\,. \qquad \text{(B20)}$$

Then (B18) and (B20) imply

$$\frac{d}{d\theta}\bar{f}(0,\theta)=\eta^{-1}\bar{f}(0,\theta) \qquad \text{(B21)}$$

or

$$\bar{f}(0,\theta)=(\text{const})\exp\left(\int^{\theta}\frac{d\theta}{\eta}\right). \qquad \text{(B22)}$$

From (B13a) for $Z=0$

$$M(\theta)=\lambda_1(0,\theta)/\bar{f}(0,\theta) \qquad \text{(B23)}$$

$$=(\text{const})\exp\left(-\int^{\theta}d\theta\,(\eta+\eta^{-1})\right), \qquad \text{(B24)}$$

where we used (B17) and (B22) in going from (B23) to (B24). Equations (B16) and (B24) imply

$$M(\theta)\,f_0^2(\theta)=(\text{const})\exp\left[\int^{\theta}\left(\eta-\eta^{-1}+\frac{1+2\rho}{\theta}\right)\right]. \qquad \text{(B25)}$$

Using (B5a) we have

$$\eta-\eta^{-1}+(1+2\rho)\,\theta^{-1}=\theta^{-1}+\frac{d}{d\theta}\ln\eta(\theta)\,, \qquad \text{(B26)}$$

so that (B25) becomes

$$f_0^2(\theta)M(\theta)\,\eta(\theta)/\theta=(\text{const})\,. \qquad \text{(B27)}$$

To obtain the constant in (B27) we evaluate the left-hand side of (B27) for $\theta\to 0$. For small θ (B2) reduces to

$$-\int_{-1}^{1}(\ln\tfrac{1}{2}\theta|x-x'|+\gamma_E)\,f(x')\,dx'=\cosh\theta x \qquad \text{(B28a)}$$

and

$$-\int_{-1}^{1}(\ln\tfrac{1}{2}\theta|x-x'|+\gamma_E)\,g(x')\,dx'=\sinh\theta x. \qquad \text{(B28b)}$$

Equations (B28) can be solved for it is known that[48] if

$$\int_{-1}^{1}u(x')[\ln 2|x-x'|+\Omega]\,dx'=w(x)\,, \qquad \text{(B29)}$$

then

$$u(x)=\frac{1}{\pi^2(1-x^2)^{1/2}}\left(P\int_{-1}^{1}dx'\,\frac{(1-x'^2)^{1/2}}{x'-x}\,\frac{dw(x')}{dx'}+\frac{1}{\Omega}P\int_{-1}^{1}dx'\,\frac{w(x')}{(1-x'^2)^{1/2}}\right) \qquad \text{(B30)}$$

In (B30) the integrals are principal-value integrals. Using (B30) we obtain for $t\to 1$

$$f(t,\theta)\sim-[\Omega\pi 2^{1/2}(1-t)^{1/2}]^{-1} \qquad \text{(B31)}$$

and

$$g(t,\theta)\sim\theta\pi^{-1}2^{-1/2}(1-t)^{-1/2}\,. \qquad \text{(B32)}$$

From the definition (B56) when $\theta\to 0$

$$\eta(\theta)\sim-\theta\Omega\,, \qquad \text{(B33)}$$

[this is how Myers derived (B6b)].

Furthermore, from (B31) and (B7a) it follows

$$f_0^2(\theta)\sim 1/2\pi^2\Omega^2\,. \qquad \text{(B34)}$$

From the definition of $\bar{f}(0,\theta)$ we have

$$\bar{f}(0,\theta)=\int_{-1}^{1}dx\,f(x)\sim-\frac{1}{\Omega}\,. \qquad \text{(B35)}$$

To obtain $M(\theta)$ we calculate $\lambda_1(0,\theta)$. To compute $\lambda_1(0,\theta)$ for $\theta\to 0$ we examine the equation

$$\Gamma\chi=e^{-Z\theta t}\sim 1\,. \qquad \text{(B36)}$$

Using (B30)

$$\chi(t,\theta)\sim-[\pi\Omega(1-t^2)^{1/2}]^{-1}\,. \qquad \text{(B37)}$$

As $\theta\to 0$, $g(t,\theta)\to 0$, hence

$$\chi(t,Z)\simeq-\lambda_1(\theta,Z)\,f(t,\theta)\,. \qquad \text{(B38)}$$

Using (B31) and (B37) we see from (B38)

$$\lim_{\theta\to 0}\lambda_1(\theta,0)=-1\,. \qquad \text{(B39)}$$

Using (B35) and (B39) we have form (B23)

$$M(\theta)\sim-\Omega\,. \qquad \text{(B40)}$$

Combining (B33), (B34), and (B40) we see that (B14) follows.

APPENDIX C

We record and derive in this appendix various formulas which are used in the text to make some expansions near T_c.

First of all the T_c condition may be written in the equivalent form of

$$\sinh 2\beta_c E_1\,\sinh 2\beta_c E_2=1 \qquad \text{(C1)}$$

or

$$z_{1c}z_{2c}+z_{1c}+z_{2c}-1=0\,, \qquad \text{(C2)}$$

with

$$z_i=\tanh\beta E_i\,,\quad i=1,\,2\,. \qquad \text{(C3)}$$

Furthermore from (C3)

$$\sinh 2\beta E_i=2(z_i^{-1}-z_i)^{-1},\quad i=1,\,2\,, \qquad \text{(C4)}$$

$$\cosh 2\beta E_i=(1+z_i^2)/(1-z_i^2)\,,\quad i=1,\,2\,, \qquad \text{(C5)}$$

and

$$\coth 2\beta E_i=\tfrac{1}{2}(z_i^{-1}+z_i)\,,\quad i=1,\,2\,, \qquad \text{(C6)}$$

and using (C2) we find

$$\sinh 2\beta_c E_1+\sinh 2\beta_c E_2=(z_{1c}+z_{2c})^2/2z_{1c}z_{2c}\,. \qquad \text{(C7)}$$

Moreover, using (C2) we have

$$(z_{2c}^{-1}+z_{2c})(1-z_{1c})=(z_{1c}^{-1}+z_{1c})(1-z_{2c})=2(z_{1c}+z_{2c}). \tag{C8}$$

Therefore as $T\to T_c^-$

$$\begin{aligned}S_\infty^{<}&\sim 4(\beta-\beta_c)(E_1\coth 2\beta_c E_1+E_2\coth 2\beta_c E_2)\\ &=2(\beta-\beta_c)\left[E_1(z_{1c}^{-1}+z_{1c})+E_2(z_{2c}^{-1}+z_{2c})\right]\\ &=4(\beta-\beta_c)(z_{1c}+z_{2c})\left[E_1(1-z_{2c})^{-1}+E_2(1-z_{1c})^{-1}\right],\end{aligned} \tag{C9}$$

where we have used (C6) and C7).

Recall the definition of R^2 as

$$\begin{aligned}R^2&=(\gamma_2/\gamma_1)^{1/2}M^2+(\gamma_1/\gamma_2)^{1/2}N^2\\ &=\left(\frac{z_1(1-z_2^2)}{z_2(1-z_1^2)}\right)^{1/2}M^2+\left(\frac{z_2(1-z_1^2)}{z_1(1-z_2^2)}\right)^{1/2}N^2\\ &=\left(\frac{\sinh 2\beta E_1}{\sinh 2\beta E_2}\right)^{1/2}M^2+\left(\frac{\sinh 2\beta E_2}{\sinh 2\beta E_1}\right)^{1/2}N^2\end{aligned} \tag{C10}$$

and

$$\begin{aligned}t^2&=(z_1z_2+z_1+z_2-1)^2\left(\frac{M^2}{z_2(1-z_1^2)}+\frac{N^2}{z_1(1-z_2^2)}\right)\\ &=(z_1z_2+z_1+z_2-1)^2\left[z_1z_2(1-z_1^2)(1-z_2^2)\right]^{-1/2}R^2.\end{aligned} \tag{C11}$$

Now at T_c we have the identity

$$(1-z_{1c}^2)(1-z_{2c}^2)=4z_{1c}z_{2c}, \tag{C12}$$

and as $T\to T_c$,

$$\begin{aligned}z_1z_2+z_1+z_2-1\sim(\beta-\beta_c)\,4z_{1c}z_{2c}&\\ \times\left[E_1(1-z_{2c})^{-1}+E_2(1-z_{1c})^{-1}\right].&\end{aligned} \tag{C13}$$

Thus we have

$$\begin{aligned}t/R\sim|\beta-\beta_c|\,2\sqrt{2}\,(z_{1c}z_{2c})^{1/2}&\\ \times\left[E_1(1-z_{2c})^{-1}+E_2(1-z_{1c})^{-1}\right],&\end{aligned} \tag{C14}$$

and so, using (C8) and (C13) we have, as $T\to T_c^-$,

$$S_\infty^{<}\sim(z_{1c}z_{2c})^{-1/8}\left[\sqrt{2}\,(z_{1c}+z_{2c})\,t/R\right]^{1/4}. \tag{C15}$$

Furthermore, since

$$\begin{aligned}\bar{S}_\infty&=\left[4z_1z_2(1-z_1^2)(1-z_2^2)\right]^{1/2}\\ &\quad\times\left[(\sinh 2\beta E_1\sinh 2\beta E_2)^{-2}-1\right]^{1/4},\end{aligned} \tag{C16}$$

we find, as $T\to T_c^+$

$$\begin{aligned}\bar{S}_\infty&\sim 4z_{1c}z_{2c}[4|\beta-\beta_c|\\ &\quad\times(E_1\coth 2\beta_c E_1+E_2\coth 2\beta_c E_2)]^{1/4}\\ &=4z_{1c}z_{2c}\{4|\beta-\beta_c|(z_{1c}+z_{2c})\\ &\quad\times\left[E_1(1-z_{2c})^{-1}+E_2(1-z_{1c})^{-1}\right]\}^{1/4}.\end{aligned} \tag{C17}$$

---

*Alfred P. Sloan Fellow, Work supported in part by Grant No. DMR73-07565 A01 of the National Science Foundation.

†Supported in part by Grant No. DID71-04010-A02 of the National Science Foundation.

‡Supported in part by Grant No. MPS75-07147 of the National Science Foundation.

§Present address: Institute for Theoretical Physics, State University of New York, Stony Brook, N. Y. 11794.

[1]A report of some of our results has been previously given in E. Barouch, B. M. McCoy, and T. T. Wu, Phys. Rev. Lett. 31, 1409 (1973); C. A. Tracy and B. M. McCoy, *ibid.* 31, 1500 (1973).

[2]E. Ising, Z. Phys. 31, 253 (1925).

[3]L. Onsager, Phys. Rev. 65, 117 (1944).

[4]C. N. Yang, Phys. Rev. 85, 808 (1952).

[5]For a complete discussion, see B. M. McCoy and T. T. Wu, *The Two-Dimensional Ising Model* (Harvard University, Cambridge, Mass., 1973).

[6]L. P. Kadanoff, Physics (N.Y.) 2, 263 (1966).

[7]M. E. Fisher, Rep. Prog. Phys. 30, 615 (1967).

[8]L. P. Kadanoff, W. Götze, D. Hamblen, R. Hecht, E. A. S. Lewis, V. V. Palciauskas, M. Rayl, J. Swift, D. Aspnes, and J. Kane, Rev. Mod. Phys. 39, 395 (1967).

[9]M. E. Fisher and R. J. Burford, Phys. Rev. 156, 583 (1967).

[10]H. B. Tarko and M. E. Fisher, Phys. Rev. Lett. 31, 926 (1973); Phys. Rev. B 11, 1217 (1975).

[11]B. I. Halperin and P. C. Hohenberg, Phys. Rev. 177, 952 (1969).

[12]M. E. Fisher and A. Aharony, Phys. Rev. Lett. 31, 1238 (1973); Phys. Rev. B 10, 2818 (1974).

[13]E. Brézin, D. Amit, and J. Zinn-Justin, Phys. Rev. Lett. 32, 151 (1974). These last two references study scaling functions by means of renormalization-group techniques, and are meant to be illustrative rather than exhaustive.

[14](a) The most recent estimate of C_{0+} and C_{1+} is M. F. Sykes, D. S. Guant, P. D. Roberts, and J. A. Wyles, J. Phys. A 5, 624 (1972). (b) For C_{0-} see J. W. Essam and D. L. Hunter, J. Phys. C 1, 392 (1968).

[15]E. W. Montroll, R. B. Potts, and J. C. Ward, J. Math. Phys. 4, 308 (1963).

[16]See also, E. W. Montroll, in *Applied Combinatoric Problems*, edited by E. F. Beckenbach (Wiley, New York, 1964).

[17]H. Cheng and T. T. Wu, Phys. Rev. 164, 719 (1967), hereafter referred to as CW.

[18]The function $S(e^{i\theta})$ is the function $\phi(e^{i\theta})$ in Refs. 5 and 17.

[19]A good discussion of Wiener-Hopf operators can be found in M. G. Krein, Am. Math. Soc. Transl. 22, 163 (1962). Also Ref. 5, Chap. IX.

[20]For numerical work the expression (2.27) for $f^{(2n)}(t)$ is not the most convenient. It is useful (to improve convergence) to change variables $y_j=\cosh\theta_j$ so that (2.27) becomes

$$f^{(2n)}(t)=(-1)^n\pi^{-2n}n^{-1}\int_0^\infty d\theta_1\cdots\int_0^\infty d\theta_{2n}\times\prod_{j=1}^{2n}\frac{e^{-t\cosh\theta_j}}{\cosh\theta_j+\cosh\theta_{j+1}}\prod_{j=1}^{n}\sinh^2\theta_{2j}\quad,$$

where $\theta_{2n+1}=\theta_1$ and n is any positive integer.

[21]See Appendix C.

[22]Our notation differs from CW in that we denote by $\bar{S}$,

$\bar{V}$, $\bar{U}$ what CW denote by $\mathcal{S}$, $\mathcal{V}$, $\mathcal{U}$.

[23]The quantity denoted here by $\bar{S}_\infty$ ($S_\infty^>$) is the same as the quantity denoted by $D(M)$ (D) in Refs. 5 and 17. Note that there is a misprint in both these references. They have $(\gamma_1\gamma_2)^{-1/2}$ in (4.19) instead of the correct value $(\gamma_1\gamma_2)^{1/2}$.

[24]The factorization $[\bar{S}(\xi)]^{-1} = \bar{P}(\xi)\bar{Q}(\xi^{-1})$ is in Refs. 5 and 17 written as $[\bar{S}(\xi)]^{-1} = P(\xi)Q(\xi^{-1})$. We have added the "bars" to distinguish from the $P(\xi)$ and $Q(\xi)$ introduced in (3.29) and (3.30), respectively.

[25]J. Stephenson, J. Math. Phys. 5, 1009 (1964).

[26]N. J. Achieser, *Theory of Approximation* (Ungar, New York, 1956).

[27]See Ref. 5, p. 261.

[28]W. L. Glaisher, Messenger Math. 7, 43 (1877). Glaisher's constant is also related to the Barnes G function; E. W. Barnes, Quart. J. Pure Appl. Math. 31 (1900).

[29]G. V. Ryazanov, Zh. Eksp. Teor. Fiz. 49, 1134 (1965) [Sov. Phys.-JETP 22, 789 (1966)].

[30]V. G. Vaks, A. I. Larkin, and Yu. N. Ovchinnikov, Zh. Eksp. Teor. Fiz. 49, 1180 (1965) [Sov. Phys.-JETP 22, 820 (1966)].

[31]T. T. Wu, Phys. Rev. 149, 380 (1966); see also Ref. 5.

[32]See, for instance, *Collected Papers of G. H. Hardy* (Oxford U.P., London, 1972), Vol. 5, pp. 259 and 476.

[33]M. Painlevé, Acta Math. 25, 1 (1902). See also, E. Ince, *Ordinary Differential Equations* (Dover, New York, 1945), Chap. 14.

[34]J. Myers, Ph.D. thesis (Harvard University, 1962) (unpublished).

[35]J. Myers, J. Math. Phys. 6, 1839 (1965).

[36]If

$$f(\xi) = \sum_{n=-\infty}^{+\infty} f_n \xi^n,$$

then

$$[f(\xi)]_+ = \sum_{n=0}^{\infty} f_n \xi^n$$

and

$$[f(\xi)]_- = \sum_{n=-\infty}^{-1} f_n \xi^n.$$

When considering integral equations on the line, the plus and minus parts of function

$$f(\zeta) = \int_{-\infty}^{+\infty} e^{i\zeta t} \hat{f}(t)\, dt$$

are defined as

$$[f(\zeta)]_+ = \int_0^{\infty} e^{i\zeta t} \hat{f}(t)\, dt$$

and

$$[f(\zeta)]_- = \int_{-\infty}^{0} e^{i\zeta t} \hat{f}(t)\, dt.$$

[37]See, for instance, Chap. 8, p. 257 of Ref. 5.

[38]F. Dyson (private communication).

[39]See, for example, Appendix B of Ref. 5.

[40]See, for example, G. H. Hardy, *Divergent Series* (Oxford U.P., London, 1949), p. 333.

[41]G. N. Watson, *A Treatise on the Theory of Bessel Functions* (Cambridge University, London, 1966), Chap. 5.

[42]C. A. Tracy and B. M. McCoy, Phys. Rev. B 12, 368 (1975). The quantity C_9 given in Table I should have a minus sign. The correct value was used in computing the entries in Table V.

[43]See also D. Abraham, Phys. Lett. A 39, 357 (1972).

[44]See Ref. 41, Chap. XIII.

[45](a) J. W. Essam and M. E. Fisher, J. Chem. Phys. 38, 802 (1963). (b) M. F. Sykes, D. S. Gaunt, J. L. Martin, S. R. Mattingly, J. W. Essam, J. Math. Phys. 14, 1071 (1973).

[46]G. Latta, J. Ration. Mech. Anal. 5, 821 (1956).

[47]Myers considers integral equations of the form

$$\Gamma f = \cos kx, \quad \Gamma g = \sin kx, \quad \Gamma\phi = e^{-ikx\cos\phi},$$

where

$$(\Gamma\psi)(x) = \int_{-1}^{1} H_0(k\,|x - x'|)\psi(x')\, dx'.$$

It is straightforward to convert his theorems to the theorems stated here.

[48]T. Carleman, Math. Z. 15, 111 (1922).

9. Reprinted article 264, listed in Chapter 10:

VOLUME 13, NUMBER 12 PHYSICAL REVIEW LETTERS 21 SEPTEMBER 1964

region of 300 Mc/sec[8] indicate a value of γ of about 2.6. The frequency distribution of the 18 showers as a function of the total effective number of sparks in the shower is shown in Fig. 3. In order to ascertain if this distribution is compatible with that to be expected for primary electrons with a geomagnetic cutoff at 4.5 BV, spectra of the type $kE^{-\gamma}$, with values of γ between 2 and 3.5, have been converted to shower-size distributions, using the mean value and spread of the distributions obtained during calibration measurements on 3, 4.5, 6, and 8 BeV/c electrons.[6] The experimental distribution is found to be compatible with the assumption that all but one [event A of Fig. 3(a)] of the electrons have momenta above the geomagnetic cutoff value and follow roughly the type of spectrum assumed. It is clear, however, that, with the present statistics, this distribution cannot be used to obtain a precise value of γ. A reliable determination of the energy spectrum requires much larger statistics and, as is planned, measurements of intensity at various latitudes during quite sun conditions.

The electron intensity measured here (which includes particles of both signs) would correspond, on the assumption of an energy spectrum of the type $kE^{-2.5}$ and of a magnetic field of 3×10^{-6} gauss, to a power of synchrotron emission of $(1.7\pm0.5)\times10^{-40}$ erg cm^{-3} Hz^{-1} sec^{-1} at a frequency of 10^9 cps, the critical frequency corresponding to 4.5 BeV/c.

We wish to thank, for their generous assistance, Professor Blamont, Mr. Regipa, and Dr. Lambert in the balloon flight; the groups from College de France, Saclay, and CERN in the calibration measurements; and Professor Labeyrie and Professor Occhialini throughout the work. The Italian group is indebted to the Consiglio Nazionale delle Ricerche for financial support.

[1]J. A. Earl, Phys. Rev. Letters 6, 125 (1961).
[2]P. Meyer and R. Vogt, Phys. Rev. Letters 6, 193 (1961).
[3]H. Alfvén and N. Herlofson, Phys. Rev. 78, 616 (1950); R. O. Kiepenheuer, Phys. Rev. 79, 738 (1950); V. L. Ginzburg, Dokl. Akad. Nauk. SSSR 76, 377 (1951).
[4]B. Agrinier, Y. Koechlin, B. Parlier, G. Boella, G. Degli Antoni, C. Dilworth, L. Scarsi, and G. Sironi, L'Ond. 432, 317 (1963).
[5]B. Agrinier, Y. Koechlin, B. Parlier, G. Boella, G. Degli Antoni, C. Dilworth, L. Scarsi, and G. Sironi, Congressino di Frascati, May 1963, Report No. INFN/63/54 (unpublished).
[6]B. Agrinier, G. Boella, G. Degli Antoni, C. Dilworth, Y. Koechlin, B. Parlier, L. Scarsi, and G. Sironi, to be published.
[7]C. J. Bland, Proceedings of the COSPAR Conference, Florence, Italy, May 1964 (to be published).
[8]J. E. Baldwin, Suppl. J. Phys. Soc. Japan 17, 173 (1962); A. J. Turtle, J. F. Pugh, S. P. Kenderdine, and I. I. K. Pauling-Toth, Monthly Notices Roy. Astron. Soc. 124, 297 (1962).

PHENOMENOLOGICAL ANALYSIS OF VIOLATION OF CP INVARIANCE IN DECAY OF K^0 AND $\bar{K}^0$†

Tai Tsun Wu* and C. N. Yang‡
Brookhaven National Laboratory, Upton, New York
(Received 18 August 1964)

1. It was recently discovered[1] that the long-lived component K_L^0 of K^0-$\bar{K}^0$ decays into the $\pi^+\pi^-$ mode. Now if CP invariance holds, the $CP=+1$ and $CP=-1$ components of K^0-$\bar{K}^0$ decay independently. The $\pi^+\pi^-$ mode in the S-wave state has $CP=1$. Hence either the short-lived component K_S^0, or K_L^0, does not decay into $\pi^++\pi^-$, in contradiction to the new discovery.

Accepting the experimental result of reference 1, one is thus forced to the conclusion that CP invariance is violated in K^0-$\bar{K}^0$ decay, as explicitly stated in reference 1. Notice that this conclusion is independent of the details of the Weisskopf-Wigner formulation[2] of decay amplitudes, as applied to the K_0-$\bar{K}_0$ case by Lee, Oehme, and Yang,[3] whose notation we shall follow.[4] (In particular, small corrections to the exponential decay rule of the formalism cannot alter the conclusion that CP invariance is violated.)

In the present note we shall analyze the decay properties of K^0-$\bar{K}^0$, mostly from the phenomenological viewpoint. Possible further experiments will be discussed for their theoretical significance.

We shall assume CPT invariance, the validity of the Weisskopf-Wigner formulation,[2,3] and that for the strong and electromagnetic interactions, separate C, P, and T invariance hold.

In the next five sections, we shall also assume that electromagnetic interactions can be neglected, and that isotopic spin is conserved for the strong interactions. We shall come back to the electromagnetic effects in Sec. 7.

2. The experimental decay rates are tabulated in Table I. (We thank P. Franzini, J. Steinberger, and W. Willis for supplying the entries.)

To analyze the decay of K^0-$\bar{K}^0$ we consider the decay matrix

$$\Gamma = \Gamma_0 + \Gamma_2 + \Gamma_l + \Gamma_{3\pi}, \tag{1}$$

and the mass matrix

$$M = M_0 + M_2 + M_l + M_{3\pi} + \cdots, \tag{2}$$

as sums of contributions from the $\pi\pi(I=0)$, $\pi\pi(I=2)$, leptonic, and 3π modes. One has, in the notation of reference 3,

$$\Gamma_0 = \begin{pmatrix} A_0^2 & A_0^2 \\ A_0^2 & A_0^2 \end{pmatrix}, \tag{3}$$

$$\Gamma_2 = \begin{pmatrix} A_2A_2^* & A_2^2 \\ A_2^{*2} & A_2A_2^* \end{pmatrix}, \tag{4}$$

$$\Gamma_l = \begin{pmatrix} \alpha_l & x_l + iy_l \\ x_l - iy_l & \alpha_l \end{pmatrix}, \tag{5}$$

and

$$\Gamma_{3\pi} = \begin{pmatrix} \alpha_{3\pi} & x_{3\pi} + iy_{3\pi} \\ x_{3\pi} - iy_{3\pi} & \alpha_{3\pi} \end{pmatrix}, \tag{6}$$

Table I. Experimental decay rates in $10^6\ \mathrm{sec}^{-1}$.

Mode	K_S^0	K_L^0
$\pi^+ + \pi^-$	$\frac{2}{3}\times 1.1\times 10^4$	2.6×10^{-2}
$\pi^0 + \pi^0$	$\frac{1}{3}\times 1.1\times 10^4$	Not known
Leptons	~11	~11
$\pi^+ + \pi^- + \pi^0$	$\lesssim 2$	~ 2
$3\pi^0$	$\lesssim 4$[a]	~ 4
All modes	1.1×10^4	~18

[a]No available experimental information; the number given is based on the assumption that the $|\Delta| = \frac{1}{2}$ rule is approximately valid.

where A_0 and A_2 are the decay amplitudes of

$$K^0 \to \pi + \pi \ (I = 0 \text{ standing wave})$$

and

$$K^0 \to \pi + \pi \ (I = 2 \text{ standing wave}),$$

respectively. We have chosen the phase of K^0 so that

$$A_0 = \text{real} > 0. \tag{7}$$

We emphasize that this choice, which is always possible, serves to define the phase of K and $\bar{K}$. The quantities p and q are given by

$$p^2 = A_0^2 + A_2^2 + x_l + iy_l + x_{3\pi} + iy_{3\pi} + iM_r - M_i,$$

and

$$q^2 = A_0^2 + A_2^{*2} + x_l - iy_l + x_{3\pi} - iy_{3\pi} + iM_r + M_i, \tag{8}$$

with the real parts of p and q chosen ≥ 0. In Eq. (8) $M_r + iM_i = M_{12}$ is an off-diagonal element of M. The eigenstates and the eigenvalues of $\Gamma + iM$ were given by Eqs. (28) and (29) of reference 3. The decay amplitudes of K_S and K_L into π-π states are easily constructed, and are tabulated in Table II. The quantity F is

$$F = \exp[i(\delta_2 - \delta_0)], \tag{9}$$

where δ_2 and δ_0 are the π-π S-wave scattering phase shifts for the $I=2$ and $I=0$ states at the energy of the rest mass of K^0. Obviously,

$$R[K_S^0 \to \pi + \pi] + R[K_L^0 \to \pi + \pi] = 2A_0^2 + 2A_2A_2^* \sim 1.1\times 10^4,$$

$$R[K_S^0 \to \text{lep}] + R[K_L^0 \to \text{lep}] = 2\alpha_l \sim 22,$$

and

$$R[K_S^0 \to 3\pi] + R[K_L^0 \to 3\pi] = 2\alpha_{3\pi} \sim 12. \tag{10}$$

The following quantities are of intrinsic experimental interest:

$$\eta_{+-} = a_{+-}^{\ L}/a_{+-}^{\ S}, \quad \eta_{00} = a_{00}^{\ L}/a_{00}^{\ S}. \tag{11}$$

We shall also use

$$\epsilon = (p - q)/p. \tag{12}$$

These quantities are useful because they are small parameters, as we shall see later.

3. The following remarks serve to orient further analysis:

(a) If $\mathrm{Im}A_2 = 0$, $y_l = 0$, $y_{3\pi} = 0$, and $M_i = 0$, then $p = q$, and Table II shows that $K_L \not\to \pi + \pi$, in con-

VOLUME 13, NUMBER 12 PHYSICAL REVIEW LETTERS 21 SEPTEMBER 1964

Table II. Decay amplitudes of K_S and K_L into π-π states.

State	Amplitudes[a]
$I=0$ (standing wave)	$a_0 = A_0(pp^* + qq^*)^{-1/2}(p \pm q)$
$I=2$ (standing wave)	$a_2 = (pp^* + qq^*)^{-1/2}(A_2 p \pm A_2^* q)$
$\pi^+ + \pi^-$ (outgoing wave)	$a_{+-} = (pp^* + qq^*)^{-1/2}\{[(\frac{2}{3})^{1/2}A_0 + (\frac{1}{3})^{1/2}A_2F]p \pm [(\frac{2}{3})^{1/2}A_0 + (\frac{1}{3})^{1/2}A_2^*F]q\}$
$\pi^0 + \pi^0$ (outgoing wave)	$a_{00} = (pp^* + qq^*)^{-1/2}\{[(\frac{1}{3})^{1/2}A_0 - (\frac{2}{3})^{1/2}A_2F]p \pm [(\frac{1}{3})^{1/2}A_0 - (\frac{2}{3})^{1/2}A_2^*F]q\}$

[a]Upper sign for K_S, lower sign for K_L.

tradiction with the experimental result of reference 1.

(b) If $\Delta Q = \Delta S$ for the leptonic decay modes of K^0-$\bar{K}^0$, then Γ_l and M_l of Eqs. (1) and (2) are both multiples of the unit matrix. Therefore, $y_l = 0$. The leptonic mode does not in this case contribute to CP violation as observed in reference 1, even though the lepton mode itself could violate CP invariance. We shall, however, not make the assumption that $\Delta Q = \Delta S$ in this paper.

(c) Phenomenologically, it is not possible to distinguish between the four M's on the right-hand side of Eq. (2). In other words, measurable quantities can only depend on M, but not on M_0, M_2, M_l, or $M_{3\pi}$ separately.

(d) If $\mathrm{Im}A_2 = 0$, then Table II gives directly

$$a_{+-}^{\ \mathrm{L}}/a_{+-}^{\ \mathrm{S}} = a_{00}^{\ \mathrm{L}}/a_{00}^{\ \mathrm{S}}. \tag{13}$$

(e) If $M_i = 0$, and $y_l + y_{3\pi} = 0$, then $p^2 - q^2 = A_2^{\ 2} - A_2^{*2}$. It will be clear in Sec. 4 that

$$|a_0^{\ \mathrm{L}}/a_2^{\ \mathrm{L}}| = O(\mathrm{Re}A_2/A_0), \tag{14}$$

provided that $|A_2|/A_0$ is small.

(f) If $\mathrm{Im}A_2 = 0$ and $M_i = 0$, then, given the experimental decay rates of K^0-$\bar{K}^0 \to$ leptons and 3π, and of $K_S \to \pi + \pi$, the rate $K_L \to \pi^+ + \pi^-$ is at most $1.75[1 + (M_\gamma/A_0^{\ 2})^2]^{-1} \times 10^{-2}$. This is too low to account for the experimental result of reference 1. We shall discuss this in more detail in Sec. 5.

Accordingly, roughly there are four ways to violate CP invariance in the decay of K^0-$\bar{K}^0$, namely, $\mathrm{Im}A_2 \neq 0$, $y_l \neq 0$, $y_{3\pi} \neq 0$, and/or $M_i \neq 0$. They correspond to CP (or T) noninvariance due to the interference between the dominant $\pi\pi(I=0)$ mode and (α) the $\pi\pi(I=2)$ mode ($\mathrm{Im}A_2/A_0 \neq 0$); ($\beta$) the lepton mode ($y_l \neq 0$), ($\gamma$) the 3π mode ($y_{3\pi} \neq 0$), and/or (δ) the off-energy shell contributions to the $K \to \bar{K}$ elements of the mass operator $M(M_i \neq 0)$. This possibility has been discussed by Sachs and Treiman.[5]

According to (f) above, (β) and (γ) together by themselves are too small to account for the magnitude of the observed effect. Thus the more important contribution to the observed CP violation has to come from (α) and/or (δ). [Theoretically it is, of course, to be expected that if any interference of the type (α), (β), or (γ) is present, then an interference of type (δ) is also present, in general.]

4. The $|\Delta I| = \frac{1}{2}$ rule is well verified in general, and for the $K \to \pi + \pi$ decay in particular. Thus, $|A_2/A_0| \ll 1$. Dropping A_2/A_0 in Table II, we obtain

$$\eta_{+-} \sim (p-q)/(p+q) = \epsilon/(2+\epsilon). \tag{15}$$

Thus the experimental small value of $|\eta_{+-}|$ shows that $|\epsilon| \ll 1$.

We proceed to expand various quantities to the lowest nonvanishing order of ϵ and A_2/A_0:

$$\eta_{+-} = \tfrac{1}{2}[\epsilon + (2)^{1/2} iF\, \mathrm{Im}A_2/A_0], \tag{16}$$

$$\eta_{00} = \tfrac{1}{2}[\epsilon - 2(2)^{1/2} iF\, \mathrm{Im}A_2/A_0], \tag{17}$$

and

$$R[K_S^{\ 0} \to \pi^+ + \pi^-] - 2R[K_S^{\ 0} \to \pi^0 + \pi^0] = 2\sqrt{2}[\mathrm{Re}A_2/A_0]\cos(\delta_2 - \delta_0) \times R[K_S^{\ 0} \to \pi + \pi(I=0)]. \tag{18}$$

Furthermore, since Γ_l is positive definite, $\alpha_l \geq |x_l|$, $\alpha_l \geq |y_l|$; or by Eq. (10),

$$|x_l| \lesssim 11, \quad |y_l| \lesssim 11. \tag{19}$$

Similarly,

$$|x_{3\pi}| \lesssim 6, \quad |y_{3\pi}| \lesssim 6. \tag{20}$$

Thus these elements are negligible compared with $A_0^{\ 2}$. {In more general cases, we can establish a somewhat better bound for y. For any mode C,

$$y_C^2 \leq R[K_S^0 \to C] R[K_L^0 \to C].\}$$

Using Eq. (8) one sees that $|\epsilon| \ll 1$ implies $M_i \ll A_0^2$. Thus one has an approximate expression for the difference of the two eigenvalues of $\Gamma + iM$:

$$\lambda_+ - \lambda_- = 2A_0^2 + 2iM_\gamma. \qquad (21)$$

Therefore,

$$M_\gamma = m_S - m_L. \qquad (22)$$

Using these one obtains from Eq. (8)

$$\epsilon = \frac{-M_i + i(y_l + y_{3\pi})}{A_0^2 + i(m_S - m_L)}. \qquad (23)$$

Equations (16)–(23) form the basis of a phenomenological analysis.

5. If $\mathrm{Im}A_2 = 0$ and $M_i = 0$, then it follows from Eqs. (16) and (23) that

$$R[K_L^0 \to \pi^+ + \pi^-]$$
$$= 6(y_l + y_{3\pi})^2 \times 10^{-5}[1 + (M_\gamma/A_0^2)^2]^{-1}. \qquad (24)$$

Statement 3(f) then follows from Eqs. (19) and (20). This is too small by a factor of 3 or 15 for[6] $|m_S - m_L| = 1/2\tau_1$ or $3/2\tau_1$. {If, moreover, we believe that the $\Delta Q = -\Delta S$ matrix element is at most 50% of that for $\Delta Q = \Delta S$ as a result of the Paris experiment,[7] then $(x_l^2 + y_l^2)^{1/2}/\alpha_l < 2(\frac{1}{2})/[1 + (\frac{1}{2})^2] = \frac{4}{5}$. Hence, $|y_l| < 9$. Thus we can strengthen the argument by about 20%.}

6. The amplitude ratios η_{+-} and η_{00} are experimentally measurable quantities. The experiment of reference 1 gives

$$|\eta_{+-}| = 2.1 \times 10^{-3}(1 \pm 0.1). \qquad (25)$$

A measurement[8] of $R[K_L^0 \to \pi^0 + \pi^0]$ would yield $|\eta_{00}|$. If one introduces further assumptions, this rate can be predicted. For example, according to Sec. 3, (d) and (e),

if $\mathrm{Im}A_2 = 0$, then $R[K_L^0 \to \pi^0 + \pi^0] = 1.3 \times 10^{-2}$;

if $M_i = y_l + y_{3\pi} = 0$, then

$$R[K_L^0 \to \pi^0 + \pi^0] = 5.2 \times 10^{-2}. \qquad (26)$$

Existing experiments yield only a rough upper bound:

$$R[K_L^0 \to \pi^0 + \pi^0] < (\sim 1), \quad |\eta_{00}| < (\sim 2 \times 10^{-2}).$$

Without additional assumptions, there are no theoretical arguments that $R[K_L^0 \to \pi^0 + \pi^0]$ cannot be as large as this experimental upper bound.

It is clear that in order to measure the phases of the amplitude ratios η_{+-} and η_{00}, interference between K_L and K_S decays into these modes must be studied. To obtain greater sensitivity, the intensity ratio of K_L and K_S must be such that their decay amplitudes into these modes are about comparable.[9] The relative phase θ between the K_L and K_S beams must also be known in order to determine the phase of η_{+-} (or of η_{00}). But to determine the difference of the phases of η_{+-} and η_{00} it is not necessary to know θ.

It is convenient to construct a diagram of the complex numbers η_{+-}, η_{00}, ϵ, and iF related through Eqs. (16) and (17), as shown in Fig. 1.

(a) If η_{+-} and η_{00} are completely measured, the quantities ϵ and $F\,\mathrm{Im}A_2/A_0$ are known. Thus $\delta_2 - \delta_0$ is measured up to $\pm n\pi$, and $\mathrm{Im}A_2/A_0$ is known up to a sign. If further $m_S - m_L$ is known, then through (23), M_i and $y_l + y_{3\pi}$ are determined.

(b) If $|\eta_{+-}|$, $|\eta_{00}|$, and the phase difference

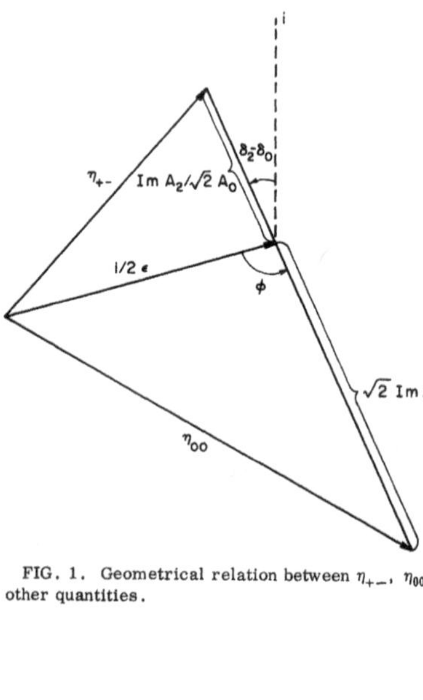

FIG. 1. Geometrical relation between η_{+-}, η_{00}, and other quantities.

of η_{+-} and η_{00} are known, then the triangle of Fig. 1 can be constructed, but its orientation relative to the real axis is known only if $\delta_2-\delta_0$ is independently obtained.

These experiments, however, do not yield any information on $\mathrm{Re}A_2/A_0$. It seems that the only experimental method of determining this ratio is to measure the rate difference on the left-hand side of Eq. (18). Existing experiment[10] gives

$$\mathrm{Re}A_2=(0\pm1)\sec(\delta_2-\delta_0). \tag{27}$$

7. In the above discussion, electromagnetic effects are completely neglected. Inclusion of these effects introduces (a) mass splits between $\pi^+\pi^-$ and $\pi^0\pi^0$ states (and related effects), and (b) additional channels like $\pi\pi\gamma$. To account for (a) one introduces two eigenstates of the S matrix for the strong and electromagnetic interactions of the $\pi\pi$ S-wave state at the K^0 mass. The resultant change comprises only small real corrections to the coefficients $(\frac{2}{3})^{1/2}$ and $(\frac{1}{3})^{1/2}$ and the phase shifts δ_2 and δ_0 in Table II.

As to (b), electromagnetic effects do not introduce CP noninvariance. Thus it is reasonable to expect additional channels, such as $\pi\pi\gamma$, not to introduce matrix elements which are imaginary in phase relative to A_0. In any case, experimentally the rates $K_{\mathrm{S,\,L}}\to\pi+\pi+\gamma$ are limited:

$$R[K_{\mathrm{S}}\to\pi^++\pi^-+\gamma]<(\sim1)$$

(from Kirsch et al.[11]) and

$$R[K_{\mathrm{L}}\to\pi+\pi+\gamma]<(\sim1)$$

(from total rate). It is reasonable to assume

$$R[K_{\mathrm{S}}\to\pi^0+\pi^0+\gamma]<(\sim1).$$

By an argument similar to that leading to Eq. (19), we have then

$$|y_{\pi\pi\gamma}|<(\sim1.5).$$

Thus electromagnetic effects are expected to be negligible in the discussion of CP noninvariances of the preceding sections.

8. We now make two supplementary remarks. One may raise the question of the evidence of T invariance in neutron[12] and Λ[13] decays and perhaps other decays. To reconcile this evidence with the CP noninvariance of reference 1, it may be that the weak interaction consists of two terms:

$$H=H_{\mathrm{strong}}+H_{\mathrm{el}}+H_{W1}+H_{W2},$$

where (a) $H_{\mathrm{strong}}+H_{\mathrm{el}}+H_{W1}$ satisfies CP invariance, (b) $H_{\mathrm{strong}}+H_{W1}$ obeys the $|\Delta I|=\frac{1}{2}$ rule, and (c) H_{W2} is weaker than H_{W1}, and violates the $|\Delta I|=\frac{1}{2}$ rule.

Another question concerns the $\pi\pi(I=2)$ decay rates of K^+ and of K^0. From Eqs. (16) and (17),

$$|\mathrm{Im}A_2|=\tfrac{1}{3}\sqrt{2}A_0|\eta_{+-}-\eta_{00}|<(\sim\tfrac{2}{3}). \tag{28}$$

If the decay amplitude of $K^+\to\pi+\pi$ $(I=2)$ is $A_2^{\ +}$, the experimental rate of this process gives

$$|A_2^{\ +}|=3.9. \tag{29}$$

Comparing Eqs. (27), (28), and (29) one concludes that more accurate measurements of $R[K_{\mathrm{S}}\to\pi^++\pi^-]-2[K_{\mathrm{S}}\to\pi^0+\pi^0]$, and information about $|\eta_{00}|$ and $\delta_2-\delta_0$ can be used to analyze the $K\to\pi+\pi$ $(I=2)$ amplitude into $|\Delta I|=\frac{3}{2}$ and $|\Delta I|=\frac{5}{2}$ components.

It is a pleasure to acknowledge the hospitality we enjoyed at Brookhaven National Laboratory where this work was done. We thank J. W. Cronin, V. L. Fitch, P. Franzini, J. Steinberger, and W. Willis for many stimulating discussions.

For other recent discussions of CP noninvariance in K^0-$\bar{K}^0$ decay, see a recent article by Sachs.[14]

---

[1]J. H. Christenson, J. W. Cronin, V. L. Fitch, and R. Turlay, Phys. Rev. Letters 13, 138 (1964).

[2]V. F. Weisskopf and E. P. Wigner, Z. Physik 63, 54 (1930); 65, 18 (1930).

[3]T. D. Lee, R. Oehme, and C. N. Yang, Phys. Rev. 106, 340 (1957).

[4]To avoid confusion we shall call the long- and short-lived components K_{L} and K_{S}. If their wave functions are called Ψ_{L} and Ψ_{S}, then, in the notation of reference 3, $\Psi_{\mathrm{L}}=\Psi_-$ and $\Psi_{\mathrm{S}}=\Psi_+$. Throughout this paper we use the unit 10^6 sec^{-1} for decay rates.

[5]R. G. Sachs and S. B. Treiman, Phys. Rev. Letters 8, 137 (1962).

[6]W. F. Fry, International Conference on Fundamental Aspects of Weak Interactions (Brookhaven National Laboratory, Upton, New York, 1963), p. 3; J. H. Christenson, J. W. Cronin, V. L. Fitch, and R. Turlay, ibid., p. 74.

[7]B. Aubert, L. Behr, J. P. Lowys, P. Mittner, and C. Pascaud, Phys. Letters 10, 215 (1964). The large ratio of e^+/e^- in the first few lifetimes of K_{S} is a strong argument that $|x|$ is quite a bit less than 1.

[8]W. Willis has been considering, independently of us, the possibility of measuring this rate.

[9]J. Steinberger has reached this same conclusion independently of us.

[10]J. L. Brown, J. A. Kadyk, G. H. Trilling, B. P. Roe, D. Sinclair, and J. C. Van der Velde, Phys. Rev. 130, 769 (1963).
[11]L. Kirsch, R. J. Plano, J. Steinberger, and P. Franzini, Phys. Rev. Letters 13, 35 (1964).
[12]M. T. Burgy, V. E. Krohn, T. B. Novey, G. R. Ringo, and V. L. Telegdi, Phys. Rev. 120, 1829 (1960).
[13]J. W. Cronin and O. E. Overseth, Phys. Rev. 129, 1795 (1963).
[14]R. G. Sachs, Phys. Rev. Letters 13, 286 (1964).

ERRATUM

AMPLIFICATION OF MICROWAVE PHONONS IN GERMANIUM. M. Pomerantz [Phys. Rev. Letters 13, 308 (1964)].

The following should appear as a footnote to the title: "This research was supported in part by the U. S. Army Electronics Laboratory, Fort Monmouth, New Jersey."

In the third paragraph from the end, the word "amplification" is to be understood to mean amplification constant, α.

10. Reprinted article 302, listed in Chapter 10:

Volume 103B, number 2 PHYSICS LETTERS 16 July 1981

SINGLE BREMSSTRAHLUNG PROCESSES IN GAUGE THEORIES☆

F.A. BERENDS and R. KLEISS
Instituut-Lorentz, Leiden, The Netherlands

P. De CAUSMAECKER[1] and R. GASTMANS[2]
Instituut voor Theoretische Fysika, University of Leuven, B-3030 Leuven, Belgium

and

Tai Tsun WU[3]
Gordon McKay Laboratory, Harvard University, Cambridge, MA 02138, USA

Received 14 April 1981

We find that, for QED and for SU(N) gauge theories, single hard bremsstrahlung cross sections are remarkably simple in the ultrarelativistic limit. They can be written as a product of two factors: one factor is connected with the lowest order, elastic process, and the second one is related to the infrared factor which describes soft quantum emission. We present explicit formulae for various processes.

In both classical and quantum physics remarkable simplifications have been found repeatedly in the limit of small wavelengths. The most striking example is optics where the solution of Maxwell's equations reduces to the determination of rays. A second example is the amplitude in the so-called high-energy scattering [1], extensively studied a quarter of a century ago. Another example is the scattering amplitude in quantum field theory at high energies, for both Regge [2] and pomeron [3] exchanges. It is the purpose of this paper to present another surprising simplification of a rather different nature for a class of *cross sections* in quantum field theory at ultrarelativistic energies.

The motivation for the present investigation comes from experiments in two different ways. On one hand it is the discovery of the gluon jet [4] at PETRA and on the other hand the required thorough knowledge of radiative corrections [5] for luminosity determinations and QED tests. At such high energies the wavelength is small compared with those characterizing the masses of the electron, the muon and even the τ lepton, while for QCD the masses of gluons vanish and those for the u, d, s and c quarks are again small. We are thus lead to study the cross sections for reactions from two bodies to three bodies for gauge theories in the ultrarelativistic limit.

More explicitly, the processes that we study involve QED and/or QCD, and include

(1) $e^+e^- \to e^+e^-\gamma$, (2) $e^+e^- \to \mu^+\mu^-\gamma$, (3) $e^+e^- \to \gamma\gamma\gamma$, (4) $q\bar{q} \to q\bar{q}g$, (5) $q\bar{q} \to q\bar{q}\gamma$,

(6) $q\bar{q} \to q'\bar{q}'g$, (7) $q\bar{q} \to q'\bar{q}'\gamma$, (8) $\bar{q}q \to ggg$, (9) $q\bar{q} \to gg\gamma$, (10) $q\bar{q} \to g\gamma\gamma$,

(11) $q\bar{q} \to \gamma\gamma\gamma$, (12) $gg \to ggg$, (13) $e^+e^- \to q\bar{q}g$, (14) $e^+e^- \to q\bar{q}\gamma$,

☆ Work supported in part by the NATO, Research Grant, No. RG 079.80.
[1] Wetenschappelijk medewerker IIKW, Belgium.
[2] Onderzoeksleider, NFWO, Belgium.
[3] Work supported in part by the US Department of Energy, contract number EY-76-S-02-3227.

 0 031-9163/81/0000-0000/$ 02.50

Volume 103B, number 2 PHYSICS LETTERS 16 July 1981

with q ≠ q′. The complete list of eighty processes is obtained from these fourteen by crossing.

Our main result is that, when no two four-momenta are nearly parallel and the masses are set to zero, these cross sections can be written as the product of two factors: one factor is associated with the two-to-two process, while the other one is a generalization of the "infrared factor" for the emission of a soft quantum [6].

Let us clarify this statement with an example: the reaction

$$e^+(p_+) + e^-(p_-) \to \mu^+(q_+) + \mu^-(q_-) + \gamma(k), \tag{1}$$

where the four-momenta of the particles are written between parentheses. We also introduce the notation:

$$s = (p_+ + p_-)^2, \quad t = (p_+ - q_+)^2, \quad u = (p_+ - q_-)^2,$$
$$s' = (q_+ + q_-)^2, \quad t' = (p_- - q_-)^2, \quad u' = (p_- - q_+)^2, \tag{2}$$

and use the metric (+ – – –) and a $2m$ spinor normalization. By explicit calculation of the hard photon bremsstrahlung process it has been found that the matrix element squared summed and averaged over the spins becomes

$$|M_\mu|^2 = e^4 [(t^2 + t'^2 + u^2 + u'^2)/ss'] S, \tag{3}$$

with

$$S = e^2 \left(\frac{s}{(p_+k)(p_-k)} + \frac{s'}{(q_+k)(q_-k)} - \frac{t}{(p_+k)(q_+k)} - \frac{t'}{(p_-k)(q_-k)} + \frac{u}{(p_+k)(q_-k)} + \frac{u'}{(p_-k)(q_+k)} \right)$$

$$= \frac{e^2}{4} \frac{2ss'(t+t') + 2tt'(s+s') + u(st + s't') + u'(st' + s't)}{(p_+k)(p_-k)(q_+k)(q_-k)}. \tag{4}$$

It can be seen that S is the factor associated with soft bremsstrahlung. If we actually take the limit $k \to 0$ we obtain the infrared factor as given by Yennie et al. [6]. The remainder of the expression reduces to the square of the nonradiative amplitude

$$\lim_{k \to 0} e^4(t^2 + t'^2 + u^2 + u'^2)/ss' = 2e^4(t^2 + u^2)/s^2 = |M_0|^2. \tag{5}$$

The fact that the result for hard bremsstrahlung is separable into two factors, one of which is actually the infrared factor without momentum conservation imposed, and the other one is formally similar to the nonradiative result, also holds true for many other reactions.

For the process

$$e^+(p_+) + e^-(p_-) \to e^+(q_+) + e^-(q_-) + \gamma(k) \tag{6}$$

we find for the square of the bremsstrahlung amplitude

$$|M_e|^2 = e^4 \{[ss'(s^2 + s'^2) + tt'(t^2 + t'^2) + uu'(u^2 + u'^2)]/ss'tt'\} S. \tag{7}$$

Again the infrared limit is obtained by setting $s' = s, t' = t, u' = u$. For

$$e^+(p_+) + e^-(p_-) \to \gamma(k_1) + \gamma(k_2) + \gamma(k_3) \tag{8}$$

we find

$$|M_\gamma|^2 = e^6 s \sum_{i=1}^{3} a_i b_i (a_i^2 + b_i^2) / a_1 a_2 a_3 b_1 b_2 b_3, \tag{9}$$

with $a_i = (p_+k_i)$, $b_i = (p_-k_i)$, $i = 1, 2, 3$. For this process the infrared factor for $k_3 \to 0$, is simply $e^2 s/(a_3 b_3)$.

The form of eq. (9) makes the symmetry among the photons manifest. The infrared limit is obtained by letting any one of k_1, k_2 or k_3 go to 0.

Volume 103B, number 2 PHYSICS LETTERS 16 July 1981

Of course, when a photon is emitted with a three-momentum parallel to the three-momentum of a charged particle, the above expressions are no longer valid and terms of order m_e^2 (or m_μ^2) must be retained in the numerators. We shall come back to this problem in a forthcoming paper.

These results for the QED processes strongly suggest that also for QCD bremsstrahlung reactions a similar factorization is to be expected. It is again straightforward to obtain the infrared factor, and the remaining terms could then be guessed from the knowledge of the lowest-order process on the basis of symmetry arguments. These conjectured cross sections are then compared with the results of the complete calculation of the hard bremsstrahlung processes by rewriting both expressions in terms of independent kinematical variables. To this end, we make extensive use of the symbolic manipulation program REDUCE [7].

We list our results for $|M|^2$ for some of the reactions. This time, we also sum (average) over all the colour of freedom for the final (initial) states.

For quark–quark scattering with different flavours,

$$\mathrm{q}_1(p_+) + \mathrm{q}_2(p_-) \to \mathrm{q}_1(q_+) + \mathrm{q}_2(q_-) + \mathrm{g}(k), \tag{10}$$

we have

$$\begin{aligned}|M|^2 = {}& \tfrac{1}{8} g^6 [(s^2 + s'^2 + u^2 + u'^2)/tt'] [(p_+k)(p_-k)(q_+k)(q_-k)]^{-1} \\ & \times \{C_1 [(u+u')(ss'+tt'-uu') + u(st+s't') + u'(st'+s't)] \\ & - C_2 [(s+s')(ss'-tt'-uu') + 2tt'(u+u') + 2uu'(t+t')]\}.\end{aligned} \tag{11}$$

Here, g is the SU(N) gauge coupling constant, and

$$C_1 = (N^2-1)^2/4N^3, \quad C_2 = (N^2-1)/4N^3. \tag{12}$$

The expressions inside the square brackets together with the denominators $(p_+k)(p_-k)(q_+k)(q_-k)$ are again infrared factors, but, because of the colour structure of the theory, some of the terms acquire weights which are no longer only ± 1, in contradistinction to the QED case [eq. (4)].

For quark–quark scattering with identical flavours, the formula is somewhat more complicated:

$$\begin{aligned}|M|^2 = {}& \tfrac{1}{8} g^6 [(p_+k)(p_-k)(q_+k)(q_-k)]^{-1} \{\{C_1 [(u+u')(ss'+tt'-uu') + s(tu+t'u') + s'(tu'+t'u)] \\ & - C_2 [(s+s')(ss'-tt'-uu') + 2tt'(u+u') + 2uu'(t+t')]\} [(s^2+s'^2+u^2+u'^2)/tt'] \\ & + \{C_1 [(t+t')(ss'-tt'+uu') + s(tu+t'u') + s'(tu'+t'u)] \\ & - C_2 [(s+s')(ss'-tt'-uu') + 2tt'(u+u') + 2uu'(t+t')]\} [(s^2+s'^2+t^2+t'^2)/uu'] \\ & + \{C_3 [(s+s')(ss'-tt'-uu') + 2tt'(u+u') + 2uu'(t+t')] \\ & + C_4 [(s+s')(ss'-tt'-uu') - 2tt'(u+u') - 2uu'(t+t') - 2s(tu+t'u') - 2s'(tu'+t'u)]\} \\ & \times [(s^2+s'^2)(ss'-tt'-uu')/tt'uu']\},\end{aligned} \tag{13}$$

with

$$C_3 = (N^4-1)/8N^4, \quad C_4 = (N^2-1)^2/8N^4. \tag{14}$$

For the process

$$\bar{\mathrm{q}}(p_+) + \mathrm{q}(p_-) \to \mathrm{g}(k_1) + \mathrm{g}(k_2) + \mathrm{g}(k_3), \tag{15}$$

there is more symmetry and, consequently, the formula is simpler:

$$|M|^2 = g^6 \frac{N^2-1}{4N^4}\left(\sum_1^3 a_i b_i (a_i^2+b_i^2)\Big/ a_1a_2a_3b_1b_2b_3\right)$$

$$\times\left[\frac{s}{2}+N^2\left(\frac{s}{2}-\frac{a_1b_2+a_2b_1}{(k_1k_2)}-\frac{a_2b_3+a_3b_2}{(k_2k_3)}-\frac{a_3b_1+a_1b_3}{(k_3k_1)}\right)\right.$$

$$\left.+\frac{2N^4}{s}\left(\frac{a_3b_3(a_1b_2+a_2b_1)}{(k_2k_3)(k_3k_1)}+\frac{a_1b_1(a_2b_3+a_3b_2)}{(k_3k_1)(k_1k_2)}+\frac{a_2b_2(a_3b_1+a_1b_3)}{(k_1k_2)(k_2k_3)}\right)\right], \qquad (16)$$

where the quantities a_i and b_i, $i = 1,2,3$, are given by eq. (9).

Finally, for

$$g(k_1) + g(k_2) \to g(k_3) + g(k_4) + g(k_5), \qquad (17)$$

we have

$$|M|^2 = \tfrac{1}{2} g^6 [N^3/(N^2-1)]\,[(12345)+(12354)+(12435)+(12453)+(12534)+(12543)$$

$$+(13245)+(13254)+(13425)+(13524)+(14235)+(14325)]$$

$$\times\frac{[(k_1k_2)^4+(k_1k_3)^4+(k_1k_4)^4+(k_1k_5)^4+(k_2k_3)^4+(k_2k_4)^4+(k_2k_5)^4+(k_3k_4)^4+(k_3k_5)^4+(k_4k_5)^4]}{(k_1k_2)(k_1k_3)(k_1k_4)(k_1k_5)(k_2k_3)(k_2k_4)(k_2k_5)(k_3k_4)(k_3k_5)(k_4k_5)}. \qquad (18)$$

We have introduced the notation

$$(ijlmn) = (k_ik_j)(k_jk_l)(k_lk_m)(k_mk_n)(k_nk_i), \qquad (19)$$

with $i, j, l, m, n = 1, ..., 5$. Note the relative simplicity of expression (18), which is the result of the evaluation of 25 Feynman diagrams!

All the presented cross sections have one more feature in common: the absence of double poles. For elastic processes, the squared matrix elements typically have double poles of the type s^{-2}, t^{-2}, or u^{-2}. For single bremsstrahlung processes, although individually squared Feynman diagrams may give rise to a double pole structure, the combination of all contributions with the same double pole always leads to a cancellation. The net result is a single pole structure and a numerator of lesser degree in the kinematical variables. This property is largely responsible for the simplicity of the obtained results. A preliminary investigation of double bremsstrahlung processes, such as $e^+e^- \to \mu^+\mu^-\gamma\gamma$ or $e^+e^- \to 4\gamma$, shows, however, that the absence of double poles is a property of single bremsstrahlung processes only, and that double poles reappear for higher-order processes.

Finally, it should be mentioned that all these cross sections were checked numerically against previously published results: $e^+e^- \to \mu^+\mu^-\gamma$ in ref. [8], $e^+e^- \to e^+e^-\gamma$ in ref. [9], $e^+e^- \to \gamma\gamma\gamma$ in ref. [10], and the QCD processes for $N = 3$ in ref. [11] (ignoring a symmetrization factor for our reaction (10), introduced in ref. [11]).

Our calculations show that single bremsstrahlung cross sections have a relatively simple structure, and we expect that our results will greatly simplify the confrontation of QED and QCD with experiment.

We thank W. Troost for several useful discussions, the theory group at DESY, Hamburg, and the Katholieke Universiteit Leuven for their hospitality. One of us (TTW) is also grateful to CERN, Geneva, for their hospitality, where this work was started.

References

[1] L.I. Schiff, Phys. Rev. 103 (1956) 443;
T.T. Wu, Phys. Rev. 108 (1957) 466;
D.S. Saxon and L.I. Schiff, Nuovo Cimento 6 (1957) 614;
R.J. Glauber, in: Lectures in theoretical physics, eds. W.E. Brittin and L.G. Dunham (Interscience, 1958) p. 315.

[2] T. Regge, Nuovo Cimento 14 (1959) 951;
B.W. Lee and R.F. Sawyer, Phys. Rev. 127 (1962) 2266;
J.D. Bjorken and T.T. Wu, Phys. Rev. 130 (1963) 2566;
B.M. McCoy and T.T. Wu, Phys. Rev. D13 (1976) 369.
[3] H. Cheng and T.T. Wu, Phys. Rev. Lett. 22 (1969) 666; 23 (1969) 1311; 24 (1970) 759, 1456.
[4] B.H. Wiik, Proc. Intern. Neutrino Conf. (Bergen, Norway, June 1979) p. 113;
P. Söding, Proc. EPS Intern. Conf. on High-energy physics (CERN, June 1979) p. 271;
see also the reports given in: T.B.W. Kirk and H.D.I. Abarbanel, eds., Proc. 1979 Intern. Symp. on Lepton and photon interactions at high energies (FNAL, August 1979).
[5] F.A. Berends and R. Gastmans, in: Electromagnetic interactions of hadrons, Vol. 2, eds. A. Donnachie and G. Shaw (Plenum, 1978) p. 471;
F.A. Berends and R. Kleiss, Nucl. Phys. B 177 (1981) 237.
[6] D.R. Yennie, S.C. Frautschi and H. Suura, Ann. Phys. (NY) 13 (1961) 379.
[7] A.C. Hearn, REDUCE 2, User's Manual, Univ. of Utah (1973).
[8] F.A. Berends, K.J.F. Gaemers and R. Gastmans, Nucl. Phys. B57 (1973) 381.
[9] S.M. Swanson, Phys. Rev. 154 (1967) 1601.
[10] F. Mandl and T.H.R. Skyrme, Proc. R. Soc. A215 (1952) 497.
[11] T. Gottschalk and D. Sivers, Phys. Rev. D21 (1980) 102;
these QCD processes were also calculated by:
J. Kripfganz and A. Schiller, Phys. Lett. 79B (1978) 317;
A. Schiller, J. Phys. G 5 (1979) 1329;
C.J. Maxwell, Nucl. Phys. B149 (1979) 61;
Z. Kunszt and E. Pietarinen, Nucl. Phys. B164 (1980) 45.

11. Reprinted article 397, listed in Chapter 11:

OPEN ACCESS
International Journal of Modern Physics A
Vol. 38, No. 28 (2023) 2350128 (36 pages)

DOI: 10.1142/S0217751X23501282

Concept of the basic standard model and a relation between the three gauge coupling constants

Tai Tsun Wu* and Sau Lan Wu
Department of Physics,
University of Wisconsin-Madison,
Madison, WI 53706, USA
**ttwu.harvard@gmail.com*

Received 17 August 2023
Accepted 21 August 2023
Published 9 November 2023

Among the known elementary particles, 15 are not massless. In the Lagrangian of the standard model, the Higgs particle couples to all 15 of them to give them their respective masses. In this paper, a new model is proposed, where in the Lagrangian, the Higgs particle couples only to the four heaviest known elementary particles t, H, Z and W, all with masses in the vicinity of $100\,\mathrm{GeV/c^2}$, but not to the other 11. In this model, the masses of these 11 elementary particles that do not couple directly to the Higgs particle are generated by radiative corrections. Because of the reduction of the number of terms in the Lagrangian, this model is referred to as the basic standard model. By studying how the bottom quark can acquire a mass, it is found that there must be a relation between the three gauge coupling constants of the gauge group $U(1) \times SU(2) \times SU(3)$. This relation between these coupling constants is the first prediction, on purely theoretical grounds without using any experimental input, of the basic standard model, and is studied in some detail. In particular, this new prediction is found to be in good agreement with experimental measurements.

Keywords: Quantum field theory; Yang–Mills non-Abelian gauge theory; standard model; quantum chromodynamics.

PACS numbers: 03.70.+k, 12.10.-g, 14.80.Bn, 12.28.-t

1. Introduction

The most important and influential paper in theoretical physics during the second half of the 20th century is that of Yang and Mills.[1] The importance of this paper on the non-Abelian gauge theory is due to its pointing out the direction for the later development and the understanding of particle physics.

*Corresponding author.

A major application of the Yang–Mills theory is the formulation of the standard model of Glashow,[2] Weinberg[3] and Salam,[4] where the non-Abelian group is $U(1) \times SU(2) \times SU(3)$. It takes into account all elementary particles that are well established experimentally, and is a most successful theory in its ability to make numerous accurate predictions of physical quantities.

An immediate consequence of this success is that there are a large number of dimensionless parameters in this standard model: 14 mass ratios, three gauge coupling constants, four Kobayashi–Maskawa parameters,[5] and four leptonic Kobayashi–Maskawa parameters. Note that the values of the last four parameters are at present completely unknown.

We are therefore in a rather uncomfortable situation: on the one hand, the standard model is a very successful physics theory, but, on the other hand, it seems to have too many dimensionless parameters. With this description of the dilemma, the most obvious way out is that these 25 dimensionless parameters are not independent.

How can these parameters not be independent?

This question has been raised and discussed during the celebration of Yang's hundredth birthday.[6] In this discussion, the electromagnetic interaction described by the fine-structure constant was neglected. This paper is a continuation of this paper, but the treatment here is more systematic and the electromagnetic interaction is taken into account.

Following Ref. 6, consider the parameters from the masses: the masses of the known elementary particles fall into three distinct ranges.

(1) Masses in the vicinity of $100\,\mathrm{GeV/c^2}$. There are four such elementary particles: the Higgs boson[7–9] itself, the two massive gauge bosons W and Z,[10–13] and the top quark.[14,15] When the top quark was discovered at Fermilab in 1995 at a mass of $173\,\mathrm{GeV/c^2}$, it was the heaviest known elementary particle; what is remarkable is that, by now after more than a quarter of a century, it is still the heaviest known elementary particle.
(2) Masses in the range of 1–$5\,\mathrm{GeV/c^2}$. There are three such elementary particles: the bottom and charm quarks together with τ, the heaviest lepton. Among these particles, the bottom quark[16] is the heaviest.
(3) Masses around $100\,\mathrm{MeV/c^2}$ or less. All the other elementary particles have masses in this range.

An attractive possibility is therefore that the Higgs boson interacts directly only with the elementary particles listed under (1), namely, the Higgs boson itself, the gauge bosons W and Z, and the top quark. In other words, there is no direct interaction in the Lagrangian density between the Higgs boson and the other elementary particles.[6] Instead, the masses of the particles of classes (2) and (3) are "generated" through radiative corrections.

Motivated by this possibility, we propose in this paper the model described in the following section, and study one of its consequences, the one on the gauge coupling constants.

The main result of this paper is given by Eq. (81) of Sec. 11; it is a new theoretical relation between the three gauge coupling constants of the group $U(1)\times SU(2)\times SU(3)$.

This main result has a number of immediate implications, a few of which are presented in Secs. 13 and 14. For example, for the first time an approximate theoretical determination is given for the strong coupling constant.

2. Basic Standard Model

The case to be studied here is the one where, in the Lagrangian density, the Higgs boson H is coupled to the Higgs boson itself, the W, the Z, and the top quark t only, but not to any of the other elementary particles. Because of the reduced number of couplings, this new theory shall be referred to as the basic standard model.

Since, at the beginning, there is only one massive fermion in this basic standard model, there is no Kobayashi–Maskawa parameters, nor the leptonic Kobayashi–Maskawa parameters. Therefore, there are only six dimensionless parameters — three mass ratios and three gauge coupling constants. By going from the original standard model to the basic standard model, the number of dimensionless parameters is changed from 25 to 6, a reduction by a factor of four.

Here are some initial observations on this basic standard model.

(1) The fundamental conjecture is: this basic standard model is capable of giving most, if not all, of the results of the original standard model.
(2) With the greatly reduced number of dimensionless parameters, how is this possible? The answer is, for example, in this basic standard model, the masses of the fermions besides the top quark are given by radiative corrections. The case of the bottom quark[16] will be further discussed in Sec. 4.
(3) Clearly there are many problems to be solved before this basic standard model can be considered to be a viable theory. Here is perhaps the most serious issue: how does the CP violation[17] arise in this basic standard model? This and a number of other important issues cannot yet be addressed at the moment.
(4) On the other hand, in some respect this basic standard model must necessarily go beyond the original standard model. This can be seen most directly from the fact that there are 25 dimensionless parameters in the original standard model but only six in the basic standard model. Therefore on the basis of the basic standard model there must be quite a few equalities between these 25 dimensionless parameters, perhaps 19 equalities or maybe even more.
(5) It is therefore the purpose of this paper to initiate the study by obtaining, on the basis of this basic standard model, the first such relation. Since this equality must come from high-order perturbation considerations, approximations play an important role. The resulting approximate equality is compared with experimental data in Sec. 10.
(6) In summary, while the present basic standard model and the original standard model have a great deal of similarity, the aims of these two theories are quite

different. The purpose of the basic standard model is not to make accurate predictions for physical quantities; rather, it is to gain some understanding of the 25 dimensionless parameters of the original standard model. In particular, while the original standard model deals with accurate low-order perturbation calculations, this basic standard model works with higher-order perturbation calculations — or more accurately with the approximate summation of all orders of perturbation calculations.

With this understanding, let us initiate the first attempt to study this basic standard model.

3. Formulation of the Problem

We proceed to formulate the first problem to be studied in this proposed model. This formulation is not entirely straightforward, and therefore will be motivated and described in some detail in this section.

Since the basic standard model is characterized by the absence, in the Lagrangian density, of any interaction between the Higgs particle and the fermions except the top quark, an immediate question is: how do these other fermions acquire their masses?

Which fermion should be chosen to be studied first? There are the following interconnected considerations.

(1) Since the only fermion that couples to the Higgs boson directly is the top quark because the top quark is the heaviest fermion, it is natural to consider first the next heaviest fermion.
(2) There is another reason to choose the heaviest elementary fermion besides the top quark: from the point of view of the basic standard model, the mass of this second heaviest fermion is likely to be from the first-order radiative correction. Such first-order radiative corrections are easier to treat than higher-order radiative corrections.
(3) The fermion chosen should interact directly with the top quark, meaning that this fermion should be one belonging to the third generation. Note that, in the basic standard model, all mixing of the generations comes from the interactions themselves, i.e. before the interactions are taken into account, there is no mixing between the generations.

All these considerations point to the desirability of studying first how the bottom quark acquires its mass in the basic standard model.

4. Underlying Structure and Some of the Immediate Implications

For the purpose of studying the coupling between the Higgs particle and the bottom quark through radiative corrections, it is convenient to use the R_ξ gauge of Fujikawa *et al.*[18] In this gauge, there are 10 one-loop diagrams for the $Hb\bar{b}$ three-vertex.

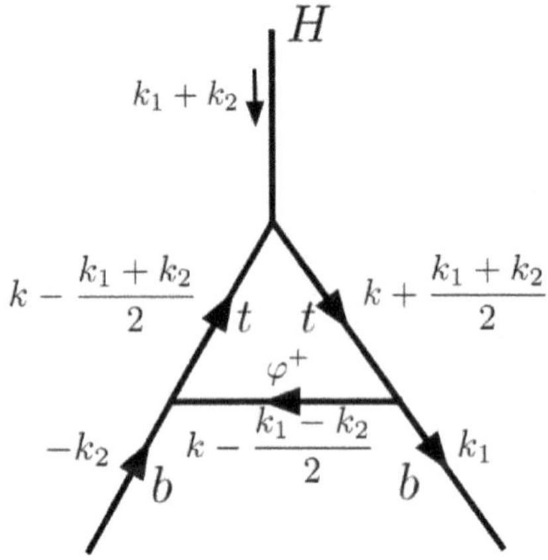

Fig. 1. The most important one-loop $Hb\bar{b}$ diagram in the R_ξ gauge.

Note that, since we are working with the basic standard model of Sec. 2, there is no $Hb\bar{b}$ three-vertex in the Lagrangian density. These 10 diagrams are shown in Figs. 1 and 2.

Consider first the diagram of Fig. 1. Since this is a perturbative diagram for the basic standard model, the two external bottom-quark lines must be taken to be massless. Under this condition, a straightforward calculation shows that the matrix element for this diagram of Fig. 1 is zero.

Similar considerations applied to the nine diagrams of Fig. 2 show that each of these diagrams gives a matrix element that is zero. Furthermore, this is also true for all higher-order $Hb\bar{b}$ diagrams in this basic standard model. There is a simple and well-known reason for this general result as follows.

Since the mass of the bottom quark is zero in the unperturbed Lagrangian density, this bottom quark has a helicity; this helicity is opposite between the bottom quark and its antiparticle. Thus the final $b\bar{b}$ state in the decay $H \to b\bar{b}$ with the initial Higgs particle at rest must have spin 1 along the direction of motion of the bottom quark and anti-bottom quark and hence cannot be the decay product of the spin-0 Higgs particle.

Since the contribution from each perturbation diagram to the coupling of the Higgs particle to the bottom quark is zero, it is necessary to include an infinite set of such diagrams in order for the bottom quark to acquire its experimentally known mass. Therefore, it is entirely unavoidable to deal with all orders of the perturbation series for the basic standard model, as stated already near the end of Sec. 2.

The higher-order terms of this perturbation series, by definition, depend on higher powers of the coupling constants, the three gauge coupling constants of the group

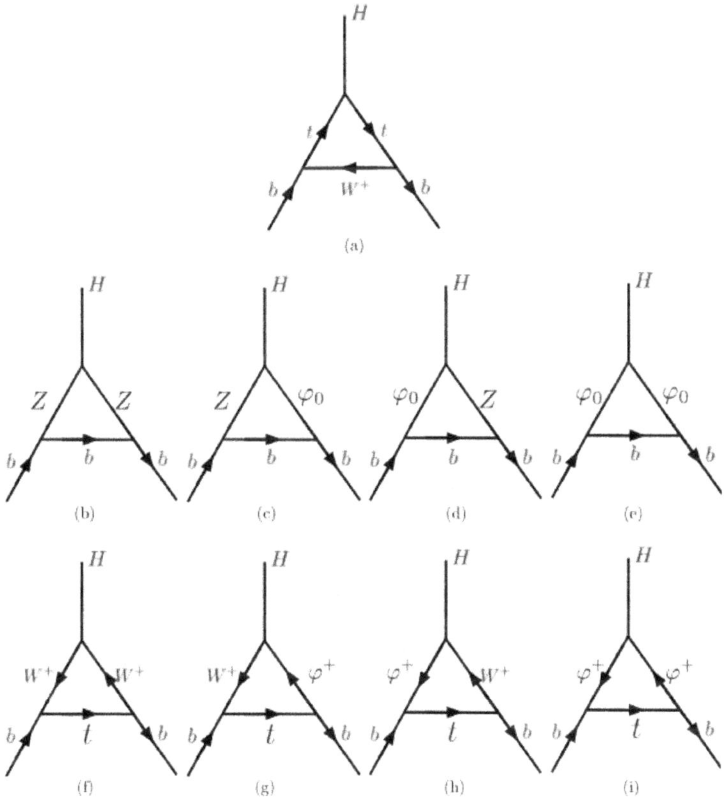

Fig. 2. Other one-loop $Hb\bar{b}$ diagram in the basic standard model when the R_ξ gauge is used.

$U(1) \times SU(2) \times SU(3)$. Accordingly, the necessity of summing approximately all orders of the perturbation series for the basic standard model has the following further implication: the three gauge coupling constants of the standard model cannot all be small. At first we found this conclusion rather startling and were rather reluctant to accept it. However, after thinking about it for some time, we now believe this conclusion to be not only acceptable but also nice in providing the first theoretical relation between the 25 dimensionless parameters of the standard model. This relation also provides a new direction for investigating and understanding elementary particle physics.

The precise statement about the specific relation between the strong coupling constant and the electroweak coupling constant is to be given in Sec. 9 where the electromagnetic interaction, being smaller, is neglected. Since the summation over all orders of the perturbation series is necessarily approximate, this relation between the two gauge coupling constant is of course also approximate. Our approximate relation is also compared with the known measured values in this Sec. 10. The further inclusion of the electromagnetic coupling constant is carried out in Sec. 11.

In order to be able to carry out this summation of the perturbation series for the basic standard model, it is essential to have some way of introducing approximations systematically. In the theoretical prediction of the increasing total cross-sections carried out more than half a century ago,[19] essential use was made of the large center-of-mass energy. For this study, a corresponding large parameter is very much needed. This leads to the considerations of next section.

5. Large m_t Approximation

Because of the way the basic standard model is formulated in Sec. 2, the obvious choice of this large parameter is m_t, the mass of the top quark, which is the heaviest among all the well-established elementary particles. This approximation shall be referred to as the "large m_t approximation".

So far as the fermions are concerned, this is certainly a good approximation. However, in order to be able to proceed with this large m_t approximation, we also have to take all the boson masses to be small compared with this m_t. Since, for example, the mass ratio $\frac{m_t^2}{m_W^2}$ is only about 4.6, this approximation cannot be expected to be very accurate. Nevertheless, it seems very difficult to do significantly better than the approximate calculation to be presented in the rest of this paper.

Let this large m_t approximation be applied to the ten one-loop triangular diagrams for the $Hb\bar{b}$ three vertices as shown in Figs. 1 and 2: that of Fig. 1 is the only one where the Feynman rules for the R_ξ gauge give a factor of m_t for each of the three vertices for a total factor of m_t^3. No such factor of m_t^3 appears in any of the nine diagrams of Fig. 2. Therefore, in this large m_t approximation, the diagram of Fig. 1 is the only one-loop diagram that needs to be kept. In applying this arguments, the external lines are not taken to be on their mass shells.

In the Feynman rules for the standard model, this mass m_t of the top quark appears in the following vertices:

$$Ht\bar{t}, \varphi_0 t\bar{t}, \varphi^+ \bar{t} b, \varphi^- t\bar{b} \tag{1}$$

again when the R_ξ gauge is used. In this list of vertices, the φ_0, φ^+ and φ^- are the Higgs ghosts.

In this large m_t approximation, all the other masses are taken to be zero, meaning that, in all propagators and all other vertices, masses other than that of the top quark are set to zero. For example, the propagator for the Higgs boson is approximated as

$$\frac{i}{p^2 - m_H^2} \sim \frac{i}{p^2}. \tag{2}$$

Furthermore, the parameter ξ in the R_ξ gauge is taken to be of order 1, and hence the propagators for the Higgs ghosts as well as the Faddeev–Popov ghosts[20] are similarly approximated as

$$\frac{i}{p^2 - \xi m_Z^2} \sim \frac{i}{p^2} \tag{3}$$

and

$$\frac{i}{p^2 - \xi m_W^2} \sim \frac{i}{p^2}. \tag{4}$$

Thus all these three propagators are the same. Since the two neutral propagators, those for H and φ_0 as given by (2) and (3), are the same, it is allowed to carry out a rotation, leading to the definition of a new particle, which we shall call φ^0, by

$$\varphi^0 = \frac{1}{\sqrt{2}}(H + i\varphi_0). \tag{5}$$

In fact, this φ^0 is the neutral component of the Higgs doublet before shifting. The propagator for this φ^0 is also given by

$$\frac{i}{p^2} \tag{6}$$

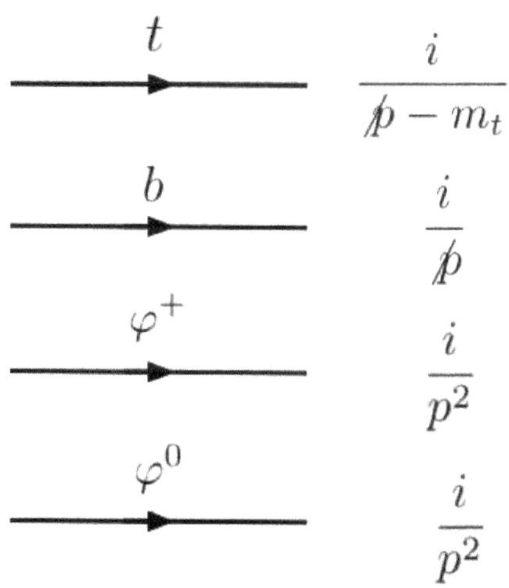

Fig. 3. Some Feynman rules for propagators in the large m_t approximation.

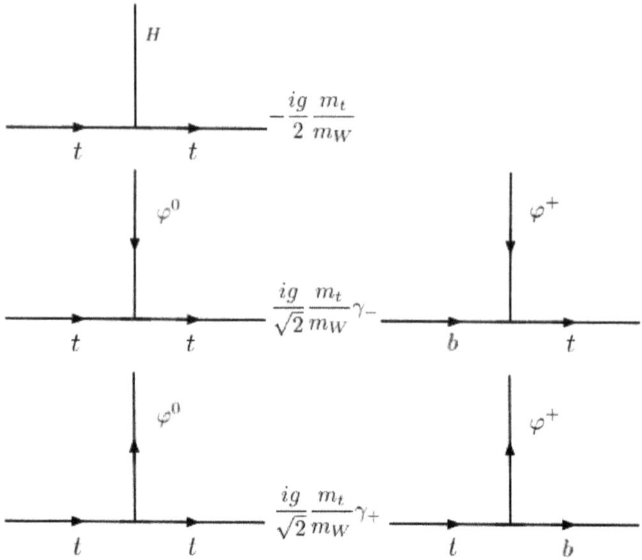

Fig. 4. Some Feynman rules for vertices in the large m_t approximation. The rules for the vertices on the right and on the left are the same.

in the large m_t approximation. Note that, although this φ^0 is neutral, its propagator comes with an arrow.

The relevant Feynman rules in this large m_t approximation are listed in Figs. 3 and 4 for the electroweak interactions.

The Feynman rules given in Fig. 4 involving the particle φ^0 defined by (5) follow from the identity

$$\{1\}\{1\} - \{\gamma_5\}\{\gamma_5\} = 2[\{\gamma_+\}\{\gamma_-\}] + [\{\gamma_-\}\{\gamma_+\}], \tag{7}$$

where

$$\gamma_\pm = \frac{1}{2}(1 \pm \gamma_5). \tag{8}$$

In these five Feynman rules of Fig. 4, the first one is to be used only when the H is an external Higgs line. That each of the other four rules involves either γ_+ or γ_- plays an important simplifying role in the analysis of this paper.

6. Choice of Diagrams

The important diagrams are those shown in Fig. 5, where the vertical line at the top is that of an external Higgs particle, the horizontal lines represent the φ particles

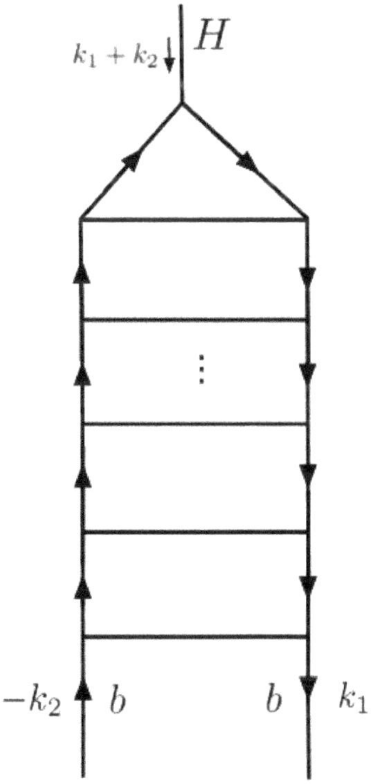

Fig. 5. Class of diagrams to be studied.

with the Feynman rules of Figs. 3 and 4, while all the other lines are fermions, with each segment either a top quark or a bottom quark.

These diagrams of Fig. 5 bear a great deal of similarity to — and also some important differences from — the tower diagrams, which have been used to predict, on purely theoretical grounds, the increasing total cross-sections.[19,21]

As already shown in Sec. 4, the matrix element of every one of the diagrams of Fig. 5 is zero when the external bottom-quark lines are both on mass shell, the mass being zero in the basic standard model. Accordingly, these diagrams are to be studied with the lowest rung removed.

As seen from the Feynman rules of Fig. 4, there is no $\varphi^0 b\bar{b}$ three-vertex in the basic standard model. Therefore, starting with the diagram of Fig. 5, the removal of the lowest rung leads to the diagram of Fig. 6(a), where the external fermion lines are those of the top quark.

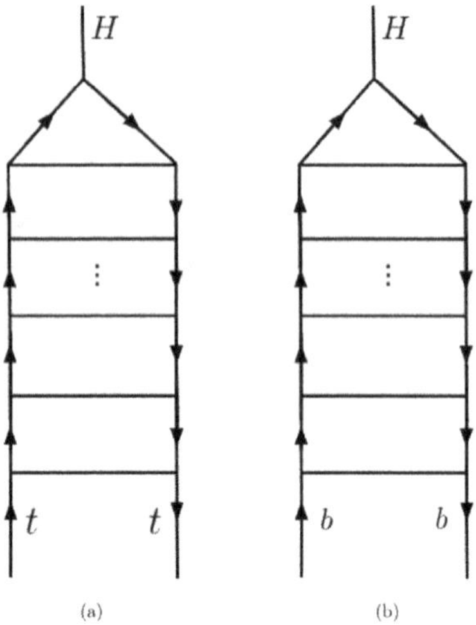

Fig. 6. Pagoda diagrams.

If these external top quark lines are replaced by external bottom quark lines, then the diagram of Fig. 6(a) becomes that of Fig. 6(b). These diagrams of Figs. 6(a) and 6(b) shall also be referred to as pagoda diagrams.

7. Summation of the Pagoda Diagrams

The next step is to carry out the summation over this set of pagoda diagrams. With electroweak interaction only, the relevant vertices are given in Fig. 4, and hence there are four types of rungs, which are shown in Fig. 7 with these four types labeled as 1, 2, 3 and 4.

From Fig. 7, it is seen that there is either a γ_+ or a γ_- at each end of the rung. This is a direct consequence of the absence of the $Hb\bar{b}$ vertex in the Lagrangian of the basic standard model. As an example of this simplification, consider the stacking of the rung 1 on top of rung 3, as shown in Fig. 8.

It is seen from Figs. 3 and 4 that the matrix element for the diagram of Fig. 8(a) contains the factor

$$\begin{aligned}&\left\{\gamma_-\frac{i}{\not{q}'-m_t}\gamma_+\right\}\left\{\gamma_-\frac{i}{\not{q}-m_t}\gamma_+\right\}\\&\quad=\frac{i}{q'^2-m_t^2}\frac{i}{q^2-m_t^2}\{\gamma_-(\not{q}'+m_t)\gamma_+\}\{\gamma_-(\not{q}+m_t)\gamma_+\}\\&\quad=\frac{i}{q'^2-m_t^2}\frac{i}{q^2-m_t^2}\{\gamma_-\not{q}'\gamma_+\}\{\gamma_-\not{q}\gamma_+\},\end{aligned}\tag{9}$$

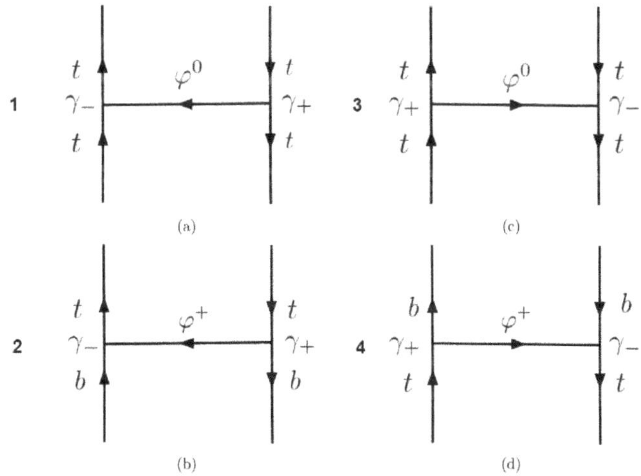

Fig. 7. The four types of rungs in the pagoda diagrams due to electroweak interactions.

where q and q' are the four-momenta of the two top-quark propagators in the middle, as indicated on Fig. 8(a). In writing down this (9), two factors of $\frac{ig}{\sqrt{2}}\frac{m_t}{m_W}$ have been omitted. The simplification comes from the trivial identity

$$\gamma_- m_t \gamma_+ = 0. \tag{10}$$

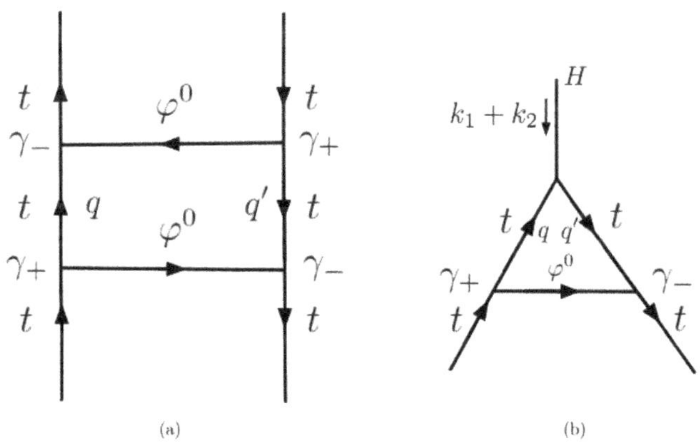

Fig. 8. (a) Rung 1 on top of rung 3; and (b) external H vertex on top of rung 3.

If the rung 1 is stacked on another rung 1 instead of the rung 3, then the corresponding factor is

$$\left\{\gamma_+ \frac{i}{\not{q}' - m_t}\gamma_+\right\}\left\{\gamma_- \frac{i}{\not{q} - m_t}\gamma_-\right\} = \frac{i}{q'^2 - m_t^2}\frac{i}{q^2 - m_t^2}\{\gamma_+ m_t \gamma_+\}\{\gamma_- m_t \gamma_-\} \quad (11)$$

as consequences of

$$\gamma_+ \not{q}' \gamma_+ = \gamma_- \not{q} \gamma_- = 0. \quad (12)$$

The diagram of Fig. 8(b) differs from that of Fig. 8(a) in that the upper rung, the rung 1, is replaced by the external H vertex. With this replacement and using the first Feynman rule of Fig. 4, Eq. (9) becomes

$$\begin{aligned}\gamma_- \frac{i}{\not{q}' - m_t}\frac{i}{\not{q} - m_t}\gamma_+ &= \frac{i}{q'^2 - m_t^2}\frac{i}{q^2 - m_t^2}\gamma_-(\not{q}' + m_t)(\not{q} + m_t)\gamma_+ \\ &= \frac{i}{q'^2 - m_t^2}\frac{i}{q^2 - m_t^2}\gamma_-(\not{q}'\not{q} + m_t\not{q} + \not{q}' m_t + m_t^2)\gamma_+ \\ &= \frac{i}{q'^2 - m_t^2}\frac{i}{q^2 - m_t^2}\gamma_- m_t(\not{q} + \not{q}')\gamma_+. \end{aligned} \quad (13)$$

Equation (13) is of central importance in this study of the gauge coupling constants in the standard model. Consider the four-momentum integration over q with

$$q' = q + k_1 + k_2; \quad (14)$$

the left-hand side of this Eq. (13) is formally of the order of $\frac{1}{q^2}$ for large q^2, while the right-hand side is of order $\frac{1}{(q^2)^{3/2}}$. This reduction of the momentum dependence is most important: since this φ^0 propagator supplies another factor of $\frac{1}{q^2}$ as seen from Fig. 3, this reduction leads to a convergent four-momentum integration over q, thanks to the factors of γ_+ and γ_-. This fact leads immediately to the general property that ALL pagoda diagrams of Figs. 5 and 6 are convergent. This is a nice property that is shared with the tower diagrams in the theoretical prediction of the increasing total cross-sections.[19]

After establishing the finiteness of the contribution from each pagoda diagram, we turn our attention to the present problem of studying the gauge coupling constants as discussed in the introduction. For this purpose, it is sufficient to consider only the special case where

$$k_1 + k_2 = 0 \quad (15)$$

— see Figs. 1 and 8(b). This relation (15) is assumed to hold throughout the rest of this paper.

Consider two ways of combining the diagrams of Figs. 8(a) and 8(b) to get two-loop diagrams, which are shown in Fig. 9. Because of Eq. (15), let the momenta of the

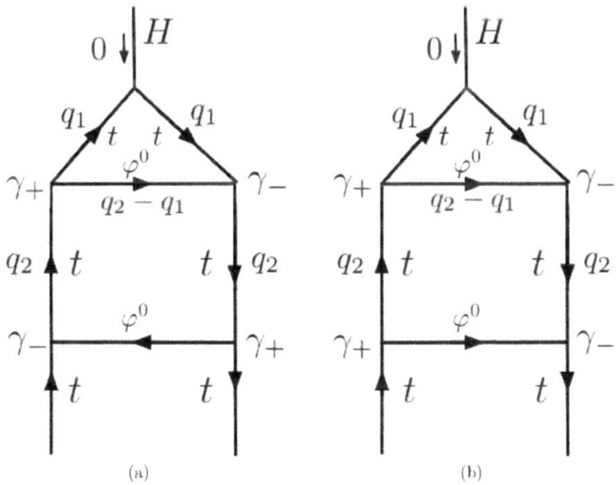

Fig. 9. Two ways to combine the diagrams of Figs. 8(a) and 8(b) to get two-loop diagrams.

propagators be chosen as those shown in that Fig. 9. Following Eqs. (9) and (13), the fermion line of Fig. 9(a) gives the factor

$$\gamma_+(\not{q}_2 + m_t)\gamma_-(\not{q}_1 + m_t)(\not{q}_1 + m_t)\gamma_+(\not{q}_2 + m_t)\gamma_-$$
$$= \gamma_+(\not{q}_2 + m_t)\gamma_- 2m_t \not{q}_1 \gamma_+(\not{q}_2 + m_t)\gamma_-. \tag{16}$$

In the integration over the four-momentum q_1, the only Lorentz index comes from the factor

$$\not{q}_1 = q_{1\mu}\gamma^\mu. \tag{17}$$

Since there is no variable besides q_1 and q_2, this $q_{1\mu}$ can be replaced by

$$q_{1\mu} \longrightarrow q_{2\mu}\frac{q_1 \cdot q_2}{q_2^2}. \tag{18}$$

Therefore

$$(16) \rightarrow \gamma_+(\not{q}_2 + m_t)\gamma_- 2m_t \frac{q_1 \cdot q_2}{q_2^2} \not{q}_2 \gamma_+(\not{q}_2 + m_t)\gamma_-$$
$$= 2m_t \frac{q_1 \cdot q_2}{q_2^2} \gamma_+ \not{q}_2 \gamma_- \not{q}_2 \gamma_+ \not{q}_2 \gamma_-$$
$$= (q_1 \cdot q_2)\gamma_+ 2m_t \not{q}_2 \gamma_-. \tag{19}$$

The importance of this result lies in the fact that the dependence on the fermions is the same for the one-loop and the two-loop cases, as seen from Eqs. (13) and (19).

The corresponding calculation for the diagram of Fig. 9(b) is only slightly different as follows:

$$\begin{aligned}
&\gamma_-(\not{q}_2+m_t)\gamma_-(\not{q}_1+m_t)(\not{q}_1+m_t)\gamma_+(\not{q}_2+m_t)\gamma_+\\
&=\gamma_-(\not{q}_2+m_t)\gamma_-2m_t\not{q}_1\gamma_+(\not{q}_2+m_t)\gamma_+\\
&\to\gamma_-(\not{q}_2+m_t)\gamma_-2m_t\frac{q_1\cdot q_2}{q_2^2}\not{q}_2\gamma_+(\not{q}_2+m_t)\gamma_+\\
&=2m_t\frac{q_1\cdot q_2}{q_2^2}\gamma_-m_t\gamma_-\not{q}_2\gamma_+m_t\gamma_+\\
&=(q_1\cdot q_2)\frac{m_t^2}{q_2^2}\gamma_-2m_t\not{q}_2\gamma_+.
\end{aligned}\tag{20}$$

Thus the major difference between Eqs. (20) and (19) is this factor $\frac{m_t^2}{q_2^2}$. In other words, the dependence on the fermions is the same for any number of loops except for the appearance of various factors like this $\frac{m_t^2}{q_2^2}$.

So far in this Sec. 7, only the cases of the rung 1 and the rung 3 have been considered, as designated in Fig. 7. It is not difficult to extend these considerations to all combinations of the four types of rungs of this Fig. 7, and the results are summarized in Table 1. In this table, the rung denoted by the index j' is placed above the one denoted by j. The entry 1 means that Eq. (19) applies in this case, while the entry β means that Eq. (20) applies instead, where β is the generalization of this ratio $\frac{m_t^2}{q_2^2}$ in this equation.

The next step is to treat the entries of Table 1 as a 4×4 matrix, and to determine its four eigenvalues. For this purpose, it is only necessary to calculate the 4×4 determinant

$$P_1(\lambda)=\begin{vmatrix}\beta-\lambda & 0 & 1 & 1\\ \beta & -\lambda & 1 & 1\\ 1 & 0 & \beta-\lambda & \beta\\ 0 & 1 & 0 & -\lambda\end{vmatrix}.\tag{21}$$

Table 1. The effect of placing rung j' above rung j — see Fig. 7 for the designation of the four types of rungs in the pagoda diagrams.

j \ j'	1	2	3	4
1	β	0	1	1
2	β	0	1	1
3	1	0	β	β
4	0	1	0	0

Here the subscript 1 means that only the electroweak interaction has been taken into account. It follows immediately that

$$P_1(\lambda) = \lambda^2[(\beta - \lambda)^2 - 2], \tag{22}$$

meaning that the four eigenvalues are

$$\lambda = 0, 0, \beta + \sqrt{2} \quad \text{and} \quad \beta - \sqrt{2}. \tag{23}$$

Since β is always positive, the largest eigenvalue is $\beta + \sqrt{2}$; this is the relevant eigenvalue.

The next task is to study the pagoda diagrams *iteratively*. Define $N_j^{(n)}(k)$, with $j = 1, 2, 3, 4$, to be the sum of the contributions from all n-loop pagoda diagrams where the lowest rung is of type j as defined in Fig. 7 with

$$k = q_n. \tag{24}$$

Then, from Table 1, the recurrence relations for these $N_j^{(n)}(k)$ are given by

$$N_1^{(n+1)}(k) = \frac{i}{2}\left(\frac{m_t}{m_W}g\right)^2 \frac{1}{(2\pi)^4}\int d^4k' \frac{1}{(k-k')^2}\frac{1}{(k'^2 - m_t^2)^2}(k\cdot k') \times \left[\frac{m_t^2}{k'^2}N_1^{(n)}(k') + N_3^{(n)}(k') + N_4^{(n)}(k')\right], \tag{25}$$

$$N_2^{(n+1)}(k) = \frac{i}{2}\left(\frac{m_t}{m_W}g\right)^2 \frac{1}{(2\pi)^4}\int d^4k' \frac{1}{(k-k')^2}\frac{1}{(k'^2 - m_t^2)^2}(k\cdot k') \times \left[\frac{m_t^2}{k'^2}N_1^{(n)}(k') + N_3^{(n)}(k') + N_4^{(n)}(k')\right], \tag{26}$$

$$N_3^{(n+1)}(k) = \frac{i}{2}\left(\frac{m_t}{m_W}g\right)^2 \frac{1}{(2\pi)^4}\int d^4k' \frac{1}{(k-k')^2}\frac{1}{(k'^2 - m_t^2)^2}(k\cdot k') \times \left[N_1^{(n)}(k') + \frac{m_t^2}{k'^2}N_3^{(n)}(k') + \frac{m_t^2}{k'^2}N_4^{(n)}(k')\right] \tag{27}$$

and

$$N_4^{(n+1)}(k) = \frac{i}{2}\left(\frac{m_t}{m_W}g\right)^2 \frac{1}{(2\pi)^4}\int d^4k' \frac{1}{(k-k')^2}\frac{1}{(k'^2)^2}(k\cdot k')N_2^{(n)}(k'). \tag{28}$$

These recurrence relation equations (25)–(28) are not to be studied any further. Instead, we proceed to include the strong interactions in addition to the electroweak interactions, leading to the recurrence relation equations (32)–(36) of the following section.

8. Inclusion of Strong Interactions

In working with the top and the bottom quarks, it is essential to include the strong interactions in addition to the electroweak interactions. In general, the contributions

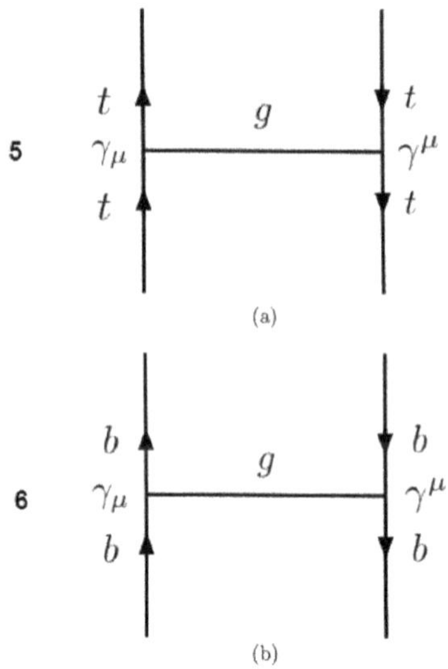

Fig. 10. The two additional types of rungs in the pagoda diagrams due to strong interactions.

from the strong interactions can be difficult to study. Fortunately, with the treatment developed in the previous sections based on the large m_t approximation, this desired inclusion of the strong interactions is complicated but manageable. We shall use Sec. 7 dealing with the electroweak interaction as a guidance for the present treatment of both the electroweak interaction and the strong interaction together.

Figure 7 shows the four types of rungs in the pagoda diagrams due to the electroweak interactions; the inclusion of strong interactions leads to two more types of rungs due to the gluon, as shown in Fig. 10. These two types 5 and 6 of rungs differ from those of Fig. 7 in not having the factors γ_+ or γ_- at each end of the rung. Since, as already seen clearly in Sec. 7, these γ_+ and γ_- play important roles in the present analysis, the first task for Sec. 8 is to rewrite the contributions from these types 5 and 6 of rungs such that these γ_+ and γ_- appear explicitly.

As a first step in generalizing the procedure of Sec. 7 to the present case, consider the diagram of Fig. 11, which is obtained from that of Fig. 8(b) replacing the rung 3 by a rung 5. The major difference between these diagrams of Figs. 8(b) and 11 is this: while, by Eq. (13), the diagram of Fig. 8(b) leads to a convergent integral, that of Fig. 11 gives a divergent integral which needs to be taken care of by renormalization.

Similarly, the lower rung of the pagoda diagram of Figs. 9(a) or 9(b) may be replaced by a rung 5. The calculation involved in arriving at Eqs. (16) and (19)

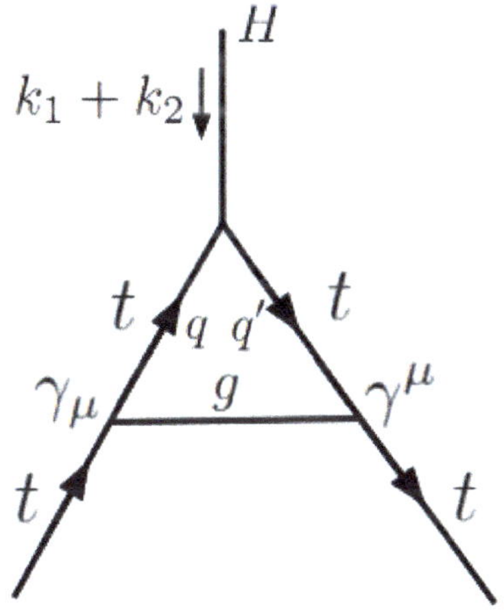

Fig. 11. One-loop diagram with rung 5.

becomes slightly more involved as follows:

$$\begin{aligned}
&\gamma^\mu(\not{q}_2 + m_t)\gamma_-(\not{q}_1 + m_t)(\not{q}_1 + m_t)\gamma_+(\not{q}_2 + m_t)\gamma_\mu \\
&\quad = \gamma^\mu(\not{q}_2 + m_t)\gamma_- 2m_t\not{q}_1\gamma_+(\not{q}_2 + m_t)\gamma_\mu \\
&\quad \to \gamma^\mu(\not{q}_2 + m_t)\gamma_- 2m_t\frac{q_1 \cdot q_2}{q_2^2}\not{q}_2\gamma_+(\not{q}_2 + m_t)\gamma_\mu \\
&\quad = 2m_t\frac{q_1 \cdot q_2}{q_2^2}(\gamma_-\gamma^\mu\not{q}_2 + \gamma_+\gamma^\mu m_t)\not{q}_2(\not{q}_2\gamma_\mu\gamma_+ + m_t\gamma_\mu\gamma_-) \\
&\quad = 2m_t\frac{q_1 \cdot q_2}{q_2^2}(\gamma_-\gamma^\mu q_2^2\not{q}_2\gamma_\mu\gamma_+ + \gamma_+\gamma^\mu m_t^2\not{q}_2\gamma_\mu\gamma_- + \gamma_+ q_2^2\gamma^\mu m_t\gamma_\mu\gamma_+ \\
&\qquad + \gamma_- q_2^2\gamma^\mu m_t\gamma_\mu\gamma_-) \\
&\quad = 2m_t\frac{q_1 \cdot q_2}{q_2^2}[-2q_2^2\gamma_-\not{q}_2\gamma_+ - 2m_t^2\gamma_+\not{q}_2\gamma_- + 4q_2^2 m_t(\gamma_+^2 + \gamma_-^2)] \\
&\quad = 2m_t\frac{q_1 \cdot q_2}{q_2^2}(-2q_2^2\gamma_-\not{q}_2\gamma_+ - 2m_t^2\gamma_+\not{q}_2\gamma_- + 4q_2^2 m_t).
\end{aligned} \tag{29}$$

Note the two factors of -2 in the first and second terms in the parentheses on the right-hand side.

Compared with Eq. (19), the new feature of this Eq. (29) is that there are three terms on the right-hand side instead of one. Of these terms, the fermion factor $\gamma_-\not{q}_2\gamma_+$ of the first term is identical to that on the right-hand side of Eq. (20), and the $\gamma_+\not{q}_2\gamma_-$ to that of Eq. (19), while the third term contains no fermion factor. This has the following implication.

For the purpose of determining the radius of convergence, the sum of the matrix elements of the n-loop $Ht\bar{t}$ pagoda diagrams shown in Fig. 6(a) is of the following form:

$$\mathcal{M}_t^{(n)}(k) = \gamma_+ 2m_t \not{k} \gamma_- N_{t1}^{(n)}(k) + \gamma_- 2m_t \not{k} \gamma_+ N_{t2}^{(n)}(k) + 2m_t^2 N_{t3}^{(n)}(k). \quad (30)$$

Equation (30) holds when the four-momentum of the external Higgs particle is taken to be zero — see Eq. (15). Since the bottom quark is massless on the level of the Lagrangian for the basic standard model, the corresponding expression is simpler for the sum of the matrix elements of the n-loop $Hb\bar{b}$ pagoda diagrams of Fig. 6(b)

$$\mathcal{M}_b^{(n)}(k) = \gamma_+ 2m_t \not{k} \gamma_- N_{b1}^{(n)}(k) + \gamma_- 2m_t \not{k} \gamma_+ N_{b2}^{(n)}(k). \quad (31)$$

Thus there are five functions of the scalar variable k^2 for each n, namely, $N_{t1}^{(n)}(k)$, $N_{t2}^{(n)}(k)$, $N_{t3}^{(n)}(k)$, $N_{b1}^{(n)}(k)$ and $N_{b2}^{(n)}(k)$.

Following the procedure of Sec. 7, the recurrence for these five functions are found to be

$$\begin{aligned} N_{t1}^{(n+1)}(k) = {} & \frac{i}{2}\left(\frac{m_t}{m_W} g\right)^2 \frac{1}{(2\pi)^4} \int d^4k' \frac{1}{(k-k')^2} \frac{k' \cdot k}{k^2} \frac{1}{(k'^2 - m_t^2)^2} \\ & \times [(m_t^2 - Ak'^2) N_{t1}^{(n)}(k') + (k'^2 - Am_t^2) N_{t2}^{(n)}(k') \\ & + 2m_t^2(1-A) N_{t3}^{(n)}(k')], \end{aligned} \quad (32)$$

$$\begin{aligned} N_{t2}^{(n+1)}(k) = {} & \frac{i}{2}\left(\frac{m_t}{m_W} g\right)^2 \frac{1}{(2\pi)^4} \int d^4k' \frac{1}{(k-k')^2} \frac{k' \cdot k}{k^2} \\ & \times \left\{ \frac{1}{(k'^2 - m_t^2)^2} [(k'^2 - Am_t^2) N_{t1}^{(n)}(k') + (m_t^2 - Ak'^2) N_{t2}^{(n)}(k') \right. \\ & \left. + 2m_t^2(1-A) N_{t3}^{(n)}(k')] + \frac{1}{(k'^2)^2} k'^2 N_{b1}^{(n)}(k') \right\}, \end{aligned} \quad (33)$$

$$\begin{aligned} N_{t3}^{(n+1)}(k) = {} & \frac{i}{2}\left(\frac{m_t}{m_W} g\right)^2 \frac{1}{(2\pi)^4} \int d^4k' \frac{1}{(k-k')^2}, \\ & \times \frac{A}{(k'^2 - m_t^2)^2} [2k'^2 N_{t1}^{(n)}(k') + 2k'^2 N_{t2}^{(n)}(k') + 2(k'^2 + m_t^2) N_{t3}^{(n)}(k')], \end{aligned} \quad (34)$$

$$\begin{aligned} N_{b1}^{(n+1)}(k) = {} & \frac{i}{2}\left(\frac{m_t}{m_W} g\right)^2 \frac{1}{(2\pi)^4} \int d^4k' \frac{1}{(k-k')^2} \frac{k' \cdot k}{k^2} \\ & \times \left\{ \frac{1}{(k'^2 - m_t^2)^2} [m_t^2 N_{t1}^{(n)}(k') + k'^2 N_{t2}^{(n)}(k') + 2m_t^2 N_{t3}^{(n)}(k')] \right. \\ & \left. + \frac{1}{(k'^2)^2} (-A) k'^2 N_{b1}^{(n)}(k') \right\} \end{aligned} \quad (35)$$

and

$$N_{b2}^{(n+1)}(k) = \frac{i}{2}\left(\frac{m_t}{m_W} g\right)^2 \frac{1}{(2\pi)^4} \int d^4k' \frac{1}{(k-k')^2} \frac{k' \cdot k}{k^2} \frac{1}{(k'^2)^2} (-A) k'^2 N_{b2}^{(n)}(k'), \quad (36)$$

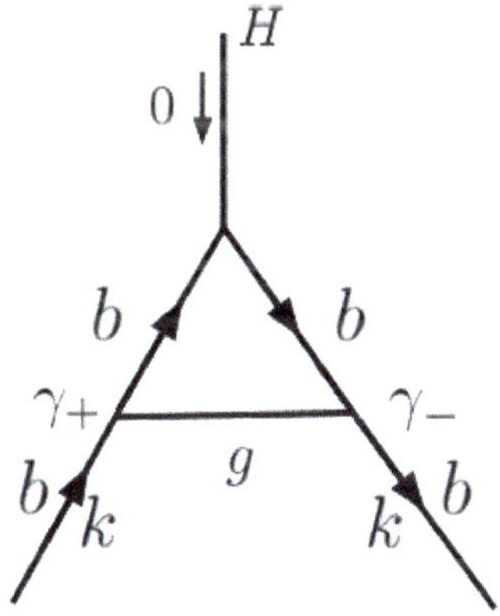

Fig. 12. One-loop diagram for $N_{b2}^{(1)}(k)$.

where A is the ratio

$$A = \left(\frac{\text{strong interactions}}{\text{electroweak interactions}}\right)^2 = \frac{6g_s^2}{\frac{1}{2}\left(\frac{m_t}{m_W}g\right)^2}. \tag{37}$$

In Eq. (37), use has been made of the fact that the gluon interacts identically with the top quark and the bottom quark; furthermore, the factor 6 in the numerator comes from the product of the three from $SU(3)$ with the two mentioned after Eq. (29), and g_s is the strong gauge coupling constant for this $SU(3)$.

Consider first Eq. (36), the simplest of the five recurrence relations given by Eqs. (32)–(36); it expresses $N_{b2}^{(n+1)}$ in terms of $N_{b2}^{(n)}$. For $n = 1$, the $N_{b2}^{(1)}(k)$ is from the diagram of Fig. 12 only. But there is no such diagram because, in the basic standard model, there is no $Hb\bar{b}$ three-vertex in the Lagrangian density. It therefore follows immediately that

$$N_{b2}^{(1)}(k) = 0. \tag{38}$$

It is therefore a consequence of the recurrence relation equation (36) that

$$N_{b2}^{(n)}(k) = 0 \tag{39}$$

for all k and all n.

It remains to study the other four recurrence relation equations (32)–(35). From these recurrence relations, it is desired to find the behavior of these four $N's$ — $N_{t1}^{(n)}(k)$, $N_{t2}^{(n)}(k)$, $N_{t3}^{(n)}(k)$ and $N_{b1}^{(n)}(k)$ — for large n.

In order to study this large-n behavior, it is essential to have a clear understanding of the most interesting and simplest case of

$$m_t \to 0. \tag{40}$$

In the large m_t approximation being used here, the meaning of this Eq. (40) is

$$|k^2| \gg m_t^2 \gg m^2, \tag{41}$$

where m is mass of any elementary particle other than the top quark. See App. B of Ref. 21 for the use of this special case Eq. (41).

In this special case, the recurrence relation equations (40)–(43) simplify to

$$N_{t1}^{(n+1)}(k) = \frac{i}{2}\left(\frac{m_t}{m_W}g\right)^2 \frac{1}{(2\pi)^4}\int d^4k' \frac{1}{(k-k')^2}\frac{k'\cdot k}{k^2k'^2}[-AN_{t1}^{(n)}(k') + N_{t2}^{(n)}(k')], \tag{42}$$

$$N_{t2}^{(n+1)}(k) = \frac{i}{2}\left(\frac{m_t}{m_W}g\right)^2 \frac{1}{(2\pi)^4}\int d^4k' \frac{1}{(k-k')^2}\frac{k'\cdot k}{k^2k'^2} \times [N_{t1}^{(n)}(k') - AN_{t2}^{(n)}(k') + N_{b1}^{(n)}(k')], \tag{43}$$

$$N_{t3}^{(n+1)}(k) = \frac{i}{2}\left(\frac{m_t}{m_W}g\right)^2 \frac{1}{(2\pi)^4}\int dk' \frac{1}{(k-k')^2}\frac{1}{k'^2}A \times [2N_{t1}^{(n)}(k') + 2N_{t2}^{(n)}(k') + 2N_{t3}^{(n)}(k')] \tag{44}$$

and

$$N_{b1}^{(n+1)}(k) = \frac{i}{2}\left(\frac{m_t}{m_W}g\right)^2 \frac{1}{(2\pi)^4}\int d^4k' \frac{1}{(k-k')^2}\frac{k'\cdot k}{k^2k'^2}[N_{t2}^{(n)}(k') - AN_{b1}^{(n)}(k')]. \tag{45}$$

It is then noted that the quantity $N_{t3}^{(n)}(k')$ appears only on the right-hand sides of Eq. (44), not in the other three recurrence relations. Thus, so far as the recurrence relations are concerned, Eq. (44) can be omitted, while the other three Eqs. (42), (43) and (45) form a set of recurrence relations for the three quantities N_{t1}, N_{t2} and N_{b1}. It is convenient and instructive to rewrite these three equations in matrix form

$$\begin{bmatrix} N_{t1}^{(n+1)}(k) \\ N_{t2}^{(n+1)}(k) \\ N_{b1}^{(n+1)}(k) \end{bmatrix} = -\frac{i}{2}\left(\frac{m_t}{m_W}g\right)^2 \frac{1}{(2\pi)^4}\int d^4k' \frac{1}{(k-k')^2}\frac{k'\cdot k}{k^2k'^2} \times \begin{bmatrix} A & -1 & 0 \\ -1 & A & -1 \\ 0 & -1 & A \end{bmatrix}\begin{bmatrix} N_{t1}^{(n)}(k') \\ N_{t2}^{(n)}(k') \\ N_{b1}^{(n)}(k') \end{bmatrix}. \tag{46}$$

Note that the 3×3 matrix

$$\begin{bmatrix} A & -1 & 0 \\ -1 & A & -1 \\ 0 & -1 & A \end{bmatrix} \tag{47}$$

on the right-hand side of Eq. (46) is real and symmetrical.

The three eigenvalues of this matrix (47) are A and $A \pm \sqrt{2}$. Therefore, the large-n behavior of these four $N's$ is controlled by the largest of these three eigenvalues, namely,

$$\lambda = A + \sqrt{2}. \tag{48}$$

The conclusion is therefore reached that, for large n, each of the four quantities $N_{t1}^{(n)}(k)$, $N_{t2}^{(n)}(k)$, $N_{t3}^{(n)}(k)$ and $N_{b1}^{(n)}(k)$ is proportional to the nth power of

$$\frac{1}{2}\left(\frac{m_t}{m_W}g\right)^2\lambda = 6g_s^2 + \frac{1}{\sqrt{2}}\left(\frac{m_t}{m_W}g\right)^2. \tag{49}$$

Besides the 3×3 matrix (47) and the overall constant, the rest of the kernel in the integral equation (46) is

$$\frac{1}{(k-k')^2}\frac{k'\cdot k}{k^2k'^2},$$

which is real and symmetric in the two four vectors k and k'. Changing from Minkowski metric to Euclidean metric, this Eq. (46) leads us to consider the scalar integral equation

$$\bar{\lambda}f(k^2) = \frac{1}{(2\pi)^4}\int d_4k'\frac{1}{(k-k')^2}\frac{k'\cdot k}{k^2k'^2}f(k'^2). \tag{50}$$

Because of rotational invariance, the arguments for f have been written as k^2 and k'^2. Using hyperspherical polar coordinated, the four-dimensional Euclidean integration is

$$d_4k' = k'^3\sin^2\theta\sin\theta' dk' d\theta d\theta' d\phi, \tag{51}$$

where $k' \geq 0$, $0 \leq \theta \leq \pi$, $0 \leq \theta' \leq \pi$ and $0 \leq \phi \leq 2\pi$. With k chosen to be the z-axis, the integral on the right-hand side of Eq. (50) is

$$\begin{aligned}&\int_0^\infty k'^3dk'\int_0^\pi d\theta\sin^2\theta\int_0^\pi d\theta'\sin\theta'\int_0^{2\pi}d\phi\frac{1}{k^2+k'^2-2kk'\cos\theta}\frac{1}{k^2}\frac{1}{k'^2}\\ &\quad\times(kk'\cos\theta)f(k'^2) = \pi^2\int_0^\infty k'dk'\frac{1}{k^2}\min\left(\frac{k^2}{k'^2},\frac{k'^2}{k^2}\right)f(k'^2).\end{aligned} \tag{52}$$

Note the overall factor of π^2.

The main question to be answered here is: what is the spectrum of the operator on $f(k'^2)$ as given by Eq. (52)?

Since we are studying the special case (41), there is no *mass scale*. Therefore, the desired spectrum can be determined by choosing $f(k'^2)$ to be

$$f(k'^2) = (k'^2)^{-\kappa} \tag{53}$$

with $\kappa > 0$. With this choice, it follows from Eqs. (50) and (52) that

$$\int d_4k'\frac{1}{(k-k')^2}\frac{1}{k^2}\frac{1}{k'^2}(k\cdot k')(k'^2)^{-\kappa} = \pi^2(k^2)^{-\kappa}\frac{1}{\kappa(2-\kappa)}. \tag{54}$$

Therefore the spectrum of $\bar{\lambda}$ is continuous and given by $[\pi^2, \infty)$.

9. Relation Between the Electroweak and the Strong Gauge Coupling Constants

With the mathematical preliminary of Sec. 8, we are now ready to return to the physics problem of obtaining an $Hb\bar{b}$ interaction through radiative connections.

In order to go from the pagoda diagrams of Fig. 6 to the $Hb\bar{b}$ diagrams of Fig. 5, it is only necessary to restore the last rung — rung 2 of Fig. 7 in the case of Fig. 6(a), and rung 6 of Fig. 10 for Fig. 6(b).

As shown in Sec. 4, because the mass of the bottom quark is zero in the Lagrangian density of the basic standard model, the matrix element for each of the $Hb\bar{b}$ diagrams of Fig. 5 is zero when both of the two external bottom-quark are on mass shell. Because all quarks are charged, they have electromagnetic interactions and hence electromagnetic masses, which have not been taken into account in the present section, but shall be in Sec. 11 entitled "Inclusion of electromagnetic interactions". Let this small electromagnetic mass for the bottom quark be called ε. With this nonzero value of ε, let $\mathcal{M}_n$ be the sum of contributions from all n-loop $Hb\bar{b}$ diagrams shown in Fig. 5. It then follows that

$$\lim_{\varepsilon\to 0} \mathcal{M}_n = 0 \tag{55}$$

for all n, and hence

$$\sum_n \lim_{\varepsilon\to 0} \mathcal{M}_n = 0. \tag{56}$$

On the other hand, in order to be able to obtain a suitable $Hb\bar{b}$ interaction through radiative corrections to give the bottom quark a nonzero mass, it is essential to have

$$\lim_{\varepsilon\to 0} \sum_n \mathcal{M}_n \neq 0. \tag{57}$$

In other words, it is essential to have the two simple operations of $\sum_n$ and $\lim_{\varepsilon\to 0}$ not to commute.

From the study of the decay

$$H \to \gamma\gamma \tag{58}$$

it is known that such lack of commutativity is not only mathematically possible but also actually happens in quantum field theory.[22] The issue is: for the present case, how does this happen?

In order to answer question, it is useful to study first the special case (41) of Sec. 8. In general, with the A defined by Eq. (37) held fixed, the sum $\sum_n \mathcal{M}_n$ is a power series in g^2. Let the radius of convergence of this series be called R_0. When (41) holds, we see from Eq. (48) that, approximately for large values of n,

$$\mathcal{M}_n \sim \left(\frac{1}{2} g^2 \frac{m_t^2}{m_W^2} \lambda\right)^n \mathcal{M}_0. \tag{59}$$

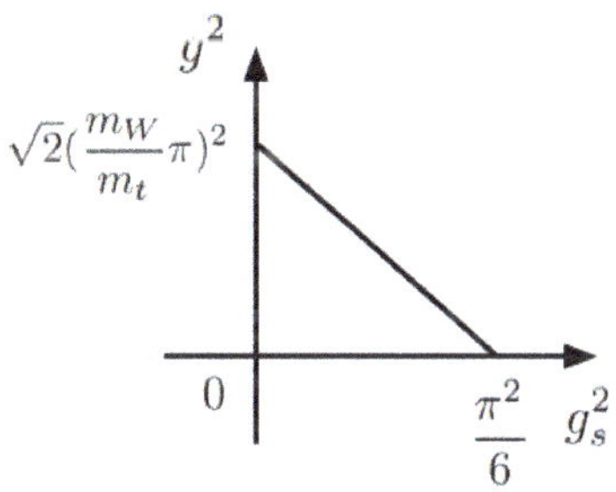

Fig. 13. Relation as given by Eq. (64) between the squares of the strong coupling constant g_s and the electroweak coupling constant g.

Whenever

$$\frac{1}{2}g^2\frac{m_t^2}{m_W^2}\lambda < R_0, \tag{60}$$

then this series $\sum_n \mathcal{M}_n$ converges exponentially, and the operations $\sum_n$ and $\lim_{\varepsilon\to 0}$ necessarily commute. Therefore, the answer to the question above is: in order for (56) and (57) to be both valid, it is necessary to have

$$\frac{1}{2}g^2\frac{m_t^2}{m_W^2}\lambda = R_0. \tag{61}$$

This result (61) holds not only for the special case (41) but also in general. The underlying reason is the same that is used for the prediction of increasing total cross-sections, namely, that the spectrum is the same for the special case

$$m = \lambda = 0 \tag{62}$$

as for the general case

$$\lambda > 0 \quad \text{and/or} \quad m > 0. \tag{63}$$

With Eq. (61), it only remains to use Eq. (49) and the result after Eq. (54) that the continuous spectrum is given by $[\pi^2, \infty)$:

$$6g_s^2 + \frac{1}{\sqrt{2}}\left(\frac{m_t}{m_W}g\right)^2 = \pi^2. \tag{64}$$

This result (64) is plotted in Fig. 13.

10. Comparison with Experiment

The following results from experimental measurements are listed in the Particle Physics Booklet:

$$m_W = 80.379\,\text{GeV}/\text{c}^2,$$

$$m_Z = 91.1876\,\mathrm{GeV/c^2},$$
$$m_t = 173\,\mathrm{GeV/c^2},$$
$$\alpha = \frac{e^2}{4\pi} = (137.036)^{-1},$$
$$\alpha_s = \frac{g_s^2}{4\pi} = 0.1181,$$

where this value for α_s is taken at the mass of Z, one of the four heaviest elementary particles listed under (1) of the introduction.

These experimental numbers imply

$$\frac{m_t^2}{m_W^2}\frac{g^2}{4\pi} = 0.1516$$

and

$$A = 9.348.$$

Therefore

$$\text{left-hand side of (64)} = 10.25 \tag{65}$$

while

$$\text{right-hand side of (64)} = 9.87. \tag{66}$$

Thus this first result equation (64) from the basic standard model is well satisfied by the experimental measurements.

11. Inclusion of Electromagnetic Interactions

In the last two sections, a relation has been obtained between the electroweak coupling constant g and the strong coupling constant g_s. What about the electromagnetic coupling constant e? Its inclusion leads only to a slight change in the development of Sec. 8.

This may be a suitable point to add the following side remark. In the standard model, it is well known that the photon and the Z are not themselves the gauge particles for the $U(1)$ and $SU(2)$ parts of the gauge group, but are instead their linear combinations. For the present theory, this fact is of no relevance because, in the large m_t approximation used here, the mass of the Z is neglected.

While the gluon couples to the top quark and the bottom quark identically, the photon couples to them differently because of their different charges. With the inclusion of the electromagnetic interactions, A of Sec. 8 has to be replaced by two different quantities

$$A_1 = \left[1 + \frac{1}{3}\left(\frac{2}{3}\frac{e}{g_s}\right)^2\right] A \tag{67}$$

for the top quark and

$$A_2 = \left[1 + \frac{1}{3}\left(\frac{1}{3}\frac{e}{g_s}\right)^2\right] A \tag{68}$$

for the bottom quark.

With all three coupling constants of the gauge group $U(1) \times SU(2) \times SU(3)$ of the standard model taken into account, the $A's$ of Eqs. (32)–(34) are replaced by $A_1's$, while those of Eqs. (35) and (36) by $A_2's$. With these replacements, Eqs. (38) and (39) remain valid, while the $A's$ of Eqs. (42) and (44) are $A_1's$, but that of (45) is A_2. Accordingly, the 3×3 matrix (47), which also appears in Eq. (46), is now

$$\begin{bmatrix} A_1 & -1 & 0 \\ -1 & A_1 & -1 \\ 0 & -1 & A_2 \end{bmatrix}. \tag{69}$$

Similar to (47), this matrix is real and symmetric.

The complication of including the electromagnetic interactions comes from the fact that the three eigenvalues of this matrix (69) are determined by the cubic equation

$$(\lambda - A_1)^2(\lambda - A_2) - (\lambda - A_1) - (\lambda - A_2) = 0, \tag{70}$$

which no longer factorizes trivially because of Eqs. (67) and (68).

Let

$$y = \lambda - A \tag{71}$$

and

$$A_0 = \frac{1}{3}\left(\frac{1}{3}\frac{e}{g_s}\right)^2 A = \frac{4}{9}\frac{m_W^2}{m_t^2}\frac{e^2}{g^2}, \tag{72}$$

then

$$y^3 - 9A_0 y^2 + (24A_0^2 - 2)y - (16A_0^3 - 5A_0) = 0. \tag{73}$$

In Eq. (73), A no longer appears. Let

$$x = y - 3A_0, \tag{74}$$

then the cubic equation for x is in the canonical form

$$x^3 + ax + b = 0, \tag{75}$$

where

$$a = -(2 + 3A_0^2)$$

and

$$b = -(A_0 - 2A_0^3). \tag{76}$$

The discriminant of this cubic equation (75) is

$$-(4a^3 + 27b^2) = 32 + 117A_0^2 + 324A_0^4 > 0, \tag{77}$$

implying that all three roots of this Eq. (75), and hence also those of Eq. (70), are always real and distinct.

With this information, the largest root of Eq. (75) is found to be given by

$$x = 2\left(\frac{2}{3} + A_0^2\right)^{1/2} \cos\theta, \tag{78}$$

where

$$\theta = \frac{1}{3}\cos^{-1}\left[\frac{1}{2}\left(\frac{2}{3} + A_0^2\right)^{-3/2}(A_0 - 2A_0^3)\right]. \tag{79}$$

In Eq. (79), the arc cosine is taken to be in the range

$$0 \le \theta < \pi. \tag{80}$$

With all three gauge coupling constants taken into account, the relation between them is thus found to be

$$6g_s^2 + \frac{1}{\sqrt{2}}\left(\frac{m_t}{m_W}g\right)^2 F\left(\frac{e^2}{g^2}\right) = \pi^2, \tag{81}$$

where

$$\begin{aligned} F\left(\frac{e^2}{g^2}\right) &= \frac{1}{\sqrt{2}}y \\ &= \frac{1}{\sqrt{2}}\left\{3A_0 + 2\left(\frac{2}{3} + A_0^2\right)^{\frac{1}{2}}\cos\frac{1}{3}\cos^{-1}\left[\frac{1}{2}\left(\frac{2}{3} + A_0^2\right)^{-\frac{3}{2}}(A_0 - 2A_0^3)\right]\right\}. \end{aligned} \tag{82}$$

In Eq. (82), A_0 is defined by Eq. (72) and is a function of the ratio $\frac{e^2}{g^2}$.

This relation equation (81) between the strong coupling constant g_s, the electroweak coupling constant g and the electromagnetic coupling constant e is the first purely theoretical result from the basic standard model without using any input from experiments. The simpler relation given by Eq. (64) is the special case when e is small. However, the smallness of e is not, at least not yet, a known consequence of the basic standard model.

12. Some Properties of the Relation between the Three Gauge Coupling Constants

This is the first time an approximate equality has been found, on purely theoretic grounds, between the three gauge coupling constants. It is therefore especially worthwhile to give some of the properties of the left-hand side of this Eq. (81).

Clearly

$$\frac{\partial}{\partial g_s^2}[\text{left-hand side of Eq. (81)}] \geq 0. \tag{83}$$

This (83) suggests that it is perhaps also true that

$$\frac{\partial}{\partial g^2}[\text{left-hand side of Eq. (81)}] \geq 0 \tag{84}$$

and

$$\frac{\partial}{\partial e^2}[\text{left-hand side of Eq. (81)}] \geq 0. \tag{85}$$

Are they?

Because of Eq. (81) and (82), the left-hand sides of these (84) and (85) are given in terms of y and A_0, which are related directly through the cubic Eq. (73). Furthermore, it is seen from either Eqs. (64) or (78) and (79) that

$$y = \sqrt{2} \tag{86}$$

when $A_0 = 0$. It follows from Eq. (73) that

$$\frac{dy}{dA_0} = \frac{9y^2 - 48A_oy + (48A_0^2 - 5)}{3y^2 - 18A_0y + (24A_0^2 - 2)}, \tag{87}$$

which can be written alternatively as

$$y - A_0\frac{dy}{dA_0} = \frac{4y - 10A_0}{3y^2 - 18A_0y + (24A_0^2 - 2)}. \tag{88}$$

From Eq. (77), the denominator is never zero; from Eq. (86), this denominator is always positive. It remains to determine the signs of the two numerators on the above right-hand sides.

Consider first the numerator of the right hand of Eq. (88). It is positive for $A_0 = 0$; can it be zero for positive values of A_0? If it is zero, then $y = \frac{5}{2}A_0$, at which the left-hand side of Eq. (73) is $\frac{27}{8}A_0^3$. In order to satisfy Eq. (73), A_0 must be zero. But at this $A_0 = 0$, by Eq. (86) y is $\sqrt{2}$, not $\frac{5}{2}A_0$. The conclusion is therefore reached this numerators of Eq. (88) is always positive, and hence

$$y - A_0\frac{dy}{dA_0} \geq 0. \tag{89}$$

This shows that the inequality Eq. (84) is indeed true.

Attention is now turned to study the third similar relation as given by (85). Similar to the case of (84), the issue is whether the numerator on the right-hand side of this Eq. (87) can be zero or not. At $A_0 = 0$, by Eq. (86), the value of this numerator is 13. Whether this numerator can be zero or not depends on the property of the 5×5 determinant

$$\begin{vmatrix} 1 & 9A_0 & 24A_0^2-2 & 16A_0^3-5A_0 & 0 \\ 0 & 1 & 9A_0 & 24A_0^2-2 & 16A_0^3-5A_0 \\ 9 & 48A_0 & 48A_0^2-5 & 0 & 0 \\ 0 & 9 & 48A_0 & 48A_0^2-5 & 0 \\ 0 & 0 & 9 & 48A_0 & 48A_0^2-5 \end{vmatrix}. \tag{90}$$

A direct evaluation of this determinant shows that it is never zero, which implies that the numerator on the right-hand side of Eq. (87) is never zero, and therefore always positive, entirely similar to the case of Eq. (88). Just like Eq. (89), this leads to

$$\frac{dy}{dA_0} \geq 0. \tag{91}$$

This shows that the three inequalities (83)–(85) are all true.

Ignoring the exceptional points where the equality sign holds, these three relations mean that the left-hand side of Eq. (81) is an increasing function of three gauge coupling constants. Furthermore, there is also the trivial property that

$$\text{left-hand side of Eq. (81)}|_{g_s^2=g^2=e^2} = 0, \tag{92}$$

which implies that it is not possible for all three gauge coupling constants to be small.

These properties are to be discussed further in Sec. 14.

13. Remarks

Equation (81) gives an approximate equality between the three gauge coupling constants of the group $U(1) \times SU(2) \times SU(3)$. Here are some simple remarks about this equality, which specifies a surface in the three-dimensional (g_s^2, g^2, e^2) space.

First, the inclusion of the electromagnetic interaction has little effect on the numerical comparison with experiment, as given in Sec. 10. The reason is that, for the present treatment of the basic standard model, the electroweak coupling constant g always appears in the combination $\frac{m_t}{m_W} g$. Therefore, the comparison of the effects of the electromagnetic interaction and the electroweak interaction is controlled by the small ratio

$$\frac{e^2}{(\frac{m_t}{m_W} g)^2} \sim \frac{1}{20}. \tag{93}$$

Second, it may be of interest to write down the intersections of the above mentioned surface with the planes or lines given by setting one or two of the three gauge coupling constants to zero.

(1) If $e = 0$, then the resulting curve is of course the one given already by Eq. (64).
(2) If $g = 0$, then $A_0 \to \infty$, $y \to 4A_0$, and the resulting curve is given by

$$6g_s^2 + \frac{8}{9}e^2 = \pi^2. \tag{94}$$

(3) If $g_s = 0$, then the curve is

$$\frac{1}{\sqrt{2}}\left(\frac{m_t}{m_W}g\right)^2 F\left(\frac{e^2}{g^2}\right) = \pi^2. \tag{95}$$

(4) If $g_s = g = 0$, then from Eq. (94)

$$\frac{8}{9}e^2 = \pi^2,$$

which means that there is an upper bound for the electromagnetic coupling constant given by

$$e^2 < \frac{9}{8}\pi^2. \tag{96}$$

(5) If $g_s = e = 0$, then from Eq. (64)

$$\frac{1}{\sqrt{2}}\left(\frac{m_t}{m_W}g\right)^2 = \pi^2,$$

which means similarly that there is an upper bound for the electroweak coupling constant given by

$$g^2 \le \sqrt{2}\left(\frac{m_W}{m_t}\right)^2 \pi^2. \tag{97}$$

(6) Finally, if $g = e = 0$, then from Eqs. (64) or (94)

$$6g_s^2 = \pi^2,$$

which means, again similarly, that there is an upper bound for the strong coupling constant given by

$$g_s^2 < \frac{\pi^2}{6}. \tag{98}$$

It should be noted that, while Eqs. (96) and (98) for the upper bounds for e^2 and g_s^2 are given with $<$ signs, that of Eq. (97) for the upper bound for g^2 has a $\le$ sign. The reason for this difference is simply that the present treatment of the basic standard model requires the electroweak coupling constant g to be nonzero.

It should be emphasized that the results of Sec. 13 so far, as well as a number of earlier sections, are completely independent of experimental measurement. They have been obtained by applying to the basic standard model the large m_t approximation and the use of the pagoda diagrams.

With this point in mind, consider the three upper bounds of (96)–(98). From the point of view of experimental measurements, the most interesting one is that of (98). From the measured value of α_s given in Sec. 10 listed in the Particle Physics Booklet,

$$\text{left-hand side of (98)} = 1.484 \tag{99}$$

and

$$\text{right-hand side of (98)} = 1.645. \tag{100}$$

Therefore this bound (98) is satisfied.

Of the many bounds given in this section, the most interesting and important one, as stated above, is the last one, that of (98) on the strength of the strong coupling constant. Since the strong interaction is the strongest among all the interactions of elementary particles, its strength g_s^2, taken at the mass of Z near $100\,\text{GeV/c}^2$, has been found on purely theoretical grounds to be given approximately by $\pi^2/6$, a new and accurate prediction, as seen from Eqs. (99) and (100). This is the first time a fundamental coupling constant has been calculated approximately.[6]

14. Discussion of First Result From the Basic Standard Model

Since what has been obtained is the first result from a new model for particle physics — the basic standard model — it seems worthwhile to have a careful discussion on what has been learned. It should also be noted that the method used to obtain this first result is also a novel one involving the innocent looking but especially tricky large m_t approximation.

One immediate feature of this first result is that the three coupling constants of the gauge group $U(1) \times SU(2) \times SU(3)$ are not independent, i.e. there is a smooth function $S(g_s^2, g^2, e^2)$ such that these coupling constants are related by

$$S(g_s^2, g^2, e^2) = 1, \tag{101}$$

where this S is given approximately by

$$S(g_s^2, g^2, e^2) \sim \frac{1}{\pi^2}\left[6g_s^2 + \frac{1}{\sqrt{2}}\left(\frac{m_t}{m_W}g\right)^2 F\left(\frac{e^2}{g^2}\right)\right] \tag{102}$$

see Eqs. (81) and (82). No such relation has been obtained before, nor conjectured.

This function S has the properties

$$\frac{\partial}{\partial g_s^2} S(g_s^2, g^2, e^2) \geq 0, \tag{103}$$

$$\frac{\partial}{\partial g^2} S(g_s^2, g^2, e^2) \geq 0, \tag{104}$$

$$\frac{\partial}{\partial e^2} S(g_s^2, g^2, e^2) \geq 0 \tag{105}$$

and

$$S(0,0,0) = 0. \tag{106}$$

See (83)–(85) and (92).

It is an immediate consequence of Eq. (106) that the three gauge coupling constants cannot all be small.

This $S(g_s^2, g^2, e^2)$ tells us not only that they cannot all be small, but also that none of these three gauge coupling constants can be too large. This conclusion comes about in the following way.

In Eq. (106), all three gauge coupling constants are taken to be zero. Suppose instead only two of the three are zero. Then, by Eq. (101), there are three cases

$$S(g_s^2, 0, 0) = 1, \tag{107}$$

$$S(0, g^2, 0) = 1 \tag{108}$$

and

$$S(0, 0, e^2) = 1. \tag{109}$$

In each of these cases, there is a unique solution; let us call these solutions

$$g_s^2 = g_{s0}^2, \tag{110}$$

$$g^2 = g_0^2 \tag{111}$$

and

$$e^2 = e_0^2, \tag{112}$$

respectively, for Eqs. (107)–(109). These solutions give the bound for the gauge coupling constants.

$$g_s^2 < g_{s0}^2, \tag{113}$$

$$g^2 \leq g_0^2 \tag{114}$$

and

$$e^2 < e_0^2. \tag{115}$$

See (96)–(98), which also imply

$$g_{s0}^2 \sim \frac{\pi^2}{6}, \tag{116}$$

$$g_0^2 \sim \sqrt{2}\left(\frac{m_W}{m_t}\right)^2 \pi^2 \tag{117}$$

and

$$e_0^2 \sim \frac{9}{8}\pi^2. \tag{118}$$

15. Going Forward

It is an important step in deriving, for the basic standard model, the relation equation (81) between the three gauge coupling constants of the group $U(1) \times SU(2) \times SU(3)$. This relation, which is new and has not even been conjectured before, is in agreement with experimental results.

This relation equation (81) has been obtained from the recurrence relations Eqs. (32)–(36) with the $A's$ suitably replaced by A_1 or A_2 of Eqs. (67) and (68) as specified in Sec. 11. These modified recurrence relations contain a great deal more physics information than Eq. (81), and a better understanding of the basic standard model shall come from the extraction of additional physics information from these recurrence relations.

The present situation has some similarity to that of the two-dimensional Ising Model seventy years ago. At that time, from the work of Onsager[23] and Kaufman,[24,25] Yang[26,27] realized that "Onsager and Kaufman had obtained much more information than just the partition function" and calculated the spontaneous magnetization. It was this result of Yang that opened up the theoretical study of the two-dimensional Ising Model.[28]

What is the analog, for this basic standard model, the spontaneous magnetization of the two-dimensional Ising Model? In view of the structure of the recurrences relations here, it is expected to be the mass ratio of the bottom and the top quarks. This will lead to the first case where a mass ratio of two elementary particles is calculated approximately on the basis of quantum field theory.

16. Summary and Outlook

A new model, referred to as the basic standard model for elementary particles, is proposed and analyzed.

The main ingredients for the formulation of this basic standard model are the following:

(1) the standard model of Glashow, Weinberg, and Salam,[2–4]
(2) the theory of increasing total cross-sections,[19] and
(3) the violation of gauge invariance.[22]

In (1), there are 25 dimensionless parameters. Of these 25, more than half of them — 14 out of 25 — come from mass ratios. A review of the measured masses of the known elementary particles show that there are four heaviest ones in the vicinity of $100\,\text{GeV/c}^2$, while that of the next heaviest one is in the expected range of radiative corrections. This suggests that the Higgs particle should interact directly only with these four heaviest ones — the top quark,[14,15] the Higgs particle,[7–9] and the gauge particles Z and W[10–13] — but not with the other elementary particles.[6] This is the idea of the basic standard model.

With this setup, the mass of the next heaviest elementary particle, the bottom quark b,[16] should be given by the one-loop $Hb\bar{b}$ diagrams. However, the matrix elements for these one-loop diagrams are all found to be zero. This by itself is not a serious problem because we have learned from (2) that it is often necessary to go beyond the one-loop approximation to obtain physics results.

Thus the calculation of the mass of the bottom quark in the basic standard model through the $Hb\bar{b}$ three-vertex cannot be limited to the one-loop diagrams, but necessarily involve multi-loop diagrams. However, it is not difficult to verify that not only the matrix elements for the one-loop diagrams but also those for all multi-loop $Hb\bar{b}$ diagrams are zero. We were stuck with this difficulty for several decades.

Then the (3) above from the violation of gauge invariance comes to our rescue: the sum of these matrix elements, each being zero, need not necessarily be zero. This sum of zeros not being zero itself gives a condition on the strengths of the gauge coupling constants. This condition equation (81) on the gauge coupling constants is a new theoretical relation, and is the main result of this paper.

Figure 14 shows schematically how the basic standard model for elementary particles has arisen from the previous theories.

A comparison with experimental measurements shows that this new result on the gauge coupling constants, derived from the basic standard model without using any experimental input, is in good agreement with such measurement. No such relation between the gauge coupling constants has been obtained before, nor even conjectured.

Many more such relations must exist, and a number of them may be expected to be found in the near future.[6,29]

When the standard model was being formulated in the nineteen sixties with the gauge group $U(1) \times SU(2) \times SU(3)$, a number of theoretical particle physicists preferred to use instead a simple group as the gauge group, because any simple gauge group leads to a single gauge coupling constant instead of three. The desire of having

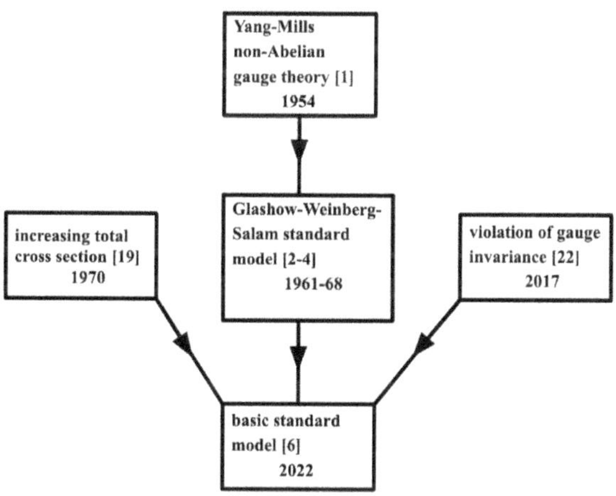

Fig. 14. Development of the basic standard model.

just one gauge coupling constant does have its appeal, but the use of a simple gauge group may not be the only way to have this achieved. Perhaps the present result from the basic standard model can be a first step in this direction, reducing the three gauge coupling constants to two independent ones. If a second relation between the three gauge coupling constants is found besides the first one studied in this paper, then there is only one independent gauge coupling constant for the gauge group $U(1) \times SU(2) \times SU(3)$.

Whether there is such a second relation or not, there are, for the basic standard model, at most two independent dimensionless parameters from the gauge coupling constants and three mass ratios from the four heaviest elementary particles t, H, Z and W for a total of five dimensionless parameters. Thus, compared with original standard model, the number of such parameters is reduced by a factor of five. It is also possible that these five parameters of the basic standard model are not independent of each other, and therefore the number of independent dimensionless parameters is less than five.

The basic standard model, or one of its modified versions, may therefore be a good candidate as a fundamental theory of elementary particles.

Acknowledgments

For the most useful discussions, we are greatly indebted to Professor Chen Ning Yang. It was many years ago when we had the first discussion with Professor Yang on possible relations between the gauge coupling constants, and we are happy that discussion has finally been brought to fruition.

ORCID

Tai Tsun Wu https://orcid.org/0009-0004-6592-2337
Sau Lan Wu https://orcid.org/0000-0001-5866-1504

References

1. C. N. Yang and R. L. Mills, *Phys. Rev.* **96**, 191 (1954).
2. S. L. Glashow, *Nucl. Phys.* **22**, 579 (1961).
3. S. Weinberg, *Phys. Rev. Lett.* **19**, 1264 (1967).
4. A. Salam, Weak and electromagnetic interactions, in *Proc. 8th. Nobel Symp.*, ed. N. Svartholm, Stockholm, Almgvist, 1968, p. 367.
5. M. Kobayashi and T. Maskawa, *Prog. Theor. Phys.* **49**, 652 (1973).
6. T. T. Wu and S. L. Wu, On the Yang-Mills non-abelian gauge theory, in *A Festschrift in Honor of the C N Yang Centenary: Scientific Papers*, eds. K. K. Phua *et al.* (World Scientific, 2022), p. 485.
7. F. Englert and R. Brout, *Phys. Rev. Lett.* **13**, 321 (1964).
8. P. W. Higgs, *Phys. Lett.* **12**, 132 (1964).
9. G. S. Guralnik, C. R. Hagen and T. W. B. Kibble, *Phys. Rev. Lett.* **13**, 585 (1964).
10. UA1 Collab. (G. Arnison *et al.*), *Phys. Lett. B* **122**, 103 (1983).
11. UA2 Collab. (M. Banner *et al.*), *Phys. Lett. B* **122**, 476 (1983).

12. UA1 Collab. (G. Arnison *et al.*), *Phys. Lett. B* **126**, 398 (1983).
13. UA2 Collab. (P. Bagnaia *et al.*), *Phys. Lett. B* **129**, 130 (1983).
14. CDF Collab. (F. Abe *et al.*), *Phys. Rev. Lett.* **74**, 2626 (1995).
15. D0 Collab. (S. Abachi *et al.*), *Phys. Rev. Lett.* **74**, 2632 (1995).
16. S. W. Herb *et al.*, *Phys. Rev. Lett.* **39**, 252 (1977).
17. J. H. Christenson, J. W. Cronin, V. L. Fitch and R. Turley, *Phys. Rev. Lett.* **13**, 138 (1964).
18. K. Fujikawa, B. W. Lee and A. I. Sanda, *Phys. Rev. Lett. D* **6**, 2923 (1972).
19. H. Cheng and T. T. Wu, *Phys. Rev. Lett.* **24**, 1456 (1970).
20. L. D. Faddeev and V. Popov, *Phys. Lett. B* **25**, 29 (1967).
21. H. Cheng and T. T. Wu, *Expanding Protons: Scattering at High Energies* (MIT Press, 1987).
22. T. T. Wu and S. L. Wu, *Nucl. Phys. B* **914**, 421 (2017).
23. L. Onsager, *Phys. Rev.* **65**, 117 (1944).
24. B. Kaufman, *Phys. Rev.* **76**, 1232 (1949).
25. B. Kaufman and L. Onsager, *Phys. Rev.* **76**, 1244 (1949).
26. C. N. Yang, *Phys. Rev.* **85**, 808 (1952).
27. C. N. Yang, *Selected Papers (1945–1980) With Comments* (World Scientific, 2005), p. 12.
28. B. M. McCoy and T. T. Wu, *The Two-Dimensional Ising Model* (Harvard University Press, Cambridge, MA, 1973) (Dover edn., 2014).
29. T. T. Wu, Alex Grossmann, scattering amplitude, Fermi pseudopotential, and particle physics, in *Theoretical Physics, Wavelets, Analysis, Genomics: An Indisciplinary Tribute to Alex Grossmann*, eds. P. Flandrin *et al.* (Birkhäuser, 2023), p. 97.